Springer Series in

# OPTICAL SCIENCES 132

*Founded by H.K.V. Lotsch*

Springer Series in
# OPTICAL SCIENCES

The Springer Series in Optical Sciences, under the leadership of Editor-in-Chief *William T. Rhodes*, Georgia Institute of Technology, USA, provides an expanding selection of research monographs in all major areas of optics: laser and quantum optics, ultrafast phenomena, optical spectroscopy techniques, optoelectronics, quantum information, information optics, applied laser technology, industrial applications, and other topics of contemporary interest.
With this broad coverage of topics, the series is of use to all research scientists and engineers who need up-to-date reference books.

The editors encourage prospective authors to correspond with them in advance of submitting a manuscript. Submission of manuscripts should be made to the Editor-in-Chief or one of the Editors. See also http://springer.com/series/624

Shuntaro Watanabe Katsumi Midorikawa
(Editors)

# Ultrafast Optics V

Shuntaro Watanabe
Institute for Solid State Physics
University of Tokyo
5-1-5 Kashiwanoha
Kashiwa-shi, Chiba 277-8581
Chiba, Japan
watanabe@issp.u-tokyo.ac.jp

Katsumi Midorikawa
Laser Technology Laboratory
RIKEN, Laser Technology Laboratory
2-1 Hirosawa, Wako-shi
Saitama 351-0198
Saitama, Japan
kmidori@postman.riken.go.jp

Library of Congress Control Number: 2007921729

ISBN 978-0-387-49117-2 ISBN 978-0-387-49119-6 (eBook)

Printed on acid-free paper.



9 8 7 6 5 4 3 2 1

springer.com

# Preface

The papers in this volume cover the major areas of recent research activity in the fields of ultrafast optics and applications of high field and short wavelength sources, and these have been selected to provide an overview of the current state of the art. The contributions were solicited from the participants of the *Joint Conference on Ultrafast Optics V and Applications of High Field and Short Wavelength Sources XI*, Nara, Japan, September 2005. Merging two historic conferences was proposed, and also supported by the common members of the two communities, because the contents of both the conferences are now getting closer. Attosecond pulses are closely related to short wavelength sources and ultrafast optics and high-field physics are on the same technology base.

Recent progress in ultrafast optics provides carrier envelope phase stabilized pulses reaching to a quasi monocycle. These pulses are amplified by a new scheme using parametric process to an optical field of which strength is comparable to the binding forces of electrons in atoms and molecules. This allows to manipulate events on atomic and molecular dynamics. The intensity of the zettawatt/cm$^2$ ($10^{21}$ W/cm$^2$) range opens the door to relativistic nonlinear optics and particle acceleration. Ultrafast intense light sources extend new wavelength range and thus make it possible generate of attosecond pulses in the XUV and soft X-ray spectral region, which enables to observe and manipulate the electron motion in atoms, molecules, and even during chemical reactions. Other technologies are also arising, such as terahertz (THz) radiation and X-ray pulse generation with applications in chemistry and biology.

We are greatly indebted to Teruto Kanai, University of Tokyo, for organizing and editing contributions to this book. We thank Japan Society for Promotion of Science, The Ogasawara Foundation for the Promotion of Science & Engineering, Nara Convention Bureau, and The Futaba Electronics Memorial Foundation for the financial support. And, of course, thanks to all the authors for allowing this book to cover the most exciting activities at the frontiers of this field.

Tokyo, Japan — Shuntaro Watanabe
Saitama, Japan — Katsumi Midorikawa

January 2006

# Contents

**Contributors** ........ **xvii**

**Part I Attosecond Metrology**

1 **Progress in Attosecond Metrology** ........ **3**
*R. Kienberger and F. Krausz*

2 **Attosecond Dynamics of Electron Wave Packets in Intense Laser Fields** ........ **15**
*K. Varjú, P. Johnsson, J. Mauritsson, R. López-Martens, E. Gustafsson, T. Remetter, and A. L'huillier*

**Part II Attosecond High Harmonic Generation**

3 **Attosecond Pulse Characterization by XUV Nonlinear Optics** ........ **27**
*Taro Sekikawa, Atsushi Kosuge, Teruto Kanai, and Shuntaro Watanabe*

4 **Spectral Interference of Attosecond Light Pulses** ........ **33**
*G. Sansone, E. Benedetti, J-P. Caumes, L. Poletto, P. Villoresi, S. Stagira, C. Vozzi, S. De Silvestri, and M. Nisoli*

5 **Frequency-resolved Autocorrelation Measurement for Characterization of an Attoseconod Pulse Train** ........ **39**
*Toshihiko Shimizu, Tomoya Okino, Hirokazu Hasegawa, Kentaro Furusawa, Yasuo Nabekawa, Kaoru Yamanouchi, and Katsumi Midorikawa*

6 **Characterization of Attosecond Pulse Trains** ........ **45**
*B. Carré, Y. Mairesse, P. Agostini, P. Breger, H. Merdji, P. Monchicourt, P. Salières, K. Varjú, P. Johnsson, J. Mauritsson, A. L'Huillier, E. Gustafsson, and L. J. Frasinski*

**Part III Carrier-Envelop Phase I**

7 **High-harmonic Generation at 100 MHz Repetition Frequency using a Femtosecond Enhancement Cavity** ........ **59**
*R. Jason Jones, Kevin Moll, Michael Thorpe, and Jun Ye*

8 **Effects of the Carrier-Envelope Phase in the Multiphoton Ionization Regime** ........ 65
*Takashi Nakajima and Shuntaro Watanabe*

9 **Carrier-Envelope Phase Detection by Interference between Surface Harmonics** ........ 73
*Atsushi Ishizawa and Hidetoshi Nakano*

10 **Coherent Control by Carrier-Envelope Phase in an Optical Poling Process** ........ 81
*Shunsuke Adachi and Takayoshi Kobayashi*

11 **Phase-coherent Spectrum from Ultrabroadband Ti:sapphire and Cr:forsterite Lasers Covering the Visible to the Infrared** ........ 87
*Jung-Won Kim, Thomas R. Schibli, Lia Matos, Hyunil Byun, and Franz X. Kärtner*

**Part IV Carrier-Envelop Phase II**

12 **Coherent Synthesis of Multicolor Femtosecond Pulses** ........ 95
*Yohei Kobayashi, Dai Yoshitomi, Masayuki Kakehata, Hideyuki Takada, and Kenji Torizuka*

13 **Quasisynchronous Pumping of Mode-locked Few-cycle Titanium Sapphire Lasers** ........ 103
*Richard Ell, Gregor Angelow, Wolfgang Seitz, Max J. Lederer, Huber Heinz, Daniel Kopf, Jonathan R. Birge, and Franz X. Kärtner*

14 **Ultrasimple Ultrabroadband Alignment-free Ultrashort Pulse Measurement Device** ........ 111
*Dongjoo Lee, Selçuk Aktürk, and Rick Trebino*

15 **Measurement of Group Velocity Distortion due to Ultrafast Index of Refraction Transients** ........ 115
*Randy A. Bartels and Klaus Hartinger*

16 **Frequency-domain Phase Conjugator for a Few-cycle and a Few-nJ Optical Pulses** ........ 119
*Hajime Nishioka, Keisuke Hayakawa, Hitoshi Tomita, and Ken-ichi Ueda*

## Part V High-field Physics and Applications I


17 **Relativistic Optics: A new Route to Attosecond Physics and Relativistic Engineering** .......... 127
*Gérard Mourou*

18 **Electron Acceleration Under Strong Radiation Damping** .......... 143
*James Koga, Sergei Bulanov, and Timur Esirkepov*

19 **Monoenergetic Electron Acceleration in a High-Density Plasma Produced by an Intense Laser Pulse** .......... 149
*E. Miura, S. Masuda, T. Watanabe, K. Koyama, S. Kato, M. Adachi, N. Saito, and M. Tanimoto*


## Part VI X-ray Plasma Sources and Applications


20 **Femtosecond Time-resolved X-ray Diffraction for Laser-excited CdTe Crystal** .......... 161
*Y. Hironaka, J. Irisawa, J. Saitoh, K. Kondo, K. Ishioka, K. Kitajima, and K. G. Nakamura*

21 **Femtosecond-laser-induced Ablation of an Aluminum Target Probed by Space- and Time-resolved Soft X-ray Absorption Spectroscopy** .......... 165
*Yasuaki Okano, Katsuya Oguri, Tadashi Nishikawa, and Hidetoshi Nakano*

22 **Observation of Ultrafast Bond-length Expansion at the Initial Stage of Laser Ablation by Picosecond Time-resolved EXAFS** .......... 173
*Katsuya Oguri, Yasuaki Okano, Tadashi Nishikawa, and Hidetoshi Nakano*


## Part VII Poster Session


23 **The Ehrenfest Theorem and Quantitative Predictions of HHG based on the Three-step Model** .......... 183
*Ariel Gordon and Franz X. Kärtner*

24 **Modeling of Ni-like Molybdenum X-ray Laser** .......... 187
*Sertan Kurnali, Arif Demir, and Elif Kacar*

25 **Enhancement and Spatial Characteristics of K$\alpha$ X-ray Emission from High-contrast Relativistic fs Laser Plasmas** ............ 193
*L. M. Chen, M. Kando, H. Kotaki, K. Nakajima, S. V. Bulanov, T. Tajima, M. H. Xu, Y. T. Li, Q. L. Dong, and J. Zhang*

26 **Optical-field Ionization by Femtosecond Laser Pulses of Time-Dependent Polarization** ............ 201
*Takamasa Ichino, Kenichi L. Ishikawa, and Masaharu Nakazawa*

27 **Transient Optical Susceptibility Induced by Nonperturbative Rotational Wave Packets** ............ 209
*Omid Masihzadeh, Mark Baertschy, Klaus Hartinger, and Randy A. Bartels*

28 **THz Wave Generation and Detection System using ~1000 nm Yb-doped Fiber Laser** ............ 217
*Atsushi Syouji, Shingo Saito, Kiyomi Sakai, Masaya Nagai, Koichiro Tanaka, Hideyuki Ohtake, Toshiaki Bessho, Toshiharu Sugiura, Tomoya Hirosumi, and Makoto Yoshida*

29 **Cavity-enhanced Optical Parametric Chirped-pulse Amplification** .. 221
*F. Ö. Ilday and F. X. Kärtner*

30 **Measuring Spatiotemporal Distortions with GRENOUILLE** ......... 225
*Xun Gu, Ziyang Wang, Selçuk Aktürk, and Rick Trebino*

31 **First-order Spatiotemporal Distortions of Gaussian Pulses and Beams** ............ 233
*Selçuk Aktürk, Xun Gu, Pablo Gabolde, and Rick Trebino*

32 **Characterization of Femtosecond Optical Pulses with Wavelet Transform of Spectral Shearing Interferogram** ............ 241
*Yuqiang Deng, Zubin Wu, Shiying Cao, Lu Chai, Ching-yue Wang, and Zhigang Zhang*

33 **Water Vapor Absorption Spectroscopy in Terahertz Range Using Wavelet-transform Analysis** ............ 249
*Yuqiang Deng, Qirong Xing, Liying Lang, Shuxin Li, Lu Chai, Ching-yue Wang, and Zhigang Zhang*

34 **Some Newly Developed Crystals for Measurement of Ultrafast Laser Pulses by Second Harmonic Generation** ............ 255
*Pathik Kumbhakar, Sunanda Chatterjee, and Takayoshi Kobayashi*

## Part VIII High-field Physics and Applications II

**35 Intensity Scalings of Attosecond Pulse Generation by the Relativistic-irradiance Laser Pulses** ............ 265
*Alexander S. Pirozhkov, Sergei V. Bulanov, Timur Zh. Esirkepov, Akito Sagisaka, Toshiki Tajima, and Hiroyuki Daido*

**36 Observation of Thin foil Preformed Plasmas with a Relativistic-intensity Ultra-short Pulse Laser by Means of Two-color Interferometer** ............ 273
*A. Sagisaka, H. Daido, K. Ogura, S. Orimo, Y. Hayashi, M. Nishiuchi, M. Mori, A. Yogo, M. Kado, T. Morita, M. Yamagiwa, A. Fukumi, Z. Li, Y. Oishi, T. Nayuki, T. Fujii, K. Nemoto, S. Nakamura, A. Noda, S. V. Bulanov, T. Zh. Esirkepov, A. S. Pirozhkov, and T. Utsumi*

## Part IX Terahertz (THz) and Nonlinear Processes I

**37 Generation of Ultrasmooth Broadband Spectra by Gain-assisted Self-phase Modulation in a Ti:sapphire Laser** ............ 281
*H. Crespo, M. V. Tognetti, M. A. Cataluna, J. T. Mendonça, and A. dos Santos*

## Part X Terahertz (THz) and Nonlinear Processes II

**38 Terahertz (THz) Pigtail Assembly Utilizing a Lens Duct for Effective Coupling of THz Radiation into Teflon Photonic Crystal Fiber Waveguide** ............ 293
*Alex Quema, Gilbert Diwa, Elmer Estacio, Romeric Pobre, Glenda Delos Reyes, Carlito Ponseca Jr., Hidetoshi Murakami, Shingo Ono, and Nobuhiko Sarukura*

**39 Terahertz Radiation from Photoconductive Switch Fabricated from a Zinc Oxide Single Crystal** ............ 301
*Hidetoshi Murakami, Elmer Estacio, Alex Quema, Glenda Delos Reyes, Shingo Ono, Nobuhiko Sarukura, Yo Ichikawa, Hiraku Ogino, Akira Yoshikawa, and Tsuguo Fukuda*

**40 Action Spectra of GaAs/AlGaAs Multiple Quantum Wells Exhibiting Terahertz Emission Peak at Excitation Energies Below the Bandgap** ............ 307
*E. Estacio, A. Quema, G. Diwa, G. De los Reyes, H. Murakami, S. Ono, N. Sarukura, A. Somintac, and A. Salvador*

**41 Ultrabroadband Detection of Terahertz Radiation from 0.1 to 100 THz with Photoconductive Antenna .................... 317**
*H. Shimosato, M. Ashida, T. Itoh, S. Saito, and K. Sakai*

**42 Improvement of the Photoluminescence Decay Response Characteristics of an Oxide-Confined Vertical Cavity Surface Emitting Laser Probed by Femtosecond Laser Pulses.................. 325**
*F. D. Recoleto Jr., J. N. Mateo, M. S. Dimamay, E. S. Estacio, A. S. Somintac, and A. A. Salvador*

**Part XI X-ray Lasers**

**43 Investigations on a Soft X-Ray Laser in the GRIP Geometry.......... 335**
*K. A. Janulewicz, J. Tümmler, H. Stiel, H. Legall, A. A. Bjeoumikhov, N. Langhoff, and P. V. Nickles*

**Part XII High-Intensity Lasers and Applications**

**44 Characterization of a $10^{20}$ W · cm$^{-2}$ Peak Intensity by Focusing Wavefront-Corrected 100-TW, 10-Hz Laser Pulses.......... 343**
*Yutaka Akahane, Jinglong Ma, Yuji Fukuda, Makoto Aoyama, Hiromitsu Kiriyama, Koichi Tsuji, Yoshiki Nakai, Yoichi Yamamoto, Koichi Yamakawa, Julia V. Sheldakova, and Alexis V. Kudryashov*

**45 Shaping Ultrafast Laser Pulses with Transient Optical Media........ 353**
*Randy A. Bartels and Klaus Hartinger*

**Part XIII Atoms and Molecules in High Field**

**46 Quantum Interference in Aligned Molecules.............................. 361**
*C. Vozzi, E. Benedetti, F. Calegari, J-P. Caumes, G. Sansone, S. Stagira, M. Nisoli, R. Torres, E. Heesel, N. Kajumba, J. P. Marangos, C. Altucci, and R. Velotta*

**47 Ellipticity Dependence of High-order Harmonics Generated in Aligned Molecules........................................................ 367**
*Tsuneto Kanai, Shinichirou Minemoto, and Hirofumi Sakai*

**48 Dissociative Ionization of Ethanol Using 400- and 800-nm Femtosecond Laser Pulses.................................................. 371**
*H. Yazawa, T. Shioyama, F. Kannari, R. Itakura, and K. Yamanouchi*

**49 Coincidence Momentum Imaging of Two- and Three-Body Coulomb Explosion of Ethanol in Intense Laser Fields ................ 375**
*Ryuji Itakura, Takahiro Teramoto, Akiyoshi Hishikawa, and Kaoru Yamanouchi*

**50 Temporal Coherent Control of Two-Photon Ionization by a Sequence of Ultrashort Laser Pulses ........................................ 377**
*Kenichi L. Ishikawa and Kiyoshi Ueda*

**Part XIV Postdeadline Paper**

**51 100-Attosecond Synchronization of Two-Color Mode-Locked Lasers by use of Optical Phase Locking .................................. 389**
*Dai Yoshitomi, Yohei Kobayashi, Masayuki Kakehata, Hideyuki Takada, and Kenji Torizuka*

**52 Complete Automatic Phase Compensation for the Generation of A-Few-Cycle Pulses ........................................................ 397**
*Keisuke Narita, Muneyuki Adachi, Ryuji Morita, and Mikio Yamashita*

**53 Carbon Nanotube Based Saturable Absorber Mirrors and Their Application to Ultrashort Pulse Generation ...................... 403**
*T. R. Schibli, K. Minoshima, H. Kataura, E. Itoga, N. Minami, S. Kazaoui, K. Miyashita, M. Tokumoto, and Y. Sakakibara*

**54 Generation of Extreme Ultraviolet Continuum Radiation Driven by Sub-10-fs Two-Color Field ...................................... 413**
*M. Kaku, Y. Oishi, A. Suda, F. Kannari, and K. Midorikawa*

**55 Ultrawideband Regenerative Amplifiers via Intracavity Acousto-optic Programmable Gain Control ............................... 421**
*Thomas Oksenhendler, Daniel Kaplan, Pierre Tournois, Gregory M. Greetham, and Frédéric Estable*

**56 Demonstration of Pulse Switching with $>10^{11}$ Prepulse Contrast by Cascaded Optical Parametric Amplification ............. 427**
*Constantin Haefner, Igor Jovanovic, Benoit Wattelier, and C. P. J. Barty*

## Part XV High-Order Harmonic Generation


57 **Generation of Exceptionally Strong Harmonics from He in an Orthogonally Polarized Two-color Laser Field** ........................ 435
*Chang Hee Nam, I Jong Kim, Chul Min Kim, Hyung Taek Kim, and Gae Hwang Lee*

58 **Single-shot Measurement of Fringe Visibility of 13-nm High-order Harmonic** ........................ 443
*Yutaka Nagata, Kentaro Furusawa, Yasuo Nabekawa, and Katsumi Midorikawa*

59 **High-order Harmonic Generation from the Laser Plasma Produced on the Surface of Solid Targets** ........................ 451
*R. A. Ganeev, M. Suzuki, M. Baba, and H. Kuroda*

60 **Scaling of keV HHG Photon Yield with Drive Wavelength** ........................ 459
*Ariel Gordon, Christian Jirauschek, and Franz X. Kärtner*

61 **Self-compression of an Ultraviolet Optical Pulse Assisted by Raman Coherence Induced in the Transient Regime** ........................ 465
*Yuichiro Kida, Shin-ichi Zaitsu, and Totaro Imasaka*


## Part XVI Ultrafast Amplifier


62 **Generation of High-energy Few-cycle Pulses by Filament Compression** ........................ 473
*C. P. Hauri, J. P. Rousseau, F. Burgy, G. Chériaux, and R. López-Martens*

63 **High Average Power, 7.5-fs Blue Source at 5 kHz** ........................ 481
*Xiangyu Zhou, Teruto Kanai, Taro Sekikawa, and Shuntaro Watanabe*

64 **Downchirped Regenerative Amplification of Femtosecond Laser Pulses at 100 kHz Repetition Rate** ........................ 493
*Kyung-Han Hong, Tae Jun Yu, Sergei Kostritsa, Jae Hee Sung, Il Woo Choi, Young-Chul Noh, Do-Kyeong Ko, and Jongmin Lee*

65 **Temporal Contrast Enhanced to $10^{-10}$ for Petawatt Class Femtosecond Lasers** ........................ 503
*A. Jullien, O. Albert, J. Etchepare, F. Augé-Rochereau, J.-P. Chambaret, G. Chériaux, S. Kourtev, N. Minkovski, and S. M. Saltiel*

## Part XVII OPCPA


66 **Infrared Optical Parametric Chirped Pulse Amplifier for High Harmonic Generation** ........ **513**
*T. Fuji, N. Ishii, Th. Metzger, C. Y. Teisset, L. Turi, A. Baltuška, N. Forget, D. Kaplan, A. Galvanauskas, and F. Krausz*

67 **Double-passed, High-energy Quasiphase-matched Optical Parametric Chirped-pulse Amplifier** ........ **521**
*Igor Jovanovic, Nicolas Forget, Curtis G. Brown, Christopher A. Ebbers, C. Le Blanc, and C.P.J. Barty*

68 **Control of Amplified Optical Parametric Fluorescence in Hybrid Chirped-pulse Amplification** ........ **527**
*Y. Hama, K. Kondo, H. Maeda, A. Zoubir, R. Kodama, K. A. Tanaka, and K. Mima*

69 **Optical Synchronization for OPCPA Chains** ........ **535**
*C.Y. Teisset, N. Ishii, T. Fuji, T. Metzger, S. Köhler, R. Holzwarth, A. Baltuška, A.M. Zheltikov, and F. Krausz*

70 **OPCPA Systems for the Amplification of Ultrashort Pulses Up to Millijoules Level** ........ **547**
*A. Renault, F. Augé-Rochereau, R. Lopez-Martens, and G. Chériaux*

71 **A Novel Method of Ultrabroadband (White light) Femtosecond Optical Parametric Amplification** ........ **553**
*Chao-Kuei Lee, Zing-Yung Zhang, J. Y. Huang, and Ci-Ling Pan*

72 **Study of Optical Parametric Chirped Pulse Amplification at 1064 and 780 nm** ........ **559**
*Xiaoyan Liang, Yuxin Leng, Ruxin Li, and Zhizhan Xu*

# Contributors

**M. Adachi** Hiroshima University, 1-3-1 Kagamiyama, Higashi-Hiroshima, Hiroshima 739-8530, Japan and Department of Applied Physics, Hokkaido University, Kita-13, Nishi-8, Sapporo, 060-8628 Japan

**Shunsuke Adachi** Department of Physics, Faculty of Science, University of Tokyo, 7-3-1 Hongo, Bunkyo, Tokyo 113-0033, Japan

**P. Agostini** CEA/DSM/DRECAM/SPAM, Bât. 522, CEA-Saclay, 91191 Gif-sur-Yvette, France

**Yutaka Akahane** Advanced Photon Research Center, KANSAI Research Establishment, Japan Atomic Energy Research Institute, Kizu, Soraku-gun, Kyoto 619-0215, Japan

**Selçuk Aktürk** School of Physics, Georgia Institute of Technology, Atlanta, Ga 30332-0430, USA

**O. Albert** Laboratoire d'Optique Appliquée, Ecole Nationale Supérieure de techniques avances, Ecole Polytechnique, Centre National de la Recherche Scientifique UMR 7639, 91761 Palaiseau Cedex, France

**C. Altucci** Coherentia–CNR–INFM and Dipartimento di Scienze Fisiche, Università di Napoli "Federico II," Napoli, Italy

**Gregor Angelow** NanoLayers, Optical Coatings GmbH, Maarweg 30, 53619 Rheinbreitbach, Germany

**Makoto Aoyama** Advanced Photon Research Center, KANSAI Research Establishment, Japan Atomic Energy Research Institute, Kizu, Soraku-gun, Kyoto 619-0215, Japan

**M. Ashida** Graduate School of Engineering Science, Osaka University, 1-3 Machikaneyama-cho, Toyonaka, Osaka, 560-8531, Japan

**F. Augé-Rochereau** Laboratoire d'Optique Appliquée, Ecole Nationale Supérieure de techniques avances, Ecole Polytechnique, Centre National de la Recherche Scientifique UMR 7639, 91761 Palaiseau Cedex, France

**M. Baba** The Institute for Solid State Physics, The University of Tokyo, 5-1-5 Kashiwanoha, Kashiwa 277-8581, Japan

**Mark Baertschy** Department of Physics, University of Colorado at Denver, Denver, CO 80217-3364, USA

**A. Baltŭska** Max-Planck-Institut für Quantenoptik, Hans-Kopfermann-Straße 1, D-85748 Garching, Germany

**Randy A. Bartels** Department of Electrical and Computer Engineering, Colorado State University, Fort Collins, CO 80523-1320, USA

**C. P. J. Barty** Lawrence Livermore National Laboratory, Mail Code L-470, 7000 East Avenue, Livermore, California 94550, USA

**E. Benedetti** National Laboratory for Ultrafast and Ultraintense Optical Science, CNR – INFM, Dipartimento di Fisica, Politecnico, Milano, Italy

**Toshiaki Bessho** Aisin Seiki Co., Ltd., Kojiritsuki, Hitotsugi-cho, Kariya 448-0003, Japan

**Jonathan R. Birge** Department of Electrical Engineering and Computer Science, Research Laboratory of Electronics, Massachusetts Institute of Technology, Cambridge, Massachusetts 02139-4307, USA

**A. A. Bjeoumikhov** IfG—Institute for Scientific Instruments GmbH, Rudower Chaussee 29/31, D-12489 Berlin, Germany

**C. Le Blanc** Lawrence Livermore National Laboratory, Mail Code L-470, 7000 East Avenue, Livermore, California 94550, USA and Laboratoire pour l'Utilisation des Lasers Intenses, École Polytechnique, Route de Saclay, 91128 Palaiseau cedex, France

**Curtis G. Brown** Lawrence Livermore National Laboratory, Mail Code L-470, 7000 East Avenue, Livermore, California 94550, USA

**S. V. Bulanov** Advanced Photon Research Center, Kansai Research Establishment, Japan Atomic Energy Research Institute, 8-1 Umemidai, Kizu, Kyoto 619-0215, Japan and A. M. Prokhorov Institute of General Physics of the Russian Academy of Sciences, 38 Vavilov Street, 119991 Moscow, Russia and Moscow Institute of Physics and Technology, 9 Institutskiy pereulok, 141700 Dolgoprudny, Moscow Region, Russia

**Sergei Bulanov** Japan Atomic Energy Agency, 8-1 Umemidai, Kizu, Souraku, Kyoto 619-0215, Japan and A. M. Prokhorov Institute of General Physics of the Russian Academy of Sciences, Vavilov Street 38, 119991 Moscow, Russia

**F. Burgy** Laboratoire d'Optique Appliquée, ENSTA-Ecole Polytechnique, Chemin de la Hunière, F-91761 Palaiseau Cedex, France

**Hyunil Byun** Department of Electrical Engineering and Computer Science and Research Laboratory of Electronics, Massachusetts Institute of Technology, Cambridge, Massachusetts 02139, USA

**F. Calegari** National Laboratory for Ultrafast and Ultraintense Optical Science, CNR– INFM, Dipartimento di Fisica, Politecnico, Milano, Italy

**Shiying Cao** Ultrafast Laser Laboratory, School of Precision Instrument and Optoelectronics Engineering, Tianjin University, Tianjin 300072, China

**B. Carré** CEA/DSM/DRECAM/SPAM, Bât. 522, CEA-Saclay, 91191 Gif-sur-Yvette, France

**M. A. Cataluna** GoLP/Centro de Física de Plasmas, Instituto Superior Técnico, Av. Rovisco Pais 1, 1049-001 Lisboa, Portugal and School of Physics and Astronomy, University of St. Andrews, St. Andrews KY16 9SS, Scotland

**J-P. Caumes** National Laboratory for Ultrafast and Ultraintense Optical Science, CNR – INFM, Dipartimento di Fisica, Politecnico, Milano, Italy

**Lu Chai** Ultrafast Laser Laboratory, School of Precision Instrument and Optoelectronics Engineering, Tianjin University, Tianjin 300072, China

**J.-P. Chambaret** Laboratoire d'Optique Appliquée, Ecole Nationale Supérieure de techniques avances, Ecole Polytechnique, Centre National de la Recherche Scientifique UMR 7639, 91761 Palaiseau Cedex, France

**Sunanda Chatterjee** Department of Physics, National Institute of Technology, Durgapur (Deemed University), Durgapur 713209, India

**L. M. Chen** Japan Atomic Energy Agency, 8-1 Umemida, Kizu, Kyoto 619-0215, Japan

**Il Woo Choi** Femto-Science Group, Advanced Photonics Research Institute, Gwangju Institute of Science and Technology, 1 Oryong-dong Buk-gu Gwangju 500-712, Republic of Korea,

**G. Chériaux** Laboratoire d'Optique Appliquée, ENSTA-Ecole Polytechnique, Chemin de la Hunière, F-91761 Palaiseau Cedex, France and Laboratoire d'Optique Appliquée ENSTA, École Polytechnique, CNRS UMR 7639, 91761 Palaiseau cedex, France and Laboratoire d'Optique Appliquée, Ecole Nationale Supérieure de techniques avances, Ecole Polytechnique, Centre National de la Recherche Scientifique UMR 7639, 91761 Palaiseau Cedex, France

**H. Crespo** CLOQ/Departamento de Física, Faculdade de Ciências, Universidade do Porto, R. do Campo Alegre 687, 4169-007 Porto, Portugal

**H. Daido** Advanced Photon Research Center, Kansai Research Establishment, Japan Atomic Energy Research Institute, 8-1 Umemidai, Kizu, Kyoto 619-0215, Japan and Advanced Photon Research Center, Japan Atomic Energy Research Institute, 8-1 Umemidai, Kizu-cho, Soraku-gun, Kyoto 619-0215, Japan

**Arif Demir** University of Kocaeli Department of Physics 41380 Kocaeli, Turkey and University of Kocaeli Laser Technologies Research and Application Center 41380 Kocaeli, Turkey

**Yuqiang Deng** Ultrafast Laser Laboratory, School of Precision Instrument and Optoelectronics Engineering, Tianjin University, Tianjin 300072, China

**M. S. Dimamay** National Institute of Physics, University of the Philippines, Quezon City, Philippines

**G. Diwa** Institute for Molecular Science (IMS), Myodaiji, Okazaki 444-8585, Japan

**Gilbert Diwa** Institute for Molecular Science (IMS), Myodaiji, Okazaki 444-8585, Japan

**Q. L. Dong** Institute of Physics and China Academy of Sciences, Beijing 100080, China

**Christopher A. Ebbers** Lawrence Livermore National Laboratory, Mail Code L-470, 7000 East Avenue, Livermore, California 94550, USA

**Richard Ell** NanoLayers, Optical Coatings GmbH, Maarweg 30, 53619 Rheinbreitbach, Germany and Department of Electrical Engineering and Computer Science, Research Laboratory of Electronics, Massachusetts Institute of Technology, Cambridge, Massachusetts 02139-4307, USA

**T. Zh. Esirkepov** Advanced Photon Research Center, Kansai Research Establishment, Japan Atomic Energy Research Institute, 8-1 Umemidai, Kizu, Kyoto 619-0215, Japan and Moscow Institute of Physics and Technology, 9 Institutskiy pereulok, 141700 Dolgoprudny, Moscow Region, Russia

**Timur Esirkepov** Japan Atomic Energy Agency, 8-1 Umemidai, Kizu, Souraku, Kyoto 619-0215, Japan and Moscow Institute of Physics and Technology, Institutskiy pereulok 9, 141700 Dolgoprudny, Moscow Region, Russia

**Frédéric Estable** AMPLITUDE TECHNOLOGIES, 2, rue du Bois Chaland, 91029 EVRY, France

**E. Estacio** Institute for Molecular Science (IMS), Myodaiji, Okazaki 444-8585, Japan

**E. S. Estacio** National Institute of Physics, University of the Philippines, Quezon City, Philippines

**J. Etchepare** Laboratoire d'Optique Appliquée, Ecole Nationale Supérieure de techniques avances, Ecole Polytechnique, Centre National de la Recherche Scientifique UMR 7639, 91761 Palaiseau Cedex, France

**N. Forget** Fastlite, Bat 403, Ecole Polytechnique, 91128 Palaiseau, France

**Nicolas Forget** Lawrence Livermore National Laboratory, Mail Code L-470, 7000 East Avenue, Livermore, California 94550, USA and Laboratoire pour l'Utilisation des Lasers Intenses, École Polytechnique, Route de Saclay, 91128 Palaiseau cedex, France

**L. J. Frasinski** The University of Reading, J. J. Thomson Physical Laboratory, Whiteknights, P.O. Box 220, Reading RG6 6AF, UK

**T. Fujii** Central Research Institute of Electric Power Industry, 2-6-1 Nagasaka, Yokosuka, Kanagawa 240-0196, Japan

**T. Fuji** Max-Planck-Institut für Quantenoptik, Hans-Kopfermann-Straße 1, D-85748 Garching, Germany

**Tsuguo Fukuda** Tohoku University, 2-1-1 Katahira, Aoba-ku, Sendai 980-8577, Japan

**Yuji Fukuda** Advanced Photon Research Center, KANSAI Research Establishment, Japan Atomic Energy Research Institute, Kizu, Soraku-gun, Kyoto 619-0215, Japan

**A. Fukumi** Advanced Photon Research Center, Kansai Research Establishment, Japan Atomic Energy Research Institute, 8-1 Umemidai, Kizu, Kyoto 619-0215, Japan and National Institute of Radiological Sciences, 4-9-1, Anagawa, Inage, Chiba 263-8555, Japan

**Kentaro Furusawa** Laser Technology Laboratory, RIKEN, 2-1 Hirosawa, Wako-shi, Saitama 351-0198, Japan

**Pablo Gabolde** School of Physics, Georgia Institute of Technology, Atlanta, GA 30332-0430 USA

**A. Galvanauskas** University of Michigan,College of Engineering, Department of Electrical Engineering and Computer Science, 1301 Beal Avenue Ann Arbor, 48109-2122, Michigan, USA

**R. A. Ganeev** The Institute for Solid State Physics, The University of Tokyo, 5-1-5 Kashiwanoha, Kashiwa 277-8581, Japan

**Ariel Gordon** Department of Electrical Engineering and Computer Science, and Research Laboratory of Electronics, Massachusetts Institute of Technology, 77 Massachusetts Avenue, Cambridge, MA 02139, USA

**Gregory M. Greetham** AMPLITUDE TECHNOLOGIES, 2, rue du Bois Chaland, 91029 EVRY, France

**Xun Gu** School of Physics, Georgia Institute of Technology, Atlanta, Ga 30332-0430, USA

**E. Gustafsson** Department of Physics, Lund Institute of Technology, P. O. Box 118, SE-221 00 Lund, Sweden

**Constantin Haefner** Nevada Terawatt Facility, University of Nevada, Reno, Nevada 89506, USA

**Y. Hama** Graduate School of Engineering, Osaka University, 2-1 Yamadaoka, Suita, 565-0871, Japan and Institute of Laser Engineering, Osaka University, 2-6 Yamadaoka, Suita, 565-0871, Japan

**Klaus Hartinger** Department of Electrical and Computer Engineering, Colorado State University, Fort Collins, CO 80523-1320, USA

**Hirokazu Hasegawa** Laser Technology Laboratory, RIKEN, 2-1 Hirosawa, Wako-shi, Saitama 351-0198, Japan and Institute for Molecular Science, Myodaiji, Okazaki 444-8585, Japan

**C. P. Hauri** Laboratoire d'Optique Appliquée, ENSTA-Ecole Polytechnique, Chemin de la Hunière, F-91761 Palaiseau Cedex, France

**Keisuke Hayakawa** Institute for Laser Science, University of Electro-Communications, 1-5-1 Chofugaoka, Chofu, Tokyo 182-8585, Japan

**Y. Hayashi** Advanced Photon Research Center, Kansai Research Establishment, Japan Atomic Energy Research Institute, 8-1 Umemidai, Kizu, Kyoto 619-0215, Japan

**E. Heesel** The Blackett Laboratory, Imperial College London, Prince Consort Road, London SW7 2BW, UK

**Huber Heinz** High Q Laser Production GmbH, Kaiser-Franz-Josef-Strasse 61, A-6845 Hohenems, Austria

**Y. Hironaka** Materials and Structures Laboratory, Tokyo Institute of Technology, Nagatsuta, 4259 Midori, Yokohama 226-8503, Japan

**Tomoya Hirosumi** Aisin Seiki Co., Ltd., Kojiritsuki, Hitotsugi-cho, Kariya 448-0003, Japan

**Akiyoshi Hishikawa** Department of Chemistry, School of Science, The University of Tokyo, 7-3-1 Hongo Bunkyo-ku, Tokyo 113-0033

**R. Holzwarth** Max-Planck-Institut für Quantenoptik, Hans-Kopfermann-Straße 1, D-85748 Garching, Germany and Menlo Systems GmbH, Am Klopferspitz 19, D-85152 Martinsried, Germany

**Kyung-Han Hong** Femto-Science Group, Advanced Photonics Research Institute, Gwangju Institute of Science and Technology, 1 Oryong-dong Buk-gu Gwangju 500-712, Republic of Korea

**J. Y. Huang** Department of Photonics and Institute of Optoelectronic Engineering, National Chiao Tung University, Taiwan, ROC

**Yo Ichikawa** Nagoya Institute of Technology, Showa, Gokiso, Nagoya 466-8555, Japan

**Takamasa Ichino** Department of Quantum Engineering and Systems Science, School of Engineering, University of Tokyo, Hongo 7-3-1, Bunkyo-ku, Tokyo 113-8656, Japan

**F. Ö. Ilday** Department of Electrical Engineering and Computer Science, and Research Laboratory of Electronics, Massachusetts Institute of Technology, 77 Massachusetts Avenue, Cambridge, MA 02139, USA

**Totaro Imasaka** Department of Applied Chemistry, Graduate School of Engineering, Kyushu University, 6-10-1, Hakozaki, Higashi-ku, Fukuoka 812-8581, Japan

**J. Irisawa** Materials and Structures Laboratory, Tokyo Institute of Technology, Nagatsuta, 4259 Midori, Yokohama 226-8503, Japan

**N. Ishii** Max-Planck-Institut für Quantenoptik, Hans-Kopfermann-Strasse 1, D-85748, Garching, Germany

**Kenichi L. Ishikawa** Department of Quantum Engineering and Systems Science, School of Engineering, University of Tokyo, Hongo 7-3-1, Bunkyo-ku, Tokyo 113-8656, Japan

**K. Ishioka** National Institute for Materials Science, Tsukuba 305-0047, Japan

**Atsushi Ishizawa** NTT Basic Research Laboratories, NTT Corporation, 3-1 Morinosato Wakamiya, Atsugi, Kanagawa 243-0198, Japan

**R. Itakura** Department of Chemistry, School of Science,The University of Tokyo, Japan and Department of Chemistry, School of Science, The University of Tokyo, 7-3-1 Hongo Bunkyo-ku, Tokyo 113-0033

**E. Itoga** National Institute of Advanced Industrial Science and Technology, AIST, 1-1-1 Umezono, Tsukuba, Ibaraki 305-8563, Japan

**T. Itoh** Graduate School of Engineering Science, Osaka University, 1-3 Machikaneyama-cho, Toyonaka, Osaka, 560-8531, Japan

**K. A. Janulewicz** Max-Born-Institut fr Nichtlineare Optik und Kurzzeitspektroskopie, Max-Born-Strasse 2A, D-12489 Berlin, Germany

**Christian Jirauschek** Department of Electrical Engineering and Computer Science, and Research Laboratory of Electronics, Massachusetts Institute of Technology, 77 Massachusetts Avenue, Cambridge, MA 02139, USA

**P. Johnsson** Department of Physics, Lund Institute of Technology, P. O. Box 118, SE-221 00 Lund, Sweden

**R. Jason Jones** JILA, National Institute of Standards and Technology and University of Colorado, 440 UCB Boulder, CO 80309

**Igor Jovanovic** Lawrence Livermore National Laboratory, Mail Code L-470, 7000 East Avenue, Livermore, California 94550, USA

**A. Jullien** Thales Laser, RD 128, Domaine de Corbeville, 91400 Orsay, France and Laboratoire d'Optique Appliquée, Ecole Nationale Supérieure de techniques avances, Ecole Polytechnique, Centre National de la Recherche Scientifique UMR 7639, 91761 Palaiseau Cedex, France

**Elif Kacar** University of Kocaeli Department of Physics 41380 Kocaeli, Turkey

**M. Kado** Advanced Photon Research Center, Kansai Research Establishment, Japan Atomic Energy Research Institute, 8-1 Umemidai, Kizu, Kyoto 619-0215, Japan

**N. Kajumba** The Blackett Laboratory, Imperial College London, Prince Consort Road, London SW7 2BW, UK

**Masayuki Kakehata** National Institute of Advanced Industrial Science and Technology (AIST) Tsukuba Central 2, 1-1-1 Umezono, Tsukuba 305-8568, Japan

**M. Kaku** RIKEN, 2-1 Hirosawa, Wako-shi, Saitama 351-0198, Japan

**Teruto Kanai** Institute for Solid State Physics, University of Tokyo, 5-1-5 Kashiwanoha, Kashiwa 277-8581, Japan *and* Department of Physics, Graduate School of Science, The University of Tokyo, 7-3-1 Hongo, Bunkyo-ku, Tokyo 113-0033, Japan

**M. Kando** Japan Atomic Energy Agency, 8-1 Umemida, Kizu, Kyoto 619-0215, Japan

**F. Kannari** Department of Electronics and Electrical Engineering, Keio University, Japan

**D. Kaplan** Fastlite, Bat 403, Ecole Polytechnique, 91128 Palaiseau, France

**Franz X. Kärtner** Department of Electrical Engineering and Computer Science, and Research Laboratory of Electronics, Massachusetts Institute of Technology, 77 Massachusetts Avenue, Cambridge, MA 02139, USA *and* Department of Electrical Engineering and Computer Science, Research Laboratory of Electronics, Massachusetts Institute of Technology, Cambridge, Massachusetts 02139-4307, USA *and* Department of Electrical Engineering and Computer Science and Research Laboratory of Electronics, Massachusetts Institute of Technology, Cambridge, Massachusetts 02139, USA

**H. Kataura** National Institute of Advanced Industrial Science and Technology, AIST, 1-1-1 Umezono, Tsukuba, Ibaraki 305-8563, Japan

**S. Kato** National Institute of Advanced Industrial Science and Technology (AIST), Tsukuba Central 2, 1-1-1 Umezono, Tsukuba, Ibaraki 305-8568, Japan

**S. Kazaoui** National Institute of Advanced Industrial Science and Technology, AIST, 1-1-1 Umezono, Tsukuba, Ibaraki 305-8563, Japan

**Yuichiro Kida** Department of Applied Chemistry, Graduate School of Engineering, Kyushu University, 6-10-1, Hakozaki, Higashi-ku, Fukuoka 812-8581, Japan

**R. Kienberger** Max-Planck Institut für Quantenoptik, Hans-Kopfermann-Straße 1, D-85748 Garching, Germany.

**Chul Min Kim** Department of Physics and Coherent X-ray Research Center, KAIST, 373-1, Kuseong-dong, Yuseong-gu, Daejeon 305-701, Korea

**Hyung Taek Kim** Department of Physics and Coherent X-ray Research Center, KAIST, 373-1, Kuseong-dong, Yuseong-gu, Daejeon 305-701, Korea

**I Jong Kim** Department of Physics and Coherent X-ray Research Center, KAIST, 373-1, Kuseong-dong, Yuseong-gu, Daejeon 305-701, Korea

**Jung-Won Kim** Department of Electrical Engineering and Computer Science and Research Laboratory of Electronics, Massachusetts Institute of Technology, Cambridge, Massachusetts 02139, USA

**Hiromitsu Kiriyama** Advanced Photon Research Center, KANSAI Research Establishment, Japan Atomic Energy Research Institute, Kizu, Soraku-gun, Kyoto 619-0215, Japan

**K. Kitajima** National Institute for Materials Science, Tsukuba 305-0047, Japan

**Do-Kyeong Ko** Femto-Science Group, Advanced Photonics Research Institute, Gwangju Institute of Science and Technology, 1 Oryong-dong Buk-gu Gwangju 500-712, Republic of Korea

**Takayoshi Kobayashi** Department of Physics, Faculty of Science, University of Tokyo, 7-3-1 Hongo, Bunkyo, Tokyo 113-0033, Japan

**Yohei Kobayashi** National Institute of Advanced Industrial Science and Technology (AIST) Tsukuba Central 2, 1-1-1 Umezono, Tsukuba 305-8568, Japan

**R. Kodama** Graduate School of Engineering, Osaka University, 2-1 Yamadaoka, Suita, 565-0871, Japan and Institute of Laser Engineering, Osaka University, 2-6 Yamadaoka, Suita, 565-0871, Japan

**James Koga** Japan Atomic Energy Agency, 8-1 Umemidai, Kizu, Souraku, Kyoto 619-0215, Japan

**S. Köhler** Max-Planck-Institut für Quantenoptik, Hans-Kopfermann-Straße 1, D-85748 Garching, Germany

**K. Kondo** Graduate School of Engineering, Osaka University, 2-1 Yamadaoka, Suita, 565-0871, Japan and Institute of Laser Engineering, Osaka University, 2-6 Yamadaoka, Suita, 565-0871, Japan and Materials and Structures Laboratory, Tokyo Institute of Technology, Nagatsuta, 4259 Midori, Yokohama 226-8503, Japan

**Daniel Kopf** High Q Laser Production GmbH, Kaiser-Franz-Josef-Strasse 61, A-6845 Hohenems, Austria

**Sergei Kostritsa** Femto-Science Group, Advanced Photonics Research Institute, Gwangju Institute of Science and Technology, 1 Oryong-dong Buk-gu Gwangju 500-712, Republic of Korea

**Atsushi Kosuge** Institute for Solid State Physics, University of Tokyo, 5-1-5 Kashiwanoha, Kashiwa 277-8581, Japan

**H. Kotaki** Japan Atomic Energy Agency, 8-1 Umemida, Kizu, Kyoto 619-0215, Japan

**S. Kourtev** Faculty of Physics, University of Sofia, 5 J. Bourchier Boulevard, BG-1164, Sofia, Bulgaria

**K. Koyama** National Institute of Advanced Industrial Science and Technology (AIST), Tsukuba Central 2, 1-1-1 Umezono, Tsukuba, Ibaraki 305-8568, Japan

**F. Krausz** Max-Planck Institut für Quantenoptik, Hans-Kopfermann-Straße 1, D-85748 Garching, Germany and Ludwig-Maximilans-Universität München, Coulombwall 1, D-85748 Garching, Germany.

**Alexis V. Kudryashov** Moscow State Open University, Sudostroitelnaya Str.18, Bld.5, Moscow, 115407 Russia

**Pathik Kumbhakar** Department of Physics, National Institute of Technology, Durgapur (Deemed University), Durgapur 713209, India

**Sertan Kurnali** Zonguldak Karaelmas University Department of Physics 67100 Zonguldak, Turkey

**H. Kuroda** The Institute for Solid State Physics, The University of Tokyo, 5-1-5 Kashiwanoha, Kashiwa 277-8581, Japan

**Liying Lang** Ultrafast Laser Laboratory, School of Precision Instrument and Optoelectronics Engineering, Tianjin University, Tianjin 300072, China

**N. Langhoff** IfG—Institute for Scientific Instruments GmbH, Rudower Chaussee 29/31, D-12489 Berlin, Germany

**Max J. Lederer** High Q Laser Production GmbH, Kaiser-Franz-Josef-Strasse 61, A-6845 Hohenems, Austria

**Chao-Kuei Lee** Institute of Electro-Optical Engineering, National Sun Yat-Sen University, Taiwan, ROC

**Dongjoo Lee** School of Physics, Georgia Institute of Technology, Atlanta, Ga 30332-0430, USA

**Gae Hwang Lee** Department of Physics and Coherent X-ray Research Center, KAIST, 373-1, Kuseong-dong, Yuseong-gu, Daejeon 305-701, Korea

**Jongmin Lee** Femto-Science Group, Advanced Photonics Research Institute, Gwangju Institute of Science and Technology, 1 Oryong-dong Buk-gu Gwangju 500-712, Republic of Korea

**H. Legall** Max-Born-Institut fr Nichtlineare Optik und Kurzzeitspektroskopie, Max-Born-Strasse 2A, D-12489 Berlin, Germany

**Yuxin Leng** State Key Laboratory of High Field Laser Physics, Shanghai Institute of Optics and Fine Mechanics, Chinese Academy of Sciences, PO Box 800-211, Shanghai 201800, China.

**A. L'Huillier** Department of Physics, Lund Institute of Technology, P. O. Box 118, SE-221 00 Lund, Sweden

**Ruxin Li** State Key Laboratory of High Field Laser Physics, Shanghai Institute of Optics and Fine Mechanics, Chinese Academy of Sciences, PO Box 800-211, Shanghai 201800, China

**Shuxin Li** Ultrafast Laser Laboratory, School of Precision Instrument and Optoelectronics Engineering, Tianjin University, Tianjin 300072, China

**Y. T. Li** Institute of Physics and China Academy of Sciences, Beijing 100080, China

**Z. Li** Advanced Photon Research Center, Kansai Research Establishment, Japan Atomic Energy Research Institute, 8-1 Umemidai, Kizu, Kyoto 619-0215, Japan and National Institute of Radiological Sciences, 4-9-1, Anagawa, Inage, Chiba 263-8555, Japan

**Xiaoyan Liang** State Key Laboratory of High Field Laser Physics, Shanghai Institute of Optics and Fine Mechanics, Chinese Academy of Sciences, PO Box 800-211, Shanghai 201800, China

**R. Lopez-Martens** Laboratoire d'Optique Appliquée ENSTA, École Polytechnique, CNRS UMR 7639, 91761 Palaiseau cedex, France and Laboratoire d'Optique Appliquée, ENSTA-Ecole Polytechnique, Chemin de la Hunière, F-91761 Palaiseau Cedex, France *and* Department of Physics, Lund Institute of Technology, P. O. Box 118, SE-221 00 Lund, Sweden

**Jinglong Ma** Advanced Photon Research Center, KANSAI Research Establishment, Japan Atomic Energy Research Institute, Kizu, Soraku-gun, Kyoto 619-0215, Japan

**H. Maeda** Graduate School of Engineering, Osaka University, 2-1 Yamadaoka, Suita, 565-0871, Japan and Institute of Laser Engineering, Osaka University, 2-6 Yamadaoka, Suita, 565-0871, Japan

**Y. Mairesse** CEA/DSM/DRECAM/SPAM, Bât. 522, CEA-Saclay, 91191 Gif-sur-Yvette, France

**J. P. Marangos** The Blackett Laboratory, Imperial College London, Prince Consort Road, London SW7 2BW, UK

**Omid Masihzadeh** Department of Electrical and Computer Engineering Colorado State University, Fort Collins, CO 80523-1320, USA

**S. Masuda** National Institute of Radiological Sciences, 4-9-1 Anagawa, Inage, Chiba, Chiba 263-8555, Japan

**J. N. Mateo** National Institute of Physics, University of the Philippines, Quezon City, Philippines

**Lia Matos** Department of Physics and Research Laboratory of Electronics, Massachusetts Institute of Technology, Cambridge, Massachusetts 02139, USA

**J. Mauritsson** Department of Physics, Lund Institute of Technology, P. O. Box 118, SE-221 00 Lund, Sweden

**J. T. Mendonça** GoLP/Centro de Fisica de Plasmas, Instituto Superior Técnico, Av. Rovisco Pais 1, 1049-001 Lisboa, Portugal and Rutherford Appleton Laboratory, CCLRC, Chilton, Didcot, Oxon OX11 0QX, England

**H. Merdji** CEA/DSM/DRECAM/SPAM, Bât. 522, CEA-Saclay, 91191 Gif-sur-Yvette, France

**T. Metzger** Max-Planck-Institut für Quantenoptik, Hans-Kopfermann-Straße 1, D-85748 Garching, Germany

**Th. Metzger** Max-Planck-Institut für Quantenoptik, Hans-Kopfermann-Strasse 1, D-85748, Garching, Germany

**K. Midorikawa** RIKEN, 2-1 Hirosawa, Wako-shi, Saitama 351-0198, Japan and Laser Technology Laboratory, RIKEN, 2-1 Hirosawa, Wako-shi, Saitama 351-0198, Japan

**K. Mima** Graduate School of Engineering, Osaka University, 2-1 Yamadaoka, Suita, 565-0871, Japan and Institute of Laser Engineering, Osaka University, 2-6 Yamadaoka, Suita, 565-0871, Japan

**N. Minami** National Institute of Advanced Industrial Science and Technology, AIST, 1-1-1 Umezono, Tsukuba, Ibaraki 305-8563, Japan

**Shinichirou Minemoto** Department of Physics, Graduate School of Science, The University of Tokyo, 7-3-1 Hongo, Bunkyo-ku, Tokyo 113-0033, Japan

**N. Minkovski** Faculty of Physics, University of Sofia, 5 J. Bourchier Boulevard, BG-1164, Sofia, Bulgaria

**K. Minoshima** National Institute of Advanced Industrial Science and Technology, AIST, 1-1-1 Umezono, Tsukuba, Ibaraki 305-8563, Japan

**E. Miura** National Institute of Advanced Industrial Science and Technology (AIST), Tsukuba Central 2, 1-1-1 Umezono, Tsukuba, Ibaraki 305-8568, Japan

**K. Miyashita** National Institute of Advanced Industrial Science and Technology, AIST, 1-1-1 Umezono, Tsukuba, Ibaraki 305-8563, Japan

**Kevin Moll** JILA, National Institute of Standards and Technology and University of Colorado, 440 UCB Boulder, CO 80309, USA

**P. Monchicourt** CEA/DSM/DRECAM/SPAM, Bât. 522, CEA-Saclay, 91191 Gif-sur-Yvette, France

**M. Mori** Advanced Photon Research Center, Kansai Research Establishment, Japan Atomic Energy Research Institute, 8-1 Umemidai, Kizu, Kyoto 619-0215, Japan

**Ryuji Morita** Department of Applied Physics, Hokkaido University, Kita-13, Nishi-8, Sapporo, 060-8628 Japan

**T. Morita** Advanced Photon Research Center, Kansai Research Establishment, Japan Atomic Energy Research Institute, 8-1 Umemidai, Kizu, Kyoto 619-0215, Japan

**Gérard Mourou** Laboratoire d'Optique Appliquée, Ecole Nationale Supérieure de techniques avances, Ecole Polytechnique, Centre National de la Recherche Scientifique UMR 7639, 91761 Palaiseau Cedex, France

**H. Murakami** Institute for Molecular Science (IMS), Myodaiji, Okazaki 444-8585, Japan

**Yasuo Nabekawa** Laser Technology Laboratory, RIKEN, 2-1 Hirosawa, Wako-shi, Saitama 351-0198, Japan

**Masaya Nagai** Department of Physics, Kyoto University, Kyoto 606-8502, Japan

**Yutaka Nagata** RIKEN, Laser technology laboratory, 2-1 Hirosawa, Wako-shi, Saitama 351-0198, Japan

**Yoshiki Nakai** Advanced Photon Research Center, KANSAI Research Establishment, Japan Atomic Energy Research Institute, Kizu, Soraku-gun, Kyoto 619-0215, Japan

**K. Nakajima** Japan Atomic Energy Agency, 8-1 Umemida, Kizu, Kyoto 619-0215, Japan and High Energy Accelerator Research Organization, 1-1 Oho, Tsukuba, Japan

**Takashi Nakajima** Institute of Advanced Energy, Kyoto University, Gokasho, Uji, Kyoto 611-0011, Japan and Institute for Solid State Physics, The University of Tokyo, Kashiwa, Chiba 277-8581, Japan

**K. G. Nakamura** Institute of Molecular Science, Okazaki, 444-8585, Japan

**S. Nakamura** Advanced Photon Research Center, Kansai Research Establishment, Japan Atomic Energy Research Institute, 8-1 Umemidai, Kizu, Kyoto 619-0215, Japan and Institute for Chemical Research, Kyoto University, Gokasho, Uji, Kyoto 611-0011, Japan

**Hidetoshi Nakano** NTT Basic Research Laboratories, NTT Corporation, 3-1 Morinosato Wakamiya, Atsugi, Kanagawa 243-0198, Japan

**Masaharu Nakazawa** Department of Quantum Engineering and Systems Science, School of Engineering, University of Tokyo, Hongo 7-3-1, Bunkyo-ku, Tokyo 113-8656, Japan

**Chang Hee Nam** Department of Physics and Coherent X-ray Research Center, KAIST, 373-1, Kuseong-dong, Yuseong-gu, Daejeon 305-701, Korea

**Keisuke Narita** Department of Applied Physics, Hokkaido University, Kita-13, Nishi-8, Sapporo, 060-8628 Japan

**T. Nayuki** Central Research Institute of Electric Power Industry, 2-6-1 Nagasaka, Yokosuka, Kanagawa 240-0196, Japan

**K. Nemoto** Central Research Institute of Electric Power Industry, 2-6-1 Nagasaka, Yokosuka, Kanagawa 240-0196, Japan

**P. V. Nickles** Max-Born-Institut fr Nichtlineare Optik und Kurzzeitspektroskopie, Max-Born-Strasse 2A, D-12489 Berlin, Germany

**Tadashi Nishikawa** NTT Basic Research Laboratories, NTT Corporation, 3-1 Morinosato Wakamiya, Atsugi, Kanagawa 243-0198, Japan

**Hajime Nishioka** Institute for Laser Science, University of Electro-Communications, 1-5-1 Chofugaoka, Chofu, Tokyo 182-8585, Japan

**M. Nishiuchi** Advanced Photon Research Center, Kansai Research Establishment, Japan Atomic Energy Research Institute, 8-1 Umemidai, Kizu, Kyoto 619-0215, Japan

**M. Nisoli** National Laboratory for Ultrafast and Ultraintense Optical Science, CNR – INFM, Dipartimento di Fisica, Politecnico, Milano, Italy

**A. Noda** Institute for Chemical Research, Kyoto University, Gokasho, Uji, Kyoto 611-0011, Japan

**Young-Chul Noh** Femto-Science Group, Advanced Photonics Research Institute, Gwangju Institute of Science and Technology, 1 Oryong-dong Buk-gu Gwangju 500-712, Republic of Korea

**Hiraku Ogino** Tohoku University, 2-1-1 Katahira, Aoba-ku, Sendai 980-8577, Japan

**K. Ogura** Advanced Photon Research Center, Kansai Research Establishment, Japan Atomic Energy Research Institute, 8-1 Umemidai, Kizu, Kyoto 619-0215, Japan

**Katsuya Oguri** NTT Basic Research Laboratories, NTT Corporation, 3-1 Morinosato Wakamiya, Atsugi, Kanagawa 243-0198, Japan

**Hideyuki Ohtake** Aisin Seiki Co., Ltd., Kojiritsuki, Hitotsugi-cho, Kariya 448-0003, Japan

**Y. Oishi** Central Research Institute of Electric Power Industry, 2-6-1 Nagasaka, Yokosuka, Kanagawa 240-0196, Japan

**Yasuaki Okano** NTT Basic Research Laboratories, NTT Corporation, 3-1 Morinosato Wakamiya, Atsugi, Kanagawa 243-0198, Japan

**Tomoya Okino** Laser Technology Laboratory, RIKEN, 2-1 Hirosawa, Wakoshi, Saitama 351-0198, Japan and Department of Chemistry, Graduate School of Science, University of Tokyo, Hongo, Bunkyo-ku, Tokyo 113-0033, Japan

**Thomas Oksenhendler** FASTLITE, Bât 403, Ecole Polytechnique, 91128 Palaiseau, France

**S. Ono** Institute for Molecular Science (IMS), Myodaiji, Okazaki 444-8585, Japan

**S. Orimo** Advanced Photon Research Center, Kansai Research Establishment, Japan Atomic Energy Research Institute, 8-1 Umemidai, Kizu, Kyoto 619-0215, Japan

**Ci-Ling Pan** Department of Photonics and Institute of Optoelectronic Engineering, National Chiao Tung University, Taiwan, ROC

**A. S. Pirozhkov** Advanced Photon Research Center, Kansai Research Establishment, Japan Atomic Energy Research Institute, 8-1 Umemidai, Kizu, Kyoto 619-0215, Japan and Division of Optics, P. N. Lebedev Physical Institute of the Russian Academy of Sciences, 53 Leninskiy prospekt, 119991 Moscow, Russia *and* Advanced Photon Research Center, Japan Atomic Energy Research Institute, 8-1 Umemidai, Kizu-cho, Soraku-gun, Kyoto 619-0215, Japan and Division of Optics, P. N. Lebedev Physical Institute of the Russian Academy of Sciences, 53 Leninskiy prospekt, 119991 Moscow, Russia

**Romeric Pobre** Department of Physics, De La Salle University, 2401 Taft Avenue, Manila 1004, Philippines

**L. Poletto** Laboratory for Ultraviolet and X-Ray Optical Research – CNR-INFM, D.E.I. – Università di Padova, Padova, Italy

**Carlito Ponseca Jr.** Institute for Molecular Science (IMS), Myodaiji, Okazaki 444-8585, Japan

**A. Quema** Institute for Molecular Science (IMS), Myodaiji, Okazaki 444-8585, Japan and Department of Physics, De La Salle University, 2401 Taft Avenue, Manila 1004, Philippines

**F. D. Recoleto Jr.** National Institute of Physics, University of the Philippines, Quezon City, Philippines

**T. Remetter** Department of Physics, Lund Institute of Technology, P. O. Box 118, SE-221 00 Lund, Sweden

**A. Renault** Laboratoire d'Optique Appliquée ENSTA, École Polytechnique, CNRS UMR 7639, 91761 Palaiseau cedex, France and Amplitude Technologies, 2 rue du Bois Chaland, CE 2926, 91029 Evry cedex, France

**G. De los Reyes** Institute for Molecular Science (IMS), Myodaiji, Okazaki 444-8585, Japan

**J. P. Rousseau** Laboratoire d'Optique Appliquée, ENSTA-Ecole Polytechnique, Chemin de la Hunière, F-91761 Palaiseau Cedex, France

**A. Sagisaka** Advanced Photon Research Center, Kansai Research Establishment, Japan Atomic Energy Research Institute, 8-1 Umemidai, Kizu, Kyoto 619-0215, Japan *and* Advanced Photon Research Center, Japan Atomic Energy Research Institute, 8-1 Umemidai, Kizu-cho, Soraku-gun, Kyoto 619-0215, Japan

**N. Saito** National Institute of Advanced Industrial Science and Technology (AIST), Tsukuba Central 2, 1-1-1 Umezono, Tsukuba, Ibaraki 305-8568, Japan

**S. Saito** KARC, National Institute of Information and Communications Technology, 588-2 Iwaoka, Nishi-ku, Kobe, Hyogo, 651-2492, Japan

**J. Saitoh** Materials and Structures Laboratory, Tokyo Institute of Technology, Nagatsuta, 4259 Midori, Yokohama 226-8503, Japan

**Hirofumi Sakai** Department of Physics, Graduate School of Science, The University of Tokyo, 7-3-1 Hongo, Bunkyo-ku, Tokyo 113-0033, Japan

**K. Sakai** KARC, National Institute of Information and Communications Technology, 588-2 Iwaoka, Nishi-ku, Kobe, Hyogo, 651-2492, Japan

**Y. Sakakibara** National Institute of Advanced Industrial Science and Technology, AIST, 1-1-1 Umezono, Tsukuba, Ibaraki 305-8563, Japan

**P. Salières** CEA/DSM/DRECAM/SPAM, Bât. 522, CEA-Saclay, 91191 Gif-sur-Yvette, France

**S. M. Saltiel** Faculty of Physics, University of Sofia, 5 J. Bourchier Boulevard, BG-1164, Sofia, Bulgaria

**A. A. Salvador** National Institute of Physics, University of the Philippines, Quezon City, Philippines *and* National Institute of Physics, University of the Philippines-Diliman, Quezon City 1101, Philippines

**G. Sansone** National Laboratory for Ultrafast and Ultraintense Optical Science, CNR – INFM, Dipartimento di Fisica, Politecnico, Milano, Italy

**A. dos Santos** Laboratoire d'Optique Appliquée, École Nationale Supérieure de Techniques Avancées, École Polytechnique, F91761 Palaiseau, France

**N. Sarukura** Institute for Molecular Science (IMS), Myodaiji, Okazaki 444-8585, Japan

**T. R. Schibli** National Institute of Advanced Industrial Science and Technology, AIST, 1-1-1 Umezono, Tsukuba, Ibaraki 305-8563, Japan

**Thomas R. Schibli** Department of Electrical Engineering and Computer Science and Research Laboratory of Electronics, Massachusetts Institute of Technology, Cambridge, Massachusetts 02139, USA

**Wolfgang Seitz** High Q Laser (US), Inc., 118 Waltham Street, Watertown, MA 02472, USA

**Taro Sekikawa** Institute for Solid State Physics, University of Tokyo, 5-1-5 Kashiwanoha, Kashiwa 277-8581, Japan

**Julia V. Sheldakova** Moscow State Open University, Sudostroitelnaya Str.18, Bld.5, Moscow, 115407, Russia

**Toshihiko Shimizu** Laser Technology Laboratory, RIKEN, 2-1 Hirosawa, Wako-shi, Saitama 351-0198, Japan

**H. Shimosato** Graduate School of Engineering Science, Osaka University, 1-3 Machikaneyama-cho, Toyonaka, Osaka, 560-8531, Japan

**T. Shioyama** Department of Electronics and Electrical Engineering, Keio University, Japan

**S. De Silvestri** National Laboratory for Ultrafast and Ultraintense Optical Science, CNR – INFM, Dipartimento di Fisica, Politecnico, Milano, Italy

**A. S. Somintac** National Institute of Physics, University of the Philippines, Quezon City, Philippines *and* National Institute of Physics, University of the Philippines-Diliman, Quezon City 1101, Philippines

**S. Stagira** National Laboratory for Ultrafast and Ultraintense Optical Science, CNR – INFM, Dipartimento di Fisica, Politecnico, Milano, Italy

**H. Stiel** Max-Born-Institut fr Nichtlineare Optik und Kurzzeitspektroskopie, Max-Born-Strasse 2A, D-12489 Berlin, Germany

**A. Suda** RIKEN, 2-1 Hirosawa, Wako-shi, Saitama 351-0198, Japan

**Toshiharu Sugiura** Aisin Seiki Co., Ltd., Kojiritsuki, Hitotsugi-cho, Kariya 448-0003, Japan

**Jae Hee Sung** Femto-Science Group, Advanced Photonics Research Institute, Gwangju Institute of Science and Technology, 1 Oryong-dong Buk-gu Gwangju 500-712, Republic of Korea

**M. Suzuki** The Institute for Solid State Physics, The University of Tokyo, 5-1-5 Kashiwanoha, Kashiwa 277-8581, Japan

**Atsushi Syouji** KARC, National Institute of Information and Communications Technology, Nishi Ku, Kobe, Hyogo 651-2492, Japan

**T. Tajima** Japan Atomic Energy Agency, 8-1 Umemida, Kizu, Kyoto 619-0215, Japan *and* Advanced Photon Research Center, Japan Atomic Energy Research Institute, 8-1 Umemidai, Kizu-cho, Soraku-gun, Kyoto 619-0215, Japan

**Hideyuki Takada** National Institute of Advanced Industrial Science and Technology (AIST) Tsukuba Central 2, 1-1-1 Umezono, Tsukuba 305-8568, Japan

**K. A. Tanaka** Graduate School of Engineering, Osaka University, 2-1 Yamadaoka, Suita, 565-0871, Japan and Institute of Laser Engineering, Osaka University, 2-6 Yamadaoka, Suita, 565-0871, Japan

**Koichiro Tanaka** Department of Physics, Kyoto University, Kyoto 606-8502, Japan

**M. Tanimoto** Meisei University, 2-1-1 Hodokubo, Hino, Tokyo 191-8506, Japan

**C. Y. Teisset** Max-Planck-Institut für Quantenoptik, Hans-Kopfermann-Strasse 1, D-85748, Garching, Germany

**Takahiro Teramoto** Department of Chemistry, School of Science, The University of Tokyo, 7-3-1 Hongo Bunkyo-ku, Tokyo 113-0033, Japan

**Michael Thorpe** JILA, National Institute of Standards and Technology and University of Colorado, 440 UCB Boulder, CO 80309, USA

**M. V. Tognetti** CLOQ/Departamento de Física, Faculdade de Ciências, Universidade do Porto, R. do Campo Alegre 687, 4169-007 Porto, Portugal

**M. Tokumoto** National Institute of Advanced Industrial Science and Technology, AIST, 1-1-1 Umezono, Tsukuba, Ibaraki 305-8563, Japan

**Hitoshi Tomita** Institute for Laser Science, University of Electro-Communications, 1-5-1 Chofugaoka, Chofu, Tokyo 182-8585, Japan

**Kenji Torizuka** National Institute of Advanced Industrial Science and Technology (AIST) Tsukuba Central 2, 1-1-1 Umezono, Tsukuba 305-8568, Japan

**R. Torres** The Blackett Laboratory, Imperial College London, Prince Consort Road, London SW7 2BW, UK

**Pierre Tournois** FASTLITE, Bât 403, Ecole Polytechnique, 91128 Palaiseau, France

**Rick Trebino** School of Physics, Georgia Institute of Technology, Atlanta, Ga 30332-0430, USA

**Koichi Tsuji** Advanced Photon Research Center, KANSAI Research Establishment, Japan Atomic Energy Research Institute, Kizu, Soraku-gun, Kyoto 619-0215, Japan

**J. Tümmler** Max-Born-Institut fr Nichtlineare Optik und Kurzzeitspektroskopie, Max-Born-Strasse 2A, D-12489 Berlin, Germany

**L. Turi** Max-Planck-Institut für Quantenoptik, Hans-Kopfermann-Strasse 1, D-85748, Garching, Germany

**Ken-ichi Ueda** Institute for Laser Science, University of Electro-Communications, 1-5-1 Chofugaoka, Chofu, Tokyo 182-8585, Japan

**Kiyoshi Ueda** Institute of Multidisciplinary Research for Advanced Materials, Tohoku University, Sendai 980-8577, Japan

**T. Utsumi** Electronics and Computer Science, Faculty of Science and Engineering, Tokyo University of Science, Yamaguchi, Daigaku-dori 1-1-1, Onoda city, Yamaguchi, 756-0884, Japan

**K. Varjú** Department of Physics, Lund Institute of Technology, P. O. Box 118, SE-221 00 Lund, Sweden

**R. Velotta** Coherentia–CNR–INFM and Dipartimento di Scienze Fisiche, Università di Napoli "Federico II," Napoli, Italy

**P. Villoresi** Laboratory for Ultraviolet and X-Ray Optical Research – CNR-INFM, D.E.I. – Università di Padova, Padova, Italy

**C. Vozzi** National Laboratory for Ultrafast and Ultraintense Optical Science, CNR – INFM, Dipartimento di Fisica, Politecnico, Milano, Italy

**Ching-yue Wang** Ultrafast Laser Laboratory, School of Precision Instrument and Optoelectronics Engineering, Tianjin University, Tianjin 300072, China

**Ziyang Wang** School of Physics, Georgia Institute of Technology, Atlanta, Ga 30332-0430, USA

**Shuntaro Watanabe** Institute for Solid State Physics, The University of Tokyo, Kashiwa, Chiba 277-8581, Japan *and* Institute for Solid State Physics, University of Tokyo, 5-1-5 Kashiwanoha, Kashiwa 277-8581, Japan

**T. Watanabe** Utsunomiya University, 7-1-2 Yoto, Utsunomiya, Tochigi 321-8585, Japan

**Benoit Wattelier** Phasics, Campus de l'Ecole Polytechnique, Bat.404, 91128 Palaiseau Cedex, France

**Zubin Wu** Ultrafast Laser Laboratory, School of Precision Instrument and Optoelectronics Engineering, Tianjin University, Tianjin 300072, China

**Qirong Xing** Ultrafast Laser Laboratory, School of Precision Instrument and Optoelectronics Engineering, Tianjin University, Tianjin 300072, China

**M. H. Xu** Institute of Physics and China Academy of Sciences, Beijing 100080, China

**Zhizhan Xu** State Key Laboratory of High Field Laser Physics, Shanghai Institute of Optics and Fine Mechanics, Chinese Academy of Sciences, PO Box 800-211, Shanghai 201800, China

**M. Yamagiwa** Advanced Photon Research Center, Kansai Research Establishment, Japan Atomic Energy Research Institute, 8-1 Umemidai, Kizu, Kyoto 619-0215, Japan

**Koichi Yamakawa** Advanced Photon Research Center, KANSAI Research Establishment, Japan Atomic Energy Research Institute, Kizu, Soraku-gun, Kyoto 619-0215, Japan

**Yoichi Yamamoto** Advanced Photon Research Center, KANSAI Research Establishment, Japan Atomic Energy Research Institute, Kizu, Soraku-gun, Kyoto 619-0215, Japan

**K. Yamanouchi** Department of Chemistry, School of Science,The University of Tokyo, Japan and Laser Technology Laboratory, RIKEN, 2-1 Hirosawa, Wako-shi, Saitama 351-0198, Japan and Department of Chemistry, Graduate School of Science, University of Tokyo, Hongo, Bunkyo-ku, Tokyo 113-0033, Japan

**Mikio Yamashita** Department of Applied Physics, Hokkaido University, Kita-13, Nishi-8, Sapporo, 060-8628, Japan

**H. Yazawa** Department of Electronics and Electrical Engineering, Keio University, Japan

**Jun Ye** JILA, National Institute of Standards and Technology and University of Colorado, 440 UCB Boulder, CO 80309, USA

**A. Yogo** Advanced Photon Research Center, Kansai Research Establishment, Japan Atomic Energy Research Institute, 8-1 Umemidai, Kizu, Kyoto 619-0215, Japan

**Makoto Yoshida** Aisin Seiki Co., Ltd., Kojiritsuki, Hitotsugi-cho, Kariya 448-0003, Japan

**Akira Yoshikawa** Tohoku University, 2-1-1 Katahira, Aoba-ku, Sendai 980-8577, Japan

**Dai Yoshitomi** National Institute of Advanced Industrial Science and Technology (AIST) Tsukuba Central 2, 1-1-1 Umezono, Tsukuba 305-8568, Japan

**Tae Jun Yu** Femto-Science Group, Advanced Photonics Research Institute, Gwangju Institute of Science and Technology, 1 Oryong-dong Buk-gu Gwangju 500-712, Republic of Korea

**Shin-ichi Zaitsu** Department of Applied Chemistry, Graduate School of Engineering, Kyushu University, 6-10-1, Hakozaki, Higashi-ku, Fukuoka 812-8581, Japan

**J. Zhang** Institute of Physics and China Academy of Sciences, Beijing 100080, China

**Zhigang Zhang** Ultrafast Laser Laboratory, School of Precision Instrument and Optoelectronics Engineering, Tianjin University, Tianjin 300072, China and Institute of Quantum Electronics, School of Electronics Engineering and Computer Science, Peking University, Beijing 100871, China

**Zing-Yung Zhang** Department of Physics, Georgia Southern University, Statesboro, GA 30460, USA

**A.M. Zheltikov** Physics Department, M.V. Lomonosov Moscow State University, Moscow 11992, Russia

**Xiangyu Zhou** Institute for Solid State Physics, University of Tokyo, 5-1-5 Kashiwanoha, Kashiwa 277-8581, Japan

**A. Zoubir** Graduate School of Engineering, Osaka University, 2-1 Yamadaoka, Suita, 565-0871, Japan and Institute of Laser Engineering, Osaka University, 2-6 Yamadaoka, Suita, 565-0871, Japan

# Part I

# Attosecond Metrology

# Progress in Attosecond Metrology

R. Kienberger[1] and F. Krausz[1,2]

[1] Max-Planck Institut für Quantenoptik, Hans-Kopfermann-Straße 1, D-85748 Garching, Germany.
[2] Ludwig-Maximilians-Universität München, Coulombwall 1, D-85748 Garching, Germany.

**Summary.** Fundamental processes in atoms, molecules, as well as condensed matter are triggered or mediated by the motion of electrons inside or between atoms. Electronic dynamics on atomic length scales tends to unfold within tens to thousands of attoseconds (1 as = $10^{-18}$ s). Recent breakthroughs in laser science are now opening the door to watching and controlling these hitherto inaccessible microscopic dynamics. The key to accessing the attosecond time domain is the control of the electric field of (visible) light, which varies its strength and direction within less than a femtosecond (1 fs = 1000 as). Atoms exposed to a few oscillation cycles of intense laser light are able to emit a single XUV burst lasting less than 1 fs. Full control of the evolution of the electromagnetic field in laser pulses comprising a few wave cycles have recently allowed the reproducible generation and measurement of isolated 250-as XUV pulses, constituting the shortest reproducible events and fastest measurement to date. These tools have enabled us to visualize the oscillating electric field of visible light with an attosecond "oscilloscope" and observing the motion of electrons in and around atoms in real time. Recent experiments hold promise for the development of an attosecond hard X-ray source, which may pave the way toward 4D electron imaging with subatomic resolution in space and time.

**PACS:** 32.80.Fb; 42.55.Vc; 42.65.Ky; 42.65.Re

Efforts to access ever shorter time scales are motivated by the endeavor to explore the microcosm in ever smaller dimensions. Recently, femtosecond laser techniques have allowed control and tracing of molecular dynamics and the related motion of atoms on the length scale of internuclear separations without the need for resolving the objects of scrutiny in space [1]. Laser light consisting of a few, well-controlled field oscillations [2] extends these capabilities to the interior of atoms, which allows to control and track electron dynamics on an atomic scale of time. Measurements of ever shorter intervals of time and tracing of dynamics within these intervals rely on reproducible generation of ever briefer events and on probing techniques of corresponding resolution. The briefest events produced until recently have been pulses of visible laser light, with durations of around 5 fs [3, 4]. Traditionally, the fastest measurement techniques have used the envelope of these laser pulses for sampling [5]. Recently, subfemtosecond bunching of femtosecond (>10 fs) extreme ultraviolet light (XUV) was observed in two-color [6,7] and two-photon [8] ionization experiments, and evidence of subfemtosecond confinement of XUV emission from few-cycle-driven (ionizing) atoms was also obtained [9]. However, time-domain technique has not been capable of resolving the time structure of subfemtosecond transients.

Herein we first report measurements and applications with an apparatus, the atomic transient recorder (ATR), which allows the reconstruction of atomic processes with a resolution within the Bohr orbit time, which is around 150 as. An accurately controlled few-cycle wave of visible light takes "tomographic images" of the time–momentum distribution of electrons ejected from atoms following sudden excitation. From these images, the temporal evolution of both the emission intensity and the initial momentum of electrons can be retrieved on a subfemtosecond time scale. Probing primary (photoexcited or collisionally excited) and secondary (Auger) electrons yields insight into excitation and relaxation processes. The transients can be triggered by an isolated attosecond electron or photon burst synchronized to the probing light field oscillations. The technique draws on the basic principle of a streak camera [10]–[12] (Fig. 1), where a light

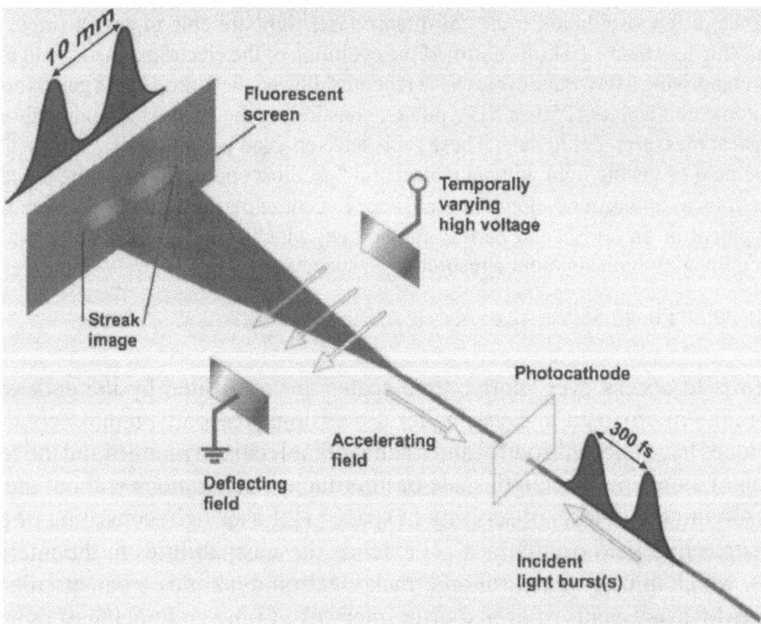


FIGURE 1. Principle of a streak camera. In its lifetime, a light flash ejects electrons from a metal plate which are then accelerated with a static electric field to a fluorescent screen. Before the electrons hit the screen, they are deflected with another field increasing linearly with time. The temporally varying deflection "streaks" the point of impact of the electrons on the screen. The spatial width of this "streak image" ($\Delta x$) is directly proportional to the duration of the electron emission, i.e., the duration of the light flash ($\Delta t$). The faster the deflecting field varies, the shorter are the pulses that can be recorded. The most modern streak cameras attain a resolution in the region of 100 fs.

pulse generates an electron bunch having exactly the same temporal structure. Deflection of the electrons in an electric field allows reconstruction of the duration of the electron bunch. By measuring the temporal evolution of the emission intensity and the momentum distribution of positive-energy electrons, the ATR [13] provides direct temporal insight into the rearrangement of the electronic shell of excited atoms on a subfemtosecond scale.

The electric field of linearly polarized femtosecond laser pulses, if sufficiently strong, induces—in a highly nonlinear interaction—gigantic dipole oscillations by pulling an electron out of the atom and smashing it back toward its core half a cycle later. The oscillations are not sinusoidal but contain very high frequency components extending into the XUV and soft-X-ray regimes [3]. In a laser field containing many oscillation cycles, the oscillations are repeated quasiperiodically, resulting in emission of a series of high-energy bursts of subfemtosecond duration and high-order harmonics of the laser radiation in the spectral domain. For a few-cycle laser driver, only a few dipole oscillations, different in amplitude, occur. The oscillation with the highest amplitude has been predicted to produce a single burst in the spectral range of the highest emitted photon energies [2]. The first evidence of the emission of an isolated subfemtosecond pulse from this interaction was reported in 2001 [9]. The implementation of the same concept—mapping the temporal distribution of a bunch of rapid particles to a spatial distribution by affecting their motion in a time-dependent manner—with photoelectrons allowed electron-optical chronoscopy by image-tube streak cameras [11, 12]. From the streaked image of the photoelectron bunch it is possible to infere with subpicosecond resolution the temporal structure of the light pulse triggering the photoemission. The time resolution of these devices is limited by the spread of the electron transit time due to a spread of their initial momentum. Herein we report temporal characterization of subfemtosecond electron emission from atoms by drawing on the same basic concept.

The three orders of magnitude improvement in time resolution results from several modifications of image-tube streak cameras: (i) the electric field of light is used for affecting the electrons' motion, (ii) the field is virtually jitter-free and (iii) applied along the direction of electron motion—implying their acceleration or deceleration instead of their deflection—(iv) directly at the location and instant of emission. While (i) implies "only" an enhanced streaking speed, the consequences of (ii)–(iv) are much more profound: (ii) allows the timing of the probing field to be systematically varied with an accuracy within the electron bunch length, and owing to (iii) this capability results in "projecting" the initial time–momentum distribution of the electron emission into a series of different final momentum distributions, whereas (iv) prevents the initial momentum spread from introducing any measurement error. As a result of (ii)–(iv), not only the spread of electron momenta stops imposing a limitation on the time resolution, but its possible temporal variation during the emission can also be captured just as that of the emission intensity.

Inspired by the physics of the first subfemtosecond experiment [9], Corkum and coworkers put forward the basic concept for ATR metrology [14], also analyzed with a comprehensive quantum theory by Brabec and coworkers [15]. Let us

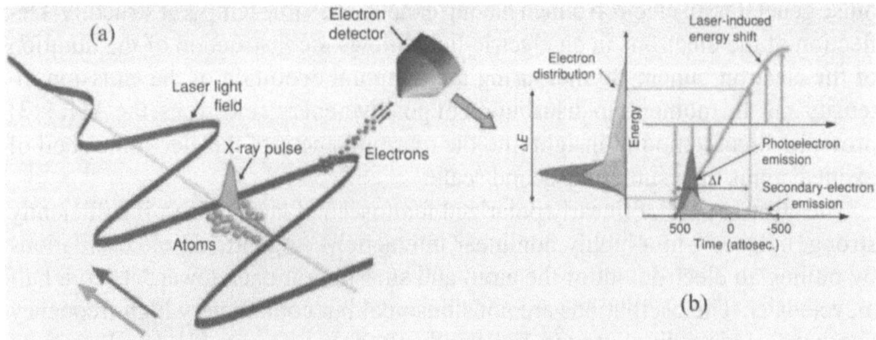


FIGURE 2. Light-field-controlled streak camera. The dark and light gray curves in (a) represent, respectively, the evolutions of the electric field of the laser light and the intensity of the attosecond X-ray flash. The latter excites the atoms, which then emit electrons. The electrons ejected in the direction of the electric field of the laser light pulse simultaneously beamed in are detected. They undergo—depending on the time of their emission within half the oscillation period of the laser light—a change of velocity (in the case illustrated the electrons emitted first are decelerated, while those released on termination of the X-ray flash are accelerated). In this manner, the successively emitted electrons are detected separately again, but this time not spatially on a screen but alongside one another on the energy scale. The width $\Delta E$ and shape of the measured energy distribution (vertical axis in (b) of the electrons reflect the duration and evolution of the electron emission, just as their spatial distribution in conventional streak imaging. In this case, however, "deflection" occurs within half a light period, which opens the way to measurement in the attosecond region.

consider electron emission from atoms exposed to a subfemtosecond X-ray burst in the presence of an intense, linearly polarized, few-cycle laser field $E_{\mathrm{L}}(t) = E_0(t)\cos(\omega_{\mathrm{L}} t + \varphi)$, where $\omega_{\mathrm{L}}$ is the frequency and $\varphi$ is the phase of the laser electric field. The momentum of the freed electrons is changed by $\Delta p = eA_{\mathrm{L}}(t)$ along the laser field vector. Here

$$A_{\mathrm{L}}(t_{\mathrm{r}}) = \int_{t_{\mathrm{r}}}^{\infty} E_{\mathrm{L}}(t)\,dt \tag{1}$$

is the vector potential of the laser field, $e$ stands for the charge of the electron and $t_{\mathrm{r}}$ is the release time of the electron. This momentum transfer (see arrows in Fig. 2b) maps the temporal emission profile into a similar distribution of final momenta $p_{\mathrm{f}} = p_{\mathrm{i}} + p$ within a time window of $T_0/2 = \pi/\omega_{\mathrm{L}}$, if the electrons' initial momentum $p_{\mathrm{i}}$ is constant in time and their emission terminates within $T_0/2$. Under these conditions, the temporal evolution of the electron emission can be retrieved from a single "streaked" momentum distribution. However, any sweep of the electrons' initial momentum revokes the unique correspondence between the electron's final momentum and release time, preventing retrieval of accurate temporal information from single streak records.

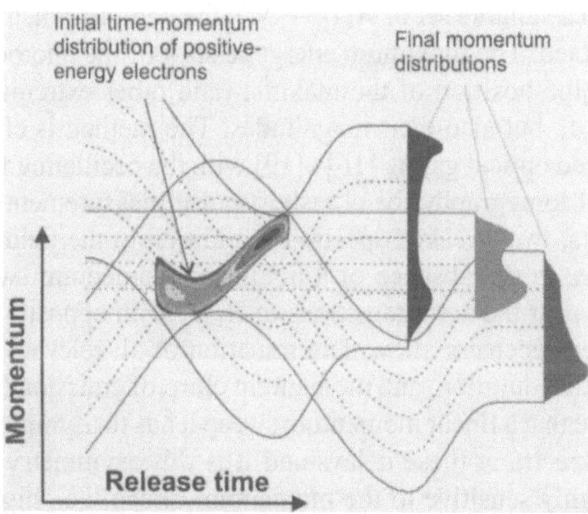


FIGURE 3. Time-dependent momentum transfer to positive-energy electrons within the laser cycle. Electrons emitted from an excited atom along the electric field vector of a strong laser field with an initial momentum $p_i$ suffer a momentum change that is proportional to the vector potential of the field $A_L(t)$ at the instant of release. In the absence of the laser field, the electron momenta do not change after detachment and accumulation of electrons with a given final momentum (mathematically, time-integration) follows lines of constant momentum (straight lines) yielding the field-free final momentum distribution (middle profile). In the presence of a probing field, accumulation of electrons with a given final momentum occurs along the lines of constant $p_i + eA_L(t)$ (sinusoidal lines). The corresponding projections (streaked spectra) are represented in dark gray. From a suitable set of such tomographic projections, the time–momentum distribution of electron emission can be retrieved, providing direct time-domain insight into atomic dynamics triggered by an attosecond excitation (in our experiments XUV pulse) synchronized to the probing laser field.

In general, the initial momentum spectrum of electrons detached from atoms by impulsive excitation varies in time during emission. A representative time–momentum distribution $n_e(p, t)$ of the electron emission rate is depicted in Fig. 3. The final electron momentum spectrum (observed after the laser pulse has left the interaction region), $\sigma(p) = \int_{-\infty}^{\infty} n_e(p, t)\, dt$ can be viewed as the projection of the time-dependent momentum distribution on the momentum space along the lines of constant $p$. In the classical description of the freed electrons' motion in the strong laser field, the final spectra obtained in the presence of a strong laser field are generalized projections along the lines where $p_f = p_i + eA_L(t)$ is constant (see Fig. 3). By delaying the laser field with respect to the the excitation that triggers electron emission, we obtain a set of tomographic records (briefly streaked spectra),

$$\sigma_A(p) = \int_{-\infty}^{\infty} n_e(p - eA_L(t), t)\, dt, \tag{2}$$

from which, with a suitable set of $A_L(t - \Delta t)$, the complete distribution $n_e(p, t)$ can be reconstructed. The maximum energetic shift of the photoelectron spectra reveals not only the position of the maxima (and other extreme values) of the vector potential $A_L$ but also their magnitudes. The method is closely related to frequency-resolved optical gating [16]– [19] with the oscillating field as gate and other concepts of tomography for ultrashort pulse measurements [20]– [25]. In the simplest cases, two streaked spectra (in addition to the field-free spectrum) may be sufficient. In the absence of a nonlinear momentum sweep, the streak records obtained near the zero transitions of $A_L(t)$ with opposite slopes together with the field-free spectrum allow determination of all relevant characteristics: the temporal profile, duration, and momentum chirp of emission (including high-order chirp). Note that a linear momentum sweep leads to asymmetric broadening of the streaked spectra at these delays and it is this asymmetry that makes the measurement highly sensitive to the momentum sweep, i.e., highly sensitive to deviations of the pulse duration from the Fourier limit.

According to an intuitive one-active-electron model of Schafer et al. [26,27], Corkum [28], and Lewenstein et al. [29], ionization of atoms by a linearly polarized laser field is accompanied by emission of XUV radiation due to recombination of the detached electron into its original ground state upon recollision with the parent ion. These theories along with a number of numerical simulations [30]– [34] predict emission of the highest-energy (cutoff) photons to occur near the zero transition(s) of the electric field following the most intense half cycle(s) at the pulse peak in a few-cycle driver. In a cosine waveform ($\varphi = 0$) this cutoff emission is expected to be confined to a single bunch. The photoelectrons ejected in the direction of the peak electric field at this instant suffer maximum increase of their momentum and hence of their energy. The streaked spectrum at this instant (delay 1.2 fs in Fig. 4), obtained with 5-fs, 750-nm cosine pulses used for both X-ray generation and subsequent electron streaking, corroborates this prediction. The photoelectron spectrum peaking at $\hbar\omega_{\text{x-ray}} - W_b \approx 72$ eV (where $\omega_{\text{x-ray}} \approx 93.5$ eV is the center of the X-ray spectrum selected by the Mo/Si mirror and $W_b = 21.5$ eV the binding energy of the most weakly bound valence electrons in Ne) in the absence of the laser field is upshifted by some 10 eV with only a few electrons scattered outside the shifted band. The clear upshift is consistent with a single, isolated X-ray burst coinciding with the zero transition of the laser electric field. Possible satellites would appear at the adjacent zero transitions of $E_L(t)$ and suffer an energy downshift. With isolated sub-fs X-ray pulses at our disposal, atomic transients can now be triggered and their subsequent evolution be captured by probing electron emission with a synchronized wave of laser light. In the first ATR measurements presented here, the objects of scrutiny were photoelectrons. The streaking field in these experiments was produced by blocking the internal part of the laser beam with a zirconium filter (transmitting the X-ray pulse) and focusing the transmitted annular beam on the target with the external section of a Mo/Si mirror [9], which can be delayed with respect to the internal section that focuses the X-ray beam. Figure 4 shows

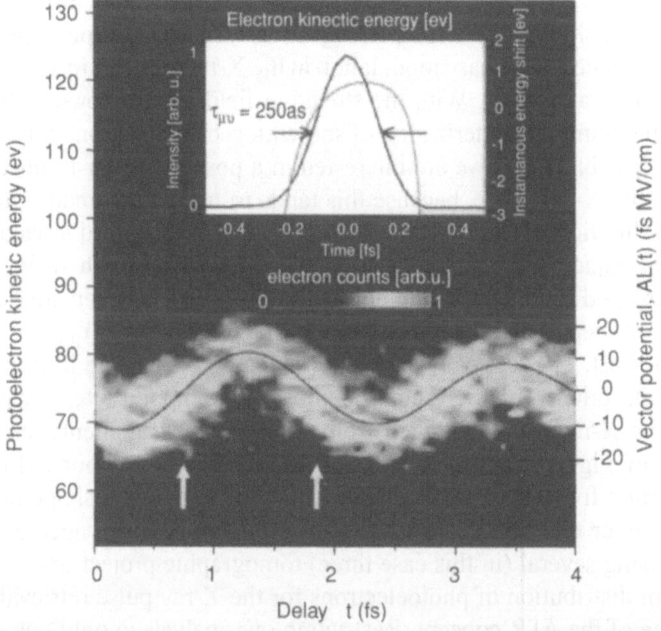


FIGURE 4. ATR measurement: A series of tomographic projections (streaked kinetic energy spectra) of the initial time–momentum distribution of photoelectrons ejected by a single subfemtosecond X-ray pulse (in false-color representation). A few-cycle laser pulse with a cosine waveform and a normalized duration of $\tau_L/T_0 = 2$ was used for both generating the single subfemtosecond excitation pulse and probing photoelectron emission in the atomic transient recorder. Black line: $A_L(t)$ of the probing field evaluated from the peak shift of the streaked spectra (see scale on the right-hand side). Inset: Temporal intensity profile and energy sweep of the subfemtosecond X-ray excitation pulse evaluated from the ATR measurements. The basis of the calculation was the unperturbed and two streaked spectra at the zero transitions of the laser electric vector potential marked by the two white arrows.

a series of streaked spectra of photoelectrons emitted from neon as a function of $\Delta t$.

If the electrons are emitted with an initial kinetic energy much larger than their average quiver energy in the laser field, and temporally confined to a fraction of the half oscillation cycle, theory predicts that their energy shift is linearly proportional to $p$ and hence to the vector potential at the instant of release of the wavepacket, $\Delta W(t_r) \approx (p_i/m)\Delta p(t_r) = (ep_i/m)A(t_r)$, where $p_i$ is the initial momentum and $m$ is the rest mass of the electron. As a consequence, $A_L(t)$ and hence $E_L(t)$ can be accurately determined from the peak shifts of the spectra—without having to analyze their detailed structure. The result is shown by the black line in Fig. 4, constituting the first direct (time-resolved) measurement of a visible light field. From this measurement we can also evaluate the half-oscillation period of the electromagnetic field as $T_0/2 \approx 1$ fs. This latter value indicates a significant blue shift to a carrier wavelength of 600 nm near the pulse peak (origin, ionization-induced self-phase modulation in the X-ray source), in agreement with previous observations [9]. With the streaking field $A_L(t)$ known, full temporal (time–momentum) characterization of subfemtosecond electron emission is now becoming feasible. First, we are interested in a possible linear frequency sweep carried by the X-ray pulse, because this tends to have a dominant effect on the shape and duration of the excitation. To this end, we analyzed streaked spectra recorded at adjacent zero transitions of $A_L(t)$ (indicated with white arrows in Fig. 4). Streaked spectra recorded at these instants are most sensitive to a linear momentum sweep. We evaluated $\tau_{\text{x-ray}} = 250\,(-5/+30)$ as and $\tau_{\text{jitter}} = 260\,(\pm 80)$ as from the best. The relatively large jitter is likely to originate from mechanical vibrations introduced by the vacuum pumps. The temporal intensity profile and chirp of the X-ray pulse obtained from the ATR measurements are shown in the insert of Fig. 4. The pulse is found to be essentially Fourier-limited (the near-quadratic frequency sweep results from the asymmetric shape of the pulse spectrum rather than from a spectral phase). The remarkable accuracy of $\tau_{\text{x-ray}}$ relies on using several (in this case three) tomographic projections of the time–momentum distribution of photoelectrons for the X-ray pulse retrieval, which is the essence of the ATR concept. Restricting our analysis to only one of the two streaked spectra—in the spirit of streak camera measurements—would only allow setting an upper limit of 500 as on the X-ray pulse duration. Extension of this analysis to a series of spectra measured within a delay range of $T_0$ revealed the absence of higher-order contributions to $\phi_{\text{x-ray}}$ as well.

Figure 5 shows a scan through a complete sub 5-fs laser pulse [35]. It demonstrates impressingly that with the tools—waveform-controlled few-cycle laser pulses in combination with single sub-fs XUV pulses—in hand one is able to fully characterize the system itself (shift of spectra gains vector potential) and to control and track electrons on an atomic time scale: each slice of the trace is a projection as figured out in Fig. 3 providing information on the attosecond pulse and the electron bunch generated by it, respectively. These measurements provide evidence of the emergence of isolated, bandwidth-limited X-ray bursts over a relative spectral band as broad as 15% (15 eV at 100 eV) from recombination emission driven by a cosine waveform with $\tau_L/T_0 = 2$. This has important implications for scaling the technology up to its limit, which lies around 100 as.

Recent experiments hold promise for the development of an attosecond hard X-ray source [36], which would allow to progress from conventional XUV photoelectron spectroscopy to X-ray diffraction providing direct information on electron density distributions. Using this powerful technique may pave the way toward 4D electron imaging with subatomic resolution in space and time.

**Acknowledgments.** We gratefully acknowledge the invaluable contributions of E. Goulielmakis, M. Uiberacker, A. Baltuska, V. Yakovlev, F. Bammer, A. Scrinzi, Th. Westerwalbesloh, U. Kleineberg, U. Heinzmann, and M. Drescher to the

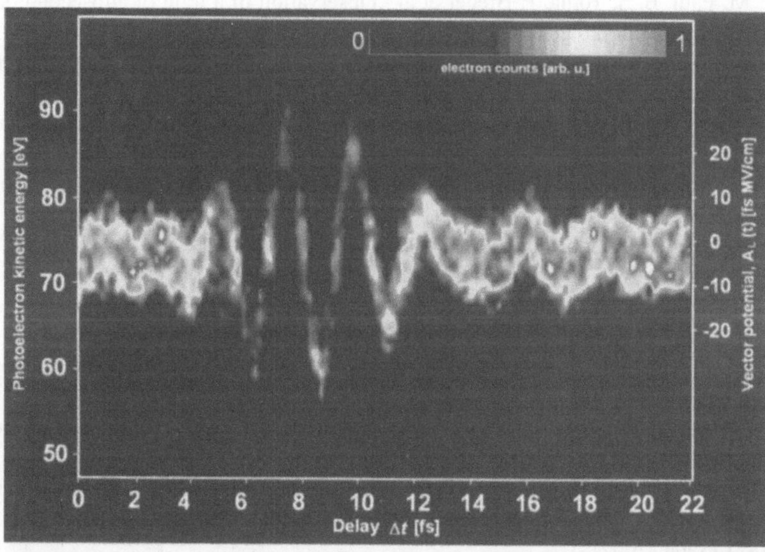


FIGURE 5. Trace of the electric field of a sub-5-fs laser pulse sampled by subfemtosecond XUV pulses. The trace is a set of some 120 spectra measured with a dressing field of phase-stabilized cosine-waveform laser pulses. The method can be used to fully characterize all parameters of the laser pulse, in particular an instantaneous frequency sweep of the pulse resulting from self-phase modulation in the target gas.

experiments reviewed in this paper. Sponsored by Fonds zur Förderung der wissenschaftlichen Forschung (Austria, grants Z63 and F016), Deutsche Forschungsgemeinschaft (Germany, grants SPP1053, HE1049/9, and KL1077/1), and the European Union's Human Potential Programme under contract HPRN-2000-00133 (Atto). R. K. is funded by an APART fellowship of the Austrian Academy of Sciences. Correspondence and request for materials should be addressed to F.K. (e-mail: ferenc.krausz@mpq.mpg.de).

## References

1. A. Zewail: Femtochemistry: atomic-scale dynamics of the chemical bond (adapted from the Nobel Lecture). J. Phys. Chem. **A104**, 5660–5694 (2000).
2. A. Baltuska, T. Udem, M. Uiberacker, et al.: Attosecond control of electronic processes by intense light fields. Nature **421**(6923), 611–615 (2003).
3. T. Brabec and F. Krausz: Intense few-cycle laser fields: frontiers of nonlinear optics. Rev. Mod. Phys. **72**, 545–591 (2000).
4. U. Keller: Recent developments in compact ultrafast lasers. Nature **424**, 831–838 (2003).
5. M. Drescher, M. Hentschel, R. Kienberger, et al.: X-ray pulses approaching the attosecond frontier. Science **291**, 1923–1927 (2001). Published online February 15, 2001; 10.1126/science.1058561.

6. P. M. Paul, E. S. Toma, P. Breger et al.: Observation of a train of attosecond pulses from high harmonic generation. Science **292**, 1689–1692 (2001).
7. Y. Mairesse, A. de Bohan, L. J. Frasinski, et al.: Attosecond Synchronisation of High-Harmonic Soft X-Rays. Science **302**, 1540–1543 (2003).
8. P. Tzallas, D. Charalambidis, N. A. Papadogiannis, K. Witte, and G. D. Tsakiris: Direct observation of attosecond light bunching. Nature **426**, 267–271 (2003).
9. M. Hentschel, R. Kienberger, Ch. Spielmann et al.: Attosecond metrology, Nature **414**, 509–513 (2001).
10. C. Wheatstonel Phil. Mag. **6**, 61 (1835).
11. D. J. Bradley, B. Liddy, and W. E. Sleat: Opt. Commun. **2**, 391 (1971).
12. M. Schelev, Ya, M. C. Richardson, and A. J. Alcock: Image-converter streak camera with picosecond resolution. Appl. Phys. Lett. **18**, 354 (1971).
13. R. Kienberger, E. Goulielmakis, M. Uiberacker et al.: Atomic transient recorder. Nature **427**, 817–821 (2004).
14. J. Itatani, F. Quéré, G. L. Yudin, M. Yu. Ivanov, F. Krausz, and P. B. Corkum: Attosecond streak camera, Phys. Rev. Lett. **88**, 173903 (2002).
15. M. Kitzler, N. Milosevic, A. Scrinzi, F. Krausz, and Th. Brabec: quantum theory of attosecond XUV pulse measurement by laser dressed photoionization, Phys. Rev. Lett. **88**, 173904 (2002).
16. D. J. Kane and R. Trebino: Characterization of arbitrary femtosecond pulses using frequency-resolved optical gating. IEEE J. Quantum Electron. **29**, 571–579 (1993).
17. T. Sekikawa, T. Katsura, S. Miura, and S. Watanabe: Measurement of the intensity-dependent atomic dipole phase of a high harmonic by frequency-resolved optical gating. Phys. Rev. Lett. **88**, 193902 (2002).
18. Vampouille et al.: J. Opt. (Paris) **15**, 385 (1984).
19. M. Kaufman et al.: Appl. Phys. Lett. **64**, 270 (1994).
20. M. Beck, M. G. Raymer, I. A. Walmsley, and Wong: Opt. Lett. **18**, 2041–2043 (1993).
21. I. A. Walmsley and V. Wong: J. Opt. Soc. Am B. **13**, 2453 (1996).
22. C. Dorrer and I. Kang: Opt. Lett. **28**, 1481 (2003).
23. M. Drescher, M. Hentschel, R. Kienberger, et al.: Time-resolved atomic inner-shell spectroscopy. Nature **419**, 803–807 (2002).
24. A. L'Huillier and P. Balcou: High-order harmonic generation in rare gases with a 1-ps 1053-nm laser. Phys. Rev. Lett. **70**, 774–777 (1993).
25. J. J. Macklin, J. D. Kmetec, and C. L. Gordon, III: High-order harmonic generation using intense femtosecond pulses. Phys. Rev. Lett. **70**, 766–769 (1993).
26. K. J. Schafer, J. L. Krause, and K. C. Kulander: Int. J. Nonlinear Opt. Phys. **1**, 245 (1992).
27. K. J. Schafer, B. Yang, L. F. DiMauro, and K. C. Kulander: Above threshold ionization beyond the high harmonic cutoff. Phys. Rev. Lett. **70**, 1599–1602 (1993).
28. P.B. Corkum: Plasma perspective on strong-field multiphoton ionization. Phys. Rev. Lett. **71**, 1994–1997 (1993).
29. M. Lewenstein, Ph. Balcou, M. Ivanov, Yu, A. L'Huillier, and P. B. Corkum: Theory of high-harmonic generation by low-frequency laser fields. Phys. Rev. A **49**, 2117–2132 (1994).
30. P. Salieres, B. Carre, L. Le Deroff, et al.: Feynman's path integral in the light of intense-laser-atom experiments. Science **292**, 902–904 (2001).
31. I. P. Christov, M. M. Murnane, and H. C. Kapteyn: High-harmonic generation of attosecond pulses in the "single-cycle" regime. Phys. Rev. Lett. **78**, 1251–1254 (1997).

32. C. Kan, N. H. Burnett, C. E. Capjack, and R. Rankin: Coherent XUV generation from gases ionized by several cycle optical pulses. Phys. Rev. Lett. **79**, 2971–2974 (1997).
33. A. de Bohan, P. Antoine, D. B. Milosevic, and B. Piraux: Phase-dependent harmonic emission with ultrashort laser pulses. Phys. Rev. Lett. **81**, 1837–1840 (1998).
34. G. Tempea, M. Geissler, and T. Brabec: Phase sensitivity of high-order harmonic generation with few-cycle laser pulses. J. Opt. Soc. Am. B. **16**, 669–674 (1999).
35. E. Goulielmakis, M. Uiberacker, and R. Kienberger, et al.: Direct measurement of light waves. Science, **305**, 1267–1269 (2004).
36. J. Seres, E. Seres, A. J. Verhoef, et al.: Nature **433**, 596 (2005).

# Attosecond Dynamics of Electron Wave Packets in Intense Laser Fields

K. Varjú, P. Johnsson, J. Mauritsson, R. López-Martens, E. Gustafsson, T. Remetter, and A. L'huillier

Department of Physics, Lund Institute of Technology, P. O. Box 118, SE-221 00 Lund, Sweden
**Katalin.Varju@fysik.lth.se**

## 1 Introduction

The continuous progress in the performances of light sources as well as detection techniques allows us to investigate and control the states of matter in even finer details. Light sources, ranging from the infrared (IR) to the extreme ultraviolet (XUV), are becoming increasingly coherent, intense, well characterized, and controlled. The shortest available light pulses are now significantly shorter than 1 fs [1–4], thus offering unique promise for studies of ultrafast electron dynamics.

The production of short light pulses requires a large spectral bandwidth where all the spectral components of the radiation add constructively during a very short instant of time. More than 10 years ago the process of high-order harmonic generation in intense laser fields [5], which provides a large spectral bandwidth in the XUV range, was proposed as a possible candidate for the production of attosecond pulse trains (APT) [6, 7]. This method has also been used to demonstrate the production of single attosecond pulses, with duration down to 250 as [2, 8, 9].

In the present paper, we review the recent studies [1, 10–12] performed on the high-harmonic line at the Lund Laser Center in collaboration with our colleagues A. S. Morlens, O. Boyko, Ph. Balcou, Ph. Zeitoun, and C. Valentin (Laboratoire d'Optique Appliquée); Y. Mairesse, H. Wabnitz, and P. Salières (Centre d'Etudes de Saclay); S. Kazamias (Laboratoire d'Interaction du rayonnement X Avec la Matière); and M. B. Gaarde and K. J. Schafer (Louisiana State University).

In this paper, we describe our method of producing on-target attosecond pulses, through filtering of high-order harmonic radiation. We use the attosecond XUV pulses—through a single photon ionisation step—to produce electron wave packets (EWPs) of comparable duration and chirp. These EWPs are generated in the presence of a synchronized IR laser field that allows for pulse characterization (perturbative probe) as well as observation of strong field effects (strong probe). Our arrangement allows us to shape the attosecond pulses corresponding to the plateau spectral region and hence the generated EWPs.

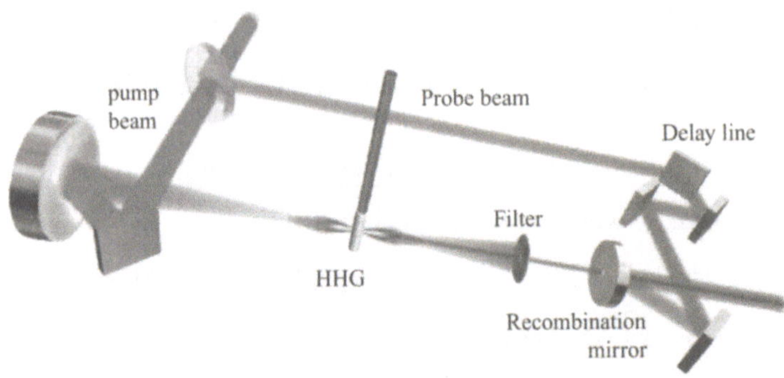


FIGURE 1. Experimental setup for the generation, postcompression, and characterization of attosecond pulses.

## 2 Experimental Setup

Figure 1 briefly shows how the XUV attosecond pulses are generated. (More details about the setup can be found, for example, in [13].) The Ti:S laser system provides 35 fs, 2 mJ pulses at 1 kHz. The laser pulse is split in two using a thin beam splitter. The (pump) pulse is focused into a 3-mm static argon gas cell (30 mbar) with a 50-cm spherical mirror. The generated harmonics are spectrally filtered by an aluminium filter (the importance of filtering is described in Section 3.2). This filter also serves the purpose of eliminating the fundamental IR light. The harmonics are passed through a 1.5-mm hole in the middle of a spherical mirror placed under vacuum. The probe pulse is sent through a variable delay stage and is recombined with the harmonic pulse after reflection on the holed mirror. The curvature of the spherical mirror is chosen such that the wavefront of the reference beam matches that of the harmonic beam. The XUV and the synchronized laser beams are then focused into the sensitive region of a magnetic bottle electron time-of-flight spectrometer, or a velocity map imaging spectrometer, to induce two-color ionization of atoms or molecules. The focusing is achieved either by a grazing incidence toroidal mirror or a zero-incidence multilayer mirror.

Recent experiments performed on the system include measurements of the relative atomic phase, refractive index of XUV-transparent materials [13] and group delay of XUV-chirped mirrors [12] (Section 3.3), and studies of generation and strong field dynamics of attosecond EWPs [10, 11].

## 3 Generation of On-target Attosecond Pulse Trains

### 3.1 High-order Harmonic Generation

High-order harmonics are generated as a nonlinear response of atoms to intense laser field. When a short laser pulse is focused into a gas of rare atoms, EWPs

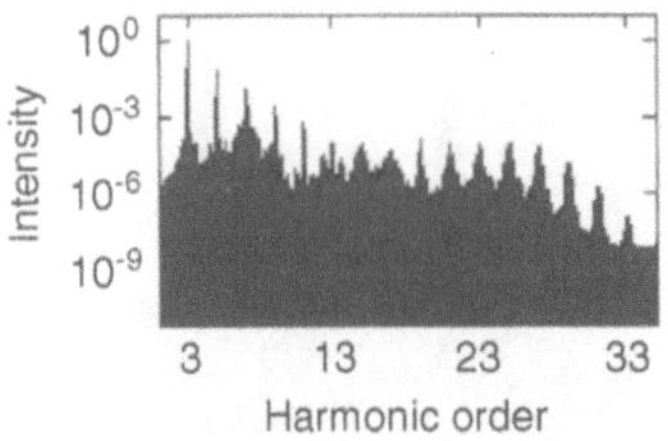


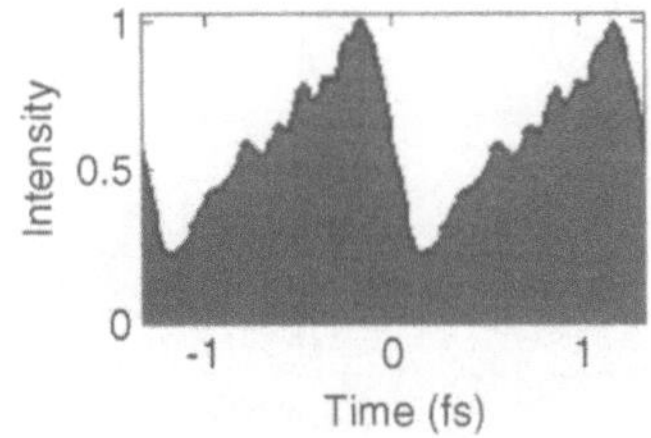


FIGURE 2. Spectral and temporal characteristics of the generated field, including all harmonics produced.

are formed twice per cycle tunneling through the barrier formed by the atomic potential lowered by the electromagnetic field and is launched into the continuum where it propagates under the influence of the laser field. Part of the wave packet is driven back to the core after approximately a half laser cycle and recombines releasing the excess energy gained in the continuum in the form of a photon. In the time domain, this leads to the emission of short XUV pulses separated by 1.3 fs (half a cycle of 800 nm radiation), while in the spectral domain discrete peaks, at odd multiples of the laser frequency, appear as a result of the periodicity of the process [14].

A typical harmonic spectrum from argon, using a laser intensity of $1.4 \times 10^{14}$ $\mathrm{W \cdot cm^{-2}}$, is shown in Fig. 2b [1]. The spectrum is calculated by integration of the time-dependent Schrödinger equation (TDSE) followed by solving Maxwell's wave equations for the generated fields in conditions close to our experimental ones. In the plateau region (harmonics 11–29), the narrow spectral peaks for the short path and lower and broader structures on both sides for the long path can be identified [14, 15].

The temporal structure within one laser cycle, corresponding to the generated harmonic spectrum introduced above, is shown in Fig. 2c. The broad structures seen are characteristic of the beating between the two dominant harmonics of the spectrum, namely the third and the fifth. The synthesis of short attosecond pulses thus obviously requires spectral filtering, achievable for example by the use of metallic filters [1, 16] or multilayer mirrors [8, 9, 17, 18].

## *3.2 Filtering the High-harmonic Radiation for APT Generation*

The production of attosecond light pulses using high-order harmonics requires to limit the spectrum to those harmonics being synchronized and comparable in amplitude [1]. A recent systematic analysis of plateau harmonics generated in argon and neon shows that the frequency components of the XUV radiation are not exactly synchronized and that the XUV pulses consequently exhibit a significant chirp (frequency variation in time) [19]. This intrinsic chirp comes from the fundamental electron dynamics responsible for high-order harmonic generation.

Our experimental setup (Fig. 1) includes an aluminium filter, which efficiently cuts the low-order harmonics till the 11th. Combined with the spectral cutoff

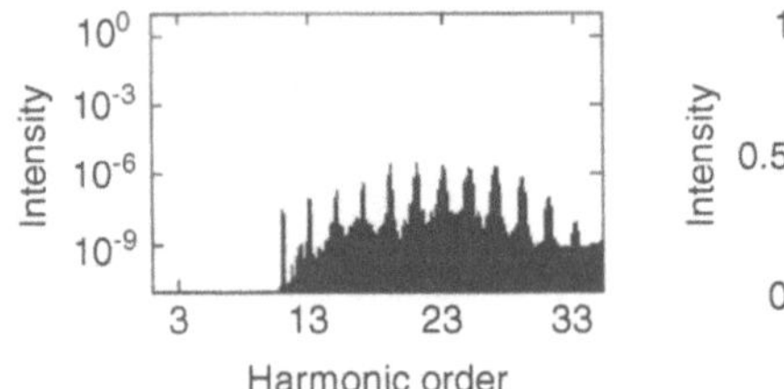


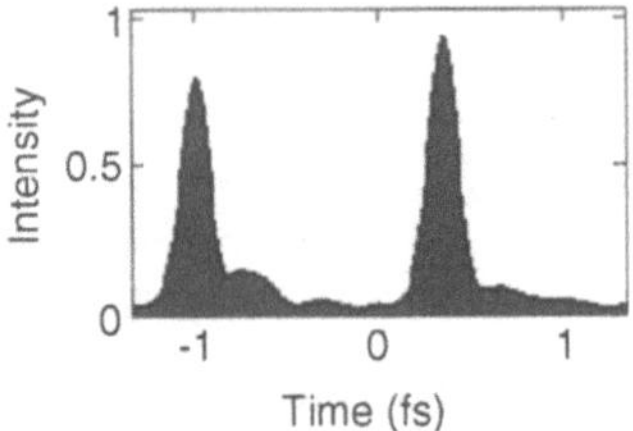


FIGURE 3. Spectral and temporal characteristics of the harmonic field, following spectral, spatial, and dispersive filtering.

in argon at higher frequencies, it then acts as a bandpass filter, which selects a spectrum centred at 30 eV with a bandwidth of 15 eV FWHM. Also, the transmitted pulses acquire a negative chirp, opposite in sign to the dispersion of the generation process [1, 16]. Another element in our experimental setup which is important for the generation of trains of short attosecond pulses is the holed recombination mirror, which cleans the transmitted harmonic beam from unwanted spectral substructures originating from longer quantum paths with higher spatial divergence [15, 20].

Figure 3 shows the theoretical result of filtering by a 600-nm thick aluminium filter and using a hard aperture to select predominantly the most collimated contribution, which corresponds to the shortest quantum path [20]. After the aperture, the individual harmonic peaks get cleaner, with strongly suppressed side structures from the more divergent long quantum paths. In the temporal domain, only one of the contributions to the previous structure remains, leaving us with short, clean pulses. We have demonstrated experimentally [1, 13] that the dispersive properties of thin aluminium filters allow us to produce on-target 170-as pulses, containing only 1.2 optical cycles of the carrier component. By varying the thickness of the aluminium filter, the chirp and hence the duration of the attosecond bursts can be taylored.

Close to the absorption peak there is a spectral region, where materials exhibit negative group delay dispersion. This makes them an ideal candidate to achieve both the spectral filtering, and dispersion compensation of the harmonic radiation. On the other hand, in terms of the wavelength tunability there is a very limited versatility of the available materials; therefore other methods, such as multilayer mirrors described in the next section, have to be investigated.

## 3.3 XUV Optics for APTs

XUV mirrors required to reflect and focus attosecond pulses should have a high reflectivity over a large spectral region and also preserve the phase-locked nature of high harmonics, responsible for the coherent buildup of attosecond pulses [17, 18]. In the XUV and soft X-ray regions the reflectivity of surfaces is very low, except for grazing incidence. A new design of multilayer structure allows us to achieve high reflectivity at normal incidence across a considerable bandwidth

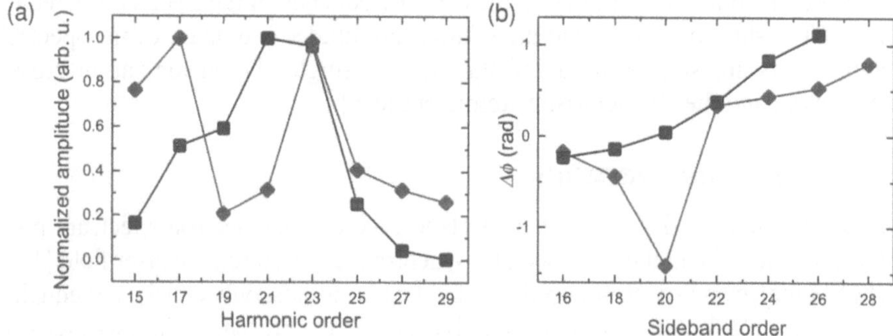


FIGURE 4. (a) Spectral amplitudes measured after focusing by a grazing incidence platinum toroidal mirror (black curve) and a normal incidence spherical multilayer mirror (red curve). For the multilayer mirror we observe a reflection minimum between harmonics 19 and 21; above that the mirror is designed to act as a bandpass filter. We do not observe the total bandwidth of the mirror as a consequence of the high-harmonic cutoff. (b) Group delay measurement of the two mirrors. The quasilinearly varying delay in case of the platinum mirror (black curve) is characteristic of the HHG process. The delay curve with the XUV mirror (red curve) is more complex: we observe the phase jump at the spectral component 20, corresponding to the reflectivity minimum of the mirror. Above this sideband order, the mirror act as a negative group delay dispersive element.

[17]. We here present measurements of the phase and reflectivity of such a mirror (designed and manufactured at the Institute d'Optique) that provides a broad and uniform reflectivity around 30 nm and compensates for the intrinsic attochirp.

We use the RABITT technique [21,22], originally developed to characterize the temporal shape of attosecond pulse trains produced by high-harmonic generation. This technique is based on measuring the group delay of the spectral components, and this is what we have exploited in the current work. For reference we have also taken data when focusing with a grazing incidence platinum mirror, which has a flat group delay response. Figure 4 shows the comparison of the harmonic amplitudes and the sideband phases for the two mirrors. The results compare well with simulated data [12].

# 4 Generation of Attosecond Electron Wave Packets

## 4.1 Single Photon Ionization

When an APT is focused to interact with an atomic gas, a train of temporally localized EWPs is generated through single-photon ionization of the atoms [23]. In this case the properties and the timing of the EWPs are determined only by the XUV pulses and the single-photon ionization parameters of the atomic gas (ionization potential, atomic phase, and cross-section).

Due to the photoionization step the carrier frequency of the EWP compared to the APT is shifted by $I_p/\hbar$ and the relative amplitudes and phase of the spectral components are slightly modified through the single-photon ionization cross-section $\sigma_q$, and the characteristic atomic phase $\phi_q^{\mathrm{at}}$.

### 4.2 Pulse Characteristics

To determine the APT pulse characteristics, we use the electron spectrum produced by the APT in the presence of a synchronized, perturbative laser field [13]. The photoelectron spectrum produced by the harmonics, only consists of equally spaced (3.1 eV for 800 nm fundamental) peaks. In the presence of a low-intensity ($10^{11}$ W $\cdot$ cm$^{-2}$) IR pulse, sidebands appear in the photoelectron spectra, due to the absorption of an harmonic and an IR photon, as well as to the absorption of the next consecutive harmonic and emission of an IR photon. The sidebands, which correspond to even harmonic orders, oscillate as a function of delay, carrying information on the relative phase of the contributing harmonics. The phase of the sidebands oscillation is $(\Delta\phi_{q+1}^{\mathrm{at}} + \Delta\Phi_{q+1})$. Combining these phases with the easily measurable amplitudes, $\sigma_q F_q$, it is possible to reconstruct the average temporal structure of the pulses in the train [19,21,22]. Interestingly we characterize here the EWPs, not directly the APT. Accounting for the atomic phase difference, and the cross-section of the ionization process, obtained from TDSE calculations [23] allows to retrieve the APT structure. For characterization of the on-target APT, care must be taken to cover the full spectrum of the harmonics, and not to impose spectral selection by the ionization potential and cross-section of the detection gas [1]. In our characterization the whole spectral range was included by using different detection gases: argon and neon. The ionization potential of argon allows for detecting all the lower order harmonics, transmitted by the aluminium filter, and neon is used to characterize the higher orders, not available in the argon scan because of its Cooper minimum. Thus our characterization indeed applies to the on-target radiation.

## 5 Interaction of Electron Wave Packets With a Strong Laser Field

An EWP injected into the continuum in the presence of an IR field will exchange momentum with the field. The nature of this momentum exchange is crucially dependent on the initial properties of the EWP, the IR intensity and the phase of the IR field at the time of injection. To be able to see the change with respect to the injection time in the photoelectron spectrum, it is further necessary that the EWP is temporally confined to a time shorter than the period of the IR field since the effect is otherwise smeared out.

In the simplest model, the interaction between the IR field and the electron is treated using classical mechanics, ignoring effects of the ion core [24]. The final momentum of the electron is given by

$$\mathbf{p}_f(t_i) = \mathbf{p}_i + e\mathbf{A}(t_i), \tag{1}$$

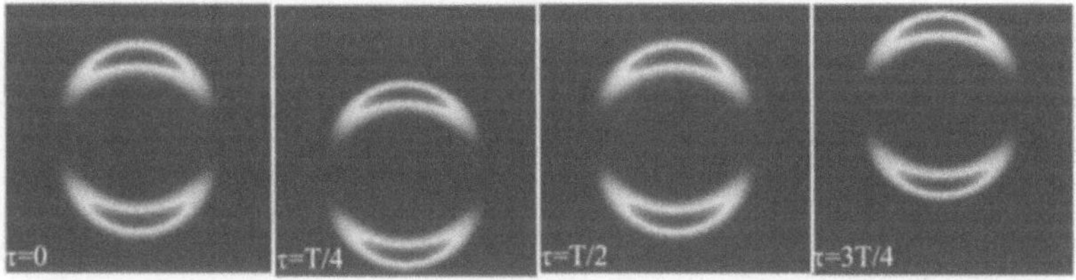


FIGURE 5. Momentum transfer between the attosecond pulse and the laser field. The shift of the photoelectron spectrum in momentum space is determined by the instantaneous value of the vector potential.

where $\mathbf{p}_i$ is the initial momentum of the electron, $e$ the electron charge, and $\mathbf{A}(t_i)$ the vector potential at the time of injection $t_i$ [10, 11]. We illustrate this effect in momentum space, using Fig. 5. If we restrict ourselves to initial momenta parallel to the IR field polarization axis by means of limiting the detection angle to 0°, we can see that the final momentum of the electron is in this case just a mapping of the vector potential of the IR field, and delay-dependent photoelectron spectra obtained using single attosecond pulses with high average energy can be used for direct measurements of electric fields [25].

## 5.1 *Momentum Transfer between the EWP and the Laser Field*

In our experiment we have used a magnetic bottle electron time-of-flight spectrometer to detect the electron spectra as a function of the delay between the ionizing APT and the strong IR field. The spectrometer is designed to collect all electrons at a $2\pi$ solid angle, but since the ionization cross-section is peaked along the polarization axis, the scalar form of Eq. 1 gives reasonably good interpretation of the results [11].

Figure 6 shows the delay-dependent photoelectron spectra when ionization takes place by a single, or a train of attosecond pulses, in the presence of a strong infinite laser field. The calculation is performed using the Strong Field Approximation [14], neglecting all effects of the atomic core [11]. The vector potential of the laser field is depicted as a red line between the figures. For the single

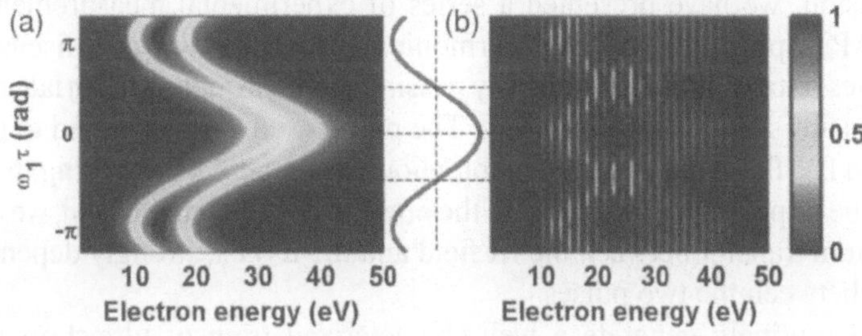


FIGURE 6. Illustrating the delay dependent momentum transfer between the laser and the EWP for a single EWP (a) and a train of EWPs spaced by half the laser period, using SFA calculation.

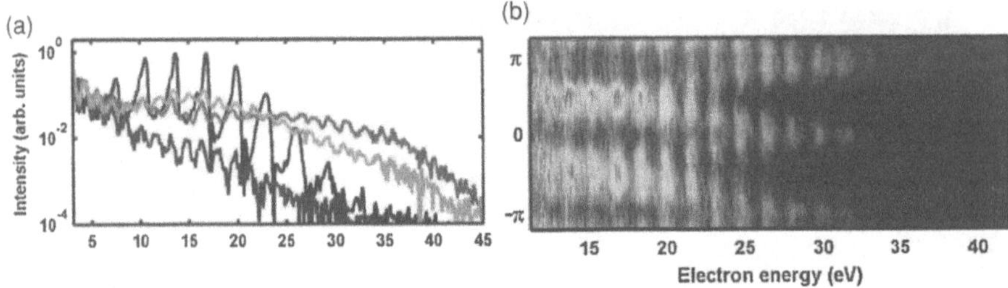


FIGURE 7. (a)Photoelectrum spectra, when Ar atoms are ionized by the IR alone (blue line), the APT alone (black line), and when both fields are present (red and green lines) at two different delays. (b) Delay-dependent electron spectrum—could have it in TOF.

pulse case we see how the electron spectrum resembles the evolution of the vector potential. Compared to the case of a single attosecond pulse, the use of an APT for the injection of the EWPs introduces two major changes. First of all the EWPs, although synchronized to the IR field, are injected twice per cycle with a phase difference of $\pi$ between them. This implies that the above-mentioned mapping of the vector potential appears twice for every delay. Secondly, the process will now be repeated with the period of the IR field, leading to the appearance of peaks spaced by the laser frequency in the photoelectron spectrum (cf. Fig. 6b).

Figure 7 shows experimental electron spectra obtained in argon as a function of the delay between the two pulses [10]. At the perfect overlap of the two pulses, the IR intensity is strong enough to steer the EWPs, and a strongly delay-dependent redistribution of the spectrum can be observed. The periodic displacement of the energy distribution toward higher energies as $\tau$ is varied is quite noticable, and electrons with energies well above the initial photoelectron distribution might be produced for certain delays.

# 6 Conclusions, Outlook

In conclusion, we have presented a series of experimental measurements concerning APTs produced via high harmonic generation. We have discussed the possibilities of tayloring these pulse by passing them through appropriate metallic filters, or using XUV chirped mirrors. The pulses were demonstrated to produce attosecond EWPs through a single photon ionization step. By tayloring the APT a chirp can be imposed on the EWPs. In the presence of a strong IR field, we observe a momentum transfer between the IR field and the EWPs, strongly dependent on the delay between the two pulses.

We can routinely generate a well-characterized train of ultrashort attosecond pulses using a commercial 2-mJ-35-fs-1-kHz Titanium Sapphire laser system, thereby showing that the attosecond time scale can be easily accessed in many laboratories. We create attosecond pulse trains in a different energy region, synthetized from plateau harmonics, than [2,8,9] and in a technologically simpler way, albeit in a train rather than as an isolated pulse.

The apparent disadvantage of the train of pulses, separated by half the period can be turned to a benefit, since the pulses separated by half a cycle will gain different phase from a synchronized laser field, and observing their interference should reveal interesting physics [26]. If for a certain experiment it is desirable to avoid the half-period repetition and the opposite in phase excitation of the system, the use of a two-color excitation produces all harmonic orders leading to a train of pulses separated by the laser period [27].

**Acknowledgments.** This research was supported by a Marie Curie Intra-European Fellowship (MEIF-CT-2004-009268), the Marie Curie Research Training Network XTRA (MRTN-CT-2003-505138), Access to Research Infrastructures activity in the Sixth Framework Programme of the EU (RII3-CT-2003-506350, Laserlab Europe), the Knut and Alice Wallenberg Foundation, and the Swedish Science Council. K. V. is on leave from Department of Optics and Quantum Electronics, University of Szeged, Szeged, Hungary. J. M. is presently at the Department of Physics and Astronomy, Louisiana State University, Baton Rouge, Louisiana 70803-4001, USA, R. L-M is currently at Laboratoire d'Optique Appliquée, ENSTA, France.

## References

1. R. López-Martens, K. Varjú, P. Johnsson, et al.: Phys. Rev. Lett. **94**, 033001 (2005).
2. R. Kienberger, E. Goulielmakis, M. Uiberacker, et al.: Nature **427**, 817 (2004).
3. P. Agostini and L. F. DiMauro: Rep. Prog. Phys. **67**, 813 (2004).
4. G. A. Reider: J. Phys. D **37**, R37 (2004).
5. M. Ferray, A. L'Huillier, X. F. Li, et al.: J. Phys. B **21**, L31 (1988).
6. Gy. Farkas and Cs. Toth: Phys. Lett. A **168**, 447 (1992).
7. S. E. Harris, J. J. Macklin, and T. W. Hänsch: Opt. Comm. **100**, 487 (1993).
8. M. Hentschel, R. Kienberger, Ch. Spielmann, et al.: Nature **414**, 509 (2001).
9. R. Kienberger, M. Hentschel, M. Uiberacker, et al.: Science **297**, 1144 (2002).
10. P. Johnsson, R. López-Martens, S. Kazamias, et al.: Phys. Rev. Lett. **95**, 013001 (2005).
11. P. Johnsson, K. Varju, T. Remetter, et al.: J. Mod. Opt. **53**, 233–245 (2006).
12. A-S. Morlens, K. Varjú, E. Gustafsson, et al.: manuscript in preparation.
13. K. Varjú, P. Johnsson, R. López-Martens, et al.: Laser Phys. **15**, 888 (2005).
14. M. Lewenstein, Ph. Balcou, M. Yu. Ivanov, et al.: Phys. Rev. A **49**, 2117 (1994).
15. P. Salières, B. Carré, L. Le Déroff, et al.: Science **292**, 902 (2001).
16. K. T. Kim, C. M. Kim, M-G. Baik, et al.: Phys. Rev. A **69**, 051805(R) (2004).
17. A-S. Morlens, Ph. Balcou, Ph. Zeitoun, et al.: Opt. Lett. **32**, 1540 (2005).
18. W. Wonisch, Th. Westerwalbesloh, W. Hachmann, et al.: Thin Solid Films **464**, 473 (2004).
19. Y. Mairesse, A. de Bohan, L. J. Frasinski, et al.: Science **302**, 1540 (2003).
20. M. Bellini, C. Lynga, A. Tozzi, et al.: Phys. Rev. Lett. **81**, 297 (1998).
21. P. M. Paul, E. S. Toma, P. Breger, et al.: Science **292**, 1689 (2001).

22. H. G. Muller: Appl. Phys. B **74**, S17 (2002).
23. J. Mauritsson, M. B. Gaarde, and K. J. Schafer: Phys. Rev. A **72**, 013401 (2005).
24. F. Quéré, Y. Mairesse, and I. Itatani: J. Mod. Opt. **52**, 339 (2005).
25. E. Goulielmakis, M. Uiberacker, R. Kienberger, et al.: Science **305**, 1267 (2004).
26. T. Remetter, P. Johnsson, J. Mauritsson, et al.: Nature Phys. **2**, 323–326 (2006).
27. J. Mauritsson, P. Johnsson, E. Gustafsson, et al.: Phys. Rev. Lett. **97**, 013001 (2006).

# Part II

# Attosecond High Harmonic Generation

# Attosecond Pulse Characterization by XUV Nonlinear Optics

Taro Sekikawa, Atsushi Kosuge, Teruto Kanai, and Shuntaro Watanabe

Institute for Solid State Physics, University of Tokyo, 5-1-5 Kashiwanoha, Kashiwa 277-8581, Japan

We demonstrate the generation of intense isolated pulses from a single harmonic (photon energy 27.9 eV) by using sub-10-fs blue laser pulses. Nonadiabatic evolution of the blue laser pulse produces a large dipole moment at the relatively low (ninth) harmonic order and the ionization of neutral gas interacting with the laser electric field shortens the harmonic pulse duration, resulting in the high-peak intensity at focus. A nonlinear optical process in the extreme ultraviolet (XUV), i.e., two-photon above-threshold ionization, was successfully observed. XUV pulses with a temporal duration of 950 as were characterized by the autocorrelation technique and were found to be close to Fourier transform limit, supported by a numerical simulation of the single-atom response to a laser electronic field.

## 1 Introduction

High harmonic generation is one of the nonlinear optical phenomena and is now drawing a great deal of attention in attosecond pulse generation and attophysics [1–4]. Attosecond pulses were generated by using the few-cycle laser [3] and by synthesizing several mode-locked harmonic orders [1]. Few-cycle laser pulses produce single, isolated attosecond pulses at the peak of the laser pulse, while the mode-locked harmonics form an attosecond pulse train with an interval of a few femtosecond. Attosecond pulses are expected to be a versatile tool to probe ultrafast phenomena such as electron motion and the Auger decay after inner-shell excitation [5, 6] in unprecedented temporal resolution.

Another interesting application of high harmonic pulses is nonlinear optics in the extreme ultraviolet (XUV) and soft X-ray regions. Compared with the synchrotron radiation (SR), high harmonic pulses have temporal coherence and are expected to cause nonlinear optical phenomena, which have never been observed by using conventional SR. So, high harmonics provide us a unique opportunity to investigate unexplored research fields. The investigation of XUV nonlinear optics by high harmonics has been continuing, since the first demonstration of the autocorrelation measurement of the ninth harmonic of a Ti:sapphire laser (photon energy 14 eV), in which the two-photon ionization of He atoms was utilized [7]. Recently, the above-threshold ionization (ATI) of He atoms was observed for the first time by using the fifth harmonic of KrF laser (photon energy 25 eV) [8]. ATI is a multiphoton-ionization process, where atoms absorb more photons than required for ionization. The cross section is one order of magnitude smaller than the normal two-photon ionization process [9]. Thus, the observation of ATI demonstrates the versatility of high harmonics in XUV nonlinear optics.

ATI caused by high harmonic pulses motivated us to characterize attosecond pulses by the autocorrelation measurement. The autocorrelation technique is the simplest type of pump and probe experiments and is a good mixture of ultrafast and nonlinear spectroscopy. In this work, the ninth harmonic (photon energy = 27.9 eV) of a blue laser was measured. The photon energy of the ninth harmonic was well above the ionization energy of target gas. In this paper, we present our approach to the autocorrelation measurement of attosecond pulses.

## 2 Autocorrelation Measurement

In the ATI process observed, He atoms absorb two ninth harmonic photons in the two-photon process and then emit photoelectrons with a kinetic energy of 31.2 eV (Fig. 1**a**). Here, it should be noted that the photon energy is larger than the ionization energy of a He atom (24. 6 eV). According to a theoretical calculation, the cross section of the two-photon ionization of He atoms at this photon energy is approximately $10^{-52}$ cm$^2$, which is one order of magnitude smaller than ever observed [9].

Experimentally, the ninth harmonic was generated by focusing the blue laser, the second harmonic of a Ti:sapphire laser, into a Ar-gas jet. The generated high

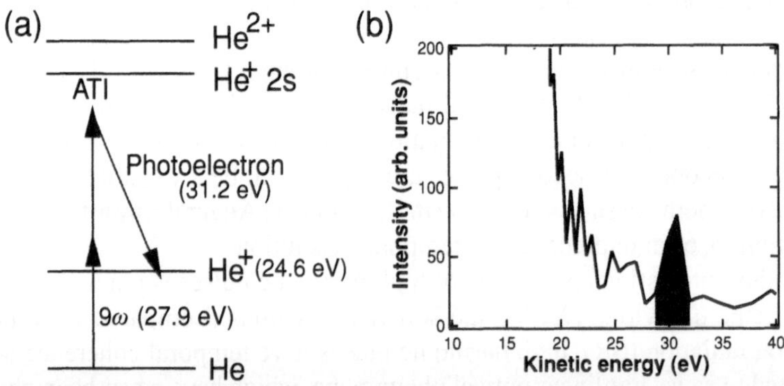


FIGURE 1. (**a**) Energy diagram of the two-photon ATI in a helium atom. 9$\omega$ indicates the ninth harmonics of the blue laser. (**b**) Photoelectron spectrum of He atoms. The hatched peak comes from the two-photon ATI by the ninth harmonic.

harmonic pulses were separated from the blue pulses by an Al filter. The ninth harmonic was selected and focused into He gas by a spherical Sc/Si multilayer mirror with a focal length of 5 cm. The pulse energy was estimated to be 1 nJ at focus. The ejected photoelectrons were collected and energy-resolved by a magnetic bottle photoelectron spectrometer with a collection efficiency of ca 50%. The flight length of electrons was 50 cm, and the retardation voltage was applied to the flight tube to improve the spectral resolution. The electron number was counted by a multichannel scaler, assuring high detection efficiency. The measured photoelectron spectrum is shown in Fig. 1b. The hatched peak comes from the ATI and was well separated from the other peaks from the one-photon process induced by the residual other harmonics.

For the autocorrelation measurement, two replicas of the ninth harmonic pulses were produced by focusing two spatially divided laser beams independently into an Ar-gas jet operated at 1 kHz in a vacuum chamber. The peak intensity of the blue laser pulse was $3.9 \times 10^{14}\ \mathrm{W \cdot cm^{-2}}$. Since the driving laser beam had high spatial coherence, assured by the spatial interference, and since the pulse energies of the divided beams were same, two harmonic pulses were almost identical. The pulse energy of one beam was 1 nJ. The time difference between the harmonic pulses was scanned by changing the relative delay time between the two blue laser pulses. The autocorrelation traces were recorded by accumulating the electron number ejected by ATI at each delay time.

The autocorrelation traces obtained by using 8.3- and 12-fs blue laser pulses are shown in Figs. 2a and 2b, respectively. The ratio of the peak to the background was almost 3:1, in good agreement with the ideal case of the autocorrelation measurement, indicating that the peak is not the coherent spike. To estimate the pulse durations of the ninth harmonics, the autocorrelation traces were fitted to Gaussian functions; the results are shown by the solid lines in Fig. 2. The

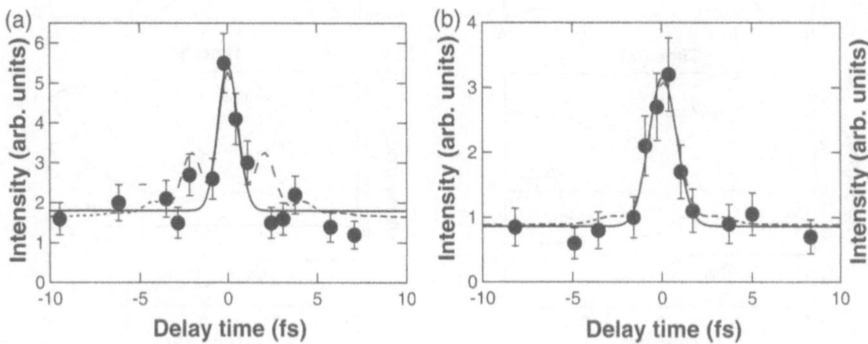


FIGURE 2. Autocorrelation traces and the spectra of the ninth harmonic of the blue laser. Autocorrelation traces of the ninth harmonic pulses generated by using (a) 8.3-fs and (b) 12-fs blue lasers. The error bars show the standard deviation of each data point. The solid lines show the fitting results. The broken lines are the calculated autocorrelation traces from the spectra of the ninth harmonic.

corresponding full widths at half maximum of the autocorrelation traces were $1.3 \pm 0.1$ and $1.8 \pm 0.1$ fs, resulting in pulse durations of $950 \pm 90$ as and $1.3 \pm 0.1$ fs, respectively.

Although we have assumed a single peak in the fitting, bumps appeared around the main peak in the 950-as pulse. Temporal profile cases were calculated by the Fourier transform of a harmonic spectrum with an assumption of a flat phase in the frequency domain and are shown in Figs. 3a and 3d. The pulse duration of the main peak is 940 as, which is close to the estimated value from the autocorrelation trace. The calculated autocorrelation traces are shown by the broken lines in Fig. 2. Both the autocorrelation traces of the 950-as and 1.3-fs pulse are reproduced well. Thus, the estimated pulse durations are valid.

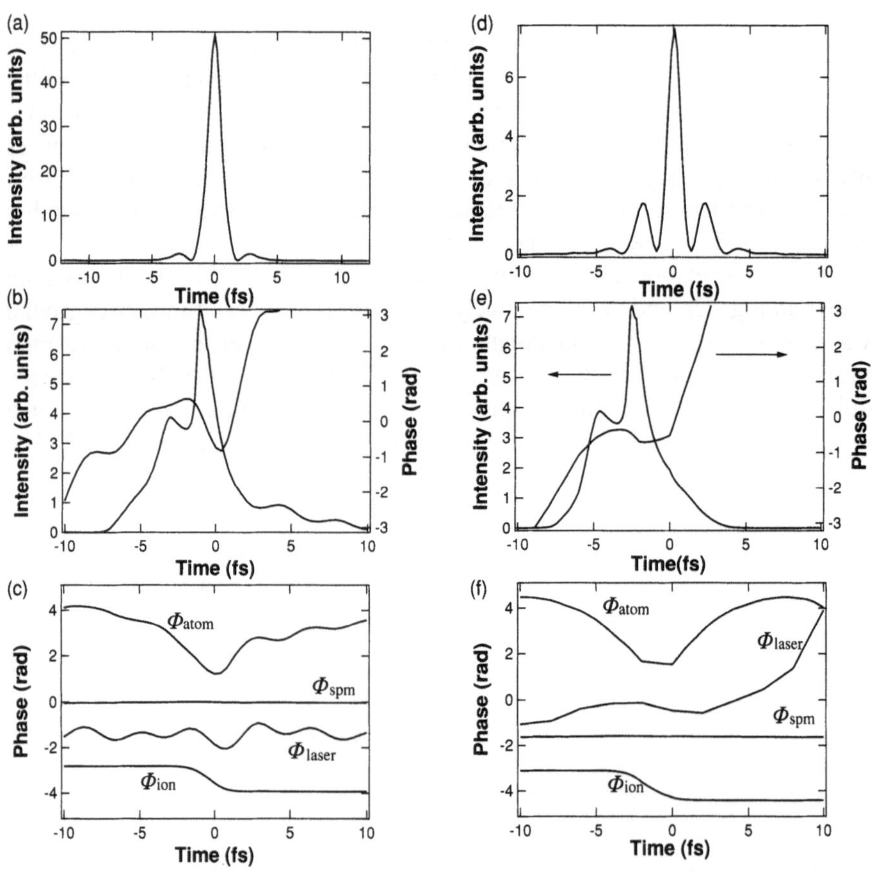


FIGURE 3. (a) Temporal profile of the ninth harmonic calculated from a spectrum with an assumption of a flat phase. (b) Temporal pulse shape and phase of the ninth harmonic obtained by a numerical simulation. (c) Temporal dependence of four phase factors: phase modulation by ionization $\phi_{\mathrm{ion}}$, atomic dipole phase $\phi_{\mathrm{atom}}$, self-phase modulation $\phi_{\mathrm{spm}}$, and the chirp from the driving laser $\phi_{\mathrm{laser}}$.

## 3 Phase of Attosecond Pulses

In the present experimental condition, the interacting gas medium was ionized efficiently and high harmonics should suffer from the phase modulation significantly. However, the experimental results suggest that the generated harmonic pulses are close to Fourier transform limit. To gain insight into the phase feature, we have made numerical simulations of high harmonic generation [10]. Herein, the single atom response was calculated by using a zero-range potential model [11], taking account of the atomic dipole phase [12], ionization, self-phase modulation, and the chirp of the fundamental pulse. The pulse shape and phase of the driving pulse used in the simulation were characterized by the self-diffraction frequency-resolved optical gating technique. Pulse propagation was not taken into account.

Results are shown in Figs. 3b and 3e. There still remain some phase dispersions. But the phase variation is almost constant or linear within the main peaks, which is consistent with the generation of nearly Fourier transform limited pulses. Figs. 3c and 3f show the phase components, $\phi_{\mathrm{ion}}$, $\phi_{\mathrm{atom}}$, $\phi_{\mathrm{spm}}$ , and $\phi_{\mathrm{laser}}$, coming from four processes, ionization, the atomic dipole phase, self-phase modulation, and the chirp of the fundamental pulse, respectively. From the simulations, we found that high harmonics are generated in the rising edge of the driving laser pulse, where the ionization occurs efficiently and $\phi_{\mathrm{ion}}$ and $\phi_{\mathrm{atom}}$ are dominant. These two components are almost linear in time, leading to Fourier transform limited pulses.

For further pulse shortening to the attosecond timescale with the use of a multicycle driving laser, one experimental approach is to generate higher-order harmonics; the lower-order harmonics rise earlier than the higher-order harmonics, resulting in a longer pulse duration. Higher-order harmonics would therefore be adequate for attosecond pulse generation, and a driving laser with shorter pulse duration is also required.

## References

1. P. M. Paul, E. S. Toma, P. Breger, et al.: Science **292**, 1689 (2001).
2. E. Goulielmakis, M. Uiberacker, R. Kienberger, et al.: Science **305**, 1267 (2004).
3. R. Kienberger, E. Goulielmakis, M. Uiberacker, et al.: Nature **427**, 817 (2004).
4. T. Sekikawa, A. Kosuge, T. Kanai, and S. Watanabe: Nature **432**, 605 (2004).
5. M. Drescher, M. Hentschel, R. Kienberger, et al.: Nature **419**, 803 (2002).
6. T. Shimizu, T. Sekikawa, T. Kanai, S. Watanabe, and M. Itoh: Phys. Rev. Lett. **91**, 017401 (2003).
7. Y. Kobayashi, T. Sekikawa, Y. Nabekawa, and S. Watanabe: Opt. Lett. **23**, 64 (1998).
8. N. Miyamoto, M. Kamei, D. Yoshitomi, et al.: Phys. Rev. Lett. **93**, 083903 (2004).
9. L. A. A. Nikolopoulos and P. Lambropoulos: J. Phys. B **34**, 545 (2001).
10. T. Sekikawa, T. Kanai, and S. Watanabe: Phys. Rev. Lett. **91**, 103902 (2003).
11. W. Becker, S. Long, and J. K. McIver: Phys. Rev. A **50**, 1540 (1994).
12. M. Lewenstein, P. Salieres, and A. L'Huillier: Phys. Rev. A **52**, 4747 (1995).

# Spectral Interference of Attosecond Light Pulses

G. Sansone,[1] E. Benedetti,[1] J-P. Caumes,[1] L. Poletto,[2] P. Villoresi,[2] S. Stagira,[1] C. Vozzi,[1] S. De Silvestri,[1] and M. Nisoli[1]

[1] National Laboratory for Ultrafast and Ultraintense Optical Science, CNR – INFM, Dipartimento di Fisica, Politecnico, Milano, Italy
mauro.nisoli@fisi.polimi.it

[2] Laboratory for Ultraviolet and X-Ray Optical Research – CNR-INFM, D.E.I. – Università di Padova, Padova, Italy

## 1 Introduction

Attosecond pulses can be produced by using the process of high-order harmonic generation in noble gases [1]. Few-cycle driving pulses allow one to confine the emission process to a single optical cycle so that single X-ray sub-femtosecond pulses can be generated [2,3]. In the case of multicycle driving pulses, trains of attosecond pulses are produced [1,4,5]. Recently, it has been demonstrated that the use of driving pulses with stable carrier-envelope phase (CEP) allows one to control the spectral distribution of the soft X-ray radiation generated by few-cycle infrared (IR) pulses [6]. From both a fundamental and a technical point of view, it is essential to understand the CEP effects on the phase of the harmonics and on the electric-field evolution of the attosecond pulses.

In this work, we present a self-referencing technique for the measurement of the effect of CEP of the driving pulses on the phase difference between consecutive harmonics [7]. Such a method is based on the measurement of the beat pattern produced by interference of attosecond pulses. The experimental results have been analyzed using the nonadiabatic saddle-point (NASP) method [8,9], which turns out to be in excellent agreement with the measurements. This numerical method provides a complete physical interpretation of the effects of the CEP on the harmonic generation process. Moreover, using the NASP method we demonstrate that, in particular experimental conditions, the temporal evolution of the electric field of the attosecond pulses can be directly controlled by the CEP of the driving pulses.

## 2 Experimental Results: Measurement of Harmonic Phase Difference

Harmonic emission has been produced focusing femtosecond light pulses (800-nm central wavelength) into an Ar jet. The gas jet was located around the position of the laser focus in order to increase the contribution of the long quantum paths due to phase-matching processes [10]. The harmonic radiation is sent to the spectrometer by a grazing-incidence toroidal mirror. The spectrometer is composed of one toroidal mirror mounted at grazing incidence for stigmatic imaging with almost unitary magnification, and one variable-line-spaced grating, mounted after the mirror [11]. The resulting spectrum is stigmatic and almost flat even in a wide spectral region. The detector is a multichannel-plate intensifier with phosphor screen optically coupled to a low-noise fast-readout CCD camera. The CEP of the IR pulses has been stabilized as described in [6] and varied by introducing in the beam path a glass plate with variable thickness.

Upon increasing the intensity of the driving pulses, the harmonic peaks broaden and eventually overlap in the spectral region between consecutive odd harmonics, where distinct spectral peaks, whose position is CEP dependent, are formed, as shown in Fig. 1. This is due to the fact that the temporal variation of the driving pulse intensity produces a harmonic chirp, which broadens the spectrum of each individual harmonic. This effect is more significant in the case of the long quantum paths. If the spectral broadening is larger than the frequency separation between consecutive harmonics, the high-frequency side of the $q$th-harmonic spectrum, generated on the leading edge of the IR pulse, overlaps the low-frequency side of the spectrum of $(q + 2)$th harmonic, generated on the pulse trailing edge, thus giving rise to the observed interference effect.

In order to analyze the CEP dependence of the interference pattern, we have then varied the CEP, $\psi$, of the driving pulses moving the glass wedges. Figure 2 shows the portion between 13th and 15th harmonics of nine spectra for different amounts $\delta z$ of glass in the beam path, corresponding to different CEPs in the range

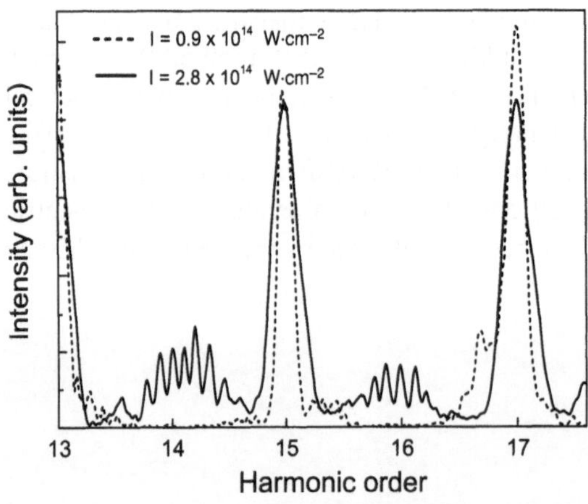


FIGURE 1. Normalized harmonic spectra generated in Ar by 35-fs pulses, with stable CEP, at two peak intensities.

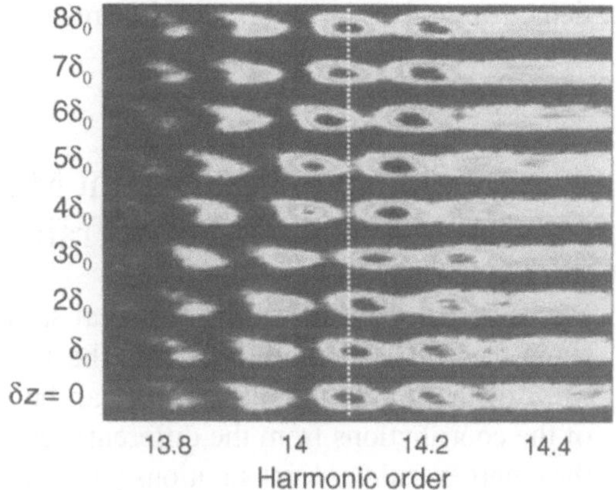


FIGURE 2. Portion between the 13th and 15th harmonics of XUV spectra generated in Ar by phase stabilized pulses, for different amounts $\delta z$ of glass in the beam path ($\delta_0 = 3.42$ μm).

$\psi_0 < \psi < \psi_0 + \pi$. Each horizontal line represents a spectrum measured at a fixed CEP. The positions of the interference peaks continuously shift by changing the pulse CEP. The beat pattern periodically changes for a CEP variation $\Delta\psi = \pi$. The same behavior was observed for all the pairs of consecutive harmonics. Using the algorithm of Fourier transform spectral interferometry [12] it is possible to retrieve, from the interference pattern, the phase difference, $\Delta\phi_q(\omega) = \phi_{q+2}(\omega) - \phi_q(\omega)$, between consecutive harmonics in the overlapping region. The inset of Fig. 3 shows the retrieved $\Delta\phi_{13}(\omega)$ for two CEP values. A CEP variation $\Delta\psi$ determines a $2\,\Delta\psi$ variation of $\Delta\phi_{13}(\omega)$. This is clearly shown in Fig. 3, which displays $\Delta\phi_{13}(\omega = 14\omega_0)$ as a function of $\Delta\psi$: the experimental points can be well fitted by a line (dashed line in Fig. 3) with slope 2. The same results are valid for all the other pairs of consecutive harmonics. From the experimental results we can conclude that the harmonic phase, $\phi_q(\omega)$, is related to the CEP of the driving pulses by the following expression: $\phi_q(\omega) = \theta_q(\omega) + q\psi$, where $\theta_q(\omega)$ does not depend on $\psi$. This conclusion is confirmed by the results of numerical simulations based on the use of the NASP method. Moreover, such calculations

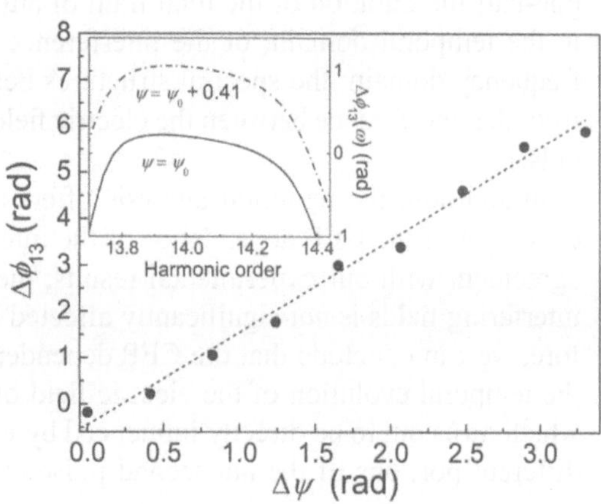


FIGURE 3. $\Delta\phi_{13}(\omega = 14\omega_0)$ as a function of the CEP variation $\Delta\psi$ calculated from the experimental spectra reported in Fig. 2; the dashed line is a linear fit $\Delta\phi_{13} = 2\Delta\psi - 0.46$. Inset: retrieved $\Delta\phi_{13}(\omega)$ for two CEP values (consecutive spectra of Fig. 2).

demonstrate that the validity of the previous conclusion can be extended also to the contributions of the short paths.

## 3 Nonadiabatic Saddle-point Method: Temporal Structure of the Attosecond Pulse Train

We have then considered the temporal structure of the emitted XUV radiation using the NASP method [9]. Using the saddle-point method, the Fourier transform of the dipole moment, $x(\omega)$, can be written as a coherent superposition of the contributions from the different electron quantum paths corresponding to the complex saddle-point solutions $(p_s, t_s, t_s')$, where $p_s$ is the stationary value of the momentum acquired by an electron which is set free at time $t_s'$ and recombines with the parent ion at time $t_s$. The sum over the relevant quantum paths can be decomposed in two terms related to the short and long quantum paths. The Fourier transform of the dipole moment can be written as

$$x(\omega) = \sum_{\text{short}} |x_s|(\omega) \exp[i\Phi_s(\omega)] + \sum_{\text{long}} |x_s|(\omega) \exp[i\Phi_s(\omega)], \tag{1}$$

where the first sum takes into account the contributions of the short quantum paths, while the second one considers the long quantum paths. $\Phi_s(\omega)$ is the phase of the complex function $x_s(\omega)$. Since the measured beat pattern is due to the long quantum paths, we have calculated the spectrum produced by the coherent superposition of the long path contributions. By taking the inverse Fourier transform, it is possible to calculate the corresponding temporal structure. A train of attosecond pulses, separated by one-half the optical period, is obtained as shown by the dashed curve in Fig. 4a. In the calculated spectrum, we have then selected the beat pattern by proper spectral windows centered in the regions between consecutive odd harmonics. The associated temporal structure (solid curve in Fig. 4a) corresponds to two trains of attosecond pulses, located one on the leading and the other on the trailing edge of the complete train, separated in time by about one-half the duration of the total train of attosecond pulses. This is the picture, in the temporal domain, of the interference effect previously described in the frequency domain: the spectral structures between the odd harmonics originate from the interference between the electric fields of such two groups of attosecond pulses.

In addition, the temporal analysis offers a simple explanation of the physical origin of the CEP dependence of the interference pattern. We found that, in agreement with our experimental results, the temporal delay, $\Delta T$, between the interfering fields is not significantly affected by the CEP of the IR pulse. Therefore, we can conclude that the CEP dependence of the beat pattern is related to the temporal evolution of the electric field of the interfering attosecond pulses, which turns out to be directly influenced by $\psi$. Indeed, Figs. 4b and 4c show two different portions of the attosecond pulse train calculated for two values of $\psi$

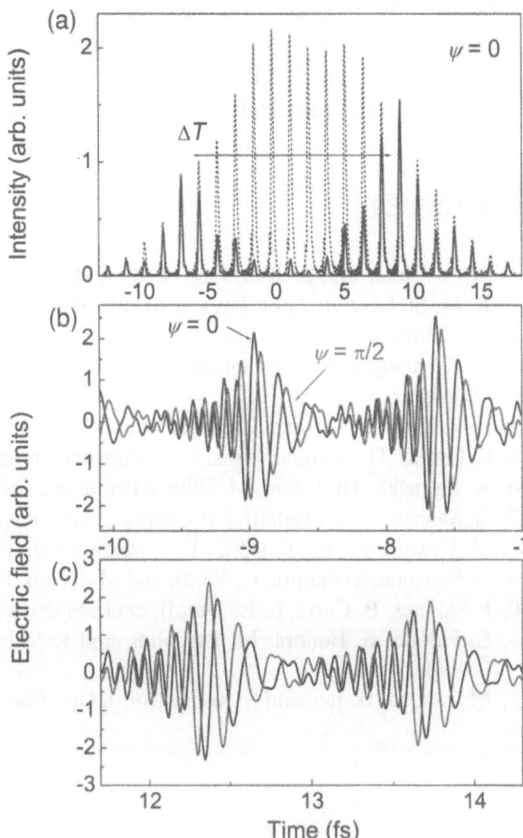


FIGURE 4. (a) Dashed curves represent intensity profile of the attosecond pulse train generated by the long paths, calculated as described in the text; solid blue curves represent calculated temporal structure associated with the interference pattern generated by the long paths. Calculated electric field of the attosecond pulses generated (b) on the leading and (c) on the trailing edge of the driving pulse, for two CEP values of the driving pulse. Driving pulse duration 30 fs; peak intensity $I = 2.6 \times 10^{14}\ \mathrm{W \cdot cm^2}$.

(this two trains have been temporally shifted to overlap the intensity envelopes of the attosecond pulses; the same shift determines the overlap of the electric fields of the corresponding IR pulses). The temporal evolution of the electric field of the attosecond pulses generated on the leading and trailing edges of the IR pulse is significantly influenced by $\psi$, while around the peak of the driving pulse the attosecond pulses (not show in Fig. 4) are not influenced by $\psi$. We have then calculated the temporal evolution of the electric field of the attosecond pulses generated by the long quantum path at low laser intensity and by the short paths. In these cases, the electric field of the attosecond pulses is not significantly influenced by the CEP of the driving pulses.

## 4 Conclusions

In conclusion, using a new self-referencing technique, based on the interference of attosecond light pulses, we have measured the phase difference between consecutive harmonics in the spectral region where they overlap. This phase difference directly depends on the CEP of the driving pulses. By using a numerical model based on the NASP method, we have shown how CEP influences the temporal evolution of the electric field of the attosecond pulses.

## References


1. (a) G. Farkas and C. Toth: Phys. Lett. A **168**, 447 (1992). (b) S. Preuss, A. Demchuk Jr, M. Stuke et al: Appl. Phys. A **61**, 33 (1995).
2. I. P. Christov, M. M. Murnane, and H. Kapteyn: Phys. Rev. Lett. **78**, 1251 (1997).
3. R. Kienberger, E. Goulielmakis, M. Uiberacker, et al.: Nature (London) **427**, 817 (2004).
4. P. M. Paul, E. S. Toma, P. Breger, et al.: Science **292**, 1689 (2001).
5. P. Tzallas, D. Charalambidis, N. A. Papadogiannis, et al.: Nature **426**, 267 (2003).
6. A. Baltuška, Th. Udem, M. Uiberacker, et al.: Nature **421**, 611 (2003).
7. G. Sansone, E. Benedetti, J-P. Caumes, et al.: Phys. Rev. Lett. **94**, 193903 (2005).
8. M. Lewenstein, Ph. Balcou, M. Y. Ivanov, et al.: Phys. Rev. A **49**, 2117 (1994).
9. G. Sansone, S. Stagira, C. Vozzi, and M. Nisoli: Phys. Rev. A **70**, 013411 (2004).
10. P. Salières, B. Carré, L. Le Déroff, et al.: Science **292**, 902 (2001).
11. L. Poletto, S. Bonora, M. Pascolini, and P. Villoresi: Rev. Sci. Instrum. **75**, 4413 (2004).
12. M. Takeda, H. Ina, and S. Kobayashi: J. Opt. Soc. Amer. **72**, 156 (1982).

# Frequency-resolved Autocorrelation Measurement for Characterization of an Attoseconod Pulse Train

Toshihiko Shimizu,[1] Tomoya Okino,[1,2] Hirokazu Hasegawa,[1,3] Kentaro Furusawa,[1] Yasuo Nabekawa,[1] Kaoru Yamanouchi,[1,2] and Katsumi Midorikawa[1]

[1] Laser Technology Laboratory, RIKEN, 2-1 Hirosawa, Wako-shi, Saitama 351-0198, Japan
[2] Department of Chemistry, Graduate School of Science, University of Tokyo, Hongo, Bunkyo-ku, Tokyo 113-0033, Japan
[3] Institute for Molecular Science, Myodaiji, Okazaki 444-8585, Japan

**Summary.** We observed photoelectron spectra with two-photon ionizations of helium excited simultaneously by multiple-order harmonics. Correlation measurements, by utilizing these photoelectron spectra, between the two replicas of the harmonic pulse including the multiple-order harmonics exhibited the pulse envelopes with cross-correlations between the different order harmonics other than the autocorrelations. This type of correlation measurement is potentially applicable to fully characterize an attosecond pulse train by improving the resolutions of photoelectron spectra and of delay.

## 1 Introduction

Attosecond science attracts growing interest since the first observation of the subfemtosecond pulse generated as a high-order harmonic pulse of a sub-10-fs Ti:sapphire laser [1]. Characterization of such a short pulse is, however, not sufficient as compared with that performed in the femtosecond regime of the visible laser pulse. Sekikawa et al. [2] demonstrated the autocorrelation measurement of the isolated attosecond pulse of the ninth harmonic of a sub-10-fs blue laser pulse by utilizing the electron yield from the above threshold ionization of a He atom, while they did not obtain the spectral phase of the pulse. López-Martens et al. [3] determined the relative phase differences between the harmonic spectra of different orders by observing the modulation of the "sideband" signal of photoelectrons generated with the harmonics and the IR laser pulse. Although they have succeeded in the reconstruction of attosecond beating, we cannot see the exact envelope of the pulse train in this experiment due to the lack of the information about the spectral phase within each harmonic spectrum. Spectral phase interferometry for direct reconstruction of the electric field (SPIDER) was achieved by Mairesse et al. [4] in the XUV regime. This method has a drawback for the specification of the relative phase difference between different order harmonics because there is no spectral fringe between them.

In this paper, we demonstrate simultaneous measurement of the correlation traces of the multiple harmonics, which are expected to be synthesized to generate an attosecond pulse train, by observing the electron spectra with two-photon absorption of the harmonic pulse. These spectra contain not only the information of the spectral phase in each harmonic but also those of the relative phase between the harmonics of different orders, because the different order harmonics can contribute the two-photon absorption. Sekikawa et al. [5] have already demonstrated the feasibility of the photoelectron yield with one harmonic as a signal of frequency resolved optical gating (FROG) technique in the vacuum ultraviolet region. Our result of the frequency-resolved correlation trace of the photoelectron spectra of the multiple harmonics can be potentially extent of this FROG technique to the attosecond pulse train, while the further improvement of the experimental condition is needed to obtain the exact result.

## 2 Experiment and Result

In the experiment, the output pulse with an energy of approximately 10 mJ and with a pulse duration of 40 fs from the CPA system of Ti:sapphire laser at a repetition rate of 10 Hz was used as a fundamental pulse for high-order harmonics generation. The laser pulse was loosely focused in a static gas cell, with a length of 10 cm, filled with xenon at a pressure of 0.2–0.25 kPa [6]. Generated harmonics and the fundamental pulse from the cell were propagated to a silicon beam separator(s) [7] placed 4-m away from the cell. Approximately half of the energy of each harmonics is reflected, and the fundamental pulse is absorbed with the beam separator(s). Two beam separators were used for dividing spatially the harmonic pulse to carry out the correlation measurement [8]. After passing through a Sn filter with a thickness of 0.1 mm, the reflected harmonics were introduced in a target chamber which included a focusing mirror made of Si/C with a radius of curvature of 100 mm, a pulsed gas jet supplying ionization targets, and a magnetic-bottle electron spectrometer.

We determined the harmonics spectra at the focus of the concave mirror by observing electron spectra yielded with one-photon ionization of Ar. Principal part of the photoelectron spectra corresponded to the 11th, 13th, and 15th order harmonics as shown in Fig. 1. We expected that three harmonics could cause two-photon ionization of He, because the ionization potential of a He atom is lower than the twice of the photon energies of these harmonics, while the harmonics at higher than or equal to 17th order would be expected to significantly ionize He atoms with one-photon absorption even if the intensities were relatively low.

Figure 2 shows a typical photoelectron spectrum of He ionized by the harmonic pulse. Horizontal axis of this figure is converted from the energies of the electrons

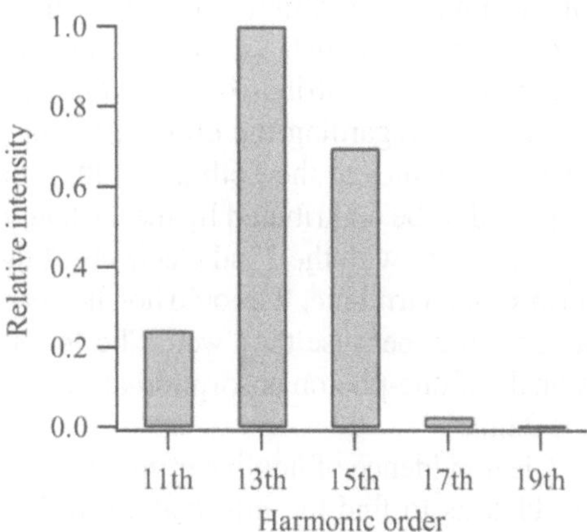


FIGURE 1. Relative intensity of harmonics.

to the corresponding harmonic orders. We note that the electrons, which are originated from the one-photon absorption of 17th- and 19th-order harmonics, with magnitude of ~10 times larger than those shown in Fig. 2, are eliminated by applying retardation voltage of 4 V to the electron spectrometer. Although there remained the electrons corresponding to the 21st-, 23rd-, and 25th-order harmonics with one-photon absorption, we could clearly resolved the electron spectra of two-photon ionization appearing at the even orders of the harmonics, with 1000-shots average of the detected signals. By taking into account the relative intensity of the harmonics shown in Fig. 1, we suppose that the most of the photoelectron at the 26th and 30th harmonic were yielded with absorption of two photons in the same order harmonic at the 13th and 15th, respectively, although the possible absorption of one photon in the 11th and one in the 15th should be considered for the 26th. In contrast to the 26th and 30th, we can easily identify the electrons at the 24th and 28th orders as results of the "cross" absorptions

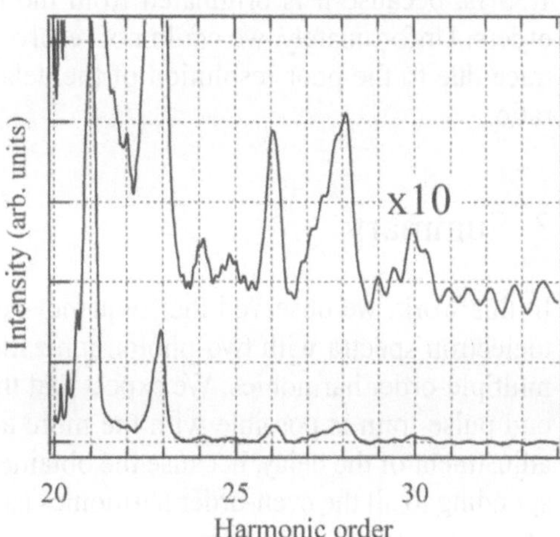


FIGURE 2. Spectrum of the photoelectrons of He.

of two photons in different orders of the harmonics, because there is no even-order harmonics such as the 12th and 14th. It is reasonable to find the 28th electrons to be contributed by the photon in the 13th-order harmonics and that in the 15th, regarding the observation of the "self" two-photon absorptions of these harmonics at the 26th and 30th order, while the 24th electrons, which is expected to be contributed by the 11th and 13th harmonics photons, ought to be accompanied with the 22nd electrons of the "self" two-photon absorption of the 11th-order harmonic. We could not, however, observe the corresponding electrons at the 22nd, because they would be fewer than those at the 24th, and the large signals of one-photon absorption at the 21st and 23rd degraded the resolution at the 22nd.

Clear evidence of nonlinear interaction of the harmonics pulse with He atoms enabled us to find the information of the temporal profile with the correlation technique. We can change the delay between two replicas of the harmonics pulse with two Si beam separator, one of which is mounted on the translation stage with a piezo actuator [8]. The observed dependences of electron yields to delay are shown in Figs. 3a, b, c, and d, corresponding to the harmonic orders of the 24th, 26th, 28th, and 30th, respectively. Each plot of filled circles in these figures was obtained from four sets of the experiment with 400-shots average of the electron signals for each delay. We can see the increases of electron yields near zero delay in each of these figures. These results should be compared with the almost flat trace of the electron yield with the one-photon ionization by the 23rd-order harmonic, shown in Fig. 3e. Fitting curves to the measured traces of the correlations with Gaussian function exhibits that the temporal profile of the harmonic pulse envelope should be confined within a duration of $\sim$30 fs, while the peak-to-background ratio of the traces are much lower than those expected. The lower signals of the correlations may be due to the incomplete spatial overlap of the two replicas of the harmonic pulse at the focal point, which should be improved for further investigation of the correlation. The correlation trace of the 28th harmonic, in principle, should include the modulation with a period of 1.35 fs, because it is originated from the two harmonics of the 13th and 15th orders. Unfortunately, we could not resolve such modulations on the correlation trace due to the poor resolution of the delay and the lower peak-to-background ratio.

## 3 Summary

In this work, we observed the frequency-resolved autocorrelation trace of photoelectron spectra with two-photon ionizations of He excited simultaneously by multiple-order harmonics. We expect that the full characterization of the attosecond pulse train is possible with the more accumulation of the data and the fine adjustment of the delay, because the obtained spectrogram of the electrons corresponding to all the even-order harmonics can be regarded as the result of a FROG of the attosecond pulse train.

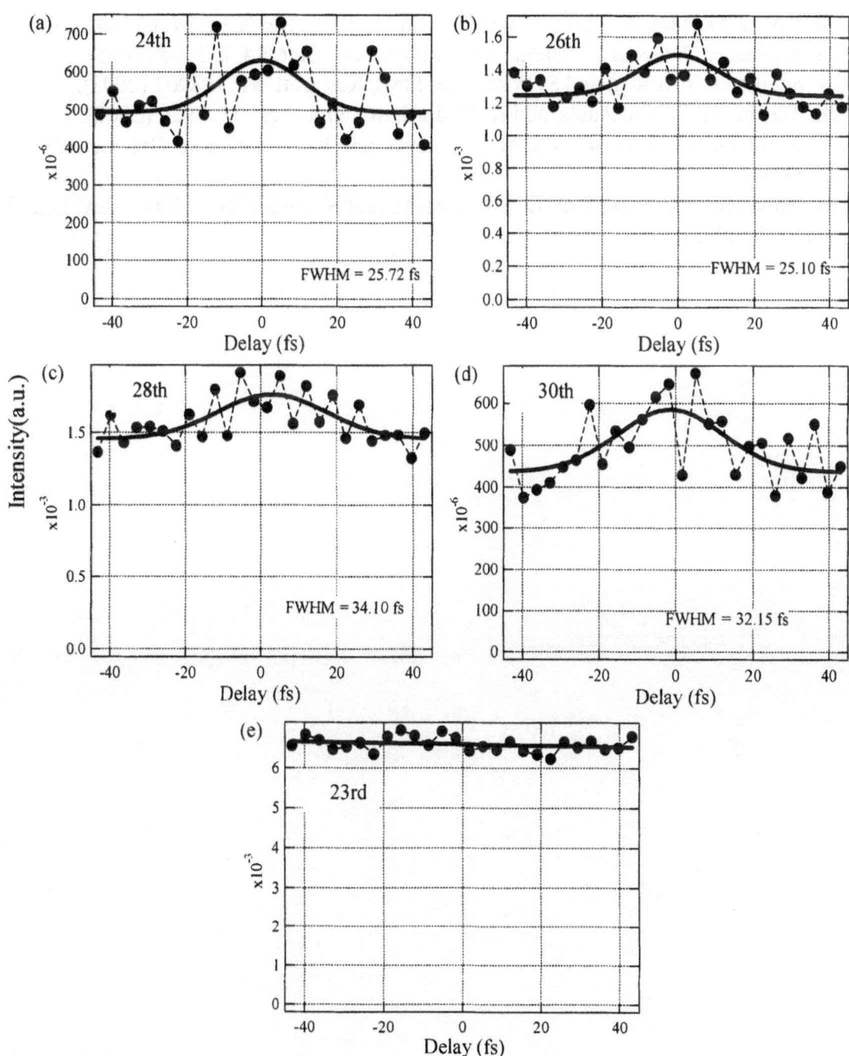


FIGURE 3. Correlation traces of the electron yield, corresponding to (a) the 24th, (b) the 26th, (c) the 28th, and (d) the 30th order. (e) Trace of the electron yield ionized by the 23rd harmonic.

**Acknowledgments.** This work was supported by the Ministry of Education, Culture, Sports, Science and Technology (MEXT) through a Grants-in-Aid for Scientific Research on Priority Area No. 140077222.

## References

1. M. Hentschel, R. Kienberger, Ch. Spielmann, et al.: Nature **414**, 509 (2001).
2. T. Sekikawa, A. Kosuge, T. Kanai, and S. Watanabe: Nature **432**, 605 (2004).

3. R. Lopez-Martens, K. Varju, P. Johnsson, et al.: Phys. Rev. Lett. **94**, 033001 (2005).
4. Y. Mairesse, O. Gobert, P. Breger, et al.: Phys. Rev. Lett. **94**, 173903 (2005).
5. T. Sekikawa, T. Kanai, and S. Watanabe: Phys. Rev. Lett. **91**, 103902 (2003).
6. E. Takahashi, Y. Nabekawa, and K. Midorikawa: Opt. Lett. **27**, 1920 (2002).
7. E. J. Takahashi, H. Hasegawa, Y. Nabekawa, and K. Midorikawa: Opt. Lett. **29**, 507 (2004).
8. Y. Nabekawa, H. Hasegawa, E. J. Takahashi, and K. Midorikawa: Phys. Rev. Lett. **94**, 043001 (2005).

# Characterization of Attosecond Pulse Trains

B. Carré,[1] Y. Mairesse,[1] P. Agostini,[1] P. Breger, H. Merdji,[1] P. Monchicourt,[1] P. Salières,[1] K. Varjú,[2] P. Johnsson,[2] J. Mauritsson,[2] A. L'Huillier,[2] E. Gustafsson,[2] and L. J. Frasinski[3]

[1] CEA/DSM/DRECAM/SPAM, Bât. 522, CEA-Saclay, 91191 Gif-sur-Yvette, France
carre@drecam.cea.fr

[2] Department of Physics, Lund Institute of Technology, P. O. Box 118, SE-221 00 Lund, Sweden
Katalin.Varju@Fysik.Lth.Se

[3] The University of Reading, J. J. Thomson Physical Laboratory, Whiteknights, P.O. Box 220, Reading RG6 6AF, UK
l.j.frasinski@reading.ac.uk

## 1 Introduction

Ultrashort light pulses, as short as 100 as, can be produced from high harmonic generation (HHG) in an atomic gas medium. This major achievement of the last 6 years in high field physics has now been demonstrated in different conditions of medium-field interaction [1].

There are two regimes for generating atto pulses. In the first regime, a single isolated atto pulse can be produced either by an "amplitude gate"—from an ultrashort driving pulse (duration ≤5fs) with well-defined carrier envelope phase (CEP) [2]—or by a polarization gate, in which the laser polarization is quickly modulated [3]. Both techniques can confine XUV emission within one laser half-cycle. The generation of isolated atto pulses opens fascinating perspectives for dynamical studies at the atto time scale, the one at which the electron wave packet evolves in the laser field. The second regime corresponds to the more standard HHG in a multicycle field, where an atto pulse train (APT) is produced. There are clearly two time scales for describing the APT. At the *atto time scale* of one half-cycle, each pulse has an envelope with a given pulse duration, a central (carrier) frequency, and a temporal phase. Moreover, the timing of the burst with respect to the laser cycle is fixed, defined by the local group delay. Now, the characteristics of the atto pulses evolve along the train, since they are generated in a time-dependent driving field. This evolution corresponds to the *femto time scale* of XUV emission. The study of the APT can be efficiently completed in the spectral domain. The atto structure is obviously related to the broadband distribution of amplitude and phase over the harmonic spectrum. Conversely, the femto structure of the APT is related to the spectral distribution within the narrow bandwidth of each individual harmonic. As the isolated pulse, the APT can serve to probe electron wave packet dynamics at the subfemto scale. A key property is that it is synchronized with the laser cycles. Measuring *the amplitude and the phase* in the XUV range has long been a challenge. This has been solved recently, both in the broad and narrow spectral ranges.

In this paper, we summarize the studies performed at Saclay and at Lund, which investigate the atto and femto structures of the APT generated by a multicycle driving pulse. To investigate the broad spectral range, we use the reconstruction of attosecond bursts by interference of two-photon transition (RABITT) technique we demonstrated in 2001 [4]. To investigate the narrow spectral range, we use two techniques, XFROG [5] and High-Harmonic SPIDER [6]. Actually, RABITT and XFROG can be viewed as FROG-type techniques. Besides those demonstrated yet for individual harmonics [7,8], or proposed for APT [9], they are well adapted to measuring the frequency modulation in HHG.

The phase properties at the atto and femto time scales can be analyzed using the adiabatic description of HHG. In this description, the instantaneous phase of the XUV field is a function of only the frequency $q\omega$ and laser intensity $I$. This double dependence determines the frequency modulation at both the atto time scale (the *atto chirp*), and the femto time scale (the *harmonic chirp*). As an important result, our studies show the relation between the two chirps, which is explained in the adiabatic description [10]. These results allow, in principle, to retrieve the complete APT generated in a multicycle laser pulse.

## 2 Description of the XUV Field : Theoretical Reminder

HHG in a driving pulse of duration 30–60 fs, i.e., including typically 10–20 optical cycles, leads to a discrete odd-harmonics spectrum. The XUV field in a finite spectral range $[q_i, q_f]$, at the exit of the generating medium, can be written as a sum of the harmonic components:

$$E(t) = \sum_{q=q_i}^{q=q_f} A_q(t) e^{-iq\omega t + i\Phi_q(t)}. \tag{1}$$

$A_q(t)$ and $-q\omega t + \Phi_q(t)$ are the amplitude and phase, respectively, of the $q$th harmonic field; $\Phi_q(t)$ is the atomic dipole phase. For the sake of simplicity, the laser phase reduces to $-\omega t$ and the transverse spatial dimension is not explicit. In (1), we assume good phase matching so that the phase of the field at $q\omega$ identifies with that of the nonlinear polarization. This means weak ionization and negligible dispersion in the generating medium. In the adiabatic description (long-laser pulse), for instance derived from the Lewenstein model [11], the atomic dipole at frequency $q$ (in units of field frequency $\omega$) is completely determined by the two variables, $q$ and field intensity $I$. Moreover, it is the sum of two main contributions, each associated to a class of quantum paths or semiclassical electron trajectories, respectively short and long. For each of the two classes, the dipole phase corresponds to the semiclassical action along the trajectory. The total differential of the phase writes

$$d\Phi = \alpha(q, I)\, dI + t_e(q, I)\, dq. \tag{2}$$

In (2), the partial derivative $t_e = \frac{\partial \Phi}{\partial q}$ denotes the group delay, equivalent to an emission time. Note that $\Phi$ depends on time through the temporal variation of the intensity. The variation of $\Phi$ is different between the two classes of trajectories. In particular, the $\alpha$ coefficient is larger (by at least a factor of 2) for the long trajectory. This offers an efficient method to discriminate between the two contributions using phase-matching and far-field spatial filtering. On the one hand, to characterize the atto structure—in pratice, after one trajectory has been selected—we should investigate the broadband variation of the phase with $q$, at fixed intensity $I$, i.e., measure the group delay $t_e(q, I)$ as a function of $q$. On the other hand, to characterize the femto structure, we should investigate the phase variation with $I$, at fixed $q$, i.e., for one individual harmonic, and measure the $\alpha(q, I)$ coefficient as a function of $I$. We will then exploit the fact that the two quantities are related so that determination of $t_e$ (conversely, of $\alpha$) already provides information on the femto (conversely, of atto) time scale.

# 3 Characterization of the Attosecond Structure of APT

## *3.1 High-harmonic Relative Phases : the RABITT Method*

To characterize the atto structure, we need to measure the amplitude and phase over a broad range of harmonics, with a rather coarse spectral resolution of $\sim$ 1eV (to span $\sim$1 fs in the time domain), so that only the average relative phases of harmonics are required. For this, we use the RABITT method [4, 12, 13]. It is based on the cross-correlation of XUV and laser pulses, in two-color *above-threshold ionization* of a gas target. The key point is that the two-photon electron signal at kinetic energy $(q+1)\omega - IP_{\text{target}}$ (sideband) results from an interference between at least two channels. In consequence, the sideband amplitude oscillates as a function of the delay $\tau$ between the XUV and laser pulses [14]; the relative phases and the emission time $t_e(q+1, I) \approx \frac{(\Phi_{q+2} - \Phi_q)}{2\omega}$ are easily determined from the sideband oscillation (the averaged laser intensity $I$ is close to the peak value). Figures 1a and b show the emission time and harmonic intensity measured in Ar at laser intensity $I = 1.2 \times 10^{14}$ W $\cdot$ cm$^2$, for the plateau harmonics $q = 11 - 25$. The contribution of the short trajectory has been selected using a far-field spatial filtering (central part of the XUV beam selected).

In Fig. 1a, the emission time, or group delay, increases linearly with order, leading to a quadratic spectral phase in Fig. 1b. From the measured phase and amplitude, we can reconstruct locally the atto pulses in the time domain, as

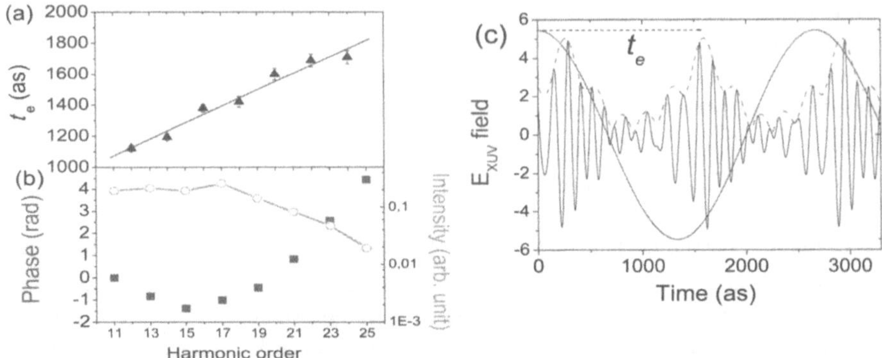


FIGURE 1. (a) Emission time of harmonics measured by RABITT in Ar; (b) Quadratic phase derived from $t_e$ (set to zero at $q = 11$) and harmonic intensity; (c) Reconstruction of the APT including harmonics 15 – 25 (the carrier envelope phase in the atto pulse is arbitrary). Laser field is in thick solid line.

illustrated in Fig. 1c. Three features are noticeable: (i) the group delay of 1600 as referred to the optical cycle corresponds to the emission time of the pulse carrier frequency $q_0 \sim 20$; (ii) the pulse is positively chirped—the red frequencies are emitted before the blue; and (iii) due to the chirp, the resulting envelope is distorted and the pulse duration is larger than the Fourier limit.

Now, it is possible to compensate the atto chirp, i.e., the positive group delay dispersion, by propagating the pulse through a medium with negative group velocity dispersion (GVD), such as a metallic foil. An Al foil of 600-nm thickness makes it possible to compensate the group delay dispersion of harmonics 13 to 35, generated in Ar. The duration of the atto pulse including these components is reduced from 480 as to 170 as, close to the Fourier limit (150 as) [15].

## 3.2 Origin of the Atto Chirp

The origin of the atto chirp can be explained from a classical picture of the electron trajectories that contribute to HHG [12, 16]. In Fig. 2a, we have plotted the trajectories $x_e(q;t)$ of the electron wavepacket along the polarization axis, for a range 11 – 21 of $q$ orders generated in Ar at intensity $I = 10^{14}$ W · cm$^{-2}$. The trajectory originates and recombines at $x = 0$, respectively, at ionization time $t_i$ and emission time $t_e$. The energy $I_p + E_{kin}$ of the XUV photon emitted in the electron-core recombination is plotted in Fig. 2b (to account for the quantum time/energy uncertainty, one should consider that emission time $t_e$ is associated to a frequency range centred on $q$). The classical picture clearly illustrates that, for each photon energy in the plateau, the short (with return time $\sim T_L/2$) and long trajectories (with return time $\sim \frac{3T_L}{4}$) contribute. If the short trajectories are selected, the emission time increases with order, as was measured in Fig. 1a; this corresponds to the positive linear chirp within the atto pulse. The variation of emission time with order derived from the classical picture is in reasonable

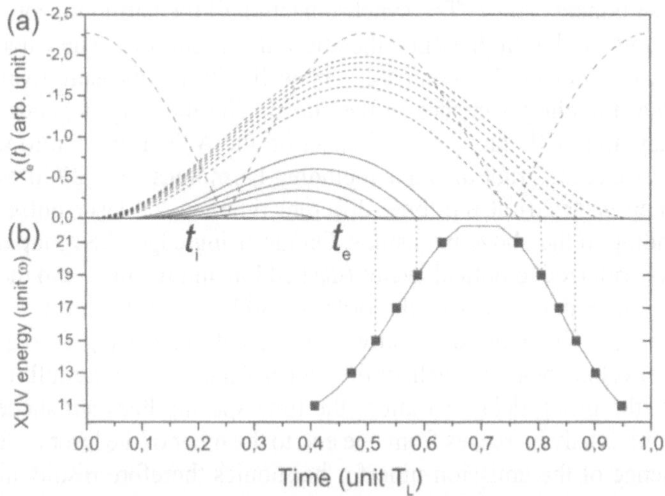


FIGURE 2. (a) Classical short (solid line) and long (dashed line) trajectories of electron in Ar at $I = 10^{14}$ W $\cdot$ cm$^{-2}$ for $q = 11 - 21$; (b) XUV photon energy versus emission time. For short (long) trajectories, emission time increases (decreases) almost linearly with order. This determines the positive (negative) *atto chirp*.

agreement with the measured group delay dispersion in Fig. 1a. Conversely, if the long trajectories are selected, the atto chirp becomes negative. In the cutoff, short and long trajectories merge and emission times converge to one single value ($\sim$0.7 $T_L$ in Fig. 2b), close to the zero of the field: the atto chirp vanishes and all the harmonics are in phase. The atto pulse structure therefore directly reflects the HHG dynamics at the scale of one optical cycle.

Figure 3a shows the variation of the emission time in the plateau region in Ne, after selection of the short trajectories, at three laser intensities $I = 2, 3, 4 \times 10^{14}$ W $\cdot$ cm$^{-2}$ [13]. Both the group delay and the group-delay dispersion (GDD)

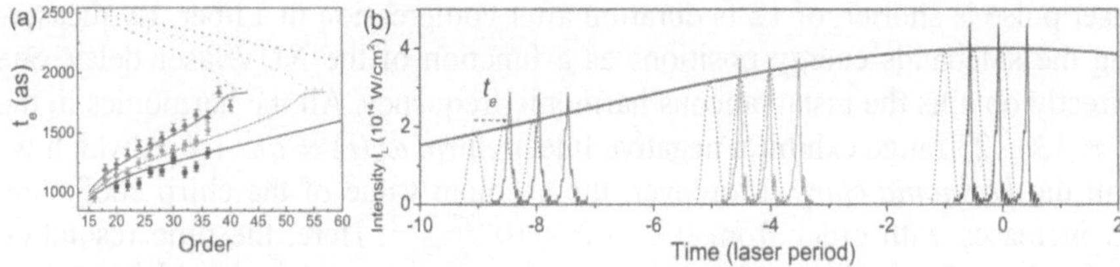


FIGURE 3. (a) Emission time measured by RABITT in Ne and calculations in Lewenstein model, for laser intensities $I = 2$ (red), 3 (green), 4 (blue) $\times 10^{14}$ W $\cdot$ cm$^{-2}$; (b) Reconstruction of APT showing variation of group delay—referred to optical cycle in dashed line—and pulse duration in the laser envelope (thick line). The nonlinear dependence of the harmonic amplitudes with $I$ is not accounted for. As a result of the group delay variation, the spacing between pulses continuously increases along the APT.

decrease with increasing $I$. The synchronization of the harmonic components is therefore improved at high $I$ (i.e., the atto pulse is shorter). The calculations in the Lewenstein model [11] compare very well with the measured values. They confirm that the chirp vanishes in the cutoff. The intensity dependence of the group delay and GDD has also an impact on the APT at the femto time scale. In Fig. 3b, we reconstruct the APT including harmonics $15 - 29$ in Ne (centred at frequency $q_0 \approx 20$) at the three different times in the laser pulse envelope, corresponding to the above intensities. On the rising edge, the group delay at $q_0$ frequency—referred to optical cycles (dashed line in Fig. 3b)—and the GDD are large; atto pulse duration is of the order of 300 as (Fourier limit 150 as). The group delay and GDD monotonically decrease with increasing $I$, to a minimum when the envelope peak is reached; they further increase on the falling edge. As a result of the group delay variation, the time spacing between successive atto pulses continuously increases from one end to the other of the laser envelope. The $I$-dependence of the emission time for harmonics therefore results in pulse-to-pulse distortions in the APT.

In summary, by studying the emission time as a function of frequency, we characterized the atto structure. By studying the emission time as a function of intensity, we add insight into the femto structure of the APT. To go further, especially to characterize the carrier frequency along the train, we need to investigate the individual harmonics and harmonic phases.

# 4 Characterization of the Femtosecond Structure of APT

## 4.1 Harmonic Temporal Phase : XFROG

To measure the time-dependent phase of one harmonic, a simple FROG-type, XFROG method has been developed in Lund, which gives direct access to the instantaneous XUV frequency [5]. As the RABITT and other XUV FROG schemes [7, 8], it is based on two-photon (XUV/laser) ionization of a rare gas target. The driving pulse has duration of 35 fs (Fourier transform—FT), thus generating $q = 13 - 23$ harmonics in Ar with pulse duration $\geq 20$ fs. The "dressing" laser pulse is shorter, of 12 fs duration after compression in a fiber. By measuring the sidebands energy positions as a function of the XUV/laser delay, one directly obtains the instantaneous harmonic frequency. All the harmonics in the $q = 13 - 23$ range exhibit a negative linear chirp, $\omega_q(t) = q\omega + b_q t$, which we call the *harmonic chirp*. Moreover, the absolute value of the chirp coefficient $b_q$ increases with order from 0.2 to $2 \times 10^{-28}$ s$^{-2}$. Here, the time resolution is fixed by the duration of the dressing laser pulse, which should be shorter than that of the XUV pulse. The intensity of the driving pulse is low enough so that the ponderomotive Stark shift of the ionization potential (in the generating medium) can be neglected. XFROG allows the straightforward monitoring of the harmonic chirp that we can manipulate. For example, by introducing a positive chirp of the laser pulse, we cancel the variation of the $b_q$ coefficient with harmonic order.

## 4.2 A Single-shot, All-optical Technique : HH-SPIDER

The SPIDER method, to completely characterize light pulse, is normally used in the visible/IR, where two replicas of the pulse, delayed in time and relatively to each other shifted in frequency, can be easily produced [17]. An extension of this technique to the XUV is challenging : beam splitter for amplitude division remains a critical optics; no process for shifting the frequency has been demonstrated yet. In the case of laser-driven HHG, a solution is to take advantage of the mutual coherence of two harmonic pulses [18, 19]. We proceed in two steps. First, two IR time-delayed replicas are produced using a DAZZLER system. Under conditions of low ionization, we check that they produce two *mutually coherent* XUV pulses. Second, the IR pulses are spectrally sheared by $\delta\omega$ using DAZZLER. The frequency shift $\sim q\delta\omega$ transferred to the XUV pulses is determined by numerically superimposing the two-component spectra. After SPIDER analysis of 11$^{\text{th}}$ harmonic, we retrieve amplitude and a quadratic spectral phase, which means a quadratic temporal phase and a negative linear chirp, in agreement with XFROG measurements. The HH-SPIDER is an all optical, single-shot method, which could be extended to broadband characterization of the full harmonic spectrum and thus the APT, provided the condition that the spectral amplitude does not go through zero, i.e., no gaps between harmonics [20]. In the case of nonidentical replicas, e.g., of strong ionization in the generating medium, SPIDER with a reference pulse can be envisaged.

## 4.3 Origin of the Harmonic Chirp

The origin of the harmonic chirp is clear from (2): the intensity-dependent dipole phase induces a temporal variation of the instantaneous frequency, $\delta\omega_q(t) = -\frac{\partial\Phi_q}{\partial t} = -\alpha_q \frac{\partial I}{\partial t}$ $(\alpha_q = \alpha(q, I))$. The generation occurs close to the peak intensity, where $I(t)$ varies quadratically in time, giving a linear chirp. Consequently, the $\alpha_q$ coefficient is directly proportional to the chirp coefficient $b_q$ which is measured. In Fig. 4, the $\alpha_q$ coefficients measured in Ar from XFROG ($1.5 \times 10^{14}$ W $\cdot$ cm$^{-2}$) are well reproduced by the TDSE calculation and the Lewenstein

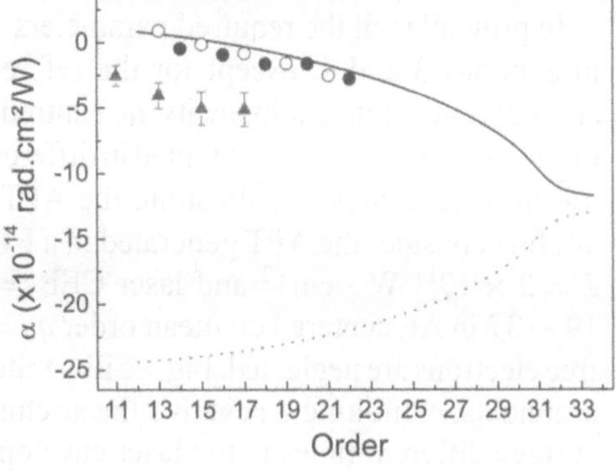


FIGURE 4. The $\alpha_q$ coefficient measured in Ar by XFROG ($I = 1.5 \times 10^{14}$ W $\cdot$ cm$^{-2}$, full circles) and HH-SPIDER ($2 \times 10^{14}$ W $\cdot$ cm$^{-2}$, triangles). Simulations from TDSE (open circles) and Lewenstein model (solid line, short trajectories; dashed line; long trajectories).

model. They are in a reasonable agreement with the results from HH-SPIDER ($2 \times 10^{14}$ W · cm$^{-2}$); the slightly different intensities may explain the different values. As in RABITT measurements, it is clear from Fig. 4 that the contribution of only the short trajectory has been selected experimentally. The $\alpha_q$ coefficient depends on $q$; we will show in the next section that this corresponds to changes at the femto time scale in the APT.

## 5 Connection Between the Harmonic and Atto Chirps

The adiabatic expression (2) of the dipole phase as a total differential implies that its partial derivatives $\alpha$ and $t_e$ should be related:

$$\frac{\partial \alpha}{\partial q} = \frac{\partial t_e}{\partial I} = \frac{\partial^2 \Phi}{\partial q \partial I}. \tag{3}$$

Unfortunately, we cannot check rigorously this equality from experimental results, since $\alpha(q, I)$ was measured in Ar (XFROG and HH-SPIDER), whereas $t_e(q, I)$ was measured in Ne (RABITT). However, it is meaningful to compare the generic behaviours of the partial derivatives in Fig. 5. Actually, although the two measurements come from two independent and very different experiments, they lead to the same generic behaviours. Both are well reproduced in the adiabatic description. It follows that $\frac{\partial \alpha}{\partial q}$ and $\frac{\partial t_e}{\partial I}$ determine equivalently the variation of the spacing between the pulses in Fig. 3. The identity of the partial derivatives of $\Phi$ in Fig. 5 shows that the atto chirp and harmonic chirps have the same physical origin, the $q$- and $I$-dependence of the dipole phase.

## 6 Reconstruction of the Atto Pulse Train

Now, we can reconstruct the dipole phase $\Phi(q, I(t))$ from (2):

$$\Phi(q, I(t)) = \Phi(q_0, I_0) + \int_{I_0}^{I(t)} \alpha(q_0, I')\, dI' + \int_{q_0}^{q} t_e(q', I(t))\, dq'. \tag{4}$$

In principle, all the required parameters are accessible from the measurements in Sections 3 and 4, except for the reference phase $\Phi(q_0, I_0)$ (the CEP of the atto pulse generated at intensity $I_0$, centred on $q_0$ frequency). However, since the measured parameters are obtained in different conditions (different gases) and thus not fully consistent, we illustrate the APT reconstruction by a simulation [21]. We first consider the APT generated in a FT laser pulse, 45 fs long, peak intensity $I = 2 \times 10^{14}$ W · cm$^{-2}$ and laser CEP = 0. The APT includes 10 harmonics, 19 – 37 in Ar, centered on mean order $q_0 = 27$. Ionization and dispersion due to free electrons are neglected. Fig. 6c illustrates the variation of the carrier frequency around $q_0$, exhibiting a negative linear chirp (solid line). Fig. 6a shows the APT at three different times in the laser envelope. We recall the main features: (i) the

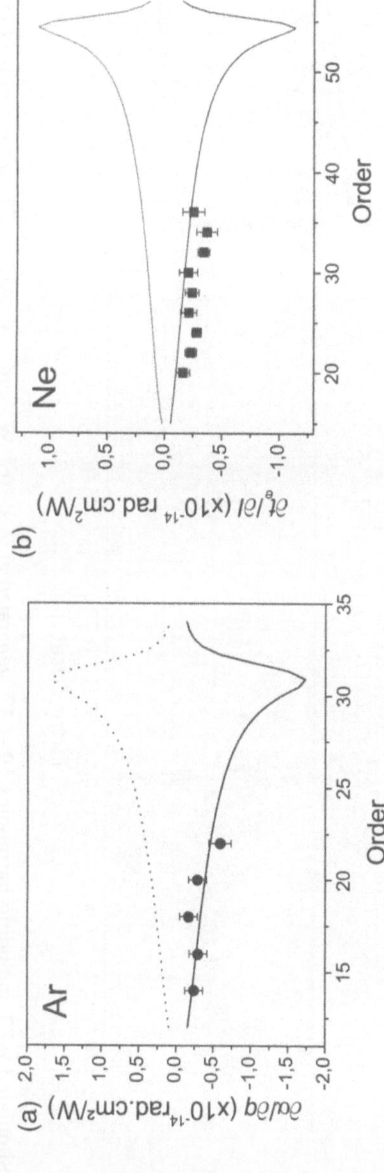


FIGURE 5. (a) Partial derivative $\frac{\partial \alpha}{\partial q}$ measured in Ar by XFROG; (b) Partial derivative $\frac{\partial t_e}{\partial I}$ measured in Ne by RABITT.

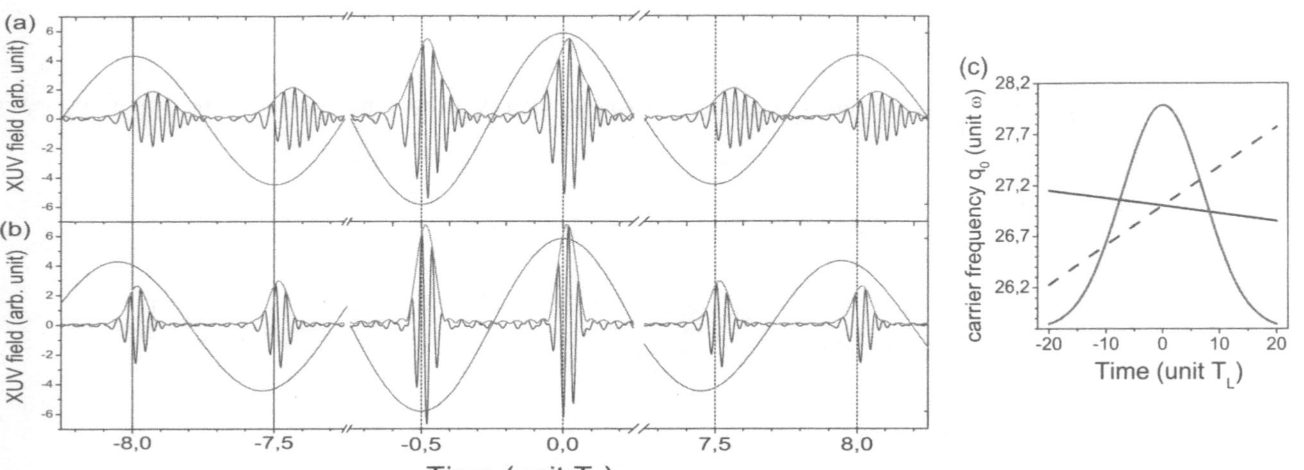


FIGURE 6. Simulated APT including harmonics 19 – 37, generated at $2 \times 10^{14}$ W · cm$^{-2}$ in Ar. (a) FT driving pulse (45 fs); (b) chirped driving pulse (cancel dispersion of harmonic chirp) and propagation through medium with negative GVD (cancel group delay dispersion of atto pulse); (c) APT carrier frequency around $q_0 = 27$ (FT laser pulse, solid line; chirped laser pulse, dashed line).

positive chirp within each atto pulse, (ii) the group delay and GDD, i.e., the atto pulse envelope, evolving from pulse to pulse with the laser intensity, and (iii) the spacing between pulses continuously increasing.

Now, we have mentioned in Sections 3 and 4 that we can control the atto and harmonic chirps. In Fig. 6b, we have reconstructed the APT when, first, the variation of the *harmonic chirp* with order is cancelled by introducing a chirp on the laser pulse ($b_L = 1.5 \times 10^{-3}$ fs$^{-2}$), and second, the positive *atto chirp* is compensated—at the peak of laser envelope—by propagating the APT through a medium with negative GVD. As a result, the atto pulses are shorter, the spacing between the pulses is the same, and the carrier frequency now increases around $q_0$ across the laser envelope (dashed line in Fig. 6c).

# 7 Conclusions

Several experimental and theoretical studies performed at Saclay and Lund have been combined into a comprehensive method for characterizing the APT from HHG in a multicycle driving pulse. Using the RABITT method, we can probe the atto time scale, giving evidence for the frequency modulation of the atto pulse, the atto chirp. Using XFROG and HH-SPIDER, we can monitor the femto time scale, giving evidence for the harmonic chirp. We have shown their compatibility and connection through the mixed partial derivatives of the emission phase. This allows the full reconstruction of the APT. Notwithstanding the advances of FROG-type methods [22], the HH-SPIDER is a promising all-optical, single-shot technique, which has the potential for characterizing XUV emission on both atto and femto times scales. This work was partially supported by the European Community under contracts MCRTN-CT-2003-505138 and RII3-CT-2003-506350 (LASERLAB Europe).

# References

1. P. Agostini and L. F. DiMauro: Rep. Prog. Phys. **67**, 813 (2004).
2. M. Hentschel, R. Kienberger, Ch. Spielmann, et al.: Nature **414**, 509 (2001).
3. O. Tcherbakoff, E. Mével, D. Descamps, et al.: Phys. Rev. A **68**, 043804 (2003).
4. P. M. Paul, E. S. Toma, P. Breger, et al.: Science **292**, 1689–1692 (2001).
5. J. Mauritsson, P. Johnsson, R. Lopez-Martens, et al.: Phys. Rev. A **70**, 021801(R) (2004).
6. Y. Mairesse, O. Gobert, P. Breger, et al.: Phys. Rev. Lett. **94**, 173903 (2005).
7. T. Sekikawa, T. Katsura, S. Miura, et al.: Phys. Rev. Lett. **88**, 193902 (2002).
8. T. Sekikawa, T. Kanai and S. Watanabe: Phys. Rev. Lett. **91**, 103902 (2003).
9. T. Shimizu, T. Okano, H. Hasegawa, et al.: UFO V/HFSW XI, Nara 2005, Book of Abstracts M2-4, p.13.
10. K. Varjú, Y. Mairesse, B. Carré, et al.: J. Mod. Opt. **52**, 379 (2005).
11. M. Lewenstein, Ph. Balcou, M. Yu. Ivanov, et al.: Phys. Rev. A **49**, 2117 (1994).
12. Y. Mairesse, A de Bohan, L. J. Frasinski, et la.: Science **302**, 1540 (2003).

13. Y. Mairesse, A. de Bohan, L. J. Frasinski, et al.: Phys. Rev. Lett. **93**, 163901 (2004).
14. V. Véniard , R. Taeb, A. Maquet: Phys. Rev. A **54**, 721 (1996).
15. R. Lopez-Martens, K. Varju, P. Johnsson, et al.: Phys. Rev. Lett. **94**, 033001 (2005).
16. S. Kazamias and Ph. Balcou: Phys. Rev. A **69** 063416 (2004).
17. C. Iaconis and I. A. Walmsley: Opt. Lett. **23**, 792 (1998).
18. M. Bellini, C. Lynga, A. Tozzi, et al.: Phys. Rev. Lett. **81**, 297 (1998).
19. P. Salières, L. Le Déroff, T. Auguste, et al.: Phys. Rev. Lett. **83**, 5483 (1999).
20. E. Cormier, I. A. Walmsley, E. M. Kosik, A. S. Wyatt, L. Corner and L. F. DiMauro: Phys. Rev. Lett. **94**, 033905 (2005).
21. K. Varju, Y. Mairesse, P. Agostini, et al.: Phys. Rev. Lett. **95**, 243901 (2005).
22. Y. Mairesse and F. Quéré: Phys. Rev. A **71**, 011401(R) (2005).

# Part III

# Carrier-Envelop Phase I

# Part III

# High-harmonic Generation at 100 MHz Repetition Frequency using a Femtosecond Enhancement Cavity

R. Jason Jones, Kevin Moll, Michael Thorpe, and Jun Ye

JILA, National Institute of Standards and Technology and University of Colorado, 440 UCB Boulder, CO 80309
rjjones@jilau1.colorado.edu

## 1 Introduction

High-harmonic generation (HHG) [1,2] provides a coherent source of vacuum ultraviolet (VUV) to soft X-ray radiation in a relatively compact system. Since its first observations [3,4], HHG has relied on high-energy, low repetition rate laser systems to provide the peak intensities needed for ionization of the gas target. The small conversion efficiency of the process, combined with the low repetition rate of amplified laser systems, results in low average powers in the VUV generation. Experiments trying to utilize these sources therefore often suffer from poor signal-to-noise levels, resulting in long data-acquisition times. Furthermore, the use of these sources as precision spectroscopic tools remains limited in comparison with sources in the visible. Ramsey-type spectroscopy has been utilized to improve spectral resolution in the VUV [5,6] but remains orders of magnitude away from the precision measurement capability available with fs laser based frequency combs in the optical and IR [7,8]. This is due to the fact that the original frequency comb structure of the laser is lost in the HHG process from the reduction of the pulse train repetition rate required to actively amplify single pulses to the needed energies/intensities.

In this work, we address these issues, namely average power, system size and cost, and spectral resolution, using an fs laser coupled to a passive optical cavity. We demonstrate coherent frequency combs in the VUV spectral region from the generation of high harmonics of the laser without any active amplification or decimation of the repetition frequency. The output from the laser is stabilized to an fs-enhancement cavity with a gas jet at the intracavity focus. The high-peak power of the intracavity pulse enables efficient HHG (Fig. 1). Since little of the fundamental pulse energy is converted, an fs-enhancement cavity is ideally suited for HHG as the driving pulse is continually "recycled" after each pass through the gas target. HHG at high repetition rates opens the door for dramatic

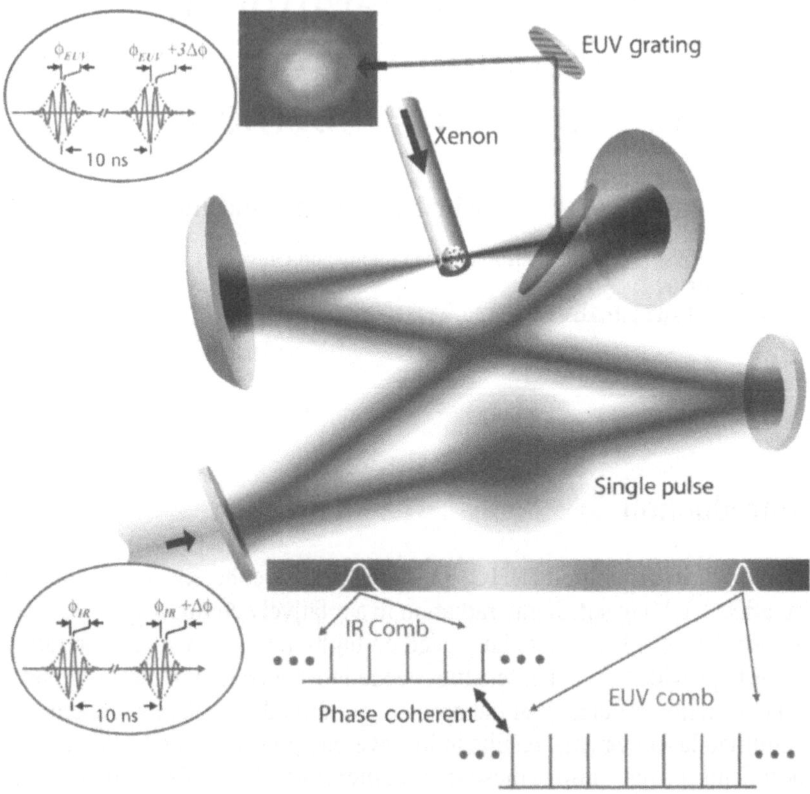


FIGURE 1. Schematic setup of intracavity HHG. A gas target at the cavity focus enables coherent HHG, resulting in a phase-stable frequency comb in the VUV spectral region. The photo inset shows the actual spatial mode profile of the third harmonic coupled out of the cavity.

improvements in average power conversion efficiency. In addition, system cost and size are greatly simplified. Optical-heterodyne-based measurements reveal that the coherent frequency comb structure of the original laser is fully preserved in the HHG process. These results lead the way for precision frequency metrology at extreme wavelengths and permit efficient HHG using only a standard laser oscillator.

## 2 Cavity Characterization

To sufficiently buildup the intracavity pulse energy, the passive optical cavity needs to demonstrate a number of important characteristics: (i) a high finesse, (ii) low round-trip group-delay dispersion (GDD) to allow ultrashort pulses to be coupled into and stored inside the cavity, and (iii) a robust servo to stabilize the two degrees of freedom of the incident pulse train to the corresponding cavity

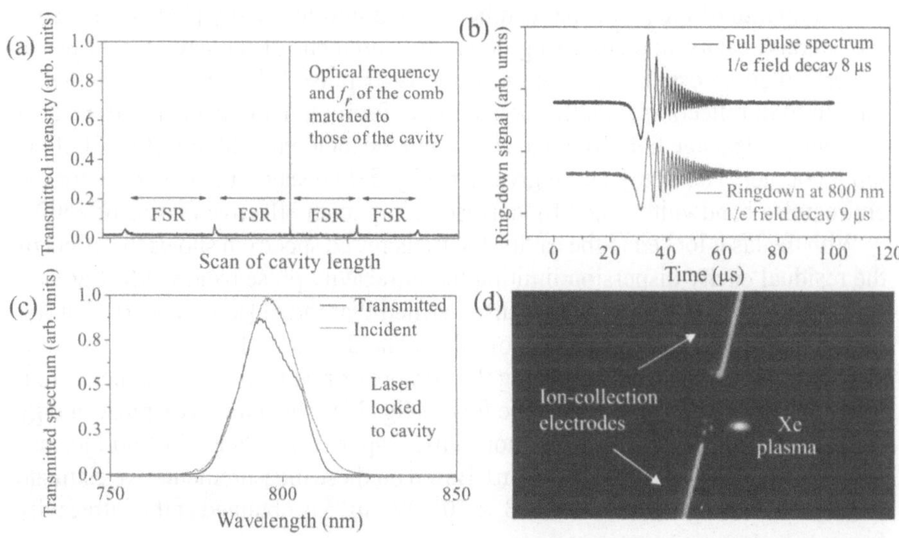


FIGURE 2. (a) Transmission through fs-enhancement cavity as the laser-cavity length is scanned. The dominant peak results from optimal alignment of frequency comb modes to resonant modes of the fs-enhancement cavity. (b) Cavity ring-down measurements for the entire pulse spectrum (top trace) and for a 5-nm region centered at 800 nm. (c) Incident and transmitted spectrum of the pulse. The pulse durations measured by frequency-resolved optical gating are 48 fs (incident) and 60 fs (transmitted). (d) Visible plasma generation at the cavity focus. A pair of electrodes for ion collection is useful for signal optimization.

resonance modes [9,10]. A standard mode-locked femtosecond Ti:Sapphire laser with a repetition frequency ($f_r$) of 100 MHz, 60-fs pulse duration, and 8-nJ pulse energy is used. The carrier-envelope offset frequency ($f_o$) and $f_r$ set the pulse-to-pulse carrier-envelope phase evolution as $\Delta\phi = 2\pi f_o/f_r$ (Fig. 1). To couple the pulse train from the mode-locked laser into the cavity, both $f_o$ and $f_r$ of the laser are adjusted such that the optical comb components are maximally aligned to a set of resonant cavity modes. For short optical pulses (<100 fs) and high finesse cavities ($F > 1000$), the comb components need to overlap with corresponding narrow-cavity resonances across a large spectral bandwidth. The power transmitted through the cavity as the length of the laser is scanned therefore shows a single sharp resonance only when both the laser carrier frequency (average comb position) and spacing ($f_r$) are optimally aligned (Fig. 2a).

To investigate the peak intensity that can be obtained with this method, an empty fs-enhancement cavity was initially characterized. The passive optical cavity has a ring geometry formed by six mirrors. All mirrors are high reflectors except the input coupler, which has a transmission of 0.1 %, nearly matching the net intracavity loss. The center wavelength of the mirror coating is 800 nm, with a bandwidth of 100 nm within which the net cavity GDD is compensated to <10 fs$^2$. However, residual GDD in the enhancement cavity causes the cavity free-spectral-range (FSR) frequency to vary as a function of wavelength, limiting the bandwidth of the pulse that can be coupled into the cavity [11]. To measure the average cavity finesse, a ring-down measurement of the cavity-field decay is performed across the bandwidth of the laser pulse. The ring-down signal is measured in reflection from the cavity, in the form of an optical heterodyne beat between the leakage field from the cavity and the incident field reflected off of the input coupler. The decay curve gives a cavity field decay time of $\approx 8$ μs for the entire pulse bandwidth (Fig. 2b), corresponding to an effective finesse of 2500.

With the laser locked to the cavity, the transmitted spectrum shows the effect of the residual cavity dispersion limiting the intracavity pulse bandwidth (Fig. 2c). Measurement of the transmitted pulse verified that the pulse was nearly Fourier transform limited with a duration of 60 fs. The duration of the pulse was minimized inside the cavity by adjusting the compressor while measuring the current produced from a plasma in Xe at the focus (Fig. 2d). The intracavity pulse energy is enhanced up to 6 μJ for these short pulses, approximately a 750-fold increase from the incident pulse energy of 8 nJ. Based on these measurements, we estimate that a peak intracavity intensity of $3 \times 10^{13}$ W·cm$^{-2}$ is obtained at the intracavity focus.

## 3 Results

To couple the HHG light out of the cavity, a 700-μm thick sapphire plate is placed at Brewster's angle (for the IR) inside the cavity (Fig. 1). The Fresnel reflection coefficient of the intracavity plate rises at shorter wavelengths, with a reflectivity of 5–10% between 40 and 100 nm. The gas target is confined within a thin, hollow brass cylinder, with a 150-μm hole to allow the intracavity pulse to pass through the sample (Fig. 1). The diffracted pattern of the HHG light, obtained with a $MgF_2$-coated aluminum grating and detected with a phosphor screen, demonstrates that at least the ninth-order harmonic has so far been generated

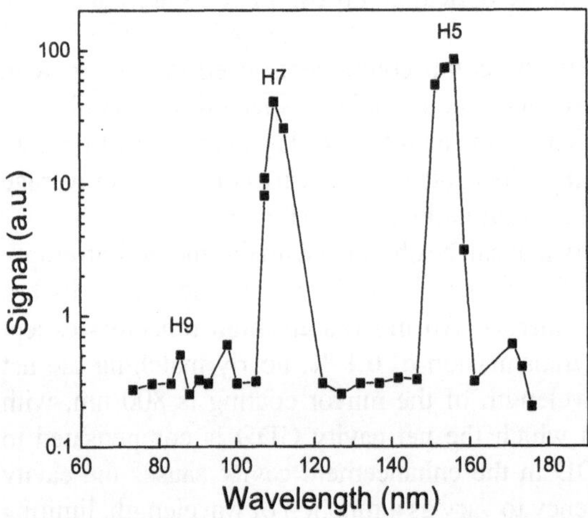


FIGURE 3. Spectrum of HHG from fs-enhancement cavity.

(Fig. 3). The average power of the third harmonic light generated inside the cavity reaches nearly 10 μW. The corresponding intracavity single-shot efficiency ( $10^{-8}$) is comparable to traditional fs amplifier based systems at similar intensity levels. This demonstrates the dramatic increase in high-harmonic power that can be accessed using a high repetition rate (100 MHz). Clearly there is significant potential to further improve this efficiency and produce harmonics far into the Extreme Ultraviolet (EUV) simply by increasing the incident pulse energy, easily allowing access to intensities $10^{14}$ W · cm$^{-2}$ at high repetition rates.

In order to utilize the precision of the fs comb in the VUV spectral region, it must be verified that no detrimental phase/frequency fluctuations in the HHG process exist. To test this, we allow the original comb to drive two independent nonlinear processes, one through the well-characterized bound electronic nonlinearities based on second harmonic and sum frequency generation via two $\beta$-$BaB_2O_4$ (BBO) optical crystals, and the other through the HHG process (Fig. 4a). Two sets of the frequency combs at 266 nm that represent the third harmonic of the fundamental IR comb are then brought together in a Mach-Zehnder interferometer geometry for beat detection, after these two separate pulse trains are temporally overlapped. A 90-MHz acousto-optic modulator is inserted in one of the interferometer arms so that the beat detection is shifted to a convenient nonzero

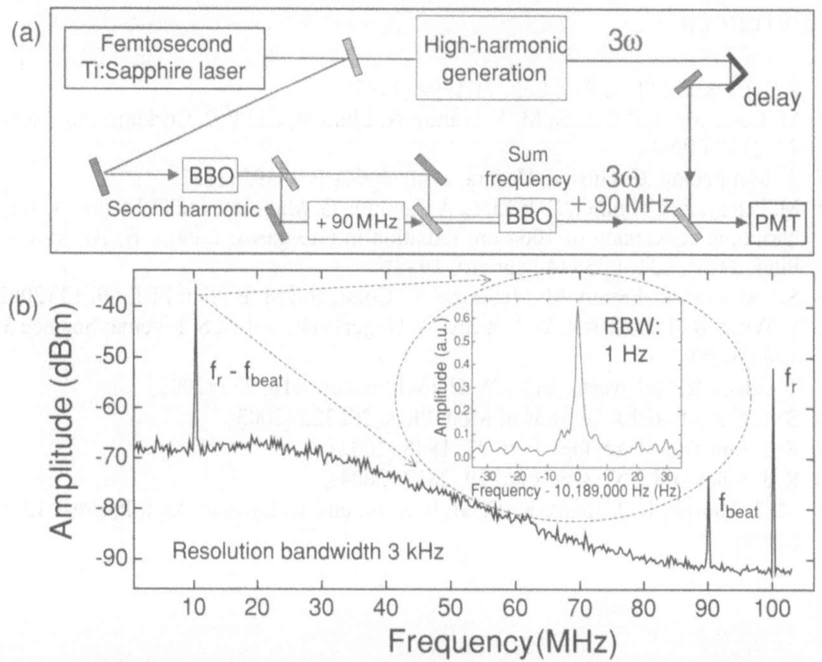


FIGURE 4. (a) Setup for coherent heterodyne detection between HHG in Xe gas and bound optical nonlinearities in BBO. (b) Measurement of the detected beat signal and repetition frequency. The linewidth shown in the inset is resolution bandwidth limited at 1 Hz.

frequency. Pairs of corresponding comb components from each spectrum produce a coherent optical beat detected by a photomultiplier. The radio-frequency spectrum of the beat note shows the clear presence of the comb structure in the UV (Fig. 4b), with a repetition frequency signal at 100 MHz and coherent beat signals both at 90 MHz and 10 MHz. The high-frequency roll-off is due to the response function of the photomultiplier amplifier. The resolution bandwidth-limited 1 Hz beat signal (inset, Fig. 4b), influenced by slow, differential path length fluctuations of the interferometer, demonstrates that the full spectral resolution and temporal coherence of the original near-IR comb has been faithfully transferred to higher harmonics in this process, limited only by the observation time period of 1 s. This type of temporal coherent interferometry of the HHG light opens the door for an exciting and unique set of precision measurements on the HHG process itself. A number of physical processes that could impact the phase coherence properties of HHG can be studied in detail, including ionization dephasing, intensity-dependent phase shifts, the effect of temporal chirp (spectral phase) of the near-IR pulse on the HHG light, etc. In the future, the presence of the frequency comb structure in the VUV and EUV spectral regions may enable similar revolutions in precision measurement, quantum control, and ultrafast science as have been recently witnessed in the visible region.

## References


1. P. B. Corkum: Phys. Rev. Lett. **71**,1994 (1993).
2. M. Lewenstein, P. Balcov, M. Y. Ivanov, A. Lhullier, and P. B. Corkum: Phys. Rev. A **49**, 2117 (1994).
3. A. Mcpherson, G. Gibson, H. Jara, et al.: JOSA B **4**, 595 (1987).
4. M. Ferray, A. L'Huillier, X. F. Li, L. A. Lompre, G. Mainfray, and C. Manus: Multiple-harmonic conversion of 1064 nm radiation in rare gases. J. Phys. B: At. Mol. Opt. Phys. **21**(3), L31–L35 (14 February 1988).
5. S. Cavalieri, R. Eramo, M. Materazzi, C. Corsi, and M. Bellini: PRL **89**, 13 (2002).
6. S. Witte, R. T. Zinkstok, W. Ubachs, W. Hogervorst, and K. S. Eikema: Science **307**, 400 (2005).
7. T. Udem, R. Holzwarth, and T.W. Hänsch: Nature **416**, 233 (2002).
8. S. T. Cundiff and J. Ye: Rev. of Mod. Phys. **75**, 325 (2003).
9. R. J. Jones and J. Ye: Opt. Lett. **27**, 1848 (2002).
10. R. J. Jones and J. Ye: Opt. Lett. **29**, 2812 (2004).
11. M. J. Thorpe, R. J. Jones, K. D. Moll, J. Ye, and R. Lalezari: Opt. Express **13**, 882 (2005).

# Effects of the Carrier-Envelope Phase in the Multiphoton Ionization Regime

Takashi Nakajima[1,2] and Shuntaro Watanabe[2]

[1] Institute of Advanced Energy, Kyoto University, Gokasho, Uji, Kyoto 611-0011, Japan
t-nakajima@iae.kyoto-u.ac.jp
[2] Institute for Solid State Physics, The University of Tokyo, Kashiwa, Chiba 277-8581, Japan
watanabe@issp.u-tokyo.ac.jp

We theoretically investigate the effects of the carrier-envelope phase (CEP) in the low-intensity regime where multiphoton ionization is the dominant mechanism for ionization. For atoms with low-ionization threshold such as Cesium, total ionization yield barely exhibits phase dependence. However, population of some bound states clearly shows dependence on the CEP. This implies that the measurement of the CEP would be possible through the photoemission between bound states without energy and angle-resolved photoelectron detection. The proposed scheme could be particularly useful to measure the CEP for a light source without an amplifier, such as a laser oscillator, which cannot provide sufficient pulse energy to induce tunneling ionization. Similar findings are also obtained for hydrogen, which shows stronger phase dependence not only for the bound state population but also for the total ionization yield.

Because of the rapid advances of laser technology and strong-field physics in the last few years, it is now well established that an intense few-cycle pulse induces various effects that are dependent on the carrier-envelope phase (CEP), in photoionization and high harmonic generation in the strong field regime where the processes are highly nonlinear and nonperturbative [1–6]. The physical picture that is most conveniently employed is tunneling ionization: once an electron escapes the Coulomb potential, its trajectory can be predicted with the use of classical mechanics. This semiclassical tunneling model essentially reproduces results obtained from the fully quantum mechanical calculations, and more importantly, agrees rather well with experimental results. Based on the CEP-dependent effects, CEP has been successfully measured by the energy- and angle-resolved photoelectron detection. The tunneling picture, however, does not tell us anything for the case in which the field is rather weak to induce tunneling ionization.

In this paper, we theoretically show that the CEP-dependent effects can be clearly seen for a few-cycle pulse even in the multiphoton ionization regime without energy- and angle-resolved photoelectron detection, and we propose a simple scheme to measure the CEP from photoemission. Specific numerical results are presented for Cs and also H atoms for comparison. We emphasize that, although the $\pm\pi$ ambiguity still remains with our scheme, the CEP can be determined by the technically much simpler manner, and the peak intensity can be much lower than that for the tunneling ionization regime. This means that the scheme considered here would be particularly useful to measure the CEP of the light pulse from a laser oscillator which obviously cannot provide a pulse energy to induce tunneling ionization.

To describe the dynamics, we solve the three-dimensional time-dependent Schrödinger equation for the Cs atom with a discretized basis set constructed in a spherical box of radius 200 au. Photon energy is assumed to be 1.55 eV, corresponding to a Ti:Sapphire laser, and the peak intensity is chosen to be $10^{11}$ W $\cdot$ cm$^{-2}$. Since the ionization potential of Cs is 3.89 eV, three photons are sufficient for photoionization. Note that even the two-photon process can also contribute to ionization due to the broad frequency spectrum of the few-cycle pulses. Keldysh parameter, $\gamma$, is 18 for the above parameter set. We define the number of cycles, $N$, for the FWHM of the vector potential, $A(t)$, which is assumed to have a Gaussian temporal envelope, i.e., $A(t) = \hat{\epsilon} A_0 \exp\left[-4\ln 2(\omega t/2\pi N)^2\right] \sin(\omega t + \phi)$, where $\hat{\epsilon}$ is a unit polarization vector and $\phi$ is a CEP. The electric field, $E(t)$, which we have used for the numerical time propagation, is defined as $E(t) = -\partial A(t)/\partial t$. In order to discuss the effects of the CEP, we now define the depth of modulation, $M$, for the total ionization yield, as $M = 2[P(\phi_{\max}) - P(\phi_{\min})]/[P(\phi_{\max}) + P(\phi_{\min})]$, where $P$ is the total ionization yield. $\phi_{\max}$ and $\phi_{\min}$ represent the CEP at which the ionization yield becomes maximum/minimum for a given photon energy, number of cycles, and peak intensity. Obviously $M$ takes a value between 0 and 2.

First we show a few representative results for the Cs atom. In Fig. 1, we present the variation of the depth of modulation, $M$, of the total ionization yield of Cs as a function of the number of cycles, $N$. It can be seen that, although we see slightly more modulation for the pulse with fewer number of cycles, the CEP dependence of the total ionization yield is extremely weak, anyway, as expected, since ionization is a very low-order process in this case, and moreover we are looking at the *total* ionization yield rather than the angle- and energy-resolved ATI photoelectrons.

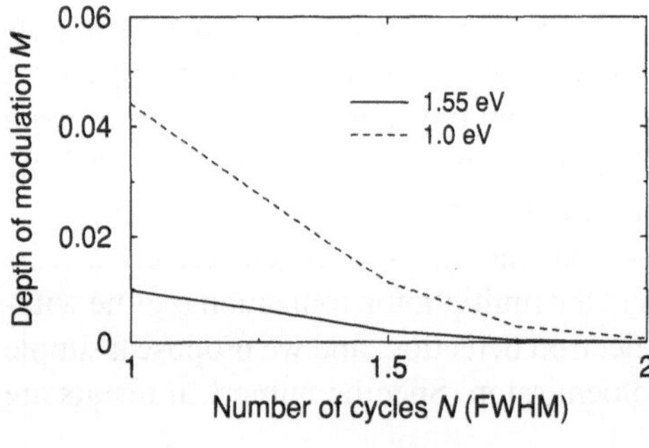


FIGURE 1. Depth of modulation for the total ionization yield of Cs as a function of the number of cycles, $N$, for the photon energies and the peak intensities of 1.55 eV and $I = 10^{11}$ W $\cdot$ cm$^{-2}$ (solid line), and 1.0 eV and $I = 4.2 \times 10^{10}$ W$\cdot$cm$^{-2}$ (dashed line). Note that the peak intensities are differently chosen so that the Keldysh parameter becomes the same value ($\gamma = 18$) for both cases.

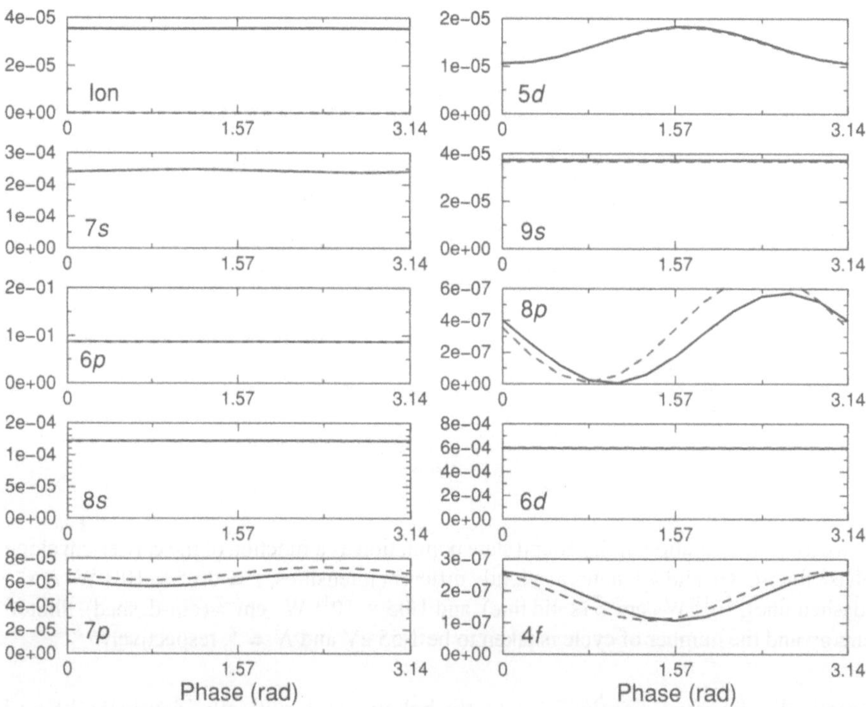


FIGURE 2. Total ionization yield and the population of some bound states of Cs as a function of CEP for 1.55 eV photon and the peak intensity of $I = 10^{11}$ W·cm$^{-2}$ with $N = 1$. Calculated results without ionization are shown by dashed lines.

When we look at the population of bound states, however, the situation is quite different. A representative result is shown in Fig. 2 for $N = 1$. Clear CEP dependence is found for the $5d$, $8p$, and $4f$ states. These states acquire small but observable population due to the extremely broad bandwidth of the pulse despite the large detuning from the resonant energy. In other words, CEP effects are not seen for the states such as $7s$, $6p$, $8s$, $9s$, and $6d$, etc., which are closer to resonance. Another interesting feature is that there are distinct phase shifts in the CEP dependence of those states. For example, modulation of the $8p$ and $4f$ states is negative sine- and cos-like, respectively, while that of the $5d$ state is negative cos-like. In order to make sure that the obtained results are not numerical artifacts, we have repeated the same calculations for different box sizes, for all of which we reproduced practically the identical results.

As a further check, we have repeated the calculation without ionization by removing all the continuum states during the time propagation, the purpose of which is to examine whether the CEP dependence of the bound state population is affected through the ionization process. The results are shown by the dashed lines in Fig. 2. For the $8p$ and $4f$ states, the calculated modulation turned out to be slightly shifted by removing the continuum. However, qualitative behavior is

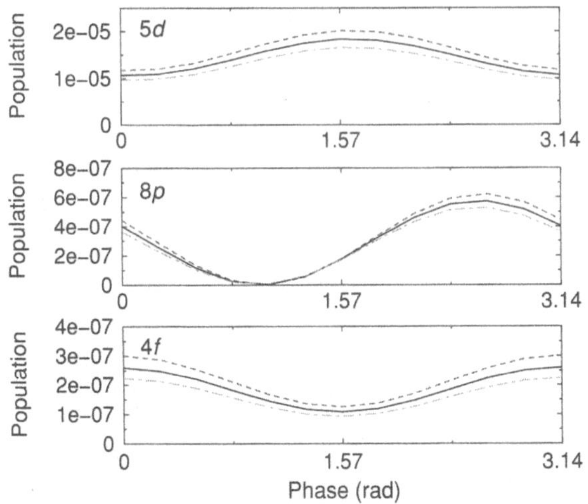


FIGURE 3. Modulation of the bound state population as a function of the carrier-envelope phase for $5d$, $8p$, and $4f$ states at slightly different intensities, $I = 0.95 \times 10^{11}$ W $\cdot$ cm$^{-2}$ (dashed line), $10^{11}$ W $\cdot$ cm$^{-2}$ (solid line), and $1.05 \times 10^{11}$ W $\cdot$ cm$^{-2}$ (dot-dashed). Photon energy and the number of cycle is taken to be 1.55 eV and $N = 1$, respectively.

practically the same. For the $5d$ state, the behavior is practically identical with and without ionization. These results clearly show that the CEP-dependent population of the bound states essentially stems from the photoabsorption processes *even without ionization*. Although we have seen that there is a clear CEP dependence on the population of some bound states, we have to check the robustness against intensity fluctuation. In Fig. 3, we show the phase-dependent modulation of the $5d$, $8p$, and $4f$ as a function of the CEP at slightly different ($\pm5\%$) intensities around $I = 10^{11}$ W $\cdot$ cm$^{-2}$. Clearly the phase-dependent modulation is quite robust against intensity fluctuation.

All of these results indicate that the CEP dependence of the bound state population can be utilized for the measurement of the CEP of few-cycle pulses through the photoemission from those states: since the lifetimes of $4f$ and $8p$ are approximately 40 ns and 300 ns, respectively, spontaneous emission takes place in a sufficiently short time. For example, population of the $4f$ state is about $10^{-7}$ by the $N = 1$ pulse at $I = 10^{11}$ W $\cdot$ cm$^{-2}$. If the number density of the Cs vapor is about $10^{14}$ cm$^{-3}$ and the interaction volume is $10^{-3}$ cm$^3$, $10^4$ atoms are excited to the $4f$ state, whose population can be easily probed from the photoemission at 1002 nm down to the $5d$ state. Photoemission from the $8p$ state down to the $5d$ state at the wavelength of 802 nm is another candidate. Note that it is rather easy to increase the photoemission signal by the slight increase of the vapor temperature or/and peak intensity.

What if the ionization potential is higher, or photon energy is smaller such that more photons are required for photoexcitation/photoionization? Based on the simple argument, we expect to see stronger CEP effects for atoms with higher

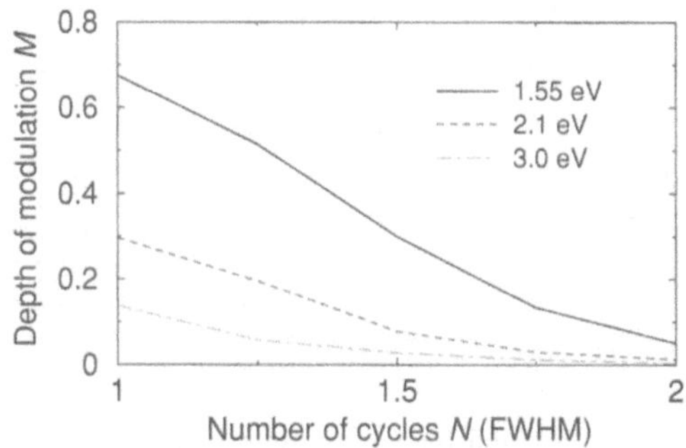


FIGURE 4. Depth of modulation for the total ionization yield of H as a function of the number of cycles, $N$, for the photon energies of 1.55 eV with the peak intensity of $I = 3 \times 10^{12}$ W $\cdot$ cm$^{-2}$ (solid line), 2.1 eV with $I = 5.4 \times 10^{12}$ W $\cdot$ cm$^{-2}$ (dashed line), and 3.0 eV with $I = 10^{13}$ W $\cdot$ cm$^{-2}$ (dot-dashed). Note that the peak intensities are differently chosen so that the Keldysh parameter takes the same value ($\gamma = 6.2$) for all cases.

ionization potential if the photon energy is chosen to be the same. Another way of saying is that, given the same atom, the CEP dependence would be smaller for bigger photons. To test our assumption, we have carried out calculations for the H atom. The photon energy has been chosen to be 1.55 eV again, and the peak intensity is set to be $3 \times 10^{12}$ W $\cdot$ cm$^{-2}$ which gives $\gamma = 6.2$. Figure 4 shows the variation of the depth of modulation, $M$, as a function of the number of cycles, $N$. As expected, the CEP-dependent modulation for H is now much larger than Cs, since the ionization process of H by 1.55 eV photon is a higher order process compared with that of Cs. As the number of cycles increases from 1 to 2, however, the value of $M$ approaches zero. For comparison, we also plot the results for the photon energies of 2.1 eV at $I = 5.4 \times 10^{12}$ W $\cdot$ cm$^{-2}$ and 3.0 eV at $I = 1.1 \times 10^{13}$ W $\cdot$ cm$^{-2}$, both of which give the same Keldysh parameter, $\gamma = 6.2$, with that for 1.55 eV at $I = 3 \times 10^{12}$ W $\cdot$ cm$^{-2}$. Regardless of the same $\gamma$, $M$ takes much smaller value for bigger photon energies simply because ionization requires a fewer photon absorption.

We now investigate how the phase-dependent modulation of the total ionization yield depends on the intensity. Figure 5a shows the variation of the modulation of the total ionization yield as a function of the CEP at different intensities. We observe an intensity-dependent phase shift. To identify the origin of the phase shift, we have performed calculations without bound states except for the ground state, which means that all intermediate states become purely virtual. The results are shown in Fig. 5b. It is clear that the intensity-dependent phase shift is due to the existence of the real bound states. This result indicates that the detail of the atomic structure significantly affects the CEP-dependent effects, in particular in the multiphoton ionization regime. We would like to point out that there must be similar effects even in the tunneling ionization regime, since, according to our numerical calculations (not shown here), calculated ionization yields are as

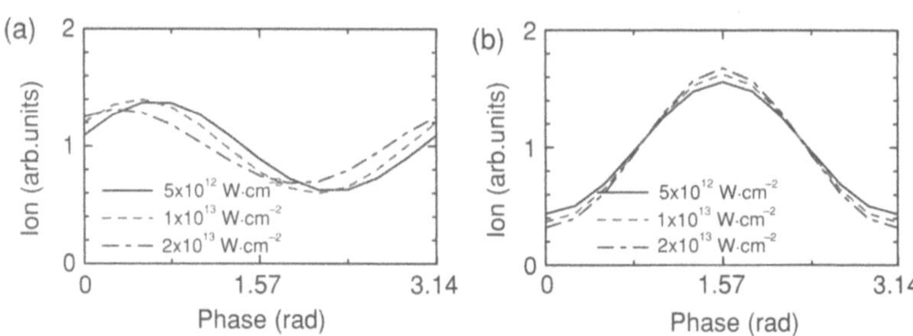


FIGURE 5. Modulation of the total ionization yield of H at different intensities for the pulse at photon energy of 1.55 eV with $N = 1$. Calculations are performed (a) with all states and (b) without bound states.

much as two orders of magnitude different with/without bound states even in the tunneling ionization regime ($\gamma \sim 1$) as far as the intensity is well below the saturation intensity. Of course, if it is near/above the saturation intensity, the difference between them is small.

Now we move on to the results and discussion on the behavior of the bound state population for H. For the 1.55-eV photon, we have already seen that, for the Cs atom, the total ionization yield barely shows phase dependence (Fig. 1), while the phase sensitivity of the population of some bound state turns out to be significant (Fig. 2). As for the H atom, we have seen (Fig. 4) that the total ionization yield shows significant phase dependence at the photon energy of 1.55 eV. The question is whether we can expect phase sensitivity for the bound state as well. From Fig. 5, we can infer that there must be strong phase dependence on the bound state population as well. Indeed, Fig. 6 shows the strong variation of the bound state population as a function of the phase. Unlike Cs, many of the bound states of H exhibits strong phase dependence, which is primarily due to the fact that the bound-excited states of H are located 10–13 eV above the ground state, implying that the excitation of those bound states necessarily requires many-photon absorption without any near-resonant states. The consequence of many-photon absorption without any real intermediate states is the enhancement of the phase sensitivity, as we have seen in Fig. 5b for the total ionization yield.

Before closing this paper, we will show how the depth of modulation, $M$, varies as a function of intensity. Calculations have been performed at the photon energy of 1.55 eV for three different pulse durations, $N = 1.0$, 1.5, and 2. The depth of modulation is larger for smaller $N$ at a given intensity, as expected. The depth modulations for the $N = 1.5$ and 2.0 pulses decrease rather monotonically as the intensity decreases. For the $N = 1.0$ pulse, however, there seems to exist an optimal intensity at which maximum depth of modulation is observed. For comparison, the variations of the depth of modulation calculated without excited bound states during the time propagation have been plotted by thin lines in Fig. 7. The depth of modulation for the $N = 1.0$ pulse increases as the intensity increases; after reaching the maximum value around $4 \times 10^{13}$ W $\cdot$ cm$^{-2}$, it decreases, and

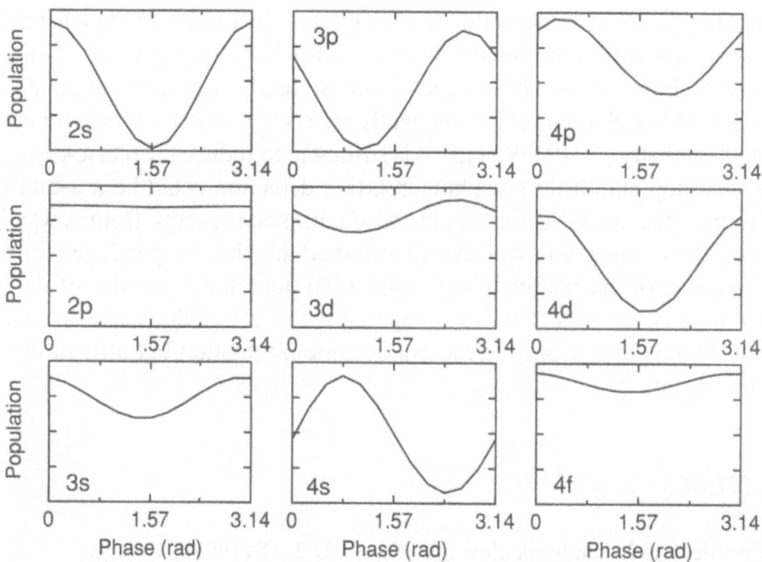


FIGURE 6. Population of some bound states of H as a function of the carrier-envelope phase for 1.55 eV photon and the peak intensity of $I = 3 \times 10^{12}$ W $\cdot$ cm$^{-2}$ with $N = 1$.

in the higher intensity limit, it approaches zero. The reason of approaching zero in the high-intensity limit is obvious. When the intensity becomes well above the intensity required for the barrier suppression ionization, the ionization process naturally becomes insensitive to the CEP.

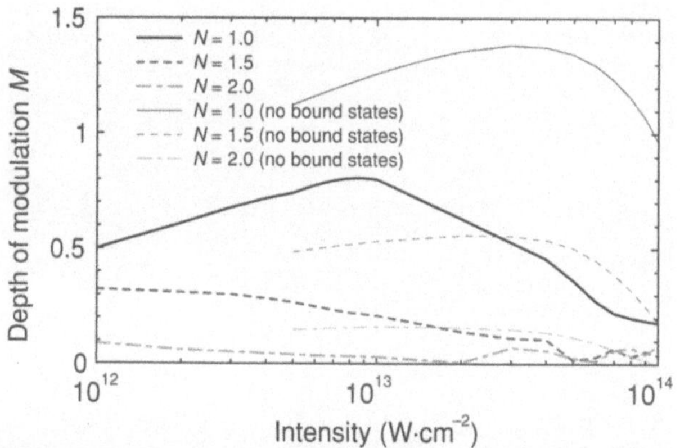


FIGURE 7. Variation of the depth of modulation, $M$, for H atom as a function of laser intensity. Photon energy is 1.55 eV. $N = 1.0$ (solid), $N = 1.5$ (dashed), $N = 2.0$ (dot-dashed) with all bound/continuum states included during the time propagation, and $N = 1.0$ (thin solid), $N = 1.5$ (thin dashed), and $N = 2.0$ (thin dot-dashed) with all excited bound states removed during the time propagation.

In summary, we have theoretically investigated the effects of the CEP at low-intensity regime (multiphoton ionization regime). For atoms with low-ionization threshold such as Cs, the phase dependence is clearly seen in some bound state population, although total ionization barely shows phase dependence. Since relatively low intensity ($\sim 10^{11}$ W $\cdot$ cm$^{-2}$) is sufficient to induce the phase-dependent bound state population for Cs, photoemission detection could be a useful technique to measure the CEP for a light pulse with small energy from, say, a laser oscillator. For comparison, we have also studied the phase dependence for the H atom. Because of the relatively high-ionization potential, the order of the excitation/ionization process is higher compared with Cs, and the phase dependence has been found for not only in the bound state population but also in the total ionization yield.

## References

1. E. Cormier and P. Lambropoulos: Eur. Phys. J. D **2**, 15 (1998).
2. I. P. Christov: Opt. Lett. **24**, 1425 (1999).
3. P. Dietrich, F. Krausz, and P. B. Corkum: Opt. Lett. **25**, 16 (2000).
4. G. G. Paulus, F. Grabson, H. Walther, et al.: Nature **414**, 182 (2001).
5. D. B. Milosevic, G. G. Paulus, and W. Becker: Phys. Rev. Lett. **89**, 15300 (2002).
6. G. G. Paulus, F. Lindncr, H. Walther, et al.: Phys. Rev. Lett. **91**, 253004 (2003).

# Carrier-Envelope Phase Detection by Interference between Surface Harmonics

Atsushi Ishizawa and Hidetoshi Nakano

NTT Basic Research Laboratories, NTT Corporation, 3-1 Morinosato Wakamiya, Atsugi, Kanagawa 243-0198, Japan
ishizawa@will.brl.ntt.co.jp

We demonstrate a measurement that provides direct information about the carrier-envelope phase (CEP) using a few-cycle laser pulse. The measurement is based on the interference between the second and third harmonics. The interference intensity is sensitive to the CEP. This result for low-energy pulses will lead to a new technique for measuring the CEP.

## 1 Introduction

Ultrashort pulse generation has been improved since the first demonstration of a laser. Pulses with only a few-cycle optical cycles at a center wavelength of 800-nm have been generated with mode-locked Ti: sapphire laser. Intense few-cycle laser pulses [1,2] have come to provide major impact on light-matter interaction, high-order harmonic generation, attosecond ($10^{-18}$ s) pulse generation, and extremely nonlinear optics. The carrier-envelope phase (CEP) is defined as the phase of the carrier wave at the maximum of the envelope. The CEP as a new parameter for characterizing a few-cycle laser pulses is of great importance for a number of modern directions of laser physics [3]. Recently, it has been observed that both above-threshold ionization [4, 5] and high-order harmonic generation [6] depend sensitively on the CEP. Almost all previous direct measurements of the CEP have relied on strong-field processes using an amplifier, which is available only at low repetition rates in a vacuum chamber with a specialized detector for measuring the CEP. Recently, CEP-sensitive phenomena in solid using an oscillator have been reported [7–9]. Müche et al. reported a rapid Rabi flopping in a gallium arsenide using an oscillator of few-cycle laser pulses [7]. The Rabi cycle is driven by the carrier wave of the laser, not by the pulse envelope, and is sensitive to the CEP. The stable CEP of an oscillator will be expected to reveal new phenomena in nonlinear optics. The energy of the pulses in our new method is 1 μJ, which is one to two orders of magnitude lower than in previous CEP measurements using an amplifier [4–6]. A recently reported Ti: sapphire oscillator provides 220-nJ pulses [10], which is one order of magnitude smaller than the energy in this experiment. However, Farkas et al. observed both even and odd harmonics up to the fifth order with efficiencies $10^{-10}$–$10^{-13}$ by focusing a gold surface at the intensity of 5 GW·cm$^{-2}$ [11]. This intensity can be achieved using a standard Ti:sapphire oscillator. Therefore, it is feasible that our method will be able to detect the CEP directly using an oscillator. Previous measurements of the CEP drift by the interference between the second and third harmonics in bulk have been proposed and demonstrated [12, 13]. Mehendale et al. proposed measuring the CEP drift of high-energy 5-fs laser pulses using the interference between the second harmonic (SH) in a $\beta$-barium borate crystal and the third harmonic (TH) from the surface of a solid [13]. Their method is good way of measuring the CEP drift. However, it is essentially the same as the conventional f-to-2f spectrum interferogram method [14]. The measured phase includes a constant offset phase component due to linear dispersion in the SHG crystal. The problem is that, since the CEP varies from pulse to pulse, it is impossible to use that method to measure the CEP directly [14]. We have devised a way of directly measuring the CEP using the interference between the SH and TH from the surface of a solid. Our paper shows how to improve the previous schemes [12, 13] so as to directly access the CEP. Our goal is to develop a method that enables us to easily determine the direct CEP. In this paper, we propose a possible way of providing direct access to the CEP in low-energy pulses (nearly 1 μJ) using the interference between the SH and TH from the surface of a solid at the same time. The interference signals are a superposition of a cos $(2\phi)$-oscillation and a cos $(3\phi)$-oscillation. The important point is that the spectra of these two harmonics overlap in the UV region. This means the interference signals do not include a constant offset phase component due to the linear dispersion in the SHG crystal and can therefore be used to measure the CEP directly. Measuring the interference intensity enables us to measure the CEP directly, which is not possible with conventional f-to-2f spectrum interferogram method. In this experiment, the interference signals from the surface were found to exhibit CEP dependence for pulses comprising only a few optical cycles. If the pulse is sufficiently short, the larger spectral width ensures that the SH and TH of the pulse overlap in spectral domain. In the overlap region, the interference intensity between the SH and TH has cyclic ups and downs for the CEP.

## 2 Experimental Setup

A schematic diagram of the experimental setup is shown in Fig. 1. We used a multipass Ti:sapphire CPA system. The CEP slip of the pulses delivered by the Ti:sapphire oscillator was controlled. The amplifier delivers 1-mJ energy in a 25-fs pulse. In our previous work, we found that by using between 2- and 2.5-atm ne gas in the hollow fiber, we could make both the pulse width and the CEP fluctuation

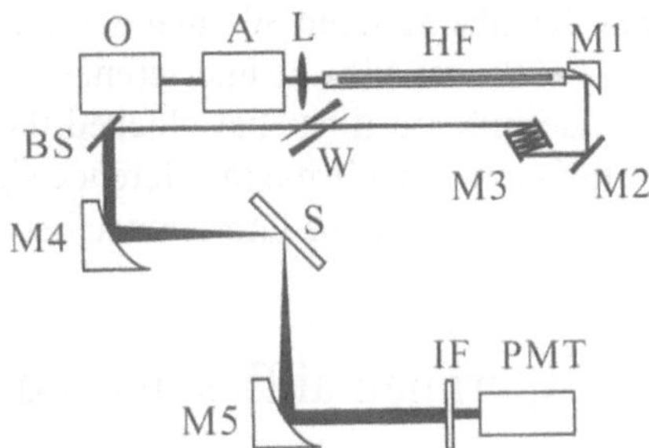


FIGURE 1. Experimental setup: O, CEP slip controlled oscillator; A, Amplifier; L, Lens; M1, spherical mirror; M2, mirror; M3, 4-pair chirp mirrors; M4 and M5, off-axis parabolic mirrors; HF, hollow fiber; W, a pair of glass wedges (apex angle 2.8°); BS, beam splitter with 5% reflection; S, sample (glass); IF, interference filter; PMT, photomultiplier.

small [15, 16]. Here, the amplified laser pulse was focused into an 1-m long hollow fiber placed in a chamber filled with 2-atm ne gas. The inner diameter of the fiber was 250 μm. The propagation in the hollow fiber broadened the spectrum more than one octave by self-phase modulation. The laser pulse width was finally compressed on reflection by the 4-pair chirp mirrors. This system delivers a 5-fs, 0.2-mJ laser pulse with a carrier wavelength of 716-nm when the gas pressure in the hollow fiber is 2 atm. The spectrum of the laser pulse is shown in Fig. 2.

In this experiment, a 5-fs, 1-μJ laser pulse was focused on a glass target at the incidence angle of 45° using an off-axis parabolic mirror. Even and odd harmonics were generated due to the broken symmetry of an interface. The out-of-focus arrangement for the low peak intensity allowed the harmonics to generate from the front interfaces of the glass. The harmonic beams reflected from the glass were collimated and detected using an UV-sensitive photomultiplier tube and an interference filter with the center wavelength of 267 nm, which blocks all other scattered photons. The SH- and TH-spectrum overlap in the UV region is shown in Fig. 2. We used color filters to verify that the signal was in fact due to the interference between the SH and TH. One (HOYA, CM-500) was transparent to

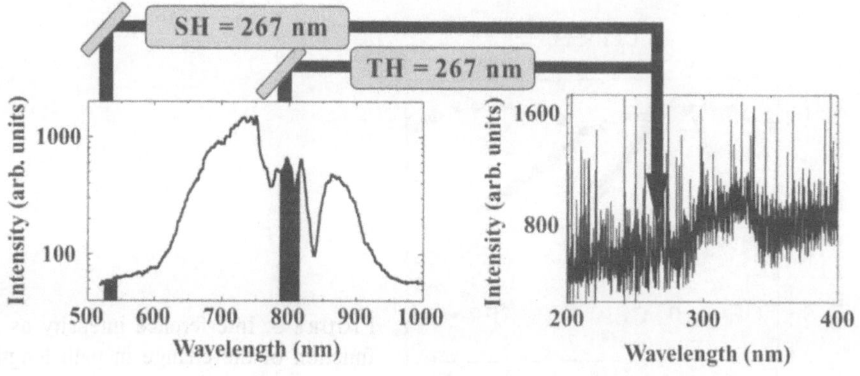


FIGURE 2. Measured spectrum of a 5-fs, 0.2-mJ laser pulse. The interference between the SH and TH were measured in the UV region.

wavelengths between 300 and 750 nm, so it blocked the SH. When we put it in front of the sample, the interference signal disappeared. The other (HOYA, IR85) was an infrared filter that blocked the TH. In this case, too, putting it in front of the sample made the interference signal disappear. This demonstrated that the signal was due to the interference between the SH and the TH.

## 3 Experimental Results and Discussion

We investigated the CEP directly using the interference signal between the SH and TH by experiment. A pair of glass wedges (apex angle of 2.8°) was placed in front of the off-axis parabolic mirror. The position of one was fixed, and the other was movable. The wedges were used to optimize dispersion and adjust the CEP. The envelope propagates with the group velocity and the carrier with the phase velocity of light. Therefore, dispersion by the glass can shift the CEP. Figure 3 shows a series of interference signals of different phases.

Each dot shows the average for a few signals. The series of interference signals could be measured within 1 min. We confirmed that both the period and amplitude of the interference signals using a silver surface agreed with those using the glass surface. It is evident that the intensity of the interference between the SH and TH generated from the surface of a solid is very sensitive to the CEP of a few-cycle laser pulses. The peaks occur at even-integer multiples of $\pi$ radians, and the valleys occur at odd-integer multiples. If we take this point to be 0 radians, then the CEP has the values shown in Fig. 3. This phenomenon is available for determining the CEP value of each laser pulse. It is possible to increase the efficiency of harmonic generation by using another kind of material as a sample [17]. We also measured the interference signal with a silver target. This yielded a stronger intensity, but there was no big difference in the clarity of the data. The higher the second-order susceptibility $\chi(2)$ and third-order susceptibility $\chi(3)$ are, the higher becomes the efficiency of harmonic generation. If the efficiency of harmonic

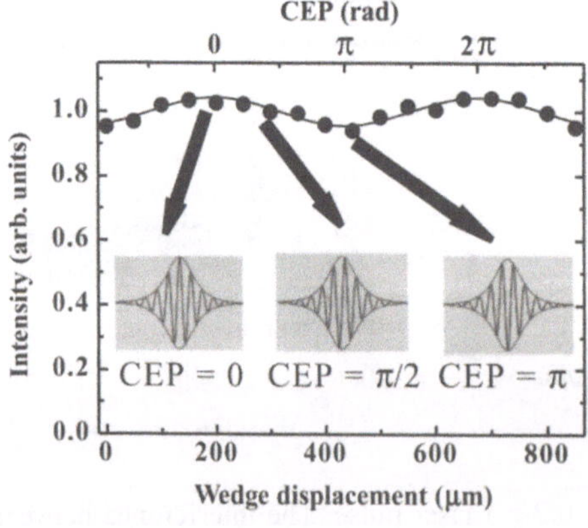


FIGURE 3. Interference intensity as a function of the change in path length through the fused silica glass wedges in Fig. 1.

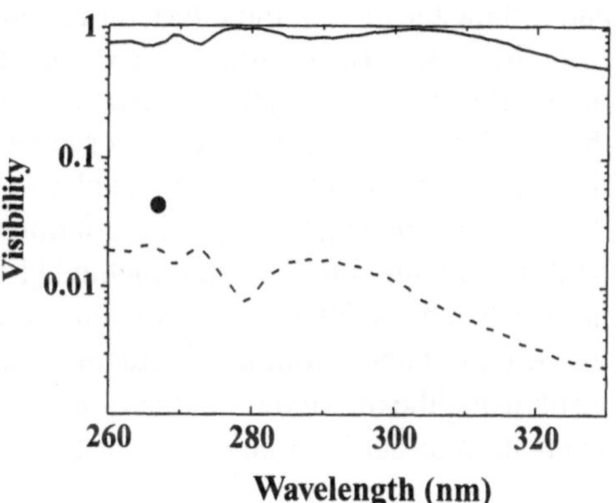


FIGURE 4. Interference visibility between the SH and TH using the laser spectrum shown in Fig. 2. The thick and dotted line shows the calculated interference visibility using a TH/SH ratios of conversion efficiency of 0.5 [11] and $10^{-5}$ [18], respectively. The dot shows experimental interference visibility.

generation can be increased, this phenomenon can be measured using even lower pulse energy, such as oscillator energy. We calculated the interference visibility between the SH and TH using the laser spectrum shown in Fig. 2. Figure 4 shows the interference visibility calculated for TH/SH intensity ratios of 0.5 [11] and $10^{-5}$ [18].

For the intensity ratio of TH/SH of 0.5, the interference signal of the wavelength of around 280- and 303-nm provides the maximum visibility. For TH/SH ratio of $10^{-5}$, the interference signal of the wavelength of around 265-nm provides the maximum visibility. The visibility of the interference signal depends on the wavelength and intensity ratio. Moreover, it has been theoretically predicted that the intensity ratio of TH/SH depends on the incident angle of a laser pulse to a target [19]. In this experiment, we chose the interference signal of 267 nm. The visibility of the interference fringe was not so high. However, it is feasible that the interference visibility can be improved by optimizing the wavelength and the incident angle of a laser pulse to a target. We estimated the CEP fluctuation in our laser pulses by using the interference signal between the SH and TH. Figures 5a

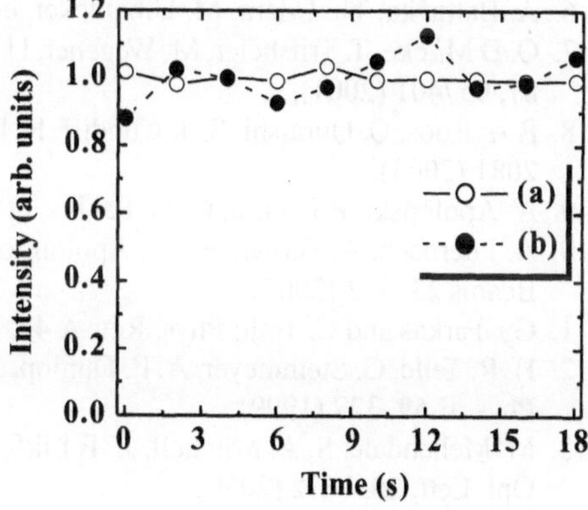


FIGURE 5. Drift of the interference signal between the SH and TH generated from the glass surface using a few-cycle laser pulse (a) with and (b) without controlling the CEP slip of the oscillator.

and b show the drift of the interference signal between the SH and TH generated from the glass surface using a few-cycle laser pulse with and without controlling the CEP slip of the oscillator, respectively. When the CEP slip was controlled, the interference signal was stable for at least 18 s. The standard deviation of the fluctuation of the interference signal was less than 3%, which corresponds to $\pm\pi/6$ CEP fluctuation using the relationship in Fig. 3. With the conventional f-to-2f spectrum interferogram method [14], a CEP fluctuation was measured to be $\pm\pi/8$ rad for 20 s in our previous work [15, 16] which almost agrees with the result obtained from the fluctuation of the interference signal in Fig. 5. This result using the interference signal was used as a cross-check to confirm the CEP fluctuation of our laser pulses in our previous work.

## 4 Summary

We experimentally demonstrated a measurement in which the interference signal between the SH and TH from the surface of a solid provides a very effective method of measuring the CEP. Moreover, it is very easy to setup the equipment and perform the measurement. We expect that various phenomena related to the CEP-sensitive interactions in solids using an oscillator will be reported. We hope this easy method using a low-energy pulse will broaden applications that are very sensitive to the CEP.

## References

1. M. Nisoli, S. De Silvestri, O. Svelto, et al.: Opt. Lett. **22**, 522 (1997).
2. G. Steinmeyer, D. H. Sutter, L. Gallmann, N. Matuschek, and U. Keller: Science **286**, 1507 (1999).
3. T. Brabec and F. Krausz: Rev. Mod. Phys. **72**, 545 (2001).
4. G. G. Paulus, F. Grasbon, H. Walther, et al.: Nature **414**, 182 (2001).
5. G. G. Paulus, F. Lindner, H. Walther, et al.: Phys. Rev. Lett. **91**, 253004 (2003).
6. A. Baltuška, Th. Udem, M. Uiberacker, et al.: Nature **421**, 611 (2003).
7. O. D Mücke, T. Tritshcler, M. Wegener, U. Morgner, and F. X. Kärtner: Phy. Rev. Lett. **87**, 057401 (2001).
8. P. A. Roos, Q. Quraishi, S. T. Cundiff, R. D. R. Bhat, and J. E. Sipe: Opt. Express. **11**, 2081 (2003).
9. A. Apolonski, P. Dombi, G. G. Paulus, et al.: Phy. Rev. Lett. **92**, 073902 (2004).
10. A. Fuerbach, A. Fernandez, A. Apolonski, T. Fuji, and F. Krausz: Laser and Particle Beams **23**, 113 (2005).
11. Gy. Farkas and C. Toth: Phys. Rev. A **46**, R3605 (1992).
12. H. R. Telle, G. Steinmeyer, A. E. Dunlop, J. Stenger, D. H. Sutter, and U. Keller: Appl. Phys. B **69**, 327 (1999).
13. M. Mehendale, S. A. Mitchell, J. -P. Likforman, D. M. Villeneuve, and P. B. Corkum: Opt. Lett. **25**, 1672 (2000).
14. M. Kakehata, H. Takada, Y. Kobayashi, et al.: Opt. Lett. **26** 1436 (2001).

15. A. Ishizawa and H. Nakano: *Ultrafast Phenomena XIV*. Chem. Phys. Vol. 79. Dordreeth: Springer, pp. 37–39 (2004).
16. A. Ishizawa and H. Nakano: Jpn. J. Appl. Phys. **44**, 6039 (2005).
17. G. Veres, S. Matsumoto, Y. Nabekawa, and K. Midorikawa: Appl. Phys. Lett. **81**, 3714 (2002).
18. A. Mishra and J. I. Gersten: Phys. Rev. B **43**, 1883 (1991).
19. A. T. Georges: Phys. Rev. A **54**, 2412 (1996).

# Coherent Control by Carrier-Envelope Phase in an Optical Poling Process

Shunsuke Adachi and Takayoshi Kobayashi

Department of Physics, Faculty of Science, University of Tokyo, 7-3-1 Hongo, Bunkyo, Tokyo 113-0033, Japan
**adachi@femto.phys.s.u-tokyo.ac.jp**

**Abstract:** We demonstrate that the efficiency of an optical poling process depends on quantum interference controlled by the self-stabilized carrier-envelope phase of the idler output from our noncollinear optical parametric amplifier system. To control quantum interference via direct control of the carrier-envelope phases instead of conventional control of the relative phases of two pulses is a new approach to coherent control.

## 1 Introduction

Carrier-envelope (c-e) phase has been attracting expanding interest because of potential applications such as controlled generations of high-harmonic generation extending to soft X-ray region [1] and above-threshold ionization [2] with high-intense field. In these experiments, the c-e phase comes only into play in the case of pulses being as short as several optical cycles. Besides, the c-e phase plays an essential role in precision optical frequency measurement [3], where c-e offset frequency is precisely shifted by controlling two frequency combs interfering with each other.

Recently, another role of the c-e phase in coherent control was demonstrated [4]. In their experiment, quantum interference of injected photocurrents in semiconductors was controlled by utilizing the c-e phase locked laser system. In this report, we will show that the efficiency of optically induced molecular orientation depends on quantum interference controlled by the c-e phase. Molecules without any cooperative interactions naturally tend to form centrosymmetric structures, which forbids any macroscopic second-order susceptibility. In an optical poling process, quasipermanent poling of the molecules is induced by light in an initially centrosymmetric material [5]. In our experiment, the efficiency of optically induced molecular orientation could be controlled by the c-e phase, utilizing noncollinear optical parametric amplifier (OPA) system developed in our group [6].

## 2 Optical Poling

In an optical poling process, two pulses at fundamental ($\omega$) and its second-harmonic(SH; $2\omega$) are incident on the molecules [5,7,8]. The molecules without an inversion center experience resonance excitation by simultaneous one-photon absorption of $2\omega$ pulse and two-photon absorption of $\omega$ pulse, corresponding to the $S_0 \rightarrow S_1$ electronic transitions. This results in a selective excitation of the molecules whose dipole is either parallel or antiparallel, depending on the polarity, caused by the coherent superposition of the two "writing" beams at $\omega$ and $2\omega$ frequencies. This so-called orientational hole-burning effect is followed by an isomerization of the molecules from the more stable *trans* form to the *cis* form. Then the reverse *cis–trans* reaction of the molecules occurs, which is thermally activated at room temperature, taking two possible pathways. Molecules that take one pathway restore the initial orientation of the molecule, while the other pathway leads to a reorientation of the molecular optical axis. Through these sequential processes of orientational hole burning and *trans–cis–trans* isomerization cycle, number of the molecules with specific orientation decreases, and consequently net macroscopic dipole moment is induced, which is embedded in polymer matrices [8,9]. The photoinduced second-order susceptibility $\chi^{(2)}_{\text{ind}}$ depends on the interference between the two fields at frequency $\omega$ and $2\omega$ and is proportional to nonzero time-average cube of the coherent superposition of $\omega$ and $2\omega$ pulses, $\langle E^3(z,t)\rangle_t$ [7,8]:

$$\begin{aligned}\chi^{(2)}_{\text{ind}} &\propto \langle E^3(z,t)\rangle_t \\ &= \langle\{E_\omega \exp\left[-i(\omega t - k_\omega z) + i\phi^{\text{c-e}}_\omega\right] \\ &\quad + E_{2\omega} \exp\left[-i(2\omega t - k_{2\omega} z) + i\Delta\phi + i\phi^{\text{c-e}}_{2\omega}\right]\}^3\rangle_t \\ &\propto \chi^{(2)}_{\text{eff}} \cos\left[\Delta k \cdot z + \Delta\phi + (\phi^{\text{c-e}}_{2\omega} - 2\phi^{\text{c-e}}_\omega)\right].\end{aligned} \tag{1}$$

Here, $\phi^{\text{c-e}}_\omega$ and $\phi^{\text{c-e}}_{2\omega}$ are the c-e phases of frequency $\omega$ and $2\omega$ pulses, respectively, $\Delta k = k_{2\omega} - 2k_\omega$ is wave-vector mismatch, $\Delta\phi$ is the relative phase between two pulses, and $\chi^{(2)}_{\text{eff}} \propto \|E^2_\omega E^*_{2\omega}\|$. We prepared one-octave spanned broadband frequency comb and employed both $\omega$ and $2\omega$ pulses from the comb that share the same c-e phase with each other, i.e., $\phi^{\text{c-e}}_{2\omega} = \phi^{\text{c-e}}_\omega$. Under this condition, the photoinduced $\chi^{(2)}_{\text{ind}}$ is probed as a SH signal of the $\omega$ pulses and by solving a classical wave equation with $\chi^{(2)}_{\text{ind}}$ [8] the signal intensity $I^{\text{SHG}}_{2\omega}$ is represented by

$$I^{\text{SHG}}_{2\omega} \propto \chi^{(2)^2}_{\text{eff}} l^2 \left\{1 + \cos\left[2\left(\Delta\phi - \phi^{\text{c-e}}_\omega\right)\right]\right\}, \tag{2}$$

where a sample thickness $l$ is assumed to be much shorter than the coherence length $l_c = 2\pi/\Delta k$ of the sample. In the experiment, relative phase $\Delta\phi$ (= delay) was fixed, and only the c-e phase $\phi^{\text{c-e}}_\omega$ was controlled to demonstrate that the efficiency of optical poling process depends on it.

## 3 Experimental

Figure 1 shows the setup for the c-e phase control and the optical poling experiment. An experimental setup of the noncollinear OPA system and the detail of the c-e phase control are described in Refs. [10] and [11], respectively. An aperture A1 extracted $\omega$:1600 nm (1600 nm component of the idler), while A2 extracted $2\omega$:800 nm (800 nm component of the idler). Both $\omega$ and $2\omega$ pulses were focused on the sample for optical poling. The sample film was prepared by spin coating tetrahydrofuran (THF) solution of Disperse Red 1 (DR1) and polymethylmethacrylate (PMMA) on a glass substrate. The thickness of the sample was set at 1 μm, which is small enough in comparison with the coherence length $l_c = 3.9$ μm of our sample; so the assumption used in the derivation of Eq. (2) is well satisfied. The experiment was performed in two sequential steps of "writing" and "probing." First, both $\omega$ and $2\omega$ pulses were simultaneously incident on the

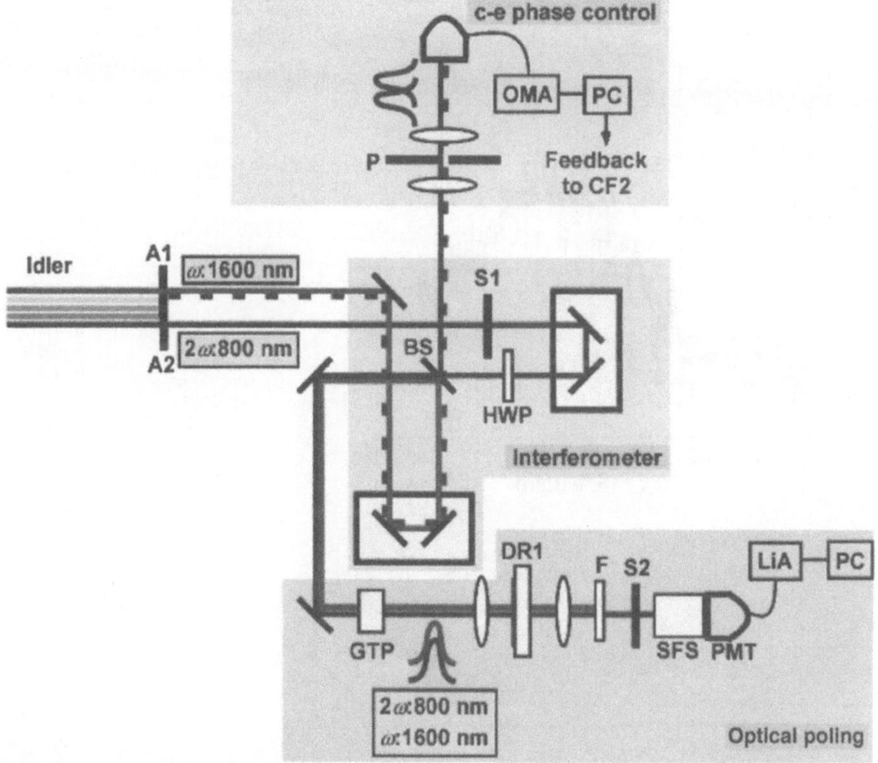


FIGURE 1. C-e control and optical poling system. A1 and A2, apertures; S1 and S2, shutters; P, pinhole; OMA, optical multichannel analyzer; GTP, Glan-Thompson poralizer; DR1, Disperse Red 1 sample; F, filter for 1600 nm; SFS, spatial filtering system; PMT, photomultiplier tube; LiA, lock-in amplifier.

sample. This writing process is then intermittently stopped by closing a shutter S1. This permitted reading the photoinduced $\chi^{(2)}_{\rm ind}$ by frequency doubling only with $\omega$ pulses incident on the sample. The generated SH signal was detected by a photomultiplier tube (PMT; H5784-20, Hamamatsu) behind a spatial filtering system (SFS). The PMT signal was acquired with a lock-in amplifier that worked synchronously with the repetition of the laser system. A shutter S2 protected the PMT from any damage that might be induced by the direct incident $2\omega$ seeding pulses, and it was opened only when S1 was closed.

## 4 Results and Discussion

Figure 2a shows the real-time growth of the SH signal with fixed c-e phases for the first experiment. At first, only $\omega$ pulses with a fixed c-e phase $\phi^{\rm c-e}_{\omega} = \phi_0$ were injected to the sample by closing S1 for several tens minutes, and SH signal was not detected because no $\chi^{(2)}_{\rm ind}$ could be induced without $2\omega$ pulses. Then both $\omega$

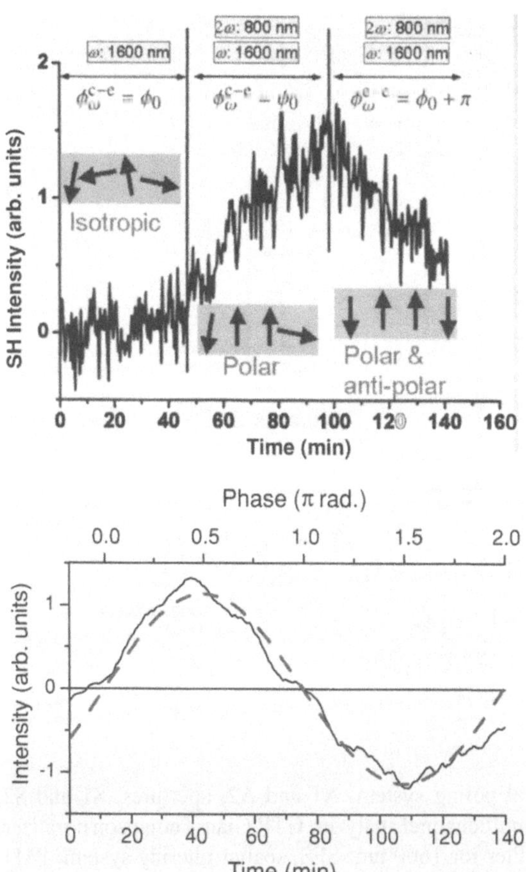


FIGURE 2. Real-time evolution of the SH signal with (a) fixed c-e phases and (b) a gradual c-e phase change. Dash curve in (b) describes the c-e phase.

and $2\omega$ pulses started to be injected, and progressive growth of the SH signal were observed. It is known that the time constant of disorientation process of the molecules of the sample, which are embedded into the PMMA matrices, is longer than 10 h [12]. Therefore, induced $\chi^{(2)}_{\mathrm{ind}}$ was stable and could be accumulated over the whole building duration. After the building duration of several tens minutes, the c-e phase of the idler was controlled into $\phi^{\mathrm{c-e}}_{\omega} = \phi_0 + \pi$. The sign of $\chi^{(2)}_{\mathrm{ind}}$ (i.e., the orientation of the molecules) is reversed when $\phi^{\mathrm{c-e}}_{\omega}$ is shifted from $\phi_0$ to $\phi_0 + \pi$ [12] though these two phases give the same signal intensity as seen from Eq. (2). Since superposition of two oriented molecules with opposite directions does not contribute to the breaking of centrosymmetry, which is the origin of $\chi^{(2)}_{\mathrm{ind}}$, the SH signal decreased with the same rate as that of the growth process.

In the second experiment, the c-e phase of the idler was scanned step by step over about 2 h. The derivation of the signal was calculated in order to obtain the optical poling efficiency per unit time and is shown in Fig. 2b with the c-e phase change. The figure clearly shows that the optical poling efficiency was synchronized with the c-e phase change.

## 5 Conclusion

We have shown that the efficiency of optical poling process was controlled by the c-e phases of two writing pulses. For the experiment, we employed our noncollinear OPA system for the self-stabilization of the c-e phase, with the $f$-to-$2f$ spectral interferometry system to control the c-e phase. To control quantum interference via direct control of the c-e phase instead of usual control of the phases of $\omega$ and $2\omega$ pulses is a new approach to quantum control. Besides, the present phenomenon has an impact on the fields of optical frequency metrology and precision optical frequency measurement, because it is possible to measure the c-e phase evolution using optical poling signal. If an optical poling experiment is performed with a pump-probe type measurement with dye molecules in solution [7], the optical poling signal intensity and the value $\phi^{\mathrm{c-e}}_{\omega}$ are related with each other on a one-to-one basis. This enables the signal to be utilized for locking the c-e phase as a substitute for $f$-to-$2f$ spectral interferometry technique.

## References


1. A. Baltuška, T. Udem, M. Uiberacker, et al.: Nature **421**, 611 (2003).
2. G. G. Paulus, F. Grasbon, H. Walther, et al.: Nature **414**, 182 (2001).
3. D. J. Jones, S. A. Diddams, J. K. Ranka, et al.: Science **288**, 635 (2000).
4. T. M. Fortier, P. A. Roos, D. J. Jones, and S. T. Cundiff: Phys. Rev. Lett. **92**, 147403 (2004).
5. N. B. Baranova and B. Ya. Zeldvich: JETP Lett. **45**, 717 (1987).
6. A. Baltuška, T. Fuji, and T. Kobayashi: Phys. Rev. Lett. **88**, 133901 (2002).

7. F. Charra, F. Devaux, J.-M. Nunzi, and P. Raimond: Phys. Rev. Lett. **68**, 2440 (1992).
8. C. Fiorini, F. Charra, J.-M. Nunzi, and P. Raimond: Nonlinear Optics **9**, 339 (1995).
9. C. Fiorini, F. Charra, and J.-M. Nunzi: J. Opt. Soc. Am. B **11**, 2347 (1994).
10. A. Baltuška, T. Fuji, and T. Kobayashi: Opt. Lett. **27**, 306 (2002).
11. S. Adachi and T. Kobayashi: Phys. Rev. Lett. **94(15)**, 153903 (2005).
12. C. Fiorini, F. Charra, J.-M. Nunzi, and P. Raimond: J. Opt. Soc. Am. B **14**, 1984 (1997).

# Phase-coherent Spectrum from Ultrabroadband Ti:sapphire and Cr:forsterite Lasers Covering the Visible to the Infrared

Jung-Won Kim,[1] Thomas R. Schibli,[1] Lia Matos,[2] Hyunil Byun,[1] and Franz X. Kärtner[1]

[1]Department of Electrical Engineering and Computer Science and Research Laboratory of Electronics, Massachusetts Institute of Technology, Cambridge, Massachusetts 02139, USA

**jungwon@mit.edu, kaertner@mit.edu**

[2]Department of Physics and Research Laboratory of Electronics, Massachusetts Institute of Technology, Cambridge, Massachusetts 02139, USA

## 1 Introduction

Few-cycle optical pulses open up new possibilities to investigate various extreme light-matter interactions, which depend on the light electric-field oscillation itself, for example, high-harmonic generation (HHG) for extreme ultraviolet and soft X-ray pulse generation [1]. For further investigations and applications of such phase-sensitive phenomena, single-cycle optical pulse, comprised of only one cycle of the light oscillation, is currently pursued in many laboratories.

Recently, single-cycle pulse trains spanning the visible to ultraviolet range have been generated by cascaded difference frequency generation of two Q-switched lasers with a vibrational transition in $D_2$ [2]. Optical broadband parametric amplification in beta barium borate (BBO) has been used to generate close- to single-cycle optical pulses at 1 μm [3]. External compression techniques based on spectral broadening and adaptive dispersion compensation also generated sub-4-fs optical pulses approaching the single-cycle regime [4,5].

The combination of phase-stabilized lasers has been long recognized as a way to synthesize ultrashort optical pulses with durations down to half an optical cycle using multiple single-frequency or mode-locked lasers (see Ref. [6] and references therein). To synthesize single-cycle optical pulses, several groups [7–10] have worked on the synthesis of optical combs spanning the visible to the infrared range from mode-locked Ti:sapphire and Cr:forsterite lasers.

As one step closer toward the single-cycle pulse synthesis, in this paper we present the generation of a phase-coherent spectrum spanning 600–1500 nm from a long-term stable Ti:sapphire and Cr:forsterite laser system with a residual timing and phase rms jitter of less than 400 as and $2\pi$ rad, respectively.

## 2 Experiments and Results

Figure 1 shows the schematic of the synchronization setup. Ultrabroadband Ti:sapphire and Cr:forsterite lasers are combined at the broadband 50:50 beam splitter [11] (BS1 in Fig. 1). For a timing synchronization, a balanced cross-correlator [7] is used. Once a tight timing synchronization is obtained, a heterodyne beat signal between the two lasers in the overlapped spectral range is obtained. This beat signal is locked to a local oscillator ($f_{LO}$ in Fig. 1) by modulating the pump power of the Ti:sapphire laser with an AOM. In the next sections, each of the key components and techniques will be explained.

### 2.1 *Ti:sapphire and Cr:forsterite Lasers with Overlapping Spectra*

Building stable and broadband mode-locked lasers with overlapping spectra is the first and still most important step in the pulse synthesis. To ensure long-term

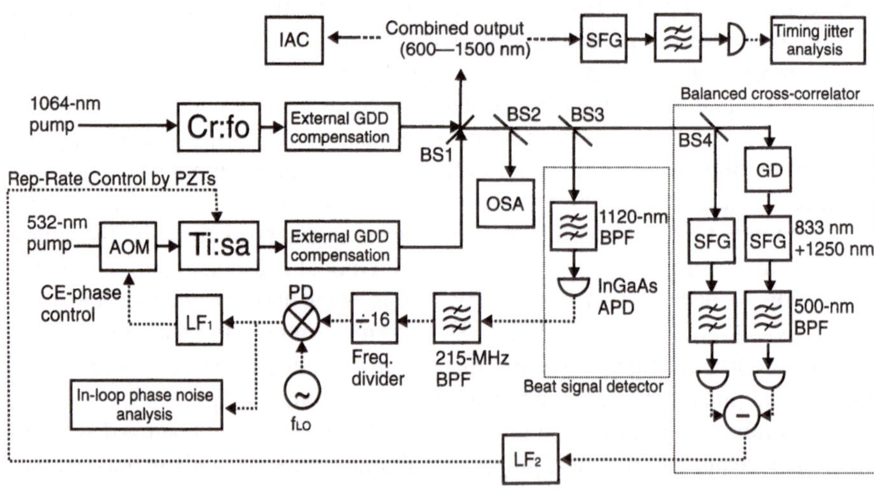


FIGURE 1. Schematic outline of the synchronization setup. IAC, interferometric autocorrelator; SFG, sum-frequency generation crystal; Cr:fo, Cr:forsterite laser oscillator; GDD, group-delay dispersion; BS1 and BS4, broadband 50:50 beam splitter with matched GDD (described in Ref. [11]); BS2, tapping beam splitter for monitoring optical spectrum; BS3, bandpass beam splitter (transmits 833 nm and 1250 nm; reflects 1120 nm); AOM, acousto-optic modulator; Ti:sa, Ti:sapphire laser oscillator; OSA, optical spectrum analyzer; GD, group delay element; BPF, bandpass filter; APD, Avalanche photodiode; PD, digital phase detector; $f_{LO}$, RF synthesizer; LF, loop filter.

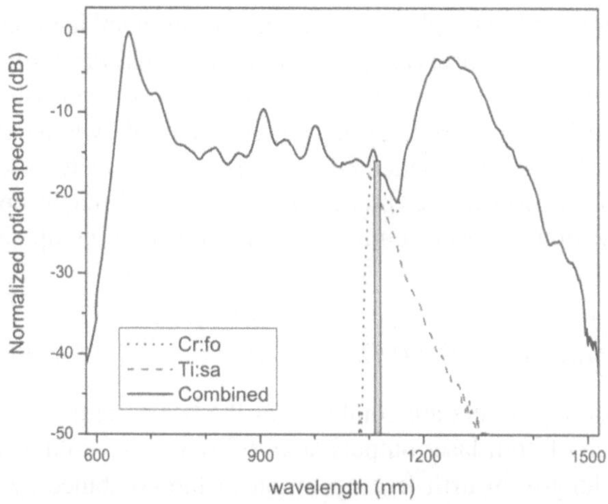


FIGURE 2. Combined optical spectra from Ti:sapphire and Cr:forsterite lasers. The gray region (15-nm wide, centered at 1120 nm) indicates the spectral range where the heterodyne beatnote between the two lasers is measured.

stable and broadband operation, both lasers were constructed as prismless cavities. The intracavity dispersions were compensated by the combination of double-chirped mirrors (DCMs) [12] and thin wedges of glass material.

The Ti:sapphire laser design is similar to the previous prismless designs in Refs. [13] and [14]. The 82-MHz Kerr-lens mode-locked Ti:sapphire laser emits ~70 mW average power when pumped with 4.5 W from a 532-nm pump laser. The optical spectrum spans from 600 nm to over 1200 nm range (see Fig. 2) with a transform-limited pulsewidth of ~4 fs.

The starting and stabilization of Cr:forsterite mode-locked laser is known as difficult due to strong thermal loading and low gain of Cr:forsterite crystal [15]. To improve the stability of laser, a broadband InGaAs saturable absorber on a large area, high-index contrast AlGaAs/$Al_xO_y$ mirror [16] is employed. The Cr:forsterite crystal was cooled to 5°C. The laser emits ~40 mW of average power when pumped with 6 W from a 1064-nm pump laser. The 3-dB bandwidth is about 90 nm (see Fig. 2), corresponding to a transform-limited pulsewidth of ~21 fs. Figure 2 shows the optical spectra of both lasers.

The combined spectrum spans 1.5 octaves from 600 nm to 1500 nm at −35 dB level measured from the peak. There is a strong spectral overlap in wavelength range 1100–1200 nm, which enables a simple direct detection of the difference in carrier-envelope offset frequency without using $1f$–$2f$ or $2f$–$3f$ interferometers [10]. The transform-limited pulse width of the combined spectra is about 3 fs. At the center wavelength, 900 nm, this corresponds to about one cycle of the light oscillation.

For a pulse synthesis in the time domain, the extracavity dispersion must be compensated. In this system, the dispersion is precompensated before the pulses from each laser are combined. This enables easier dispersion compensation by use of chirped mirrors and glass plates than in the overall combined output over 1.5 octaves. To make sure the lasers are generating near-transform-limited pulses and the extracavity dispersion compensations are properly set, we measured the pulses with the IAC (in Fig. 1) of each laser. The retrieved pulsewidths of Ti:sapphire and Cr:forsterite lasers were about 5.8 fs and 24 fs, respectively. They are slightly longer than the transform-limited pulsewidths calculated from the optical spectra. Further optimization of laser conditions and extracavity compensation are in progress.

## 2.2 Ultrabroadband 50:50 Beam Splitter with Matched GDD

The pulses from both lasers are combined at the beam splitter (BS1 in Fig. 1). The combination of both laser outputs at an in-loop beam splitter is of key importance for a long-term drift-free operation of the combined system. For the beam combining, it is desirable to have a very broadband constant transmission and reflection (over 600–1500 nm full range) as well as a well-defined dispersion characteristic. These conditions are fulfilled by matching the dispersion in reflection from the coating with that of the single pass transmission through the substrate while keeping the coating reflection constant over the interested spectral range. A broadband 50:50 beam splitter (ranging 600–1500 nm) with matched GDD is designed, fabricated, and used for combining and splitting pulses at BS1 and BS4 in Fig. 1. A more detailed description of this beam splitter can be found in Ref. [11].

Furthermore, this 50:50 splitting ratio will also enable later the locking of the difference carrier-envelope frequency ($\Delta f_{ceo}$) to zero via homodyne detection at the beam splitter BS1, which stabilizes the relative carrier-envelope phase in a long-term stable way.

## 2.3 Timing Synchronization with Balanced Cross-correlator

Synchronizing the pulse envelopes from each laser with subcycle timing fluctuation is one of the most important tasks in pulse synthesis. To achieve subfemtosecond timing jitter, a balanced cross-correlator described in Ref. [7] is used. The main improvement here is achieved by using the broadband 50:50 beam splitter (Ref. [11]) to avoid potential cross-talk between rep-rate lock and carrier-envelope phase lock due to the unbalanced intensity changes at the cross-correlator. The measured out-of-loop timing jitter was about 375 as $\pm$ 203 as over 10-mHz to 2.3-MHz bandwidth. The error is given by the amplitude noise in the cross-correlation signal.

## 2.4 Phase Locking by Direct Photodetection of Overlapped Spectra

As soon as the tight-timing synchronization is obtained, we observe a strong beat signal in the overlap region of the optical spectrum (indicated as the gray

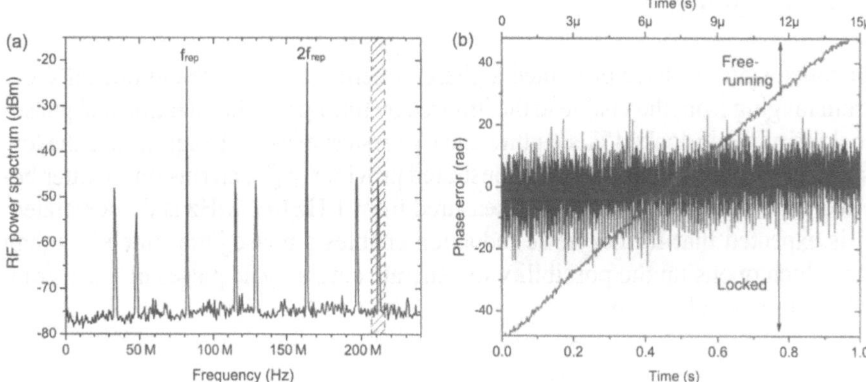


FIGURE 3. (a) RF spectrum (RBW 30 kHz) of the beat signals from the InGaAs APD output in Fig. 1. Red-shadowed region indicates the bandpass filter centered at 215 MHz that selects the beat signal used for phase locking. (b) In-loop phase error signal from the digital phase detector (PD in Fig. 1). Red line shows the free-running phase error signal over 15 μs time scale. Blue line shows the residual in-loop phase error for 1 s when it is locked. Note that the phase error shown here are already calibrated to the full range, i.e., $[-16\pi, +16\pi]$ range.

region in Fig. 2). To lock the difference in carrier-envelope offset frequency ($\Delta f_{\text{ceo}}$) between the two lasers, we filter out a 15-nm-wide part of the spectrum in the overlap spectral range at 1120 nm and detect it with a high-speed InGaAs Avalanche photodetector. Figure 3a shows the RF spectrum of the output from the photodetector. Heterodyne beatnotes ($mf_{\text{R}} \pm \Delta f_{\text{ceo}}$, $m =$ integer) show 30 dB signal-to-naire ratio measured with a 30-kHz resolution bandwidth. The beatnote component at 215 MHz is bandpass filtered, amplified, and frequency-divided by a factor of 16 to increase the locking range. This beatnote is locked to an RF synthesizer, $f_{\text{LO}}$, at ~13.43 MHz via modulating intracavity energy of the Ti:sapphire laser with an AOM (see Fig. 1).

Figure 3b shows the output from the digital phase detector for free-running and locked states with two different time scales (15 μs and 1 s, respectively). When the carrier-envelope difference phase is locked, the in-loop integrated rms phase noise is $2.2\pi$ ($1.7\pi$) radians measured from 1 (10) Hz to 1 MHz. This is equivalent to 3.6-fs rms phase jitter from 1 Hz to 1 MHz.

Further improvements are clearly possible by optimization of the phase-locked loop and noise reduction in the Cr:forsterite laser, which seems to be the main reason for the large remaining carrier-envelope phase noise. This large noise currently prevents us from locking the difference carrier-envelope frequency to zero by homodyne detection and synthesizing a single-cycle pulse. Note that this cannot be achieved with conventional offset locking in a long-term stable way since offset locking introduces an uncontrolled path difference between the two laser outputs.

## 3 Conclusion

In conclusion, we have presented a phase-coherent ultrabroadband optical spectrum ranging from the visible to the infrared by timing synchronization and phase-locking of mode-locked Ti:sapphire and Cr:forsterite lasers. In addition to the low subfemtosecond timing jitter demonstrated previously [7], an rms phase jitter between the two lasers of $2.2\pi$ rad measured from 1 Hz to 1 MHz is demonstrated. It is expected that further noise reduction enables a phase jitter much less than $2\pi$ which opens up the possibility to generate single-cycle pulses at 1 μm from a long-term stable setup.

## References


1. A. Baltuska, T. Udem, M. Uiberacker, et al.: Nature **421**, 611 (2003).
2. M. Y. Shverdin, D. R. Walker, D. D. Yavuz, G. Y. Yin, and S. E. Harris: Phys. Rev. Lett. **94**, 033904 (2005).
3. S. Adachi, P. Kumbhakar, T. Kobayashi: Opt. Lett. **29**, 1150 (2004).
4. K. Yamane, Z. Zhang, K. Oka, R. Morita, M. Yamashita, A. Suguro: Opt. Lett. **28**, 2258 (2003).
5. B. Shenkel, J. Biegert, U. Keller, et al.: Opt. Lett. **28**, 1987 (2003).
6. T. W. Hänsch: Opt. Commun. **80**, 71 (1990).
7. T. R. Schibli, J. Kim, O. Kuzucu, et al.: Opt. Lett. **28**, 947 (2003).
8. A. Bartels, N. R. Newbury, I. Thomann, L. Hollberg, and S. A. Diddams: Opt. Lett. **29**, 403 (2004).
9. D. Yoshitomi, Y. Kobayashi, H. Takada, M. Kakehata, and K. Torizuka: Opt. Lett. **30**, 1408 (2005).
10. Y. Kobayashi, D. Yoshitomi, M. Kakehata, H. Takada, and K. Torizuka: Opt. Lett. **30**, 2496 (2005).
11. J. Kim, J. R. Birge, V. Sharma, et al.: Opt. Lett. **30**, 1569 (2005).
12. F. X. Kärtner, N. Matuschek, T. Schibli, et al.: Opt. Lett. **22**, 831 (1997).
13. T. R. Schibli, O. Kuzucu, J. Kim, E. P. Ippen, J. G. Fujimoto, F. X. Kaertner: IEEE J. Sel. Top. Quant. Elec. **9**, 990 (2003).
14. L. Matos, D. Kleppner, O. Kuzucu, et al.: Opt. Lett. **29**, 1683 (2004).
15. Z. Zhang, K. Torizuka, T. Itatani, K. Kobayashi, T. Sugaya, and T. Nakagawa: IEEE J. Quant. Elec. **33,** 1975 (1997).
16. S. N. Tandon, J. T. Gopinath, H. M. Shen, et al.: Opt. Lett. **29**, 2551 (2004).

# Part IV

# Carrier-Envelop Phase II

# Part IV

# Coherent Synthesis of Multicolor Femtosecond Pulses

Yohei Kobayashi, Dai Yoshitomi, Masayuki Kakehata, Hideyuki Takada, and Kenji Torizuka

National Institute of Advanced Industrial Science and Technology (AIST)
Tsukuba Central 2, 1-1-1 Umezono, Tsukuba 305-8568, Japan
y.kobayashi@aist.go.jp

## 1 Introduction

The attosecond pulse generation is realized in XUV region by using high-order harmonics [1,2]. The pulse duration in visible and infrared region is approaching to the monocycle limit [3]. For further pulse shortening in the visible and infrared region, the coherent addition of the different-color pulses is the candidate for the generation of the monocycle pulse. A subfemtosecond pulse train may also be generated by using Fourier synthesis of multicolor pulses [4,5]. The schematic of Fourier synthesis is shown in Fig. 1. Fourier synthesis of multicolor pulses may generate not only attosecond pulse train but also desired shape of the electric field such as triangle or rectangle shapes. This can be called as an optical function generator. The optical phase relation of different-color pulses have to be locked in order to realize it.

In this report, We have generated optical phase-coherent multicolor femtosecond pulses by two methods in order to realize coherent synthesis of different-color femtosecond pulses. One is the optical phase locking among a pump, signal, and idler pulses in a femtosecond optical parametric oscillator, and the other is that in a two-wavelength laser consisting of mode-locked Ti:sapphire and Cr:forsterite lasers. Timing-synchronized multicolor pulses were generated from a two-wavelength laser or an optical parametric oscillator (OPO). In both cases, the optical phase relation is locked tightly for long time. The fluctuation of the locked optical frequency is suppressed to 1 MHz level for hours. The superimposed electric field could be a subfemtosecond pulse train. This technique would also be useful for an ultrawide frequency comb generation in the frequency domain application.

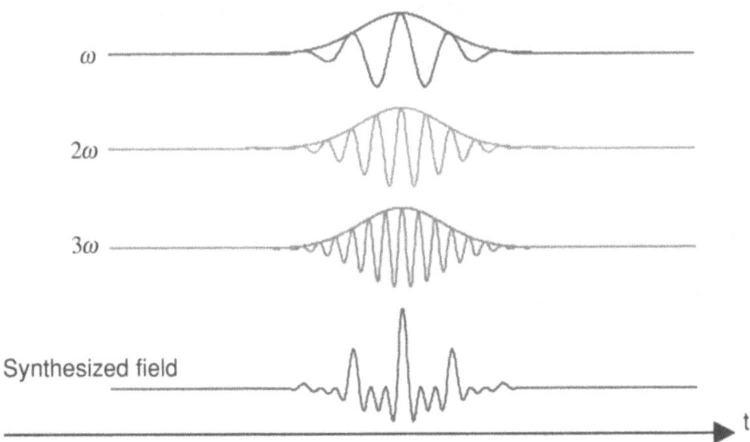


FIGURE 1. Schematic of Fourier synthesis.

## 2 Optical Phase Locking in OPO

We have developed a femtosecond OPO to generate subharmonic pulses. This femtosecond OPO produces timing-synchronized pump, signal, and idler pulses whose optical frequencies are $3\omega$, $2\omega$, and $\omega$, respectively. This OPO also produces sum frequencies of them ($4\omega_1$, $4\omega_2$, $5\omega$, and $6\omega$). The second harmonic of the signal ($4\omega_1$) and the sum frequency between the pump and the idler ($4\omega_2$) have the spectrum overlap so that we can obtain the heterodyne beat by superposing them. This beat signal shows the optical-phase relation among the pump, signal, and the idler pulses. The beat signal can be controlled by varying the cavity-length difference between the OPO and the Ti:sapphire oscillator that is used for the pump pulse. The optical-phase relation among these multicolor pulses thus can be fixed by a feedback system. The schematic of the experimental setup is shown in Fig. 2.

Figure 3 shows the power spectral density and integrated phase error of the locked beat signal. This shows that the integrated phase error from 1 mHz to 1 MHz is 0.24 rad, which guarantees tight phase locking among different-color pulses [6]. In order to superimpose multicolor pulses with fixed phase relation, the beat signal should be fixed to zero. We applied DC-locking technique to realize it. The interferences between two $4\omega$ pulses can be observed by eyes or a video (Fig. 4) in this condition. The phase locking can be kept about 60 h (Fig. 5). The measured frequency fluctuation between two optical pulses is 0.3 mHz in the standard deviation in 60 h. These measurements guaranteed phase-coherent six-color femtosecond pulses generation for long time.

## 3 Optical Phase Locking in Two-wavelength Laser

We are developing passively timing-synchronized Ti:sapphire and Cr:forsterite lasers to realize Fourier synthesis of them and their sum frequencies. The coherent

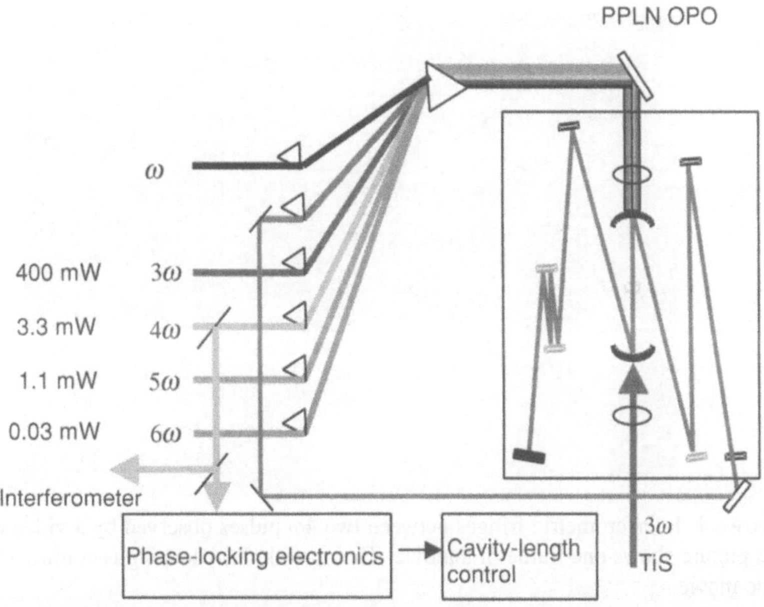


FIGURE 2. Schematic of the subharmonic OPO.

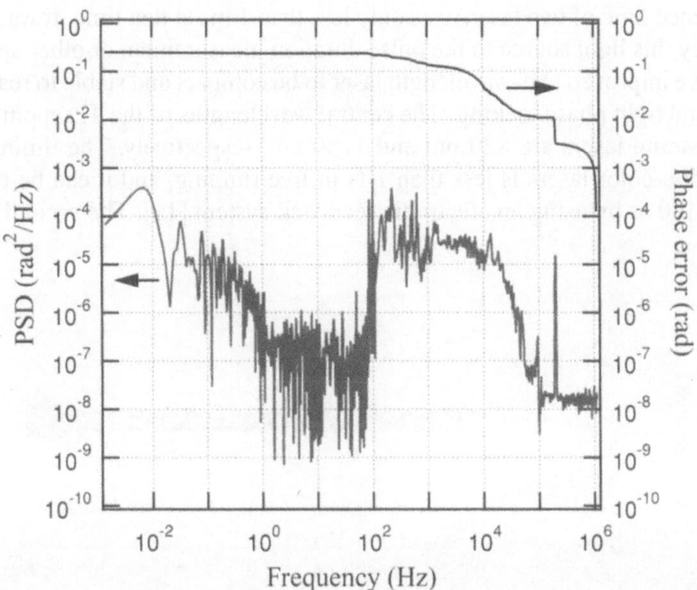


FIGURE 3. Power spectral density and integrated phase noise of the locked beatnote. The integrated phase error from 1 mHz to 1 MHz is 0.24 rad.

FIGURE 4. Interferometric fringes between two 4$\omega$ pulses observed by a video camera. This picture shows one frame of a movie. The optical phase locking is confirmed by the video movie.

superposition of spectrum-overlapped two laser pulses could generate shorter pulse duration. The combination of Ti:sapphire and Cr:forsterite lasers is the candidate to achieve over one-octave spectrum by this technique [7–9]. We have demonstrated the phase-relation control of these two lasers [10]; however, the coherence time of two lasers was only less than 1 ms at that time. It was too short to apply this light source to the pulse duration measurement or other application. We have improved two-wavelength laser to be compact and stable to realize long-term and tight phase locking. The central wavelengths of the Ti:sapphire and the Cr:forsterite lasers are 820 nm and 1230 nm, respectively. The timing jitter of these two-color lasers is less than 1 fs in free running, and it can be reduced to about 100 as by using an electronic feedback system [11]. The second harmonic

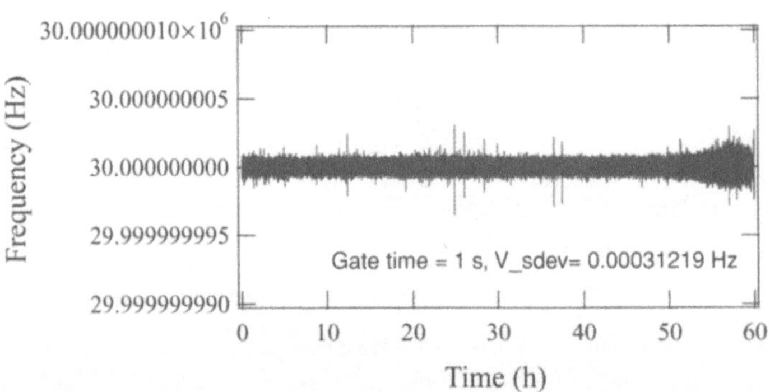


FIGURE 5. Frequency fluctuation of the locked beat signal measured by 1-s gated counter. The standard deviation of 0.3 mHz in 60 h shows stable phase locking for long time.

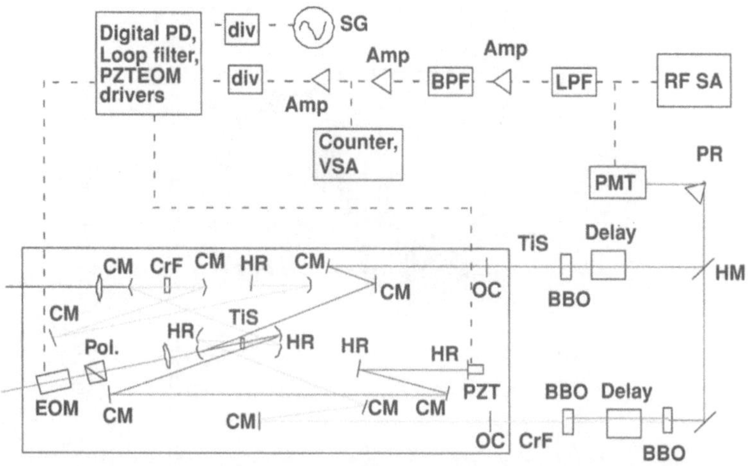


FIGURE 6. Experimental setup of two-wavelength laser and optical phase locking between Ti:sapphire and Cr:forsterite lasers. HR, high reflector; CM, chirp-compensation mirror; CrF, Cr:forsterite crystal; TiS, Ti:sapphire crystal; OC, output coupler; HM, half mirror; RF SA, rf spectrum analyzer; LPF, low-pass filter; BPF, band-pass filter; Amp, amplifier; VSA, vector signal analyzer; div, dividers; SG, signal generator; PD, phase-frequency comparator; PR, prism.

of the Ti:sapphire and the third harmonic of the Cr:forsterite laser pulses overlap in the frequency domain so that we can obtain the heterodyne beat between them. The optical phase relation can be obtained and controlled from this beat signal in the same manner as described above. The experimental setup is shown in Fig. 6.

The optical phase locking can be kept for many hours. The frequency fluctuation of the locked beat signal was 0.32 mHz in 3400 s (Fig. 7). The locked beat signal has the line width of less than 1 mHz, and the phase error was 0.43 rad from

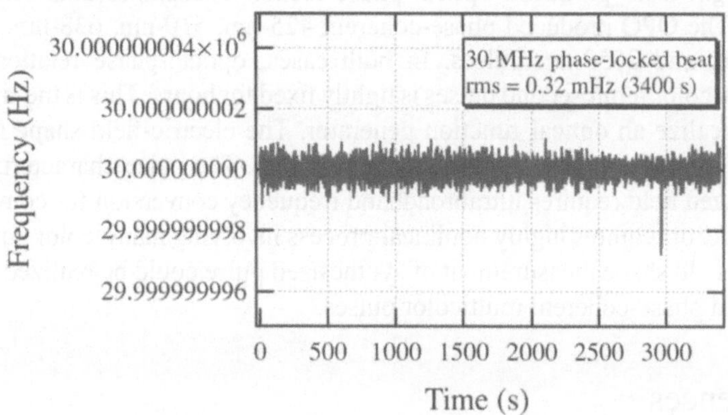


FIGURE 7. Frequency fluctuation of the locked beat note measured by 1-s averaged counter. The standard deviation in 3400 s is 0.32 mHz.

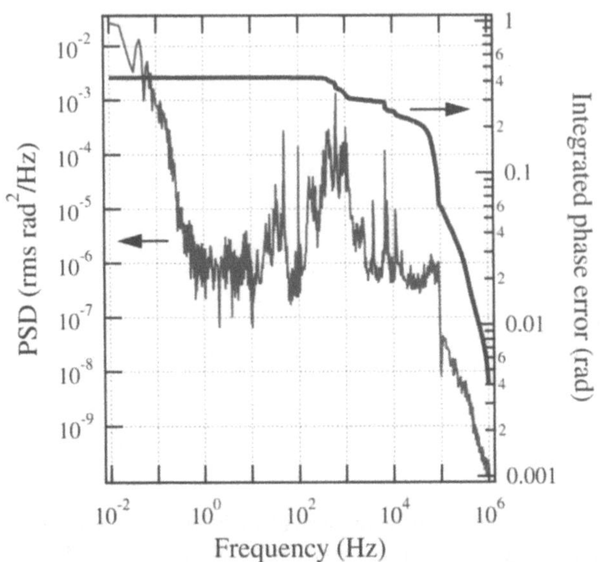


FIGURE 8. Power spectral density of the locked beat signal and integrated phase error. Accumulated phase error from 10-mHz to 1-MHz integration is 0.43 rad, which shows the tight phase locking.

10-mHz to 1-MHz integration (Fig. 8). These measurements guarantee the long-term and tight optical phase locking between Ti:sapphire and Cr:forsterite lasers.

## 4 Conclusion

We have generated phase-coherent multicolor pulses by using two systems. Two-wavelength laser produced optical phase locked Ti:sapphire and Cr:forsterite pulses. The OPO produced phase-coherent 425-nm, 510-nm, 638-nm, 850-nm, 1275-nm, and 2550-nm pulses. In both cases, optical-phase relation among different-color femtosecond pulses is tightly fixed for hours. This is the significant step to realize an optical function generator. The electric-field shape measurement of the superimposed pulses is the next step. The pulse characterization of synthesized field requires ultrabroadband frequency conversion for conventional technique, or requires highly nonlinear process involving many-color pulses. The electric-field shape measurement of synthesized pulse could be realized by using amplified phase-coherent multicolor pulses.

## References

1. M. Hentschel, R. Kienberger, Ch. Spielmann, et al.: Nature **414**, 509 (2001).
2. T. Sekikawa, A. Kosuge, T. Kanai, and S. Watanabe: Nature **432**, 605 (2004).

3. K. Yamane, T. Kito, R. Morita, and M. Yamashita: In Conf. Lasers and Electro-Optics, Optical Society of America, Washington, DC, (2004), postdeadline paper PDC2.
4. T. W. Hänsch: Opt. Comm. **80**, 71 (1990).
5. K. Shimoda: Jpn. J. Appl. Phys. 34, 3566 (1995).
6. Y. Kobayashi, H. Takada, M. Kakehata, and K. Torizuka: Appl. Phys. Lett. **83**, 839 (2003).
7. A. Bartels, N. R. Newbury, I. Thomann, L. Hollberg, and S. Diddams: Opt. Lett. **29**, 403 (2004).
8. J. Kim, T. R. Schibli, L. Matos, H. Byunn, and F. X. Kärtner: In Joint Conference on Ultrafast Optics V and Applications of High Field and Short Wavelength Sources XI (2005), paper M3-5.
9. T. R. Schibli, J. Kim, O. Kuzucu, et al.: Opt. Lett. **28**, 947 (2003).
10. Y. Kobayashi, K. Torizuka, and Z. Wei: Opt. Lett. **28**, 746 (2003).
11. D. Yoshitomi, Y. Kobayashi, H. Takada, M. Kakehata, and K. Torizuka: Opt. Lett. **30**, 1408 (2005).

# Quasisynchronous Pumping of Mode-locked Few-cycle Titanium Sapphire Lasers

Richard Ell,[1,2] Gregor Angelow,[1] Wolfgang Seitz,[3] Max J. Lederer,[4] Huber Heinz,[4] Daniel Kopf,[4] Jonathan R. Birge,[2] and Franz X. Kärtner[2]

[1] NanoLayers, Optical Coatings GmbH, Maarweg 30, 53619 Rheinbreitbach, Germany
**richard.ell@nanolayers.de**

[2] Department of Electrical Engineering and Computer Science, Research Laboratory of Electronics, Massachusetts Institute of Technology, Cambridge, Massachusetts 02139-4307, USA
**kaertner@mit.edu**

[3] High Q Laser (US), Inc., 118 Waltham Street, Watertown, MA 02472, USA
**wolfgangseitz@highq-us.com**

[4] High Q Laser Production GmbH, Kaiser-Franz-Josef-Strasse 61, A-6845 Hohenems, Austria
**danielkopf@highqlaser.at**

## 1 Introduction

Dispersion-managed mode-locked (DM-KLM) (Ti:sapphire) lasers [1] are the work horses in the domain of sub-10 fs laser pulses. Since the first observation of Kerr-lens mode-locking (KLM) in 1991 [2], continuous laser development has lead to the generation of octave-spanning spectra and 5-fs pulses directly from the oscillator [3, 4]. DM-KLM lasers exploit the intensity dependent non-linear refractive index in conjunction with a careful management of the distribution of discrete dispersive elements inside the cavity. In the time domain, this leads to self-phase modulation and hence additional spectral broadening, whereas in the transverse spatial beam dimensions, the buildup of a Kerr-lens together with a suitable resonator geometry enables efficient gain modulation to favor pulsed operation in comparison to the continuous wave (CW) operation. However, KLM lasers with pulse durations below a few tens of femtoseconds are generally not, and usually require, external (mechanical) perturbations to initiate mode-locking.

In this work, we report on the detailed dynamics of DM-KLM lasers when quasisynchronously pumped by a mode-locked picosecond $Nd:YVO_4$ laser. We investigate the self-starting behavior and residual power modulation, demonstrating that self-starting is achieved for a wide range of pulse durations and optical bandwidths.

## 2 General Aspects of the Starting Behavior of Kerr-lens Mode-locked Lasers

Though KLM allows for a large modulation of the effective gain up to several tens of percent, it is generally not self-starting. A parameter that characterizes the self-starting ability is the so-called mode-locking driving force and is defined as $(\partial(\Delta g)/\partial I)$ for $I \to 0$, with $I$ the intensity and $\Delta g$ the gain modulation [5]. In sub-10-fs lasers, the mode-locking driving force is designed to be small in order not to overdrive the KLM when the laser transitions from CW operation to pulsed operation.

So far, three different approaches have been published to overcome the self-starting problem in KLM lasers. One approach is to maximize the mode-locking driving force by proper cavity alignment [6–8]. Cerullo et al. showed that a particular resonator design enables a maximization of the nonlinear mode variation and consequently dynamic loss modulation, achieving self-starting in a KLM Ti:sapphire when operating close enough to the stability edge [7]. This is only possible down to pulse durations of about 40 fs until KLM is overdriven, resulting in a noncontinuous mode-locking. Alternatively, one can use a semiconductor saturable absorber mirror (SESAM) [9] or saturable Bragg reflector (SBR) [10] inside the cavity. The laser then exhibits self-starting mode-locking because SESAMs/SBRs provide large mode-locking driving forces in the initial pulse buildup phase. Saturation of the SBR after pulse buildup does not harm the laser dynamics, because KLM is taking over the pulse shaping. Another advantage of this method is a relaxed cavity alignment in contrast to purely DM-KLM lasers. A drawback is the bandwidth limitation introduced by these devices that can only be overcome by using nonconventional fabrication procedures which are still under development [11, 12].

A third approach to attain self-starting DM-KLM is (quasi-) pumping [13, 14]. The term "quasi" accounts for the fact that precise of the Ti:sapphire laser is not necessary. Siders et al. have already demonstrated self-starting 30-fs pulses with quasisynchronous pumping [14]. However, they used hardaperture KLM not suitable for few-cycle pulses. Such pulses possess optical bandwidths of several hundred nanometers up to one octave, and since the mode size varies strongly with wavelength, an efficient use of an hard aperture is not possible.

We extend this work to the sub-10-fs regime and provide detailed information on the starting transients, as well as residual Radio Frequency (RF) sideband modulation. We also show that only a very small gain modulation induced by the pulsed pump laser is sufficient to initiate soft-aperture mode-locking of the shortest pulses directly generated from an oscillator.

# 3 Motivation for (quasi-) Synchronous Pumping

Synchronous or quasisynchronous pumping has many interesting applications. First of all, self-starting is achieved, which is most important when considering the poor starting performance of high repetition-rate DM-KLM lasers. For repetition frequencies of several hundred MHz up to a few GHz [15], synchronous pumping will significantly improve the starting behavior. We also predict that pumping at a subharmonic of the fundamental repetition frequency will still lead to self-starting of the DM-KLM process. As will be illustrated later on, the residual output power modulation is only on the order of −30 dBc and can even be reduced further by pumping with a higher repetition-rate source due to stronger filtering of the gain medium. Active synchronization of the mode-locked pump and Ti:sapphire lasers completely removes the residual output power modulation and is very appealing for many pump-probe experiments, because the system simultaneously delivers timing-controlled femtosecond near–IR pulses, picosecond green pulses, and picosecond IR pulses [16]. In the context of the emerging field of optical, phase-synchronized mode-locked pump and Ti:sapphire lasers offer exciting new perspectives by accessing new spectral regions not yet covered by phase-controlled frequency combs [17]. Phase synchronization of the picoseond $Nd:YVO_4$ laser with the DM-KLM Ti:sapphire laser is facilitated by the fact that our broadband sub-10-fs pulses have enough optical power at the wavelength of the picosecond laser, around 1064 nm. An interference experiment of the two synchronized lasers provides the difference of the two offset frequencies [17]. Together with the appropriate scheme for the control of the carrier-envelope offset frequency of the Ti:sapphire laser and a suitable feedback loop, one is readily able to provide fully phase-coherent mode combs for time domain applications and frequency metrology, either by using the available wavelength ranges directly or by extending the mode combs to the IR or visible spectral regions by nonlinear frequency mixing.

# 4 Experimental Results

The experimental setup for the quasisynchronous pumping is depicted in Fig. 1. The pump source is a frequency doubled (532 nm), picosecond (7 ps), saturable absorber mirror mode-locked $Nd:YVO_4$ laser delivering up to 6.5 W of pump power at a repetition rate of 100 MHz and with a diffraction-limited beam $M^2 < 1.2$ (picoTrain, High Q Laser Production GmbH, Austria). The Ti:sapphire laser is sketched below and is a compact, z-folded, prismless oscillator with optics for spectra (VENTEON OS, NanoLayers Optical Coatings GmbH, Germany). To control the cavity length of the Ti:sapphire laser, one end mirror was mounted on a piezo-controlled translation stage.

After manually equalizing the resonator lengths to a common repetition frequency of approximately 100 MHz, the Ti:sapphire laser immediately starts mode-locking. Changing the total intracavity dispersion by moving or completely

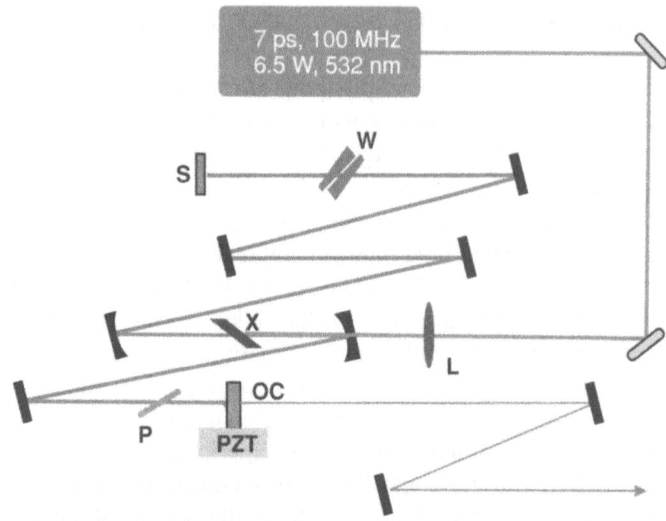


FIGURE 1. Setup of the quasisynchronously pumped Titanium–Sapphire laser. All cavity mirrors except for the two end mirrors are dispersion-compensating mirrors. S, silver mirror; W, $BaF_2$ glass wedges; X, laser crystal (2 mm path length); P, $BaF_2$ glass plate; OC, 1% output coupler; PZT, piezo-controlled mirror mount; L, pump lens.

removing the glass wedges (and the plate), we had access to different pulse duration regimes and observed reliable self-starting behavior for 100-fs pulses down to 6-fs pulses and octave-spanning spectra. In the sub-10-fs regime of interest, a typical cavity detuning of about 10 μm, corresponding to 500 Hz, is tolerated while still maintaining self-starting. Both lasers are built on water-cooled breadboards, and therefore the passive stability is sufficient to keep the lasers within the self-starting range for many hours without cavity-length adjustments. To study the transition dynamics from CW operation to mode-locked operation, a chopper wheel was inserted into the intracavity beam of the Ti:sapphire laser. The average fundamental power was detected with a photodetector (10 MHz bandwidth) as well as the second-harmonic (SH) power when the Ti:sapphire laser is mode-locked ( 1 GHz bandwidth). The signals were analyzed using a 1.5-GHz LeCroy oscilloscope. Both traces are given in Fig. 2a. The orange upper trace represents the fundamental Ti:sapphire power, and the blue trace below is the SH power of the Ti:sapphire. Since the detector for the fundamental power is slow, the signal shows an average over the individual laser pulses. The graph illustrates reliable self-starting behavior. Fig. 2b shows a closer look on the temporal evolution of the fundamental laser power and reveals some detail on the complex laser dynamics that occurs during the buildup of the 6-fs pulse from the continuous wave running laser over a time span of about 5–10 ms (independently confirmed by the onset of the SH signal, not shown in magnification). This is about one to two orders of magnitude slower than typical times for standard KLM lasers [7] or SESAM/SBR mode-locked lasers [5]. When the system is operated near the edge of the self-

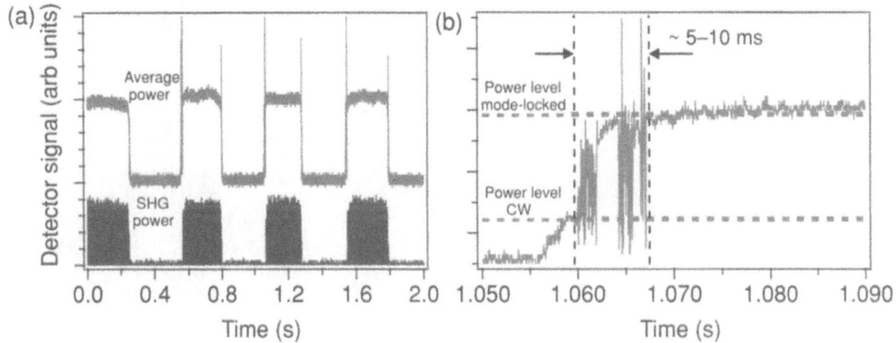


FIGURE 2. (a)Fundamental and SH average power of the self-starting mode-locked Ti:sapphire laser with a chopper wheel in the intracavity beam illustrating the fast and reliable self-starting behavior. (b) Magnification of the temporal evolution of the fundamental average power. The transient from CW operation toward mode-locked operation typically lasts 5–10 ms.

starting limits, the self-starting becomes unreliable and stochastic with respect to buildup time.

For quasisynchronous pumping, it is important to characterize the output power modulation induced by the mismatch of the pulse repetition rates of the two lasers. When the repetition frequencies are manually adjusted to be equal, no modulation in the output power is observed. If the lasers are not synchronous, the output power of the Ti:sapphire will be modulated with the difference frequency of the repetition rates of both lasers. Figure 3 displays the RF power spectrum of the Ti:sapphire output power with a pump detuning of 5 kHz. It shows sidebands slightly below

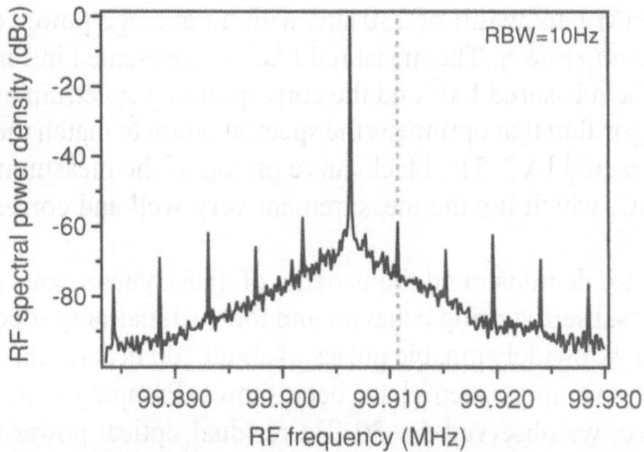


FIGURE 3. Radio-frequency spectrum of the quasisynchronously pumped Ti:sapphire. The repetition frequency of the ps-pump laser is detuned by about 5 kHz and matches the gray dotted line. The output power of the Ti:sapphire laser is modulated by only −30 dBc, corresponding to a 60 dB suppression in the radio–frequency power spectrum.

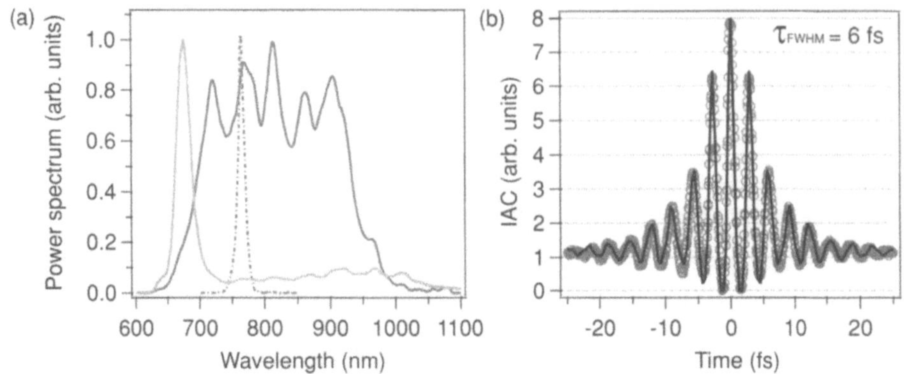


FIGURE 4. (a) Measured optical power spectra displaying the wide self-starting range. The spectrum represented by the gray curve is "octave-spanning" (see text) and the narrow dotted spectrum gave tunable 100-fs pulses. The spectrum in orange corresponds to the measured autocorrelation. (b) Measured (orange circles) and calculated (black curve) IAC traces revealing a pulse duration of 6 fs.

−60 dBc resulting in a 0.1% power modulation of the optical output. This 60 dB suppression is observable over a wide detuning range from within the self starting limits up to several MHz.

In this mode of operation, the laser produces a clean train of mode-locked pulses characterized by its spectrum and interferometric autocorrelation (IAC). Figure 4a shows several spectra for which self-starting is observed, from a narrow bandwidth (tunable) 100-fs pulse to octave-spanning spectra, important for a direct and simple carrier-envelope offset frequency stabilization scheme [18]. The spectrum for which the IAC was measured is represented by the orange curve, and it has a FWHM bandwidth of 230 nm, with an average power of typically 150 mW at 5 W pump power. The measured IAC is represented in Fig. 4b by the orange circles. The measured IAC and the corresponding spectrum are used in a phase retrieval algorithm that optimizes the spectral phase to match the calculated IAC with the measured IAC. The black curve on top of the measurement data is the calculated IAC, which fits the measurement very well and corresponds to a pulse duration of 6 fs.

In conclusion, we demonstrated the benefits of quasisynchronous pumping in terms of an universal self-starting behavior and low residual-output power modulation. Self-starting of widely tunable pulses of about 100 fs duration down to 6 fs pulses and octave-spanning spectra have been shown. Pumping with a 100-MHz picosecond source, we observed a −30 dBc residual optical power modulation when both lasers are not perfectly synchronized. These proof-of-principle experiments open up the possibility of a more general use of mode-locked lasers to pump few-cycle oscillators. Not only do mode-locked pump lasers potentially offer advantages in terms of reduced system complexity and costs (because of the ease of single-pass external doubling) but a completely mode-locked and synchronized pump-oscillator system can also prove very valuable for pump-probe experiments, optical frequency metrology, and phase-controlled few-cycle laser pulses.

## References


1. Y. Chen, F. X. Kärtner, U. Morgner, et al.: J. Opt. Soc. Am. B **16**, 1999 (1999).
2. D. E. Spence, P. N. Kean, and W. Sibbett: Opt. Lett. **16**, 42 (1991).
3. R. Ell, U. Morgner, F. X. Kärtner, et al.: Opt. Lett. **26**, 373 (2001).
4. L. Matos, D. Kleppner, O. Kuzucu, et al.: Opt. Lett. **29**, 1683 (2004).
5. L. R. Brovelli, U. Keller, and T. H. Chiu: J. Opt. Soc. Am. B **12**, 311 (1995).
6. K. Tamura, J. Jacobson, E. P. Ippen, H. A. Haus, and J. G. Fujimoto: Opt. Lett. **18**, 220 (1993).
7. G. Cerullo, S. De Silvestri, and V. Magni: Opt. Lett. **19**, 1040 (1994).
8. J. Solis, J. Siegel, C. N. Afonso, N. P. Barry, R. Mellish, and P. M. W. French: Opt. Commun. **123**, 547 (1996).
9. D. H. Sutter, G. Steinmeyer, L. Gallmann, et al.: Opt. Lett. **24**, 631 (1999).
10. S. Tsuda, W. H. Knox, S. T. Cundiff, W. Y. Jan, and J. E. Cuningham: IEEE J. Sel. Top. in Quantum Elect. **2**, 454 (1996).
11. R. Fluck, I. D. Jung, G. Zhang, F. X. Kärtner, and U. Keller: Opt. Lett. **21**, 743 (1996).
12. S. N. Tandon, J. T. Gopinath, H. M. Shen, et al.: Opt. Lett. **29**, 2551 (2004).
13. Ch. Spielmann, F. Krausz, T. Brabec, E. Wintner, and A. J. Schmidt: Opt. Lett. **16**, 1180 (1991).
14. C. W. Siders, E. W. Gaul, M. C. Downer, A. Babine, and A. Stepanov: Rev. Sci. Instrum. **65**, 3140 (1994).
15. A. Bartels and H. Kurz: Opt. Lett. **27**, 1839 (2002).
16. W. Seitz, T. R. Schibli, U. Morgner, et al.: Opt. Lett. **27**, 454 (2002).
17. R. Ell, W. Seitz, U. Morgner, T. R. Schibli, and F. X. Kärtner: Opt. Commun. **220**, 211 (2003).
18. O. D. Mücke, R. Ell, A. Winter, et al.: Opt. Expr. **13**, 5163 (2005).

# Ultrasimple Ultrabroadband Alignment-free Ultrashort Pulse Measurement Device

Dongjoo Lee,[1] Selçuk Aktürk,[1] and Rick Trebino[1]

School of Physics, Georgia Institute of Technology, Atlanta, Ga 30332-0430, USA
**akturk@socrates.physics.gatech.edu**

## 1 Introduction

The measurement of ultrashort laser pulses has traditionally experienced difficulties. Even now, newly proposed techniques for doing so are often extremely complex, containing as many as a dozen sensitive alignment parameters. As a result, such techniques may be more likely to introduce a distortion than to measure it.

Recently, we developed a technique based on the second-harmonic-generation frequency-resolved optical gating (SHG FROG) method, which is extremely simple and requires no alignment. We called this method GRating-Eliminated No-nonsense Observation of Ultrashort Incident Laser Light E-fields (GRENOUILLE) [1]. GRENOUILLE involves replacing the beam-splitter, delay line, and beam-combining optics with a single optical component called a Fresnel biprism, which simplified the device tremendously, and has no sensitive alignment parameters. Thus, it is inexpensive and simple to use. It has also been quite successful, providing measurement capability for visible and IR pulses over a wide range of pulse lengths energies, and repetition rates.

Unfortunately, GRENOUILLE, like all SHG-based pulse-measurement devices, yields symmetrical traces and hence an ambiguity in the direction of time. Also, like all SHG-based devices, it has a limited wavelength range. In particular, it cannot measure pulses in the UV because SHG crystals become opaque at the second harmonic (<200 nm).

Optical parametric amplifiers (OPAs) now yield microjoule- or higher-energy ultrashort pulses over broad ranges of wavelengths. Measuring OPA output pulses, especially in the visible and UV, however, is very difficult with conventional pulse-measurement devices that usually use SHG because the second harmonic of these pulses falls into UV, where the responses of most detectors drop significantly. In addition, because of higher dispersion at shorter wavelengths, these pulses suffer significantly from group velocity mismatch in the nonlinear crystal, requiring the use of extremely thin crystals, which reduces the sensitivity. OPA output pulses also suffer from a wide range of temporal and spatiotemporal distortions, which calls for an ultrasimple pulse-measurement device that could measure pulses over a broader range of wavelengths.

## 2 Ultrasimple Transient-grating FROG Setup

As a result, we present such a device here. In order to remove the direction-of-time ambiguity and achieve UV pulse measurement, a third-order nonlinear-optical beam geometry (such as self-diffraction, polarization gating, or transient grating) is required. The most versatile, sensitive, and accurate of the three is the Transient-Grating (TG) geometry. However it is extremely complex, requiring splitting the pulse into three beams, overlapping two of them in space and time to generate a TG in a $\chi^{(3)}$ medium that is probed by the third beam, and then spectrally resolving the diffracted four-wave-mixing output pulse. The two grating-excitation beams must be overlapped perfectly in time, while the third must experience a variable delay with respect to the other two. [2] However, we utilize two simple tricks that significantly simplify this device, yielding an ultrasimple device that, like GRENOUILLE, has no sensitive alignment parameters. The device, which is easy to setup, is automatically aligned, and inherently extremely broadband. A single device can operate from the UV to the IR, with no change of optics. It operates on a single-shot basis and also yields a very sensitive measurement of the pulse-front tilt, very useful for OPAs, which usually also suffer from this distortion.

Figure 1 schematically shows a picture in which the input beam is split into three beams by a simple mask with three holes in it. A cylindrical lens focuses the beams to lines in the $\chi^{(3)}$ medium. A Fresnel biprism then crosses and overlaps the beams in space and time. Two of the beams overlap perfectly in space and time across the glass medium and form the TG. The other pulse crosses the grating at an angle, thus varying the relative delay, between the pulses. The diffracted pulse emerges from the interaction region as the fourth beam in the rectangle formed by the three input pulses. The input mask is imaged onto another mask after the $\chi^{(3)}$ medium, which eliminates the three input beams, passing only the diffracted signal pulse, which is then spectrally resolved by a simple grating/lens spectrometer, whose input slit is the line focus in the $\chi^{(3)}$ medium. As a result, there is no sensitive alignment parameters.

## 3 Experiment

We demonstrated our device for ~400-fs long, 800-nm amplified Ti:sapphire pulses of about 100 μJ.

The configuration of our alignment-free TG-FROG is shown in Fig. 1. The input beam is expanded by a 4× telescope, and is then split by an input mask into three

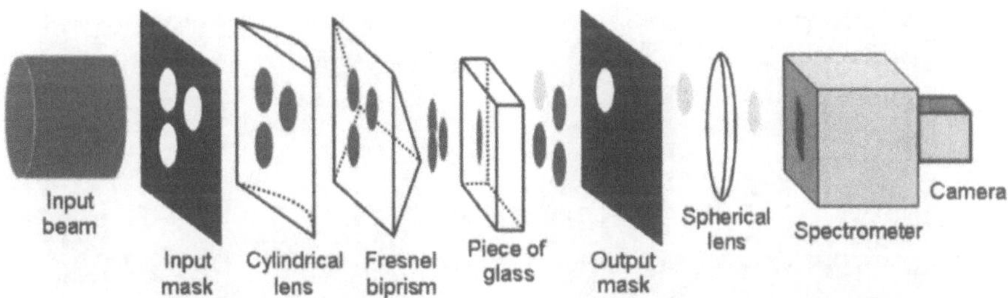


FIGURE 1. The input mask splits the input beam into three beams, which are then overlapped in the $\chi^{(3)}$ medium (ZnSe in this initial measurement, but fused silica in a more broadband device). The cylindrical lens yields line foci, which maps delay onto transverse (vertical) position, allowing single-shot measurement. The upper two beams cross and form a TG in the crystal. The lower beam is diffracted by the TG and generates an autocorrelation signal beam in the other corner of the rectangle. The line focus then acts as the entrance slit to a homemade spectrometer consisting of a collimating lens, diffraction grating, and focusing lens, yielding a single-shot TG FROG trace with delay running vertically and wavelength horizontally.

beams. Two beams are overlapped in space and time at the third-order nonlinear material by a cylindrical lens, producing a line-shaped transient refractive-index grating. The third beam is overlapped and delayed by the Fresnel biprism, whose diffraction by the induced grating produces a signal pulse. For the third-order nonlinear material, we use a 3-mm thick ZnSe, but fused silica can be used for broader-band (UV, visible, and IR) operation [3].

We demonstrate our alignment-free TG-FROG on pulses from a 1-kHz rep rate 800-nm regenerative Ti:sapphire amplifier. The focal length of the cylindrical lens is 150 mm. The Fresnel biprism has an apex angle of 174.7°, yielding a delay range of 0.7 ps. The crossing angle of the two pump beams is about 3.8°, resulting in a transient grating of about 6-μm spatial period.

The use of a very thick crystal (3 mm) is permitted because the TG process is automatically phase matched, and the group-delay dispersion is negligible [4]. The generated TG-FROG signal is collimated and spectrally resolved using a simple alignment-free spectrometer in which the line focus at the ZnSe acts as the input slit and which uses a 600-line/mm groove-density grating. The spectrally resolved signal is captured by a CCD camera. The Femtosoft FROG code is then used to retrieve the intensity and phase of the pulse from the measured trace. The agreement between the measured and retrieved traces (Fig. 2) is very good. The FROG error is about 0.6%.

To verify the accuracy of the results, we measured the same pulse, using a commercial GRENOUILLE. The retrieved intensity and phase of the pulse are shown in Fig. 2 for both measurements, and the agreement is good. The pulse's FWHM duration and bandwidth that TG-FROG retrieved are 420-fs and 4.58 nm, respectively. GRENOUILLE measured a 406-fs pulse, in comparison. This device should be able to measure amplified pulses over an extremely broad range

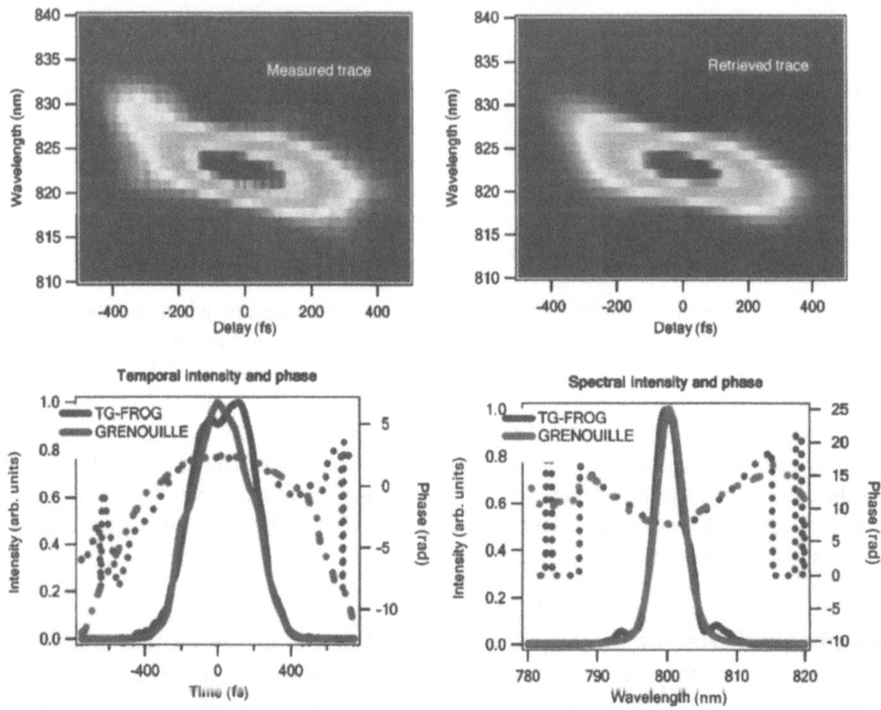


FIGURE 2. Measured and retrieved TG-FROG traces and retrieved intensity and phase of the pulse from TG-FROG and from GRENOUILLE for comparison. On the left are the spectral intensity and phase; on the right are the temporal intensity and phase.

of wavelengths with no change of components if a nonlinear medium such as fused silica is used because no component is wavelength sensitive. While OPAs yield weaker pulses in the UV than in the visible and near-IR, third-order nonlinearities increase in the UV, so the device sensitivity scales appropriately. Finally, while the ultrasimple TG-FROG device described here utilizes transmissive components, an essentially all-reflective device can be designed for extremely short pulses.

## References


1. P. O'Shea, M. Kimmel, X. Gu, and R. Trebino: Highly simplified device for ultrashort-pulse measurement. *Opt. Lett.* **26(12)**:932–934 (2001).
2. R. Trebino: *Frequency-Resolved Optical Gating*. Kluwer Academic Publishers, Boston: (2002).
3. R. W. Boyd: *Nonlinear Optics*. San Diego: Academic Press (2003).
4. J. N. Sweetser, D. N. Fittinghoff, and R. Trebino: Transient-grating frequency-resolved optical gating. *Opt. Lett.* **22**:519–521 (1997).

# Measurement of Group Velocity Distortion due to Ultrafast Index of Refraction Transients

Randy A. Bartels and Klaus Hartinger

Department of Electrical and Computer Engineering, Colorado State University, Fort Collins, CO 80523-1320, USA
**Randy.Bartels@colorstate.edu**

## 1 Introduction

The interaction of a linearly polarized laser pulse with an anisotropic molecular gas exerts a torque on molecules not initially aligned with the polarization direction [1]. Laser pulses shorter than the molecular rotation time will impulsively excite a rotational wave packet composed of a superposition of rotational eigenstates. The rotational wave packet periodically collapses and revives, producing ultrafast transients in the index of refraction [2,3]. A time-delayed ultrafast probe pulse propagating in the ultrafast index transient accumulates a phase modulation that depends on the particular shape of the rotational revival. Strong temporal phase modulation may be applied to the probe pulse by extending the interaction length with a hollow core fiber. This large phase modulation leads to substantial spectral changes in the probe pulse and has been applied to produce transform-limited pulses both temporally compressed [4] and stretched [5].

Pure phase modulation as described above is the lowest-order effect that occurs when a probe pulse is propagating in the ultrafast index transient. When the propagation length of the probe pulse in the transient index becomes large, additional second-order effect begin to distort the pulse propagation. We have measured changes in the propagation time of probe pulses in an ultrafast index transient that can be attributed to two second-order physical effects that distort the group velocity of the probe pulse. The first distortion effect is an interplay between the dispersion in the experiment and a shifting of the central frequency of the probe pulse, i.e., a frequency shift results in a change of the group-velocity due to high-order spectral dispersion in the gas. The second effect is directly due to the index transient and originates from the second-order expansion of the transient index source term in the wave equation describing the propagation of the probe pulse. In this paper, we report on the first direct measurements of this ultrafast index transient-induced group delay. We measure changes in the fiber transit time of >200 fs and determined that the dominant mechanism of this measured group velocity distortion can be attributed to the direct modification of the group velocity by the index transient, rather than a combination of dispersion and frequency shifting.

## 2 Experiment

In the experiment shown in Fig. 1a, the polarization of a linearly polarized 0.8-mJ, 50-fs transform-limited pulse is rotated with a half-wave plate and sent into a broad bandwidth-polarizing beam splitter forming the input of a orthogonal polarization Mach-Zehnder interferometer. By adjusting the waveplate, approximately 90% of the pulse energy is directed into the pump arm of interferometer and the remaining 10% of the energy is in the probe pulse. The probe pulse is then sent into a computer-controlled optical delay line that allows control over the delay between the pump and probe pulses. Before the pump and probe pulses are recombined, 50% of the probe arm is split off with a reflective beam splitter to serve as a reference pulse. By splitting off the reference pulse after the motorized stage, an adjustment of the pump–probe delay does not influence the relative probe-reference delay. The remaining probe energy is combined with the orthogonally polarized *pump* pulse and focussed into a 150-μm-diameter hollow-core fiber filled with $CO_2$ gas. At the exit of the hollow-core fiber, the probe pulse is collimated and isolated from the pump with two CVI reflective thin-film polarizers. The path length of the reference pulse is adjusted so that the probe and reference pulses are nearly temporally overlapped when both are coupled into a spectrometer to record the spectral interferometry (SI) between the probe and

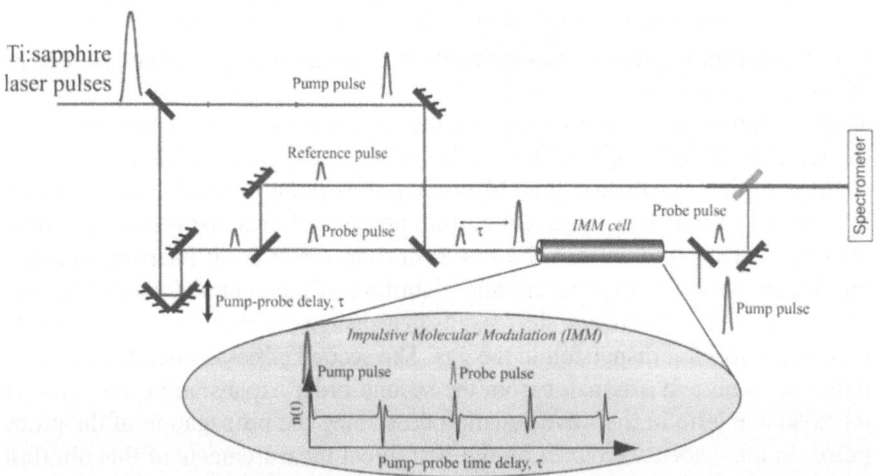


FIGURE 1. Schematic of experimental setup.

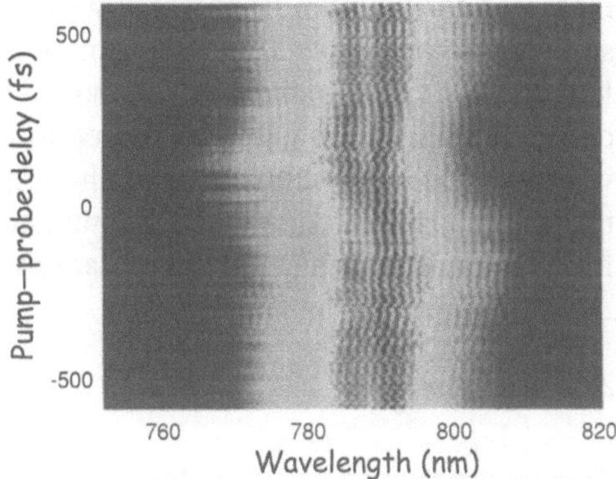


FIGURE 2. Interference spectrum recorded between the probe and reference pulses as the probe pulse is scanned through a quarter revival of $CO_2$.

reference pulses. Modulations of the probe-reference delay are recorded SI signal as the probe pulse is scanned through the quarter rotational revival of $CO_2$.

## 3 Results

Figure 2 shows the SI data as the probe pulse is scanned through the quarter revival of $CO_2$ at a pressure of 450-Torr in a 10-cm long fiber. Figure 3 plots the sideband due to the reference-probe pulse interference that appears in the Fourier transform of the spectral interference as the probe pulse is scanned through the quarter revival. As discussed earlier, the possible mechanisms for group delay distortion of the probe pulse is a combination of frequency shifting due to temporal phase modulation and linear dispersion or directly due to the index transient. At ≈450-Torr, the group velocity dispersion (GVD) of the hollow fiber and $CO_2$ cancel one another, leading to weak dispersion. Near zero GVD, the dispersion-induced group velocity distortion should be minimal. The spectrum of the probe pulse was

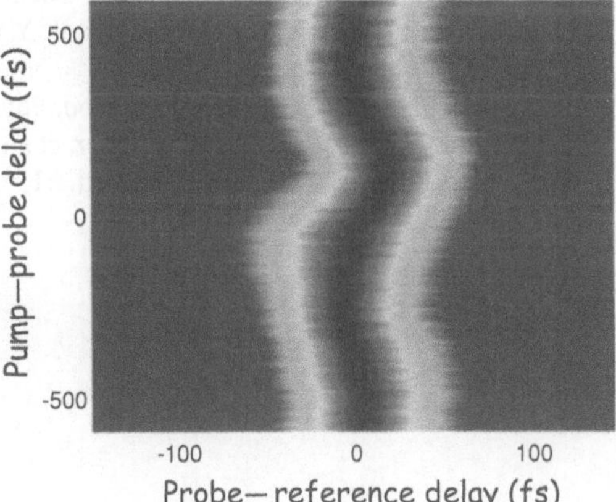


FIGURE 3. Interference spectrum recorded between the probe and reference pulses as the probe pulse is scanned through a quarter revival of $CO_2$.

measured for the scan from which we observe a maximum central wavelength shift during the scan of 10 nm. Assuming a spectral shift of 10 nm over the entire length of the fiber, we find an upper bound of 0.05 fs of dispersion-induced probe delay. The dispersion calculation used to estimate the dispersion-induced delay was independently confirmed in another experiment. From this, we may conclude that the propagation distortion is dominated by the transient index of refraction from the transiently aligned molecular gas.

## 4 Summary

To summarize, we have, to our knowledge, performed the first direct measurement of distortions of the propagation of laser pulses through optical media exhibiting ultrafast transients in the index of refraction. Presently, we are measuring the group velocity distortion over a wider range of experimental conditions. We are also characterizing the distorted probe pulses. These measurements represent measurements of a new effect occurring when short laser pulses propagation in optical media with ultrafast index of refraction transients. These effects will also impact the ability to compress pulses with ultrafast index transients. An understanding of these effects will improve the understanding of ultrafast temporal phase modulation of short laser pulses. Moreover, further studies may lead to new approaches to manipulate ultrafast laser pulses.

**Acknowledgments.** The authors gratefully acknowledge support for this work from the National Science Foundation CAREER Award ECS-0348068, the Office of Naval Research Young Investigator Award, and the Beckman Young Investigator Award. One author (RAB) gratefully acknowledges generous support from a Sloan Research Fellowship.

## References

1. B. Friedrich and D. Herschbach: J. Phys. Chem. **99**, 15686 (1995).
2. C. H. Lin, J. P. Heritage, T. K. Gustafson, R. Y. Chiao, and J. P. Mctague: Phys. Rev. A **13**, 813 (1976).
3. H. Stapelfeldt and T. Seideman: Rev. Mod. Phys. **75**, 543 (2003).
4. R. A. Bartels, T. C. Weinacht, N. Wagner, et al.: Phys. Rev. Lett. **88**, 013903 (2002).
5. K. Hartinger and R. A. Bartels: Opt. Lett. **31**, 3526–3528 (2006).

# Frequency-domain Phase Conjugator for a Few-cycle and a Few-nJ Optical Pulses

Hajime Nishioka, Keisuke Hayakawa, Hitoshi Tomita, and Ken-ichi Ueda

Institute for Laser Science, University of Electro-Communications, 1-5-1 Chofugaoka, Chofu, Tokyo 182-8585, Japan
**nishioka@ils.uec.ac.jp**

Temporal phase structure of a 10-fs optical pulse has been regenerated by frequency-domain phase conjugator based on the two-photon recorded Bragg diffraction. This optically programmable diffractive device with the index modulation of $\Delta n = 0.03$ was directly written by a mode-locked laser with no cavity extension.

## 1 Introduction

The frequency domain phase conjugator (FDPC) generates time-reversed replica of a phase-modulated (frequency chirped) pulse [1]. The time-reversed replica regenerates initial pulse shape after passing though again the same GDD element. We have demonstrated FD phase conjugation by the Time Ordered Pulse Recording/Readout in a Periodic Diffractive Optics (TOPEDO), where the temporal phase structure of optical pulse is recorded as a second-order cross-correlation function in space as shown in Fig. 1. The recorded phase structure is read in reflection of a transform-limited (TL) read pulse that is the same as the gate pulse. The readout becomes time-reversed replica of the signal wave, i.e., FDPC wave. A function of the two-photon TOPEDO is an optically programmable phase-sensitive device that records, storages, and playbacks optical pulse structure in addition to the self-phase correction. In the two-photon interference, the signal and gate pulses having the same frequency components can be interacting each other with a long time duration exceeding their coherent length. Moreover, the two-photon interference efficiently produces deep amplitude modulation even with the small gate pulse energy. The first FDPC experiment has been demonstrated in semiconductor-doped colored glasses having an index modulation of

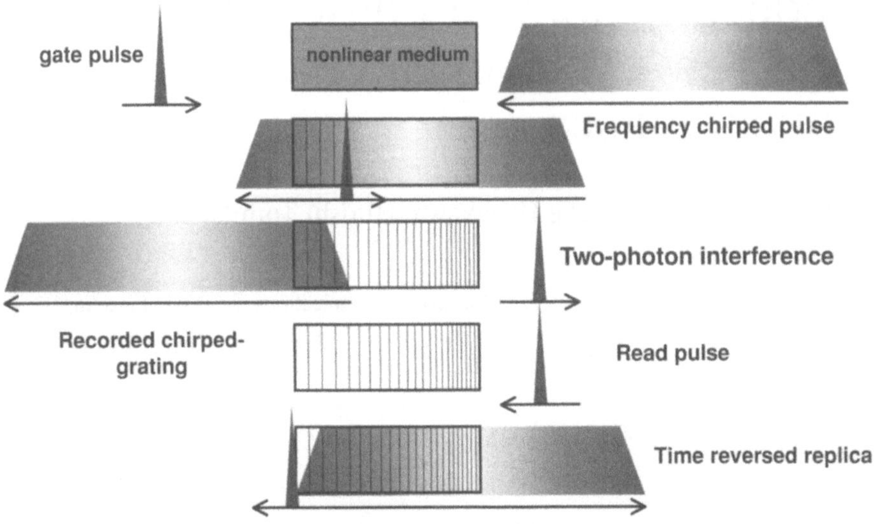


FIGURE 1. Schematics illustration of two-photon-gated pulse recording and time-reversed readout in a periodic diffractive optics.

$\Delta n = 10^{-5}$ for sub-mJ and 100-fs laser pulses [2]. For the 10-fs pulses, however, these media were inefficient. Diffraction efficiency is proportion to square of coherent length, for example, a few-cycle grating has million times lower efficiency as compared with typical 10,000 grooves gratings. An index modulation of $>10^{-2}$ is required for the few-cycle experiments. In this paper, we have demonstrated FDPC operation for a few-cycle optical pulse. Deep refractive index modulation was produced by a few-nJ laser pulse directly from a mode-locked oscillator without cavity extension.

## 2 Two-photon Gated Recoding

Phase structure of the signal pulse is recorded as a second-order cross-correlation function with a TL ultrashort pulse that has the same frequency bandwidth that of the signal pulse. Typically, the signal pulse is strongly chirped pulse so that the pulse duration of signal is much longer than the coherent time. The time-integrated second-order autocorrelation function of the chirped pulse has interference fringes in a limited time window corresponding to the coherent time as shown in Fig. 2a. In contrast, two-photon interaction between the TL pulse and signal pulse records the whole phase information of the signal pulse as shown in Fig. 2b. At this time, the interference contrast, however, is degraded as the compression factor because the TL pulse has whole frequency components. The contrast ratio optimized for normalized pulse energy and pulse duration is shown in Fig. 3. The result shows that a gate pulse energy having 1/100 fraction of

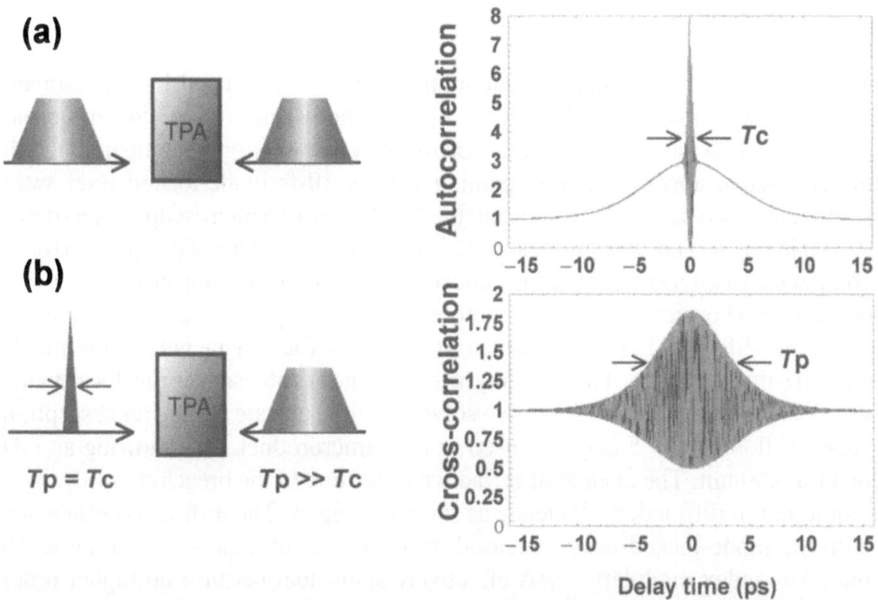


FIGURE 2. Instantaneous phase recording of a chirped pulse by two-photon interference. (a) Interaction with itself, and (b) interaction with a TL pulse. $T_p$, pulse dration; $T_c$, coherent time.

the signal is enough for the signal pulse having 10,000 times longer pulse duration. The contrast is given as the electric-field amplitude corresponding to quartic root of the two-photon absorption, therefore this method is essentially highly sensitive.

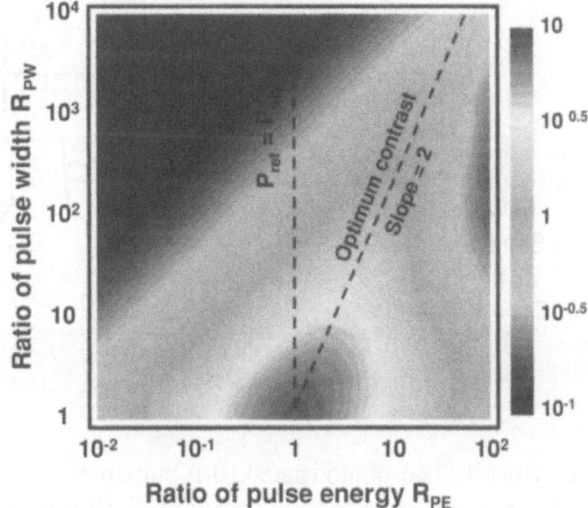


FIGURE 3. Fringe contrast optimized between signal pulse duration and energy. The horizontal and vertical axis are normalized by the gate pulse duration and energy, respectively.

## 3 Experiment

The requirement, deep index modulation, should be produced by off-resonant large absorption change. We have developed photosensitive dye-doped thermo plastic film that has $10^3$ times large absorption cross-section in comparison with the solid-state absorber. Pulses coming from a 10-fs mode-locked laser were splitted into two beams and interact at the focal plane of a microscope objective as shown in Fig. 4. The laser pulse bound 22 times between a pair of chirped mirror to compensate a GDD of 1260 $fs^2$ produced by the microscope objective. The pulse duration and Fourier phase were monitored by frequency shearing interferometory technique (SPIDER). An interference fringes at the focal plane is 13 micron wide and 207 micron thick. The spatially averaged laser intensity at the focal plane was 3.7 $GW \cdot cm^{-2}$. C307 laser dye solved in vinyl acetate having an absorption cross-section of $10^{-16}$ $cm^2$ was used as a 72-micron thick film showing an OD of 11 at 400 nm. The change of refractive index due to the breaching of dye was monitored in diffraction efficiency as shown in Fig. 5. The diffraction efficiency with the mode-locked laser increased as a function of exposure time up to 10 min. The index modulation was effectively sinusoidal because no higher order diffraction was observed. The maximum diffraction efficiency was 3.2%. The diffraction efficiency corresponds to a change in refractive index of 0.03 when the writing laser field was 10 fs (only four cycles in optical field corresponding to 8 grooves). This index change is the highest value written by two-photon process as far as our knowledge. No diffraction portion was observed when we stopped mode lock; therefore the contribution by fundamental absorption was negligible. Spectra width of diffracted beam was 150 nm centered at 800 nm, which is enough for efficient diffraction of the 10-fs pulses. A beam divergence was measured to

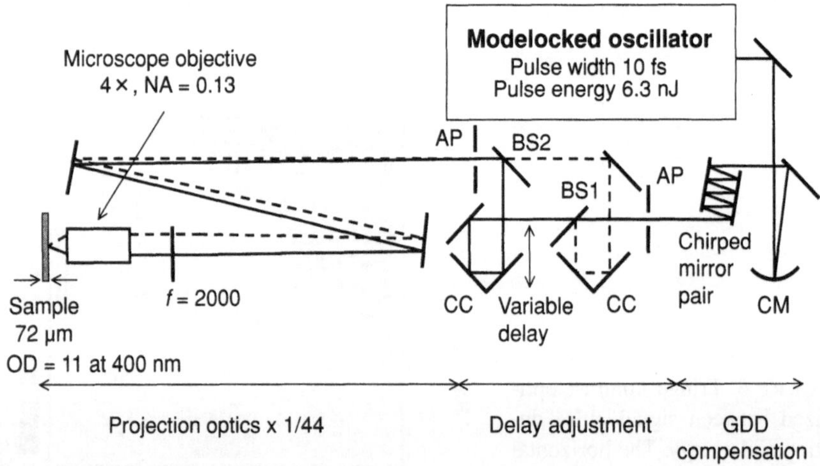


FIGURE 4. Two-photon gated 10-fs pulse recording. The GDD compensation with chirped mirrors is designed to form transform-limited pulses at the focus.

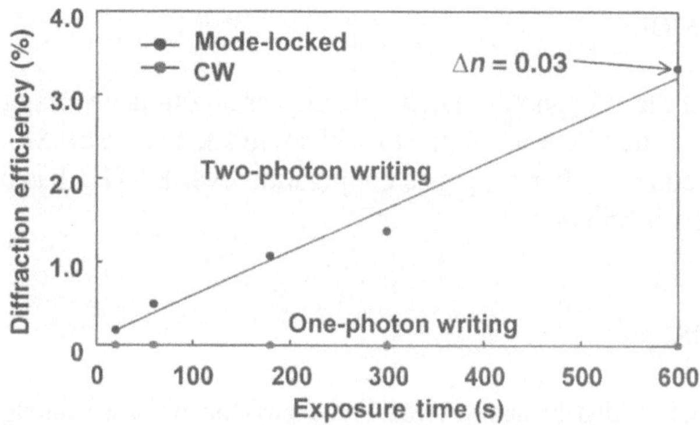


FIGURE 5. Diffraction efficiency of two-photon recorded 8-grooves grating as a function of exposure time.

be diffraction-limited number. A change of Fourier phase diffracted by the FDPC was monitored by the SPIDER. In this experiment, a 3-mm thick BK-7 glass plate was inserted to single arm, and temporal phase structure of the chirped pulse was recoded in the film. We have compared Fourier phase between the chirped signal, diffracted (FDPC), and passing though again the same glass plate as shown in Fig. 6. The FDPC operation for a bandwidth over 80 THz has been observed.

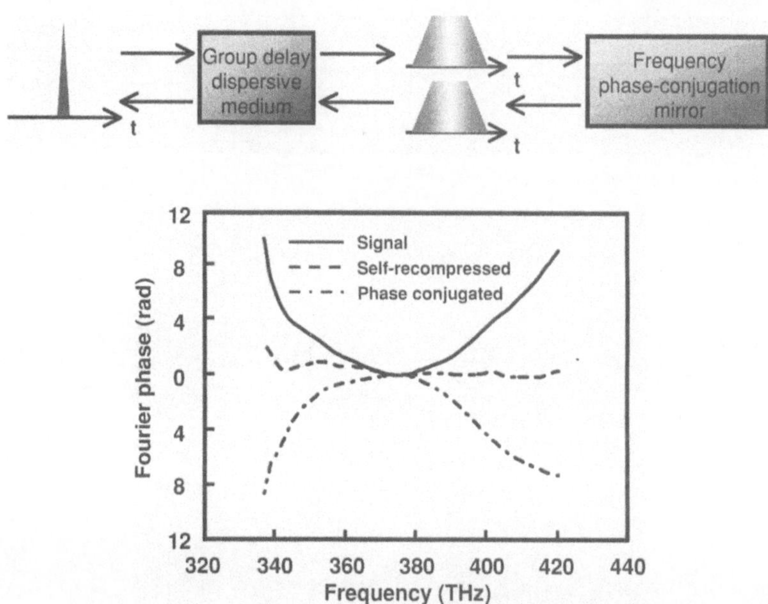


FIGURE 6. Fourier phase of the 10-fs laser pulse measured by SPIDER after 3-mm thick BK-7 (signal), phase-conjugated, and passing though again same glass plate (self-recompression).

## 4 Conclusion

In conclusion, the two-photon gated optically programmable Bragg grating has been directly written by 6-nJ IR pulses with an index change of $\Delta n = 3 \times 10^{-2}$. All-optical frequency-domain phase conjugation over 80 THz has been demonstrated with 10-fs pulses.

## References

1. H. Nishioka, S. Ichihashi, and K. Ueda: Frequency-domain phase-conjugate femtosecond pulse generation using frequency resolved cross-phase modulation. Opt. Express. **10**, 920 (2002).
2. H. Nishioka and K. Ueda: Femtosecond pulse recoding and regeneration by a two-photon gated periodic diffractive optics, Springer Ser. Chem. Phys. **79**, Springer-Verlag 777 (2004).

# Part V

# High-field Physics and Applications I

# Part V

# Relativistic Optics: A new Route to Attosecond Physics and Relativistic Engineering

Gérard Mourou

Laboratoire d'Optique Appliquée, Ecole Nationale Supérieure de techniques avances, Ecole Polytechnique, Centre National de la Recherche Scientifique UMR 7639, 91761 Palaiseau Cedex, France
**Gerard.Mourou@ensta.fr**

**Summary.** The advent of ultraintense lasers capable of producing intensities such that laser–matter interaction is governed by the electron relativistic behavior is the gateway to a new type of nonlinear optics where the electron in the laser field has a relativistic character. In contrast to the nonrelativistic regime, the laser field is capable of moving matter much more effectively. Contrary to the bound electron optics, which is producing radiation typically in the eV range, relativistic optics is producing radiation and particles with much higher characteristic energies in the keV to the GeV. This energy is bound to go up when higher power will become available. Because these radiations/particles are produced over length of tens of micrometers, relativistic optics opens the field of relativistic microelectronics/photonics engineering. One of the unpredicted surprise is the possibility to produce attosecond pulses of radiations/particles efficiently and well synchronized with the laser pulse. This is opening the possibility to produce intensity at the level close to the Schwinger intensity thus getting access to the nonlinear QED regime.

## 1 Introduction

Available focused intensity is as high as $10^{22}$ W · cm$^{-2}$. We will reach soon $10^{23}$ W · cm$^{-2}$ (Fig.1). These intensities are well above the regime where the electron motion starts to be relativistic. If we consider optics as the science of light–electron interaction, it is natural to call *relativistic optics* when the light–electron interaction is dominated by relativistic effects. We will draw as much as possible on the similarities between relativistic optics and conventional nonlinear optics. As mentioned previously, the progresses in high-intensity lasers have been so rapid that it is necessary to redefine the different regimes of intensities. Now, we refer to high intensity when the laser field $E$ fulfills the following condition:

$$\hbar\omega < \left[ m_0 c^2 \sqrt{1 + a_0^2} - 1 \right] < m_0 c^2 a_0 \gg 1 \tag{1}$$

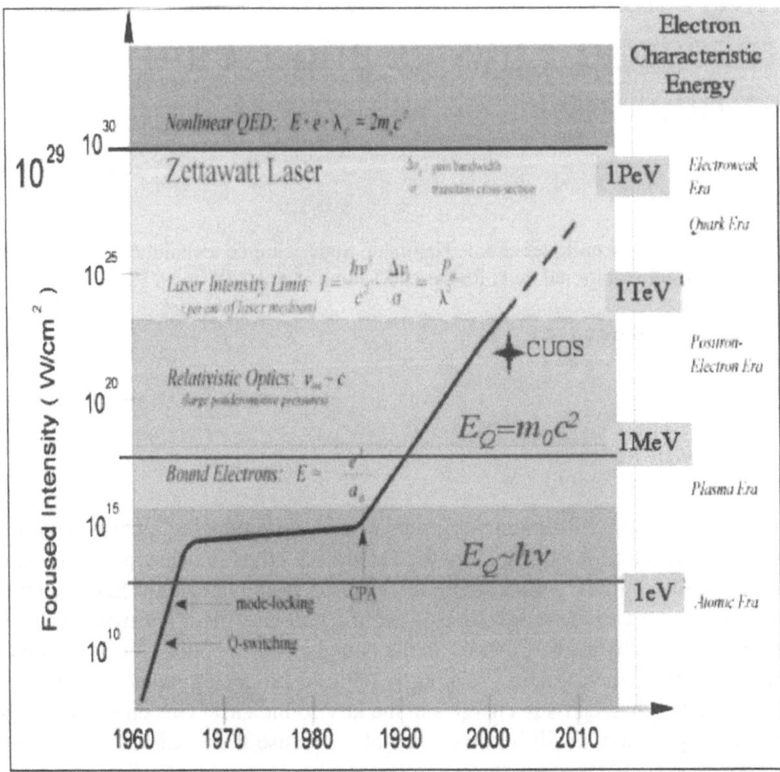


FIGURE 1. Intensity versus years, showing the rapid progresses after 1985 where it became possible to reach the relativistic intensity level that open up the field of relativistic nonlinear optics.

with the dimensionless amplitude of the laser radiation

$$a_0 = eA/m_0c^2. \tag{2}$$

Here $m_0c^2\sqrt{1+a_0^2}$ is the ponderomotive potential (in the limit of $a_0 \gg 1$ it is equal to $eE\lambda/2\pi$ and for $a_0 \ll 1$ it is $e^2E_0^2/2\omega^2m_0$), $\hbar\omega$ the photon energy and $m_0c^2$ the rest mass energy of the electron, $\lambda = 2\pi c/\omega$ the laser wavelength, and e and $m_0$ the electron charge and mass. This regime corresponds to intensities between $5.10^{14}$ and $10^{18}$ $\text{W} \cdot \text{cm}^{-2}$ for 1-μm wavelength. The ultrahigh intensity regime will be defined as the one above the $10^{18}$ $\text{W} \cdot \text{cm}^{-2}$ limit. That is where

$$eE_0\lambda > 2mc^2. \tag{3}$$

For excimer wavelength at 248 nm, the relativistic limit will be at $10^{19}$ $\text{W} \cdot \text{cm}^{-2}$ while for 10.6 μm for $CO_2$ this limit will be at $10^{16}$ $\text{W} \cdot \text{cm}^{-2}$. Finally displayed in the Fig. 1 is the nonlinear QED limit reached for the laser field $E$ such that

$$eE\lambda_c > 2m_0c^2, \tag{4}$$

where $\lambda_c = \hbar/m_0c$ is the Compton length. Relation 4 shows that the work that the field has to produce over a Compton length $\lambda_c$ set by the uncertainty principle to separate an electron positron pair must be greater than $2m_0c^2$. This regime corresponds to intensities on the order of $10^{29}$ W · cm$^{-2}$ for 1-μm light. Such a field, $m_0c^2/\lambda_c e$, is called the Schwinger field. Not surprising similar isomorphysisms of fields to breakdown a neutral atom is governed by the Keldysh field, and semiconductors. Let us recall that the laser field $E$ is related to the intensity $I$ by

$$E^2 = Z_0 I, \tag{5}$$

where $Z_0 = 377$ is the vacuum impedance in ohms.

The physics in the high-intensity regime includes tunneling ionization, harmonic generation [1], etc. It deals with bound electron nonlinear optics. This regime has been covered extensively by a number of reviews [2] and would not be addressed in this article. The ultrahigh intensity regime has already produced a wealth of scientific results that are all related to the relativistic character of the electrons. In laser–atom interactions, the work at high intensity has been based on the nonrelativistic Schroedinger equation and dipole approximation. Trying to extend the laser–atom interaction in the relativistic regime would require solving the time-dependent Dirac equation [3–7]. The laser-plasma interaction in the ultrahigh intensity regime leads to a panoply of new phenomena like X-ray generation [8–10], $\gamma$-ray generation [11], relativistic self-focusing [12–18], high harmonic generation [19–22], electron [23–34] and proton [35–45], acceleration, neutron [46, 47] and positron [48] production, as well as the demonstration of nonlinear QED [49, 50].

## 2 On the Similarities and Differences Between Bound-electron and Relativistic Nonlinear Optics (Qualitative Introduction of Relativistic Regime)

The classical treatment of classical linear and bound-electron nonlinear optics deals with the electron displacement $x(t)$ around the nucleus (Fig. 2a). This displacement gives rise to the polarizibility

$$P(t) = Nex(t), \tag{6}$$

where $N$ is the electron density. The force applied to the electron is the Lorentz force

$$F(t) = eE(t) \tag{7}$$

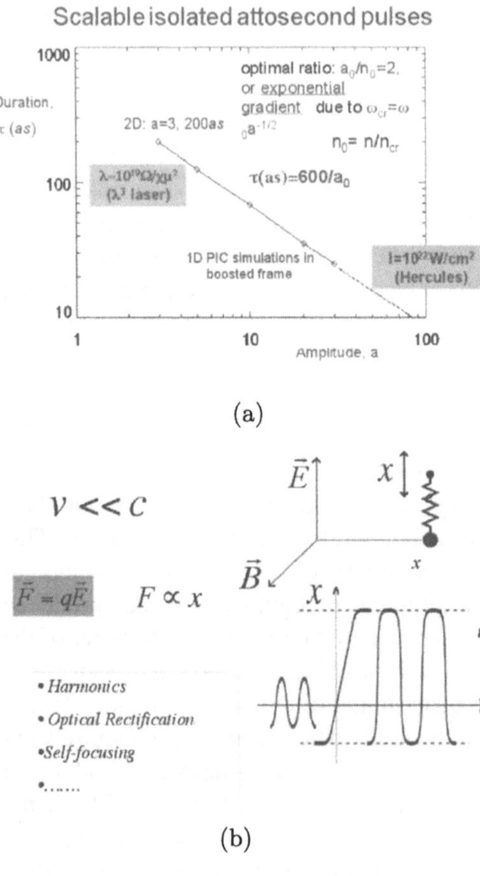


FIGURE 2. (a) In the relativistic regime, the velocity of the electron $v$ is comparable to $c$. The $v \wedge B/c$ becomes very large. The electron mass $m$ changes during one oscillation, and its main motion is forward. (b) Nonlinear optics of the bound electron, where, $F = qE$, but $F$ is not proportional to the electron displacement $x$ like in linear optics causing the nonlinear optical effects of the bound electron. Its main motion is transverse.

in which in the classical limit we neglected the magnetic field part due to the smallness of the v/c ratio. In the linear regime, for electrons bound to their nucleus, $F(t)$ is proportional to the displacement $x(t)$. As the displacement increases the proportionality between $x(t)$ and $E(t)$ is not respected anymore and is at the origin of the well-known nonlinear optical effects of the bound electrons: harmonic generation, optical rectification, etc. As the laser intensity increases to the $10^{14}$ $\mathrm{W \cdot cm^{-2}}$ level, the material will ionize or damage and the electrons will become free. It is just at the threshold where the electron is still bonded and in the process of becoming free that high harmonics are created [2]. It is also at the same intensity level that the deterministic character of the damage threshold is observed [51] in solid. At much higher intensity level $10^{18}$ $\mathrm{W \cdot cm^{-2}}$ the electron becomes unbound and its velocity will approach the speed of light. The Lorentz force applied to the electron is

$$F(t) = e\left[E(t) + \frac{1}{c} v \times B\right], \tag{8}$$

where the term $\frac{v \times B}{c}$ cannot be neglected anymore. Because of the combined action of the $E$ and $B$ fields, the electron will follow a complicated trajectory. For a linearly polarized light, this trajectory is a figure eight in the frame moving at the average electron velocity as it is explained in the *Classical Theory of Fields* by Landau and Lifshitz [52]. The normalized vector potential quantity $a_0 = eA/m_0c^2$ represents the quivering momentum normalized to $m_0c^2$. Here $A$ is the electromagnetic vector potential. The longitudinal displacement is proportional to $a_0^2$ whereas the transverse displacement scales as $a_0$. In the reference frame where the charged particle initially is at rest for $a_0 < 1$ its transverse momentum is larger than the longitudinal one, whereas for $a_0 > 1$ the situation is obviously reversed and the longitudinal momentum becomes much larger than the transverse one (Fig. 2b). This complicated electron motion is the source of the relativistic nonlinear effects like relativistic rectification, relativistic self-focusing, relativistic harmonic generation, etc.

## 3 Relativistic Rectification or Wake Field effect

The concept of wake field acceleration that we call relativistic rectification by analogy with its bound electron counterpart was first introduced by Tajima and Dawson [53]. The idea is to introduce a stable method of exciting large amplitude fast waves, whereas the previous collective acceleration suffer from instabilities involving ions. It was further theoretically studied by different groups [54–57]. In the plasma, the electrons are strongly pushed forward due to the $vB$ force (Fig. 3). They drag behind the much more massive ions setting up a large electrostatic field parallel to the laser propagation direction. This field is extremely large and of the order of magnitude of the laser transverse field. The $vB$ term *transforms* the laser field into a longitudinal electrostatic field with an amplitude equivalent to the laser transverse field. This is a remarkable result if we consider that for the longest time, laser researchers recognized the enormous amplitude of the laser transverse field and tried to flip a fraction of this field along the transverse direction using various schemes [58–60]. In the relativistic regime, this conversion performed in plasma is automatic and efficient. If we consider that harmonic generation is the hallmark of bound electron nonlinear optics,

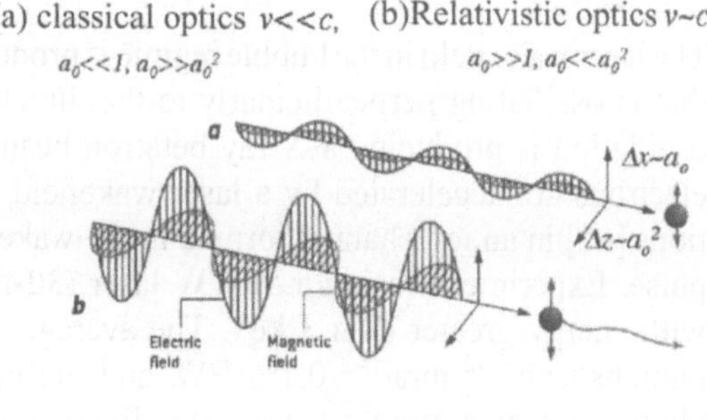


FIGURE 3. In the relativistic regime under the large $v \wedge B$ force, the electrons are pushed forward. The ions more massive stay in the back setting up a large longitudinal (electrostatic) gradient as large as the laser field. The laser electromagnetic field has been efficiently rectified. The resulting field is in the longitudinal direction.

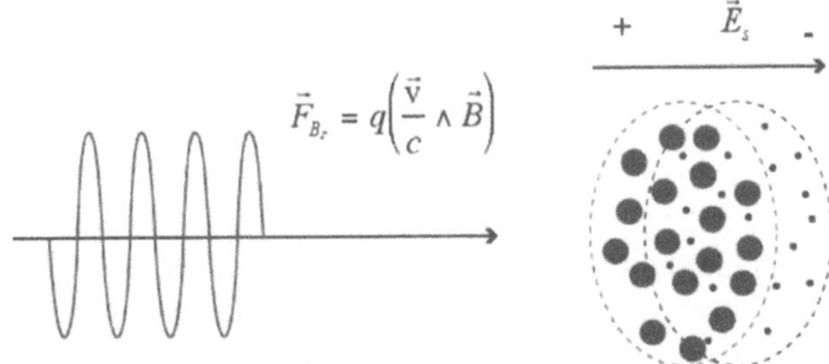


FIGURE 4. The field rectification can lead to quasimonochromatic (200 MeV) beam of superior transverse beam divergence and emittance as shown in the insert.

relativistic rectification seems to be the most prominent effect of relativistic optics. In addition, let us note that optical rectification in classical nonlinear optics is not often used. It occurs only in noncentrosymmetric crystals and is not very efficient. It is due to the fact that in a noncentrosymmetric system the charges are pushed unequally in the direction perpendicular to the propagation axis to produce a net electrostatic field perpendicular to the propagation axis. In relativistic optics, it is the converse. The rectified field is longitudinal, it is produced in centrosymmetric media—plasmas—it is efficient, and its value is of the order of the transverse field.

As a consequence, relativistic intensities can produce extremely large electrostatic fields. For example for $I = 10^{18}\ \mathrm{W \cdot cm^{-2}}$, we could generate (see Eq. (5)) an electrostatic field up to 2 TV/m and 0.6 PV/m for $10^{23}\ \mathrm{W \cdot cm^{-2}}$. These values are gargantuan. To put them in perspective, they correspond to SLAC (50 GeV) on 100-μm. It is mesmerizing to think that if we were able to maintain this gradient over 1m, a PeV accelerator using conventional technology as discussed by Fermi, which would circumvents the earth, could fit on a table. Recently, 100–200 MeV quasimonoenergetic beams [32–35] with good transverse emittance were produced by four groups, all using a novel plasma embodiment described by A. Pukhov [61] (see Fig. 4).

## 4 Transverse Motion: Betatron Oscillations

The laser wake field in the bubble regime is producing a high-energy electron beam that is oscillating perpendicularly to the direction of propagation. The electron oscillation is producing a X-ray betatron beam of keV energy. Here, energetic electrons are accelerated by a laser wakefield, and experience betatron oscillations [62] in an ion channel formed in the wake of the intense femtosecond laser pulse. Experiments using a 50-TW laser (30-fs duration) have produced X-ray with energy greater than 3 keV. The average brightness demonstrated is $5.10^6$ photons/s/mm$^{-2}$/mrad$^{-2}$/0.1% BW, and the peak spectral brightness is $2.10^{22}$ photons/s/mm$^{-2}$/mrad$^{-2}$/0.1% BW. Beyond the demonstration that compact

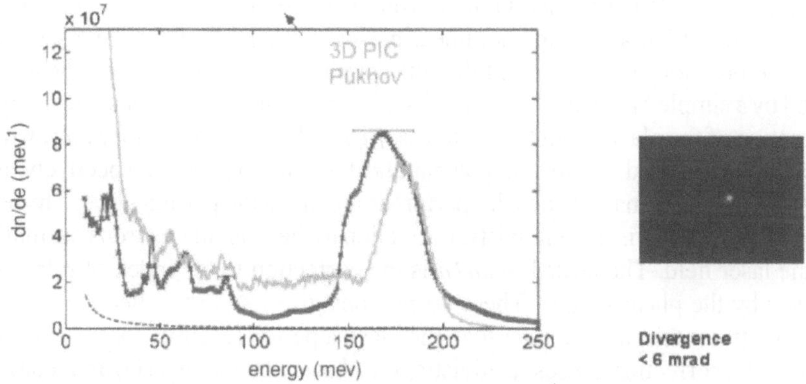


FIGURE 5. The transverse beam oscillation can lead to betatron forward emission. The energetic electron beam and the betatron emission are separated with magnets. Up to 3 keV forward X-ray, beam is produced.

synchrotron radiation can be produced with femtosecond laser systems, this radiative process opens very promising perspectives in a near future. Presently, its reasonably high flux ($5.10^6$ photons/pulse/s/0.1% BW) and its perfect synchronization with the laser system enable visible pump/X-ray probe experiments with insignificant time jitter. Bragg diffraction and absorption X-ray studies can be achieved, which will significantly extend the first X-ray diffraction studies already demonstrated in ultrafast X-ray science [63, 64]. These results pave the way to a new generation of radiation in the X-ray spectral range, with high collimation and a short pulse duration, produced by the use of a compact laser system as shown in Fig. 5.

# 5 Opportunity of the Relativistic Regime: Attosecond Physics

## *5.1 Relativistic Pulse Compression with Subcritical Plasma [65]*

At relativistic intensity, using plasmas for further compressing laser pulses is only natural: plasmas can support much higher intensity levels than classical nonlinear optics material, and relativistic laser–plasma interaction provides the nonlinearities required for laser pulse compression. Indeed, the past few years have seen the emergence of numerous theoretical and numerical studies on pulse compression in plasmas. Amongst others, one can find the superradiant amplification of colliding laser pulses [16, 66], the reflection of a laser onto a counterpropagating ionization front [17, 67], or onto a counterpropagating nonlinear plasma wave [18, 68]. Recent theoretical work has shown that it is possible to obtain pulse compression in a plasma using a single laser pulse: the interplay of relativistic self-phase modulation, group velocity dispersion (GVD), and the presence of a high-amplitude nonlinear plasma wave can lead to the laser pulse self-compression. The physics of the self-compression mechanism can be understood by a simple 1D picture. In relativistic laser–plasma interactions, the index of refraction of the plasma can be written as $\eta(\xi) = 1 - (1 + \delta n/n)\omega_p^2/\gamma\omega_0^2$, where $\xi = ct - z$, $\omega_p$ and $\omega_0$ are the plasma and laser frequencies, respectively, and $\delta n/n(\xi)$ is the normalized density perturbation due to the plasma wave. The term $\gamma = (1 + a^2/2)^{1/2}$ is the relativistic correction to the mass of electrons oscillating in the laser field. The term $1 + \delta n/n$ is the correction to the index of refraction caused by the plasma wave. These corrections give rise to a collection of well-known effects for pulses longer than the plasma period [25,69]. For instance, the relativistic correction causes relativistic self-phase modulation (1D) and relativistic self-focusing (2D). Similarly, the plasma wave gives rise to Raman forward scattering (1D) and the self-modulation instability (2D).

Here, we are interested by the case where the pulse is shorter than the plasma wavelength ($c\tau < \lambda p$). The pulse completely sits inside the first oscillation of the plasma wave; its front is red shifted whereas its back is blue shifted. In conjunction with the GVD of the plasma, this leads to pulse compression: the back of the pulse travels faster than the front of the pulse. As the pulse becomes compressed, it becomes essentially red shifted because it stays in the descending slope of the plasma wave. The physics is analogous to the soliton effect taking place in optical fibers.

The laser pulse had a $38 \pm 2$-*fs* duration at full width at half maximum (FWHM) and contained 1.2 J of laser energy at central wavelength 820 nm. It was focused onto the edge of a 3-mm cylindrical supersonic He gas jet, with a 500-μm density gradient at the front. The focal spot size was $r_0 = 21$ μm at FWHM, producing vacuum-focused laser intensities of $I = 3.2 \times 10^{18}$ W · cm$^{-2}$, for which the corresponding normalized vector potential is $a_0 = 1.26$. For these high laser intensities, the He gas was fully ionized by the foot of the laser pulse. The shortening of the laser pulse was observed for electron densities between $n_e = 510^{18}$cm$^{-3}$ and $n_e = 10^{19}$cm$^{-3}$. The highest shortening with the best reproducibility was obtained for $n_e = 6 - 7.510^{18}$cm$^{-3}$ for which the plasma wavelength ($\lambda p = 12.2 - 13.6$ μm) is comparable with the laser pulse length ($c\tau = 11$ μm). The diameter was larger than the plasma wavelength, but one may expect that self-focusing in the plasma brings it down to the matched value. Figure 6 presents the raw autocorrelation images which clearly show a possible pulse compressed down to less than 10 fs when the laser pulse interacts with the plasma corresponding to a shortening by a factor of four from the initial pulse.

## 5.2 *Overcritical Plasma: Toward Efficient Isolated Attosecond Pulses [70–73]*

The relativistic regime could provide a new avenue to the generation of efficient attosecond pulses of X and $\gamma$ radiations as well as electrons. In addition,

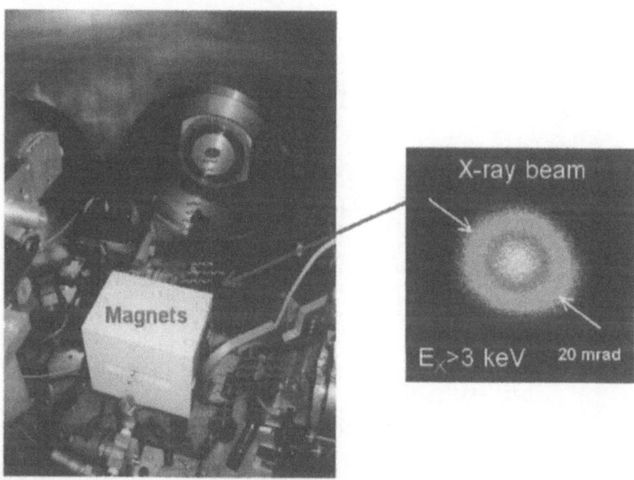


FIGURE 6. The relativistic self-phase modulation produced in a plasma can lead to pulse spectral broadening followed by pulse compression due the negative GVD in the plasma. This effect was able to compress joule level pulse from 38 fs to sub-10-fs. It offers the additional advantage that it is a grating-free technique.

the possibility to produce efficient attosecond pulses makes possible to generate pulses with intensity close to the critical field, taking the field of high-field physics near the vacuum polarization and pair generation regime.

Analysis with 2D PIC simulations of laser–plasma interaction in the relativistic $\lambda^3$ regime reveals that dense attosecond electron bunches can be formed in vacuum. The attosecond pattern appears due to the joint action of incident and generated electromagnetic fields. In addition, the bunched electrons, escaping the target with relativistic velocities, compress the reflected radiation, generating the train of attosecond pulses (Fig. 7). This illustrates a connection between the origin of attosecond electromagnetic pulses and bunched electrons, and their synchronized behavior.

The analytical model of a relativistically driven charged layer [70] confirms that an oscillating electric field at oblique incidence is sufficient to provide the pulse compression. Depending on the electric-field phase, the charged sheet moves from or toward the reflected radiation, forming stretched background radiation or prominent spikes. Indeed, both the features are observed in the reflected radiation in Fig. 7. Consequently, the self-induced relativistic motion of charges in the $\lambda^3$ regime directly generates isolated attosecond pulses by reflection, deflection, and compression.To scale this mechanism of attosecond pulse generation to higher intensities, it is necessary to keep similar slopes in electron density profiles. To do so, the plasma density should be scaled linearly with the driving field strength, maintaining the same balance between charge separation forces and forces arising from the incident fields. With higher field strength, electrons also move faster and the resulting Doppler compression leads to the formation of still shorter pulses.

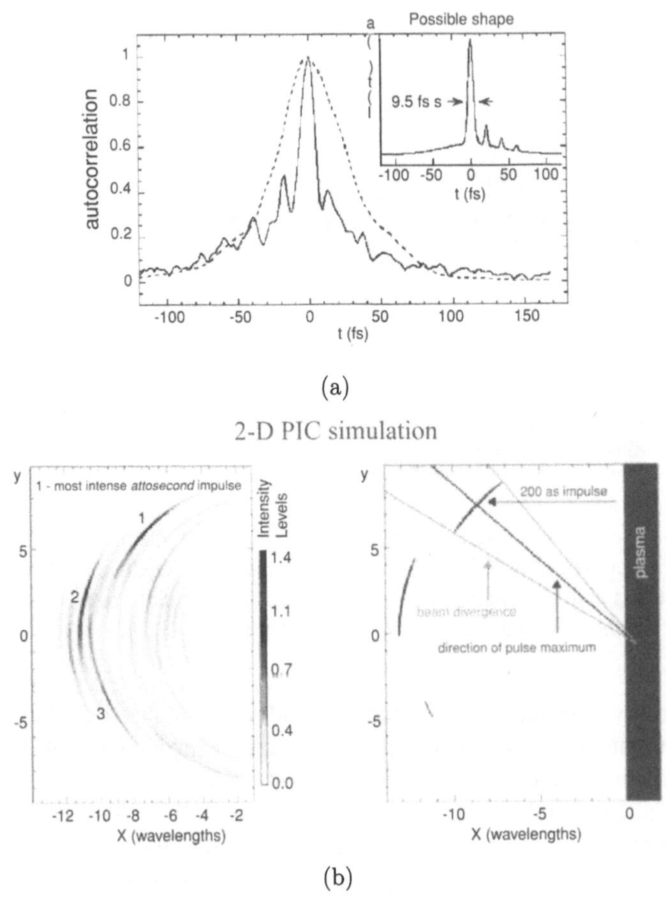


FIGURE 7. A $\lambda^3$ relativistic intensity laser with a $a_0 = 10$ amplitude is focused on a slanted target, and the interaction with the target traps some electrons that will contribute to compress the light cycle. As a result, attosecond X-ray pulses are produced in synchronism with the attosecond pulses of electrons. In the future, this technique will take advantage of the phase stabilization technique that has been recently developed [74].

Using 2D PIC simulations, it is found that the ratio of $a_0/n_0 \sim 2$, where $n_0 = n/n_{cr}$ is optimal for the production of isolated attosecond pulses. Boosted-frame 1D PIC simulations are capable of modeling oblique incidence only over a limited range of amplitude, because reflected light is bound to propagate only in the specular direction. Nevertheless, we find that attosecond pulse duration produced follows the dependence $T_{pulse} = 600\ as/a_0$ for the ratio $a_0/n_0 = 2$ in the range $3 < a_0 < 20$ (Fig. 8).

Similar results with 2D plasmas were obtained by the group of A. Pukhov [75]. The advantage of the $\lambda^3$ geometry is that it provides isolated attosecond pulses with much less input energy or higher $a_0$.

$a_0=10$, $\tau=15$ fs, f/1, $n_0=25\,n_{cr}$

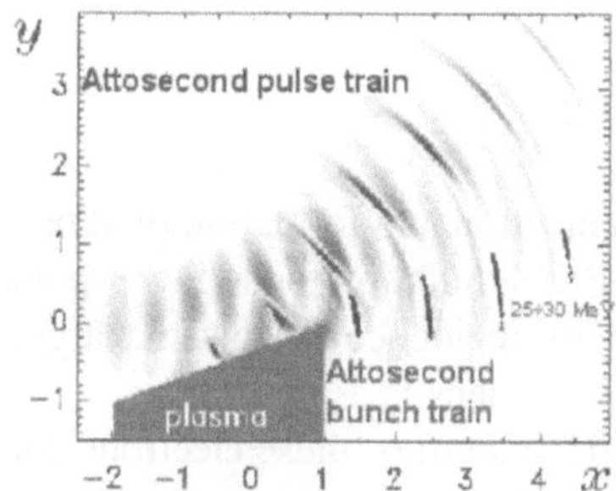


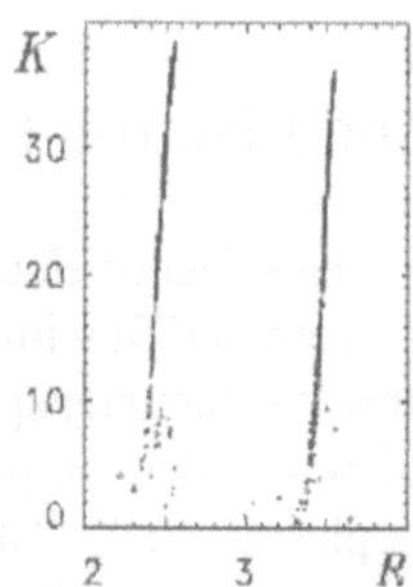

FIGURE 8. There is the hope that by increasing the intensity and keeping $a_0/n_0 \sim 2$ much shorter and efficient pulses could be generated $\tau \sim 600\ as/a_0$. This could lead to intensities close to the pair generation threshold with relatively compact setup.

## 6 The Ultra-relativistic Regime: Self-induced Focusing of Light in Reflection

In a simple view, attosecond electromagnetic pulses, emerging from a target which is curved by the laser pulse pressure, can be simultaneously focused to a much smaller volume, providing higher intensity in focus. To demonstrate this potential effect, preliminary 2D simulations were performed for the above-stated laser pulse parameters and for the steplike plasma profile.

Evaluation of the maximal value of the electromagnetic energy density in time leads to the optimal density value $n_0 = 4n_{\text{cr}}$, for which the strongest generated attosecond pulse gives the maximal intensity in its focus, after emerging from the target. With variation of the plasma density, the *focusing optics* and the attosecond pulse duration are changed. The former depends mainly on the ratio $a_0/n_0$ and the latter depends on $a_0$ and $a_0/n_0$.

The focal volume of an attosecond electromagnetic pulse generated in reflection from a plasma surface can be very small fractions of the $\lambda^3$ volume, with a commensurate increase of focal intensity. Indeed, this effect of self-induced focusing of light is similar to the self-focusing observed in under-dense plasma, when the laser pulse changes the refractive index and converges the wavefront producing higher intensity.

## 7 Pair Production [73]

If the self-focusing by reflection, discussed previously, is realized experimentally, it offers the potential for the generation of extreme fields. Under the influence of intensities available today, using current laser technology the potential exists to efficiently generate pulses with several-attosecond duration and focus them to a few nanometers. The resulting intense fields should be strong enough to extract significant numbers of electron–positron pairs from vacuum as seen in the Fig. 1.

## 8 Attosecond Electron Pulses

We demonstrate the effect of electron ejection and formation of attosecond electron bunches in vacuum [75] with 2D PIC simulation for a 15-fs p-polarized laser pulse focused to a 1-$\lambda$ diameter spot at 70° incidence on a plasma target of density of $25n_{\rm cr}$. The dimensionless amplitude in focus is $a_0 = 10$.

Escaping the plasma with relativistic velocities, these electrons compress the reflected radiation and form attosecond electromagnetic pulses while maintaining their own attosecond bunched structure (Fig. 7a). The trains of attosecond electromagnetic pulses and electron bunches are mutually synchronized, having been formed cooperatively. Estimating the number of particles in the strongest bunches, we find that they contain $10^8$ electrons for these laser–plasma parameters. The energy distribution in each bunch is chirped (Fig. 7b), and above 10 MeV the bunch thickness is $\lambda/7$. As the maximal energy in the bunch depends on the maximal amplitude in the corresponding cycle, by shaping the laser pulse it is possible to generate a single attosecond electron bunch.

## 9 Generation of Efficient $\gamma$-ray by Coherent Thomson Scattering [74]

The dense attosecond electron bunches have the ability to scatter counterpropagating electromagnetic radiation with a frequency up-shift of $4\gamma^2$. Analysis of the influence of the bunch thickness $d$ on the scattering efficiency with respect to the wave vector $k$ leads to the conclusion that near unity conversion efficiency may be obtained under the condition $kd = 1/4\gamma^2$, provided that a sufficient number of electrons is included in the bunch. The chirped energy distribution observed in simulations might be exploited to obtain high scattering efficiency from a lower charge density. Implementing this scattering would be a means of generating extremely bright attosecond X-ray pulses.

## 10 Conclusions

If the hallmark of bound electron nonlinear optics was the generation of radiation with eV energy, today the hallmark of relativistic optics is the generation of radiations and particles with characteristic energy in the MeV–GeV range. The energy will increase as the laser peak power will, to reach 100 GeV to 1 TeV levels. Relativistic optics provides the basis for extremely compact particle/radiation devices through incoherent/coherent scattering. The unexpected benefit is that as the intensity increases the pulses, duration decreases from the femtosecond to the attosecond and possibly to the zeptosecond range. This represents the onset of a new kind of technology that we could call relativistic microelectronic/photonic engineering that could replace RF-based devices today. Relativistic optics combined with the latest refinement in pulse shortening and carrier phase stabilization [76] could provide the technological leap necessary to reach intensities from compact lasers thought to be unreachable where we could observe vacuum polarization, pairs generation, and nonlinear QED.

## References

1. P. Agostini, G. Barjot, J. F. Bonnal, G. Mainfray, and C. Manus: IEEE J. Quantum. Electron. QE-4, 667 (1968).
2. C. J. Joshi, and P. B. Corkum: Phys. Today 48, 36 (1995).
3. Popov, V. S., V. D. Mur, and B. M. Karnakov: JETP Lett. 66, 229 (1997).
4. C. J. Joachain, N. J. Kylstra, and R. M. Potvliege: In Proceedings of the 2nd International Conference on Superstrong Fields in Plasmas, Varenna (2002), 2001. M. Lontano, G. Mourou, O. Svelto, and T. Tajima (eds). (Am. Inst. of Physics, New York, 2002), AIP Conf. Proc., 611, 9.
5. C. H. Keitel: Contemp. Phys. 42, 353 (2001).
6. A. Maquet, and R. Grobe: J. Mod. Opt. 49, 2001 (2002).
7. G. R. Mocken, and C. H. Keitel: Phys. Rev. Lett. 91, 173202 (2003).
8. J. C. Kieffer, J. P. Matte, H. Pepin, et al.: Phys. Rev. Lett. 68, 480 (1992).
9. J. D. Kmetec, C. L. Gordon III, J. J. Macklin, et al.: Phys. Rev. Lett. 68, 1527 (1992).
10. F. Beg, A. R. Bell, A. E. Dangor, et al.: Phys. Plasmas 4, 447 (1997).
11. P. A. Norreys, M. Santala, E. Clark, et al.: Phys. Plasmas 6, 2150 (1999).
12. C. Max, J. Arons, and B. Langdon: Phys. Rev. Lett. **33**, 209 (1974).
13. P. Sprangle, C. M. Tang, and E. Esarey: IEEE Trans. Plasma Sci. PS-15, 145 (1987).
14. A. B. Borisov, A. V. Borovskiy, O. B. Shiryaev, et al.: Phys. Rev. A **45**, 5830 (1992).
15. P. Monot, T. Auguste, P. Gibbon, F. Jakober, and G. Mainfray: Phys. Rev. Lett. 74, 2953 (1995).
16. P. Gibbon, P. Monot, T. August, and G. Mainfray: Phys. Plasmas 2, 1304 (1995).
17. S. -Y. Chen, G. S. Sarkisov, A. Maksimchuk, et al.: Phys. Rev. Lett. 80, 2610 (1998).
18. J. Fuchs, G. Malka, J. C. Adam, et al.: Phys. Rev. Lett. 80, 1658 (1998).
19. S. V. Bulanov, N. M. Naumova, and F. Pegoraro: Interaction of an ultrashort, relativistically strong laser-pulse with an overdense plasma. Phys. Plasmas **1**, 745–757 (1994).
20. R. Lichters, J. Meyerter Vehn, and A. Pukhov: Short-pulse laser harmonics from oscillating plasma surfaces driven at relativistic intensity. Phy. Plasmas **3**, 3425–3437 (1996).
21. D. Von der Linde: AIP Conf. Proc. 426, 221 (1997).
22. A. Tarasevitch, A. Orisch, D. von der Linde, et al.: Phys. Rev. E 62, 023816 (2000).
23. C. E. Clayton, C. Joshi, C. Darrow, and D. Umstadter: Phys. Rev. Lett. 54, 2343 (1985).
24. A. Modena, Z. Najimudin, A. E. Dangor, et al.: Nature 337, 606 (1995).
25. K. Nakajima, D. Fisher, T. Kawakubo, et al.: Phys. Rev. Lett. 74, 4428 (1995).
26. D. Umstadter, J. K. Kim, and E. Dodd: Phys. Rev. Lett. 76, 2073 (1996).

27. R. Wagner, S. -Y. Chen, A. Maksimchuk, and D. Umstadter: Phys. Rev. Lett. 78, 3125 (1997).
28. D. Gordon, K. C. Tzeng, C. E. Clayton, et al.: Phys. Rev. Lett. 80, 2133 (1998).
29. S.-Y. Chen, A. Maksimchuk, E. Esarey, and D. Umstadter: Phys. Rev. Lett. 84, 5528 (2000).
30. R. Bingham, J. T. Mendonca, and P. K. Shukla: Plasma Phys. and Controll. Fus. 46, R1 (2004).
31. S. Y. Tochitsky, R. Narang, C. V. Filip, et al.: Phys. Rev. Lett. **92**, 095004 (2004).
32. S. P. D. Mangles, C. D. Murphy, Z. Najmudin, et al.: Nature 431, 535 (2004).
33. C. G. R. Geddes, C. Toth, J. van Tilborg, et al.: Nature 431, 538 (2004).
34. J. Faure, Y. Glinec, A. Pukhov, et al.: Nature 431, 541 (2004).
35. K. Krushelnick, E. L. Clark, Z. Najmudin, et al.: Phys. Rev. Lett. **83**, 737–740 (1999).
36. M., Dangor, A. E., Malka, V., Neely, D., Allott, R., and C. Danson, et al.: Phys. Rev. Lett. 83, 737 (1999).
37. G. S. Sarkisov, V. Yu, Bychenkov, V. N. Novikov, et al.: Phys. Rev. E **59**, 7042 (1999).
38. A. Zhidkov, A. Sasaki, and T. Tajima: Phys. Rev. E 61, R2224 (1999).
39. T. Zh. Esirkepov, F. Sentoku, F. Califano, et al.: JETP Lett. 70, 82 (1999).
40. S. V. Bulanov, T. Zh. Esirkepov, F. Califano, et al.: JETP Lett. **71**, 407 (2000).
41. A. Maksimchuk, S. Gu, K. Flippo, D. Umstadter, et al.: Phys. Rev. Lett. 84, 4108 (2000).
42. R. Snavely, M.H. Key, S.P. Hatchett, et al.: Phys. Rev. Lett. 85, 2945 (2000).
43. D. Umstadter: J. Phy. D: Appl. Phys. 36, R151 (2003).
44. R. Bingham, J. T. Mendoncza, and P. K. Shukla: Plasma Phys. and Controll. Fus. 46, R1 (2004).
45. A. Maksimchuk, S. Gu, K. Flippo, D. Umstadter, et al.: Phys. Rev. Lett. 84, 4108 (2004).
46. G. Pretzler, A. Saemann, A. Pukhov, A. et al.: Phys. Rev. E 58, 1165 (1998).
47. L. Disdier, J. P. Garconnet, G. Malka, and J. L. Miquel: Phys. Rev. Lett. 82, 1454 (1999).
48. C. Gahn, G. D. Tsakiris, G. Pretzler, et al: Appl. Phys. Lett. **77**, 2662 (2000).
49. C. Bula, K. T. McDonald, E. I. Prebys, et al.: Phys. Rev. Lett. 76, 3116 (1996).
50. D. L. Burke, R. C. Field, G. Horton-Smith, et al.: Phys. Rev. Lett. 79, 1626 (1997).
51. A. P. Joglekar, H. Liu, G. J. Spooner, E. Meyhofer, G. Mourou, and A. J. Hunt: Appl. Phys. B 77, 25 (2003).
52. L. D. Landau, and E. M. Lifshitz: *The Classical Theory of Fields*. Oxford Pergamon Press. (1980).
53. T. Tajima, and J. M. Dawson: Phys. Rev. Lett. 43,267 (1979).
54. L. M. Gorbunov, and V. I. Kirsanov: JETP 93, 509 (1987).
55. P. Sprangle, E. Esarey, A. Ting, and G. Joyce: Appl. Phys. Lett. 53, 2146 (1988).
56. S. V. Bulanov, V. I. Kirsanov, and A. S. Sakharov: JETP Lett. 50, 176 (1989).
57. V. I. Berezhiani, and I. G. Murusidze: Phys. Lett. A 148, 338 (1990).
58. L. Schaechter: Phys. Rev. Lett. 83, 92 (1999).
59. L. Schaechter, R. L. Byer, and R. H. Siemann: In Advanced Accelerator Concepts: Tenth Workshop. AIP Conference Proceedings, Vol. 647, p. 310 (2002). (American Institute of Physics Conference Series).
60. V. V. Apollonov, A. I. Artem'ev, Y. L. Kalachev, A. M. Prokhorov, and M. V. Fedorov: JETP Lett. 47, 91 (1998).
61. A. Pukhov and J. Meyer-ter-Vehn: Appl. Phys. B, 74, 355 (2002).
62. A. Rousse, K. Ta Phuoc, R. Shah, et al.: Phys. Rev. Lett. 93, 135005 (2004).

63. A. Rousse, C. Rischel, S. Fourmaux, et al.: Nature (London) **410**, 65 (2001).
64. K. Sokolovski-Tinten, C. Blome, J. Blums, et al.: Nature (London) **422**, 287 (2003).
65. M. Manclossi, J. J. Santos, D. Batani, et al.: Phys. Rev. Lett. **96**, 125002 (2006).
66. G. Shvets, N. J. Fish, A. Pukhov, et al.: Phys. Rev. Lett. 81, 4879 (1998).
67. W. B. Mori: Phys. Rev. A 44, 5118 (1991).
68. S. V. Bulanov, T. Esirkepov, and T. Tajima: Phys. Rev. Lett. 91, 85001 (2003).
69. M. Dreher, E. Takahashi, J. Meyer-ter-Vehn, et al.: Phys. Rev. Lett. 93, 95001 (2004).
70. S. V. Bulanov, T. Esirkepov, and T. Tajima: Phys. Rev. Lett. 91, 85001 (2003).
71. F. S. Tsung, C. Ren, L. O. Silva, et al.: Proc. Nat. Acad. Sci. USA 99, 29 (2002).
72. N. M. Naumova, J. A. Nees, I. V. Sokolov, B. Hou, and G. A. Mourou: Phys. Rev. Lett. 92, 063902 (2004).
73. N. Naumova, I. Sokolov, J. Nees, A. Maksimchuk, V. Yanovsky, and G. Mourou: Phys. Rev. Lett. **93**, 195003 (2004).
74. N. Naumova, J. A. Nees, and G. A. Mourou: Phys. Of Plasmas 12, 056707 (2005).
75. P. Gordienko, and B. Shorokhov: Phys. Rev. Lett. 93, 115002 (2004); and Phys. Rev. Lett. 94, 103903 (2005).
76. A. Baltuska, M. Uiberacker, E. Goulielmakis, et al.: IEEE J. Sel. Top. Quantum Electron. **9**, 972–989 (2003).

# Electron Acceleration Under Strong Radiation Damping

James Koga,[1] Sergei Bulanov,[1,2] and Timur Esirkepov[1,3]

[1]Japan Atomic Energy Agency, 8-1 Umemidai, Kizu, Souraku, Kyoto 619-0215, Japan
koga.james@jaea.go.jp
[2]A. M. Prokhorov Institute of General Physics of the Russian Academy of Sciences, Vavilov Street 38, 119991 Moscow, Russia
bulanov.sergei@jaea.go.jp
[3]Moscow Institute of Physics and Technology, Institutskiy pereulok 9, 141700 Dolgoprudny, Moscow Region, Russia
timur.esirkepov@jaea.go.jp

## 1 Introduction

With short-pulse high-power lasers, it is possible via strong focusing to achieve irradiance levels up to $10^{22}\ \mathrm{W \cdot cm^{-2}}$ [1], and with peta-watt level lasers much higher irradiances [2]. At these irradiance levels, the effect of radiation damping on the electron motion in the intense wave can become large [3]. Even with current lasers irradiances reaching the Schwinger limit, $10^{29}\ \mathrm{W \cdot cm^{-2}}$ could be achieved by using counterpropagating laser pulses [4] and ultra-short pulses of relativistically strong electromagnetic waves reflected back at a plasma-vacuum interface [5]. In addition, through coherent radiation effects the damping could be strong even for relatively low irradiance laser pulses interacting with clusters [6]. In this paper, we numerically examine the effect of the interaction of an extremely high irradiance pulse with a high-energy electron.

## 2 Results

We calculated the interaction of a single 150-MeV electron counterstreaming with a linearly polarized gaussian plane wave laser with a pulse duration of 30-fs and wavelength $\lambda_0$ equal to 0.8 μm using various irradiances. The gaussian laser pulse is of the form

$$\boldsymbol{E}(\boldsymbol{x}, t) = \mathbf{e_z} E_0 h(\phi) \sin(\phi) \tag{1}$$
$$\boldsymbol{B}(\boldsymbol{x}, t) = -\mathbf{e_y} E_0 h(\phi) \sin(\phi), \tag{2}$$

where $\boldsymbol{E}$ is the electric field, $\boldsymbol{B}$ is the magnetic field, $\mathbf{e_z}$ is the unit vector in the $z$

direction, $\mathbf{e}_y$ is the unit vector in the $y$ direction, $\phi = \omega_0(t - x/c)$, $\omega_0$ is the laser frequency,

$$h(\phi) = \exp\left[-\left(\frac{\phi}{\omega_0 \tau}\right)^2\right], \tag{3}$$

and $\tau$ is the laser pulse duration. We advance the electron using the classical equations of motion including radiation damping [7]:

$$\frac{d\gamma\boldsymbol{\beta}}{dt} = \frac{e}{mc}(\boldsymbol{E} + \boldsymbol{\beta} \times \boldsymbol{B}) + \frac{2e^2}{3mc^2}\boldsymbol{g}_0, \tag{4}$$

and the expression for the spatial part of the damping force is [7]

$$\begin{aligned} \boldsymbol{g}_0 = & \frac{e}{mc^2}\gamma\left(\frac{\partial}{\partial t} + \boldsymbol{v}\cdot\boldsymbol{\nabla}\right)(\boldsymbol{E} + \boldsymbol{\beta} \times \boldsymbol{B}) \\ & + \left(\frac{e}{mc^2}\right)^2 c\left[(\boldsymbol{\beta}\cdot\boldsymbol{E})\boldsymbol{E} + (\boldsymbol{E} + \boldsymbol{\beta}\times\boldsymbol{B})\times\boldsymbol{B}\right] \\ & - \left(\frac{e}{mc^2}\right)^2 \gamma^2 c\boldsymbol{\beta}\left[(\boldsymbol{E} + \boldsymbol{\beta}\times\boldsymbol{B})^2 - (\boldsymbol{\beta}\cdot\boldsymbol{E})^2\right], \end{aligned} \tag{5}$$

where $\boldsymbol{\beta} = \boldsymbol{v}/c$, $\boldsymbol{v}$ is the velocity of the electron, and $\gamma = 1/\sqrt{1-\beta^2}$. These equations have been shown to be valid in the classical regime when $\lambda_0' > \lambda_C$, where $\lambda_0'$ is the wavelength of the laser in the electron's rest frame and $\lambda_C$ is the Compton wavelength $\lambda_C = h/m_e c$ [8]. The equations are numerically integrated using an adaptive Runge-Kutte scheme [9]. Figure 1 shows the percentage of energy loss for the electron interacting with laser pulses of various amplitudes, where $a_0$ is the normalized laser amplitude $eE_0/mc\omega_0$ and $E_0$ is the peak laser amplitude. The dots represent the numerical results and solid line indicates the theoretical prediction of the energy loss using the largest term in the damping equation [3,7]. They are in good agreement. We can see that for large amplitudes,

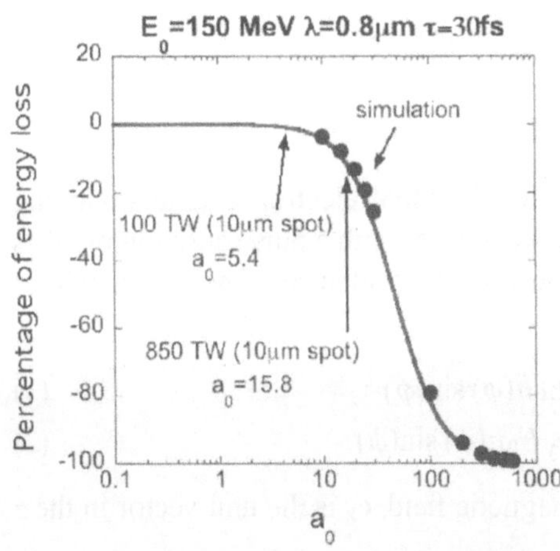


FIGURE 1. Electron energy loss as a function of the normalized laser amplitude $a_0$.

FIGURE 2. Trace of the entire electron energy trajectory represented by the relativistic factor $\gamma$ as it interacts with the laser which is propagating in the +x direction where x is normalized by the laser wavelength $\lambda_0$. The insert shows an expanded region showing the initial interaction of the electron with the laser pulse.

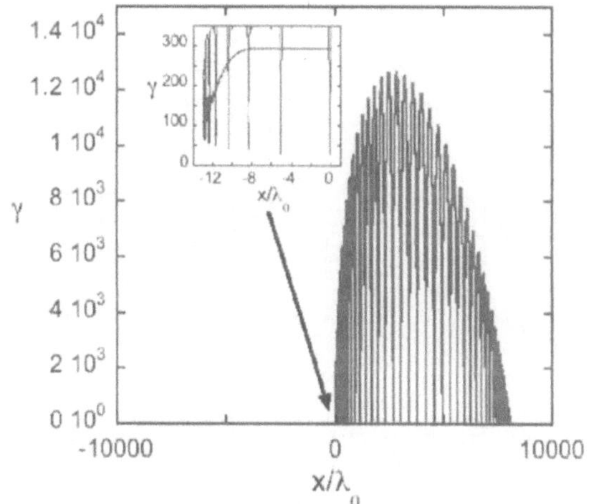


the electron looses nearly all of it's energy after the interaction. Figure 2 shows the electron energy trajectory as a function of propagation distance for a laser amplitude of $a_0 = 500$ or irradiance $5 \times 10^{23}\ \mathrm{W} \cdot \mathrm{cm}^{-2}$. The insert in the figure is an expanded region showing the initial interaction of the electron with the laser pulse. The laser is propagating from the left to the right, and the electron is initially propagating from the right to the left. During the interaction of the electron with the laser pulse, the electron's energy goes to nearly zero and then propagates in the direction of the laser pulse gaining energy. From the figure, it can be seen that the electron turns around in a distance on the order of $4\lambda_0$, which corresponds to a real distance of 3.2 μm. From the figure it can be seen that the electron's peak energy in the laser pulse exceeds 6 GeV and the final energy is about 1.5 MeV. The peak energy of the electron is less than the theoretical maximum in a plane wave of $a_0^2/2$ or about 64 GeV for an initially stationary electron. This may be due to the radiation damping. In addition, in a real focusing and defocusing laser pulse the final energy would be different. The spot size of the focus spot would also be a consideration.

We calculate the radiation scattered from the interaction between the laser pulse and electron using [10]

$$(\boldsymbol{E}_z)_{\mathrm{scat}} \propto \left[ \frac{\hat{n} \times [(\hat{n} - \boldsymbol{\beta}) \times \dot{\boldsymbol{\beta}}]}{(1 - \boldsymbol{\beta} \cdot \hat{n})^3} \right]_{\mathrm{ret}}, \tag{6}$$

where $(\boldsymbol{E}_z)_{\mathrm{scat}}$ is the scattered electric field in the polarization direction of the laser pulse, $\hat{n}$ is the unit normal vector in the direction of radiation emission, $\boldsymbol{\beta} = \boldsymbol{v}/c$, $\dot{\boldsymbol{\beta}} = \dot{\boldsymbol{v}}/c$, ret refers to the retarded time, $t'$ evaluated at $t' + R(t')/c = t$, and $R(t')$ is the distance of the electron from the observation point.

The radiation scattered in the initial propagation of the electron from the laser–electron interaction is shown in Fig. 3. The dotted line is for the case with no damping included and the solid line is with damping. The time is normalized

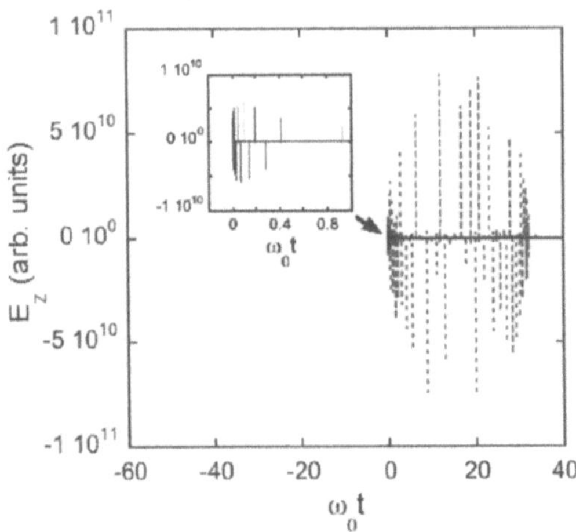


FIGURE 3. Trace of the scattered electric field from the electron laser interaction with (solid line) and without (dotted line) damping. The insert shows a close-up of the scattered radiation in the damping case only.

to the laser frequency $\omega_0$. The laser pulse length would be about $\omega_0\tau = 71$. As can be seen in the figure, with no damping the overall length of the scattered radiation is longer than in the case with damping. This is most likely due to the rapid loss of electron energy during the interaction and the fact that the electron is turned around. In the case with no damping, the electron continues to propagate in the initial propagation direction and looses no energy after the interaction. The insert in Fig. 3 shows a closeup of the scattered radiation in the damping case only. The overall length of the scattered radiation pulse is about 0.3 to 0.4 $\omega_0\tau$. This about a factor of 200 times smaller that the laser pulse giving a width of about 150 as. Figure 4 shows a expanded plot of the largest spike. The arrows in the figure indicate that the width of the spike is less than $2 \times 10^{-7}\omega_0\tau$ or about a 85-yoctosecond pulse (yocto $\equiv 10^{-24}$). In this regime by the uncertainty principle, the corresponding photon energy is about 49 MeV or very hard gammas.

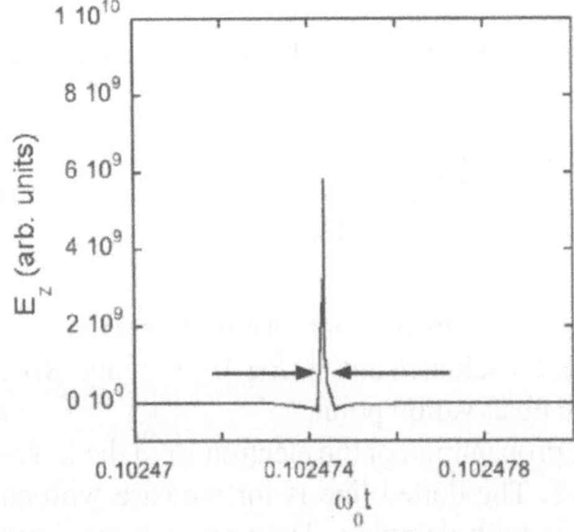


FIGURE 4. Expanded trace of the scattered electric field from the electron–laser interaction with radiation damping.

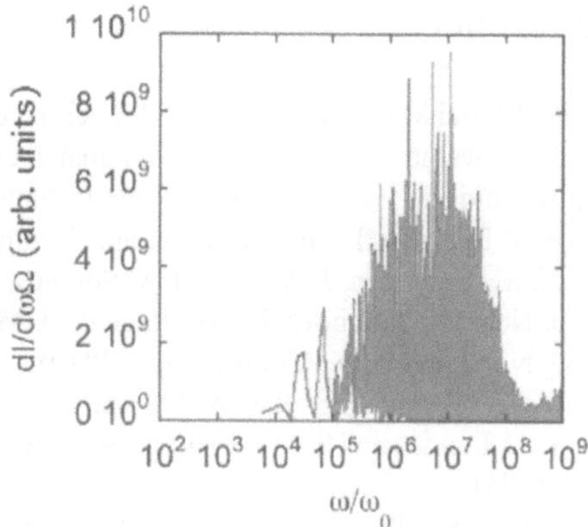


FIGURE 5. Power spectrum for $a_0 = 500$ for electron motion with damping for a 150-MeV electron interacting with a 30-fs $a_0 = 500$ laser pulse.

We have calculated the power spectra resulting from the electron–laser interaction using [10]

$$\frac{d^2 I}{d\omega d\Omega} = \frac{e^2}{4\pi^2 c}\left|\int_{-\infty}^{\infty}\left[\frac{\hat{n}\times[(\hat{n}-\boldsymbol{\beta})\times\dot{\boldsymbol{\beta}}]}{(1-\boldsymbol{\beta}\cdot\hat{n})^3}\right]_{\text{ret}} e^{i\omega t}dt\right|^2. \tag{7}$$

Since the points used to calculate the spectra are unevenly distributed in time, we use the method developed to calculate the spectra for unevenly sampled data [11–13]. Figure 5 shows the spectra of the damping case. The position of the peak in the spectrum is about $10^7\omega/\omega_0$ corresponding to photon energies on the order of 15 MeV. The spectrum is broad ranging over about three orders of magnitude. At lower laser amplitudes, the peak occurs at lower frequencies [14].

## 3 Conclusion

We have calculated the interaction of an ultrahigh irradiance laser pulse counterstreaming with a single 150-MeV electron. The numerical calculations show that at large laser amplitudes ($a_0 = 500$), the electron is turned around on distances of the order of four laser wavelengths. The electron is accelerated in the propagation direction of the laser pulse up to energies of the order of 6 GeV. Examination of the backscattered radiation field shows an overall pulse width of about 150 as with spikes on the order of 85 yocto-seconds. The shortness of these spikes indicates an energy of 49 MeV from the uncertainty principle. Power spectra of the scattered radiation shows a peak at about 15 MeV with a broadness spanning three orders of magnitude. In future work, the collective behavior of multiple electrons will be examined including a more realistic focusing laser pulse. Future work must also include quantum mechanical calculations of such an interaction.

## References

1. S.-W. Bahk, P. Rousseau, T. A. Planchon, et al.: Opt. Lett. **29**, 2837 (2004).
2. M. Aoyama, K. Yamakawa, Y. Akahane, et al.: Opt. Lett. **28**, 1594 (2003).
3. A. Zhidkov, J. Koga, A. Sasaki, et al.: Phys. Rev. Lett. **88**, 185002 (2002).
4. S. V. Bulanov, T. Zh. Esirkepov, and T. Tajima: Phys. Rev. Lett. **91**, 085001 (2003).
5. N. M. Naumova, J. A. Nees, I. V. Sokolov, et al.: Phys. Rev. Lett. **92**, 063902 (2004); J. Nees, N. Naumova, E. Power, et al.: J. Mod. Opt. **52**, 305 (2005); N. M. Naumova, J. Nees, and G. A. Mourou, Phys. Plasmas **12**, 056707 (2005).
6. S. V. Bulanov, T. Esirkepov, J. Koga, et al.: Plasma Phys. Rep. **30**, 196 (2004).
7. L. D. Landau and E. M. Lifshitz: *The Classical Theory of Fields*. New York: Pergamon. (1994).
8. J. K. Koga: Phys. Rev. E **70**, 046502 (2004).
9. W. H. Press, S. A. Teukolsky, W. T. Vetterling, et al.: *Numerical Recipes in FORTRAN*. Cambridge: Cambridge University Press. (1992).
10. J. D. Jackson: *Classical electrodynamics*. New York: Wiley. (1999).
11. N. R. Lomb: Astrophys. Space Sci. **39**, 447 (1976).
12. W. H. Press and G. B. Rybicki: Astrophys. J. **338**, 277 (1989).
13. W. H. Press, S. A. Teukolsky, W. T. Vetterling, et al.: *Numerical Recipes in FORTRAN*. Cambridge: Cambridge University Press. (1992).
14. J. Koga, T. Zh. Esirkepov, and S. V. Bulanov: Phys. Plasmas. **12**, 093106 (2005).

# Monoenergetic Electron Acceleration in a High-density Plasma Produced by an Intense Laser Pulse

E. Miura,[1] S. Masuda,[2] T. Watanabe,[3] K. Koyama,[1] S. Kato,[1] M. Adachi,[4] N. Saito,[1] and M. Tanimoto[5]

[1] National Institute of Advanced Industrial Science and Technology (AIST), Tsukuba Central 2, 1-1-1 Umezono, Tsukuba, Ibaraki 305-8568, Japan
e-miura@aist.go.jp
[2] National Institute of Radiological Sciences, 4-9-1 Anagawa, Inage, Chiba, Chiba 263-8555, Japan
[3] Utsunomiya University, 7-1-2 Yoto, Utsunomiya, Tochigi 321-8585, Japan
[4] Hiroshima University, 1-3-1 Kagamiyama, Higashi-Hiroshima, Hiroshima 739-8530, Japan
[5] Meisei University, 2-1-1 Hodokubo, Hino, Tokyo 191-8506, Japan

**Abstract:** A monoenergetic electron beam with an energy of 23 MeV has been obtained by the laser-driven plasma acceleration from a plasma with an electron density of $4 \times 10^{19}$ cm$^{-3}$ produced by a 3-TW, 50-fs, 800-nm laser pulse. The normalized amplitude of the plasma wave was estimated to be 0.7 from the forward-scattered light spectrum. The interaction length of the laser pulse with the plasma was measured to be 300–500 μm from the side-scattered light image. The plasma diagnostics based on the optical measurements suggest that the monoenergetic beam generation is due to the matching of the acceleration length to the dephasing length over which accelerated electrons outrun the acceleration phase in the plasma wave and begin to be decelerated.

## 1 Introduction

Particle acceleration via the interaction of an intense laser pulse with a plasma, i.e., laser-driven plasma acceleration, has been intensively studied to realize an advanced compact accelerator [1] since the fundamental concept was proposed [2]. The generation of electron beams with the maximum energy of up to several hundred MeV has been demonstrated [3, 4]. However, the energy spectra of accelerated electrons have so far been Maxwellian-like or power-law distributions, and the energy spreads have been nearly 100%. Therefore, the key issue in realizing a compact accelerator based on laser-driven plasma acceleration has been the generation of a monoenergetic beam. Recently, several groups have reported the monoenergetic electron beam generation [5–9]. This is a big step toward the realization of a laser-driven plasma accelerator. We have so far reported the generation of a monoenergetic electron beam with an energy of 7 MeV from a high-density plasma with the electron density of the order of $10^{20}$ cm$^{-3}$ produced by a 2-TW, 50-fs, 800-nm laser pulse [8]. For various applications, more energetic and intense beams are required. It is also necessary to achieve more stable generation of monoenergetic beams with the required energy.

The maximum electron energy $W_{max}$ in laser-driven plasma acceleration is given by [2]

$$W_{max} = 2\gamma_{\phi}^2 m_e c^2 \varepsilon, \quad (1)$$

where $\gamma_{\phi}$, $m_e$, $c$, and $\varepsilon$ are the Lorentz factor for the phase velocity of the plasma wave, the electron mass, the speed of light, and the normalized amplitude of the plasma wave, respectively. The Lorentz factor $\gamma_{\phi}$ is given by $\gamma_{\phi} \sim (n_c/n_e)^{1/2}$, where $n_c$ and $n_e$ are the critical density and the electron density. Then, the maximum energy is proportional to $n_e^{-1}$. To get a more energetic electron beam, a lower-density plasma should be used. Here, it is noted that the maximum energy given by Eq. (1) can be obtained, where the acceleration length reaches the dephasing length. The dephasing length is defined as the length over which accelerated electrons outrun the acceleration phase in the plasma wave and begin to be decelerated. The dephasing length $L_d$ is given by $L_d \sim \gamma_{\phi}^2 \lambda_p$ in the linear regime, where $\lambda_p$ is the wavelength of the plasma wave. Then, the dephasing length is proportional to $n_e^{-3/2}$. In a lower-density plasma, it is also necessary to make the interaction length of a laser pulse with a plasma longer. To make the interaction length longer, the formation of a plasma channel produced by several laser pulses [6] or discharge [10] and the use of a hollow capillary [11] have been proposed. Another method is the use of a focusing optics with a longer focal length, which gives the longer Rayleigh length [5, 7].

In this paper, we report experiments to get more energetic electron beam by using a lower-density plasma of the order of $10^{19}$ cm$^{-3}$ and a focusing optics with a longer focal length compared with our previous experiment [8]. A monoenergetic electron beam with an energy of 23 MeV has been obtained from a plasma with an electron density of $4 \times 10^{19}$ cm$^{-3}$. The mechanism of the monoenergetic beam generation is discussed according to the plasma diagnostics based on the optical measurements.

## 2 Experimental Conditions

The experimental setup is shown in Fig. 1. The experiment was performed with a Ti:sapphire laser system which delivered a 200-mJ, 50-fs laser pulse with the center wavelength of 800 nm. The laser beam was focused by an off-axis parabolic mirror with the focal length of 300 mm. The spot diameter was 9 μm in full width at half maximum, which contained 40% of the laser energy. The maximum laser intensity was estimated to be $2.5 \times 10^{18}$ W $\cdot$ cm$^{-2}$ corresponding to the

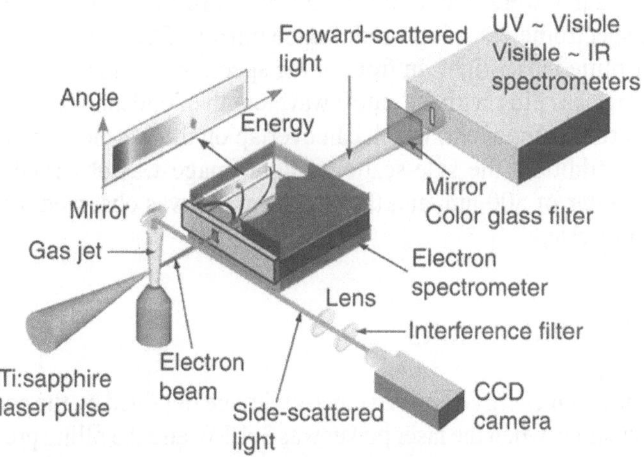


FIGURE 1. Experimental setup.

normalized vector potential $a_0 \sim 1$. The Rayleigh length was estimated to be 110 μm. A He gas jet was ejected from a supersonic nozzle with the Mach number of 5.3 attached to a pulsed gas valve (General valve, Series 9). The nozzle exit diameter was 0.7 mm. The laser beam was focused at 1 mm above the nozzle exit. The density profile of the gas jet was measured with Ar gas by Jamin interferometer, because the refractive index of He was too low. It is expected that an Ar gas jet gives a density profile similar to that of He gas jet, because the specific-heat ratio of Ar is equal to that of He. The atomic density of the gas jet center at 1 mm above the nozzle exit was measured to be $1.6 \times 10^{19}$ cm$^{-3}$, when the filling pressure of the pulsed valve was 50 bar. The electron density was estimated to be $3.2 \times 10^{19}$ cm$^{-3}$ due to the full ionization of He under the laser intensity of our experiment [12]. The gas jet density was proportional to the filling pressure of the pulsed valve. The electron density of the plasma was controlled in the range of $(2.6 - 5.1) \times 10^{19}$ cm$^{-3}$ by varying the filling pressure. The product of the plasma frequency $\omega_p$ and the laser pulse duration $\tau_L$ was between 14 and 20. The experiment was still performed in the self-modulated laser wakefield acceleration regime ($\omega_p \tau_L \gg 2\pi$).

An energy spectrum of accelerated electrons was measured by an electron spectrometer in the range from 0.2 to 30 MeV. The magnetic-field strength was 0.2 T. The electron energy-spectrum was recorded on an imaging plate (IP) (Fuji Film: BAS-SR). To prevent exposure by scattered laser light, the IP was covered with 15-μm-thick Al foil. The sensitivity curve of the IP reported in Ref. [13] was used. A slit was set in front of the magnet. The angular divergence of an electron beam was also obtained.

To investigate the acceleration mechanism, several optical measurements were conducted simultaneously with the electron beam measurement. The forward-scattered light spectrum was observed by two spectrometers. The observed wavelength ranges were 200–850 nm and 650–1100 nm, respectively. In front of both the spectrometers, a high-reflectance mirror of 800 nm was set to attenuate the transmitted laser light. In front of the spectrometer for longer wavelength regime, a long pass filter with the cutoff wavelength of 650 nm was set to attenuate the second-order component, which can overlap on the wavelength range around 1000 nm. In addition, the side-scattered light image was observed through an interference filter of 800 nm, and the shadowgraph was observed using a 50-fs, 800-nm laser pulse.

## 3 Experimental Results

Figure 2a shows an energy-resolved electron image obtained by the accumulation of eight laser shots, when the laser power was 3.4 TW and the filling pressure of the pulsed valve was 50 bar corresponding to the electron density of $3.2 \times 10^{19}$ cm$^{-3}$. Spots observed around 23 MeV indicate the generation of a monoenergetic electron beam with small divergence. Three spots around 23 MeV suggest that a monoenergetic beam generated in three shots among eight shots. The emission directions may be diverged due to the shot-to-shot variation of the pointing of the focused beam. Figure 2b shows the electron energy spectrum obtained as the average of eight shots. The monoenergetic electron beam with an energy of 23 MeV was seen. The number of electrons of the monoenergetic beam was estimated to $10^6$ per shot. This value can be the lower limit, because the estimated electron number was the average of eight shots. The observed energy spread of

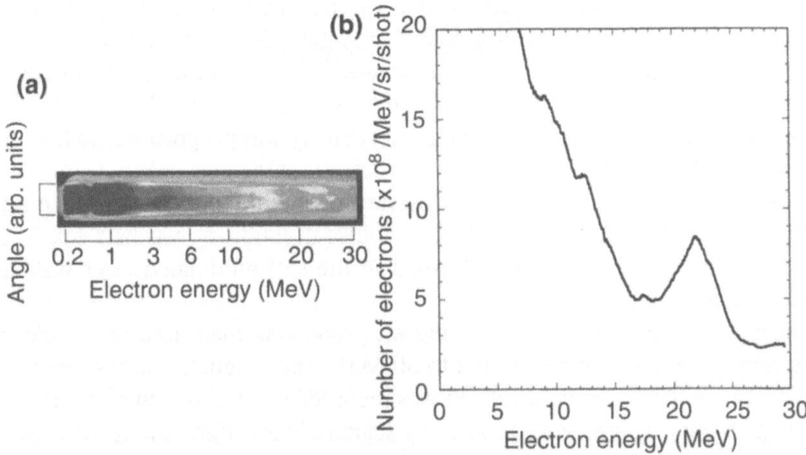


FIGURE 2. (a) Energy-resolved electron image obtained by the accumulation of eight laser shots. Three spots around 23 MeV indicate the generation of a monoenergetic electron beam with small divergence. (b) Electron energy spectrum obtained as the average of eight shots. The laser power was 3.4 TW, and the filling pressure of the pulsed valve was 50 bar corresponding to the electron density of $3.2 \times 10^{19}$ cm$^{-3}$.

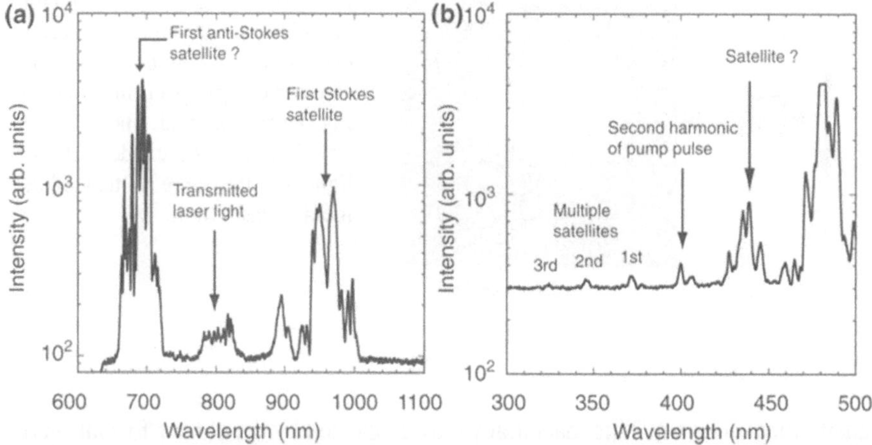


FIGURE 3. Forward-scattered light spectra in the wavelength ranges of (a) 650–1100 nm (b) and 300–500 nm, when the electron energy spectrum shown in Fig. 2 was obtained. Both spectra were obtained at the same shot. In the long wavelength region, the first Stokes satellite of the pump pulse caused by stimulated forward Raman scattering was observed. In the short wavelength range, the multiple satellites of the second harmonic of the pump pulse caused by the collective Thomson scattering by the plasma wave were observed.

4 MeV was limited by the resolution of the electron spectrometer. The angular divergence of the monoenergetic beam was estimated to be $\pm 0.4°$ from the vertical size of the spots seen in Fig. 2a. The normalized emittance $\varepsilon_n$ is given by $\varepsilon_n = \gamma\sigma\delta\theta$, where $\gamma$, $\sigma$, and $\delta\theta$ are the relativistic Lorentz factor for the electron energy, the radius of the electron source, and the angular divergence at the half width at half maximum, respectively. Assuming that the electron source radius is 4.5 μm which is the laser spot radius, the normalized emittance is estimated to be $1.5\pi$ mrad. The normalized emittance is comparable to those of present linear accelerators. The divergence of the 23-MeV beam was much smaller than that of the 7-MeV beam obtained in our previous experiments [8]. The beam divergence for the higher-energy electrons can be smaller, because the relative transverse momentum becomes smaller due to the longer acceleration length.

Figures 3a and b show typical forward-scattered light spectra observed at the same shot, when the electron energy spectrum shown in Fig. 2 was obtained. As seen in Fig. 3a, the first Stokes satellite of stimulated forward Raman scattering was observed around 950 nm. This is the evidence for the excitation of the plasma wave. The first Stokes satellite was strongly modulated. The modulation may be due to the relativistic cross-phase modulation between the first Stokes satellite and the incident laser light in a plasma [14]. The electron density was estimated to be $4 \times 10^{19}$ cm$^{-3}$ from the wavelength of the Stokes satellite. The electron density estimated from the wavelength of the satellite peak was a little bit higher than that estimated from the gas jet density. This is due to inaccuracy of the

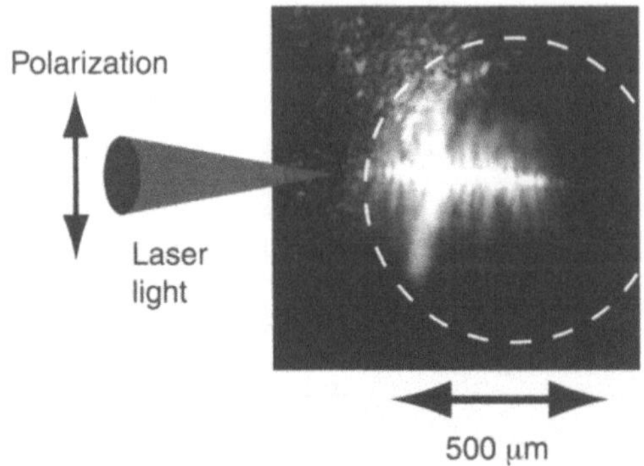


FIGURE 4. Side-scattered light image observed through an interference filter of 800 nm, when the electron energy spectrum shown in Fig. 2 was obtained. The polarization direction of the side-scattered light was the same as that of the incident laser light.

density measurement. This inaccuracy may be caused by the cluster formation of Ar in a gas jet. When the electron density was $4 \times 10^{19}$ cm$^{-3}$, the first anti-Stokes satellite can be observed around 700 nm. A spectral peak was seen around 700 nm in Fig. 3a. However, it is not clear that the peak is the anti-Stokes satellite, because the component by the ionization blue shift of the laser light can overlap in this wavelength region.

Figure 3b shows the forward-scattered light spectrum around 400 nm. The second harmonic of the pump laser pulse was observed at 400 nm due to the nonlinear interaction between the laser pulse and the plasma [15]. In addition, the multiple satellites were also observed around 350 nm. When the second harmonic of the pump pulse propagates through a plasma, collective Thomson scattering from the plasma wave causes multiple satellites. From the wavelengths of the multiple satellites, the electron density was estimated to be $4.2 \times 10^{19}$ cm$^{-3}$, which is in good agreement with that estimated from the wavelength of the first Stokes satellite of the pump pulse. Assuming that the second harmonic functions as a probe pulse, the intensity ratio of the first- and second-order satellites $P_2/P_1$ gives the normalized amplitude of the plasma wave. The normalized amplitude of the plasma wave $\varepsilon$ is given by $\varepsilon \sim (P_2/P_1)^{1/2}$ [16]. The intensity ratio $(P_2/P_1) \sim 0.5$ gives the normalized amplitude $\varepsilon \sim 0.7$.

Figure 4 shows a typical side-scattered light image, when the electron energy spectrum shown in Fig. 2 was obtained. The laser light propagated from left to right. The dotted circle shows the position of the nozzle exit. The arrow shows the polarization direction of the incident laser light. The strong side-scattered light image with a fishbone structure was observed around the nozzle center. The wavelength of the side-scattered light is close to that of the incident laser light, because the image was observed through an interference filter of 800 nm. In addition, the polarization direction of the side-scattered light is the same as that of the incident laser light. It is supposed that the side-scattered light is due to collective Thomson scattering induced by the excitation of a plasma wave [17]. From the length of the side-scattered light image, the interaction length of the laser pulse with the plasma was estimated to be 300–500 μm by taking account of the shot-to-shot variation. The observed interaction length is much longer than the

Rayleigh length of 110 μm. This may be due to the relativistic self-focusing [18]. The laser power was higher than the critical power for the relativistic self-focusing $P_{cr}$ given by $P_{cr} = 17(n_c/n_e)$ [GW].

Figures 5a and b show the electron-density dependence of the electron-energy spectrum and the forward-scattered light spectrum. Here, it is noted that the electron densities shown in the figure and the text are the values estimated from the density measurement. In this case, the laser power was 2.7 TW. As seen in Fig. 5a, the generation of a monoenergetic beam was observed at the density around $4 \times 10^{19}$ cm$^{-3}$ as well as the result shown in Fig. 2. For the lower-density ($2.6 \times 10^{19}$ cm$^{-3}$) plasma, no energetic electrons were observed. The intensity ratio of the first Stokes satellite to the transmitted laser light was also small. This shows that the amplitude of the plasma wave was small at this density. In contrast, for the higher-density ($5.1 \times 10^{19}$ cm$^{-3}$) plasma, the number of energetic electrons dramatically increased. However, no monoenergetic peaks were observed. The intensity ratio of the Stokes satellite to the transmitted laser light was large, and the spectral width of the satellite was broad. These suggest the excitation of the nonlinear wave with the large amplitude [19]. Although copious amounts of electrons are self-trapped, the energy spread becomes large due to the trapping of electrons in the wide phase range. The monoenergetic beam was obtained only in the narrow electron-density region around $4 \times 10^{19}$ cm$^{-3}$. Furthermore, the intensity and the spectral width of the first Stokes satellite in the forward-scattered light spectrum depend on the electron density and correlate with the electron energy spectrum. These results are very similar to those in our previous experiment with a high-density plasma of the order of $10^{20}$ cm$^{-3}$ [20].

## 4 Discussion

Here, we discuss the mechanism of the monoenergetic beam generation. The monoenergetic beam was obtained only at the electron density around $4 \times 10^{19}$ cm$^{-3}$. From Eq. (1), the normalized amplitude of the plasma wave of 0.7 gives the maximum electron energy of 30 MeV at $4 \times 10^{19}$ cm$^{-3}$. The energy of the monoenergetic beam of 23 MeV is close to the maximum energy estimated from the normalized amplitude of the plasma wave (30 MeV). The dephasing length at $4 \times 10^{19}$ cm$^{-3}$ is estimated to be 230 μm in the linear regime. The dephasing length is also close to the interaction length of 300–500 μm measured from the side-scattered light image. Considering that the interaction length gives the maximum acceleration length, it is reasonable to suppose that the monoenergetic beam generation was due to the matching of the acceleration length, to the dephasing length, which causes the monoenergetic peak formation due to the self-bunching of electrons entering the deceleration phase [5,6]. No monoenergetic peaks in the higher-density ($5.1 \times 10^{19}$ cm$^{-3}$) plasma can be explained as follows. The dephasing length becomes shorter in the higher-density plasma. However, the length of the side-scattered light image was nearly constant in the electron-density region

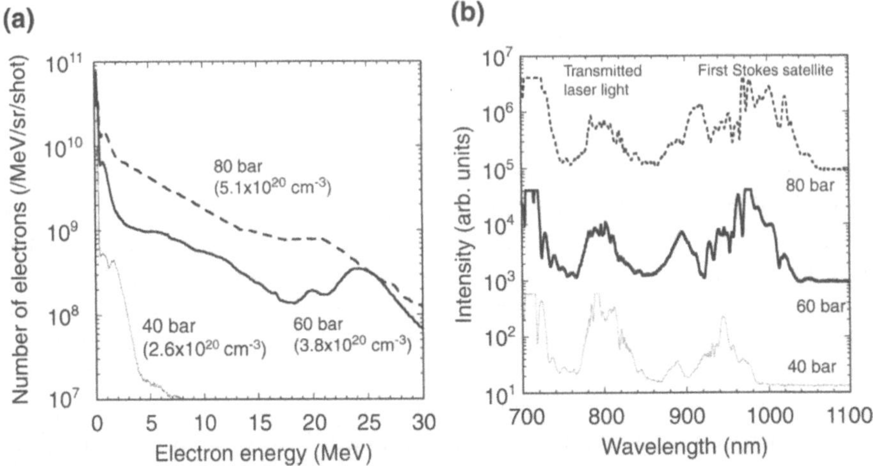


FIGURE 5. Electron density dependence of the (a) electron energy spectrum and (b) the forward-scattered light spectrum. The laser power was 2.7 TW. A monoenergetic beam was generated only in the narrow electron-density range around $4 \times 10^{19}$ cm$^{-3}$. The electron energy spectrum correlates with the intensity and width of the first Stokes satellite.

higher than $4 \times 10^{19}$ cm$^{-3}$. Then, the matching of the acceleration length to the dephasing length cannot be kept in the higher-density plasma.

As shown in Figs. 5a and b, the intensity and the spectral width of the first Stokes satellite in the forward scattered light spectrum depend on the electron density and correlate with the electron energy spectrum. This result shows that the excitation of the plasma wave strongly depends on the electron density. To obtain a monoenergetic beam, electrons should be preferentially trapped at the narrow phase range in a plasma wave with a single period. For the generation of a monoenergetic beam by self-trapping electrons in a plasma, the condition of the excitation of the plasma wave can be extremely limited. Considering that the excitation of the plasma wave strongly depends on the electron density, it is reasonable to obtain a monoenergetic beam only in the narrow electron-density range.

## 5 Summary

Generation of a monoenergetic electron beam with an energy of 23 MeV has been demonstrated in laser-driven plasma acceleration by using a 3-TW, 50-fs, 800-nm laser pulse. The number of electrons of the monoenergetic beam was $10^6$, and the divergence was $\pm 0.4°$. The monoenergetic beam was obtained only in the narrow electron-density range around $4 \times 10^{19}$ cm$^{-3}$. The normalized amplitude of the plasma wave was estimated to be 0.7 from the forward-scattered light spectrum. The interaction length was measured to be 300–500 μm from the side-scattered light image. These results suggest that the monoenergetic beam generation is due to the matching of the acceleration length to the dephasing length. The monoenergetic beam generation in the narrow electron-density range was supported by the density dependence of the intensity and the spectral width of the first Stokes satellite in the forward scattered light spectrum.

**Acknowledgement.** A part of this study was financially supported by the Budget for Nuclear Research of the Ministry of Education, Culture, Sports, Science and Technology (MEXT), based on the screening and counseling by the Atomic Energy Commission, and the Advanced Compact Accelerator Development of the MEXT.

# References

1. D. Umstadter: J. Phys. D **36**, R151 (2003); R. Bingham, J. T. Mendonça, and P. K. Shukla: Plasma Phys. Control. Fusion **46**, R1 (2004), and references therein.
2. T. Tajima and J. M. Dawson: Phys. Rev. Lett. **43**, 267 (1979).
3. V. Malka, S. Fritzler, E. Lefebvre, et al.: Science **298**, 1596 (2002).
4. S. P. D. Mangles, B. R. Walton, M. Tzoufras, et al.: Phys. Rev. Lett. **94**, 245001 (2005).
5. S. P. D. Mangles, C. D. Murphy, Z. Najmudin, et al.: Nature **431**, 535 (2004).
6. C. G. R. Geddes, Cs. Toth, J. van Tilborg, et al.: Nature **431**, 538 (2004).
7. J. Faure, Y. Glinec, A. Pukhov, et al.: Nature **431**, 541 (2004).
8. E. Miura, K. Koyama, S. Kato, et al.: Appl. Phys. Lett. **86**, 251501 (2005).
9. A. Yamazaki, H. Kotaki, I. Daito, et al.: Phys. Plasma **12**, 093101 (2005).
10. T. Hosokai, M. Kando, H. Dewa, et al.: Opt. Lett. **25**, 10 (2000).
11. Y. Kitagawa, Y.Sentoku, S. Akamatsu, et al.: Phys. Rev. Lett. **92**, 205002 (2004).
12. S. Augst, D. Strickland, D. D. Meyerhofer, S. L. Chin, and J. H. Eberly: Phys. Rev. Lett. **63**, 2212 (1989).
13. K. A. Tanaka, T. Yabuuchi, T. Sato, et al.: Rev. Sci. Instrum. **76**, 013507 (2005).
14. S. Chen, P. Zhang, W. Theobald, et al.: Observation of relativistic cross-phase modulation in high intensity laser-plasma interactions. In *Conference on Lasers and Electro-Optics 2005*, (Optical Society of America, Washington DC, 2005) JThD6.
15. K. Krushelnick, A. Ting, H. R. Burris, A. Fisher, C. Manka, and E. Esary: Phys. Rev. Lett. **75**, 3681 (1995).
16. S. P. Le Blanc, M. C. Downer, R. Wagner, et al.: Phys. Rev. Lett. **77**, 5381 (1996).
17. S.-Y. Chen, M. Krishnan, A. Maksimchuk, and D. Umstadter: Phys. Plasma **7**, 403 (2000).
18. P. Sprangle, C.-M. Tang, and E. Esary: IEEE Trans. Plasma Sci. **PS-15**, 145 (1987).
19. A. Modena, Z. Najmudin, A. E. Dangor, et al.: Nature **377**, 606 (1995).
20. E. Miura, K. Koyama, S. Masuda, et al.: Monoenergetic electron beam generation from a high-density plasma produced by a 2-TW, 50-fs laser pulse. In: *Superstrong field in plasmas: The third international conference on Superstrong field in plasmas*,ted by M. Lontano (American Institute of Physics, New York, 2006) to be published.

# Part VI

# X-ray Plasma Sources and Applications

# Femtosecond Time-resolved X-ray Diffraction for Laser-excited CdTe Crystal

Y. Hironaka,[1] J. Irisawa,[1] J. Saitoh,[1] K. Kondo,[1] K. Ishioka,[2] K. Kitajima,[2] and K. G. Nakamura[3]

[1]Materials and Structures Laboratory, Tokyo Institute of Technology, Nagatsuta, 4259 Midori, Yokohama 226-8503, Japan
hironaka@msl.titech.ac.jp
[2]National Institute for Materials Science, Tsukuba 305-0047, Japan
[3]Institute of Molecular Science, Okazaki, 444-8585, Japan

## 1 Introduction

Recent developments of high-power femtosecond laser system enable to generate the ultrashort X-ray pulses, which completely synchronize with the laser pulse. Time-resolved X-ray diffraction for laser-excited materials has been extensively performed [1–7]. Since the X-ray diffraction signal is sensitive to a change of lattice d-spacing with an accuracy of milliangstrom, this technique is ideal for measuring the rapid and small changes in crystal structures. Recently, Sokolowski-Tinten et al. [8] measured the longitudinal coherent optical phonon in semimetal (Bi) using ultrafast X-ray diffraction with a new analysis. This technique was performed to detect. Time-resolved X-ray diffracted signal taking within a much shorter period than the oscillation of coherent phonon showed alteration of its intensity due to the coherent atomic deviation from the crystallographic atomic position. In this paper, we performed femtosecond time-resolved X-ray diffraction (FTXRD) to detect the coherent optical phonon in CdTe (111) single crystal.

## 2 Method

Classical coherent atomic vibration $\delta_j$ in the crystal will be expressed as follows:

$$\boldsymbol{\delta}_j = \boldsymbol{u}_j \sin\left(\omega t + \phi_j\right), \tag{1}$$

where $\boldsymbol{u}_j$ and $\phi_j$ are the maximum deviation vector and initial phase of oscillation for $j$th atom in the unit cell and $\omega$ is the frequency of phonon. For easy consideration, the symmetrical vibration mode is taken into account. We assumed that

$\phi_j = 0$ for all the atoms and summation of $u_j$ in the unit cell culminates in zero. In this assumption, the crystal structure factor will be described as follows:

$$\boldsymbol{F} = \sum_{j}^{N} f_j \exp\{-i\boldsymbol{G}(r_j + u_j \sin \omega t)\}, \tag{2}$$

where $f_j$ and $r_j$ are the atomic scattering factor and crystallographic atomic position in the unit cell, respectively. $N$ and $G$ are the total number of atoms in the unit cell and reciprocal lattice vector, respectively. Equation (2) is the time-dependent structure factor for the crystal in which the coherent phonon is stationary exited. First-order Taylor expansion of Eq. (2) for the case of small amplitude of coherent phonon is written as

$$\boldsymbol{F} = \boldsymbol{F}_0 - \left\{ i \sum_{j}^{N} f_j G u_j \exp\left(-iGr_j\right) \right\} \sin \omega t. \tag{3}$$

where $F_0$ is the crystallographic structure factor (time independent). Equation (3) gives sensuous understanding to the modulation of X-ray diffracted signal because the X-ray reflectivity is proportional to $|F|^2$. Figure 1 shows the calculated $|F|^2$ for the case of CdTe single crystal. The reflected plane of (111) and (222) are plotted for various atomic deviations from the crystallographic atomic position. As the Eq. (3) includes inner product of $G$ and $u_j$, the modulation of X-ray reflection will be measured only for the phonon motion, which has a parallel component to the reciprocal lattice vector. Although zincblend structure like CdTe has two branches of optical phonon mode at the center of Brillion zone in the [111] direction, only longitudinal mode will be exposed to view in the signal.

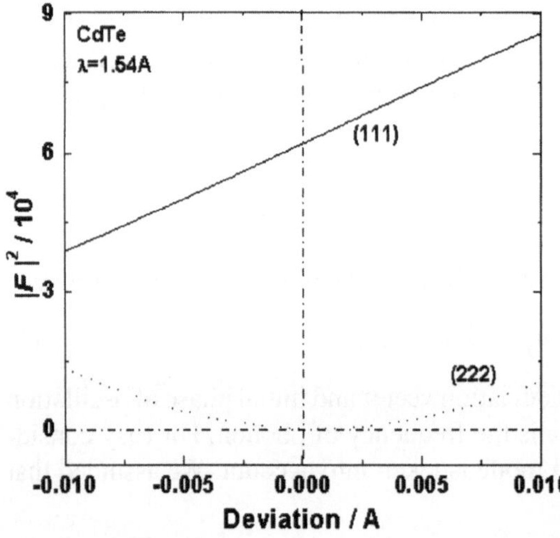


FIGURE 1. Calculated $|F|^2$ for various coherent atomic deviations using Eq. (2). Solid line shows the reflection from the lattice plane of CdTe (111). Dotted line shows the reflection of (222) plane, which is the X-ray forbidden reflection condition. Phonon oscillation will also modulate $|F|^2$, while $F_0$ is larger than a second term of Eq. (3). If the X-ray forbidden reflection is applied for the measurement of large amplitude of phonon, the X-ray intensity will be modulated with $2\omega$.

## 3 Experimental Results

Figure 2 shows schematic drawings of FTXRD experiment. A table-top terawatt laser system [1], which consists of a Ti:sapphire laser and chirped pulse amplification (CDA) system, operating at 10 Hz is used to generate Cu K $\alpha$ from the copper tape target (30 m in thickness, and 500-m long ). A compressed laser pulse (60 fs) with an incident angle of 60 is focused on the Cu tape target with a focal spot of 46 m in diameter by a parabolic mirror. Another femtosecond laser pulse was used for the excitation of coherent optical phonon in CdTe single crystal. The arrival time of pump laser and probe X-ray to the crystal surface were controlled by the optical delay line with the temporal resolution of 20 fs. FTXRD experiment was performed with the symmetrical Bragg reflection geometry (B = 11.88°). Diffracted X-ray signal was detected by X-ray CCD camera. Both of the diffracted signal from the laser perturbed and unperturbed surface of CdTe were detected at the same time, and normalized intensity was used for the reflectivity analysis. The pump laser pulse was focused on CdTe with the power density of 161 GW $\cdot$ cm$^{-2}$. Fourier transform was applied to the FTXRD data to obtain the phonon spectrum. Figure 3 shows the Fourier power spectrum of the experimental result. The peak of 5.07 THz is obtained, and it is equivalent to the LO phonon frequency at the Brillion zone center.

In summery, we performed femtosecond time-resolved X-ray diffraction experiment and successfully observed coherent LO phonon in CdTe single crystal excited by femtosecond laser irradiation. It is expected that this technique will become powerful tools to analyze phonon dynamics from the aspect of solid state physics and material sciences.

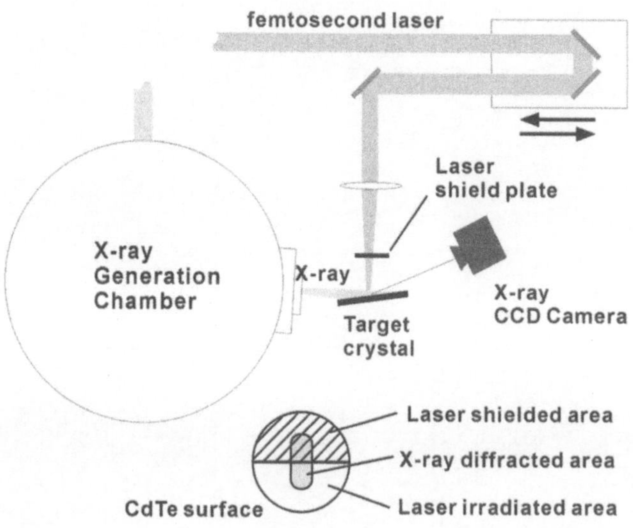


FIGURE 2. Schematic drawings of FTXRD experiment.

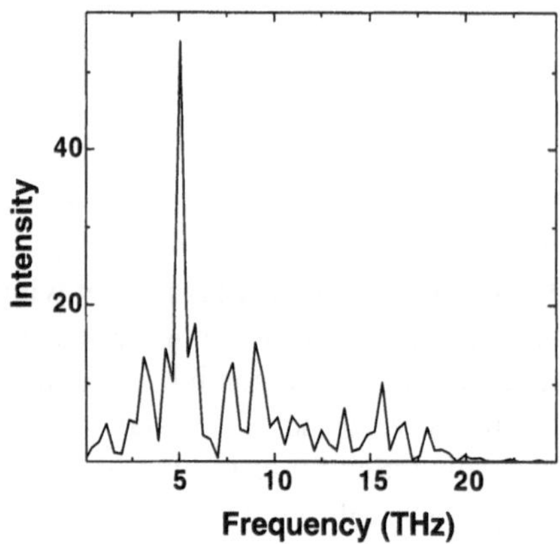


FIGURE 3. Fourier power spectrum of time-dependent X-ray diffraction signal. The peak of 5.07 THz is equivalent to the LO phonon frequency at the Brillouin zone center of CdTe [111] direction.

# References


1. Y. Hironaka, et al.: Appl. Phys. Lett.**77**, 1967 (2000).
2. A. H. Chin, et al.: Phys. Rev. Lett. **83**, 336 (1999).
3. C. W. Siders, et al.: Science **286**, 1340 (1999).
4. A. M. Lindenberg, et al.: Phys. Rev. Lett. **84**, 111 (2000).
5. A. Rousse, et al.: Nature **410**, 65 (2001).
6. K. Sokolowski-Tinten, et al.: Phys. Rev. Lett. **87**, 5701(2001).
7. A. Cavalleri, et al.: Phys. Rev. Lett. **87**, 237401 (2001).
8. K. Sokolowski-Tinten, et al.: Nature **422**, 287 (2003).

# Femtosecond-laser-induced Ablation of an Aluminum Target Probed by Space- and Time-resolved Soft X-ray Absorption Spectroscopy

Yasuaki Okano, Katsuya Oguri, Tadashi Nishikawa, and Hidetoshi Nakano

NTT Basic Research Laboratories, NTT Corporation, 3-1 Morinosato Wakamiya, Atsugi, Kanagawa 243-0198, Japan
okano@will.brl.ntt.co.jp

## 1 Introduction

Laser ablation involves phase transitions that produce an ablation plume consisting of various states of matter, namely solid, liquid, vapor, and plasma. This plume plays an important role in the deposition of thin films [1], the formation of nanoparticles [2], and other processes. It is important to investigate this ablation plume at the atomic level not only for a basic study of laser ablation dynamics but also to provide a better understanding of such processes. Because the plume evolves spatially over time, space- and time-resolved measurements are also required. One atomic-level diagnostic method is X-ray absorption spectroscopy (XAS). This involves an element-specific probe of the local structure in a material and provides information about its electronic and atomic structure. An important advantage of this technique is that a wide variety of solid, liquid, and gaseous samples can be examined directly and nondestructively. Furthermore, we can combine this technique with an ultrafast X-ray probe generated by femtosecond-laser-induced plasma [3–6], and this makes time-resolved XAS suitable for the study of laser ablation processes.

In this study, we investigated the expansion process of an ablation plume induced by irradiating aluminum foil with a femtosecond laser, using a space- and time-resolved XAS system that we developed. The ablation of vaporized aluminum and condensed particles, such as droplets, following the expansion of the plasma was observed as a shift of the aluminum $L$-shell photoabsorption edge in space- and time-resolved absorbance spectra.

## 2 Soft X-ray Imaging System for Time-resolved XAS

A schematic illustration of the XAS system we developed is shown in Fig. 1. The system consists of a laser-plasma X-ray source, an X-ray microscope with critical illumination, and a spectrometer. The key feature of this system is that a Kirkpatrick-Baez (K-B) microscope [7], which consists of two spherical mirrors, is used for imaging. The system is arranged for the picosecond time-resolved study of photoexcited materials by the pump-probe method. It employs a femtosecond Ti:sapphire laser system with a pulse width of 100–120 fs at a central wavelength of 790 nm. In this experiment, to enhance the X-ray intensity [8,9] we added an artificial prepulse, which was generated by a regenerative amplifier in the laser system. The prepulse had a contrast ratio to the main pulse of approximately $10^{-1}$ or less, which was estimated using a photodiode.

The output laser pulse is divided into two parts by a beam splitter. One is used for probe X-ray generation, and the other is used as a pump pulse that excites a sample. An optical delay line placed on the pump laser path adjusts the arrival time between the pump and probe pulses at the sample position for the time-resolved measurement. The transferred image is spectrally resolved with a transmission grating (TG) placed between the K-B mirrors and the MCP. The TG is mounted on a slit with a transfer stage to select a proper portion of the backlit image. The TG is made of SiN bars supported by a thin SiN membrane and has a grating period of 833 nm (1200 bars/mm).

### 2.1 Ultrafast X-ray Probe

The X-ray probe pulse is generated by irradiating a tantalum-foil target with the laser pulses at a laser intensity of $10^{16}$ W/cm$^2$. The X-ray target is held on a drive apparatus to expose a fresh surface for each laser shot. The temporal resolution of the system is determined by the pulse width of the probe X-rays. Spectrally

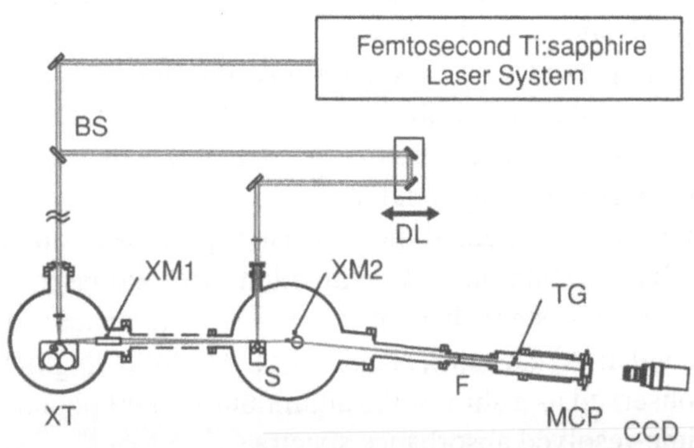


FIGURE 1. Schematic illustration of the experimental setup for the space- and time-resolved X-ray absorption measurement. BS, beam splitter; DL, optical delay line; XT, X-ray target; XM1, condenser mirror; XM2, KB mirrors; S, sample; F, filter; TG, transmission grating.

FIGURE 2. Temporal profiles of the probe X-ray pulses. (a) Spectrally resolved image. (b) Average intensity of the X-ray pulse.

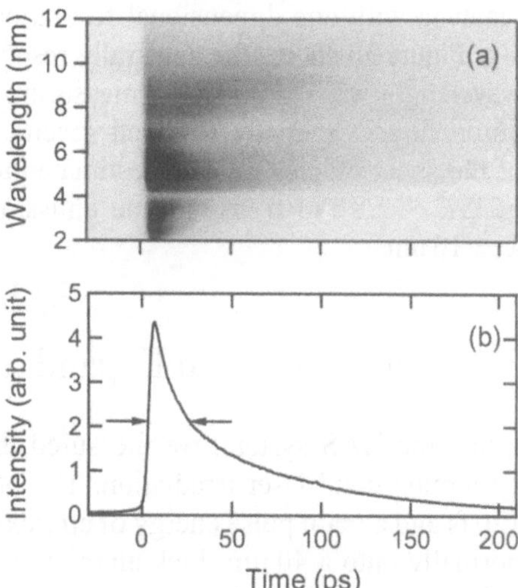


resolved temporal profiles of the pulses were measured using an X-ray streak camera with a transmission grating [8]. Figure 2a is a typical spectrally resolved streak image of the X-ray probe. There are large structures in the spectrum due to the photon energy dependence of the quantum yield of the CsI photocathode of the streak camera [10]. The average intensities in the soft X-ray region between 2 and 12 nm, excluding the quantum yield, are plotted in Fig. 2b. The full width at half maximum of the X-ray pulse was 23 ps, which was the temporal resolution in this experiment.

## 2.2 Spatially Resolved Spectroscopy

The spatial resolution was estimated by measuring images of several fine grids. Figure 3a shows a typical backlit image of a 2000-mesh through-hole copper grid, which has a period of 12.5 μm. Part of the transferred image is spectrally

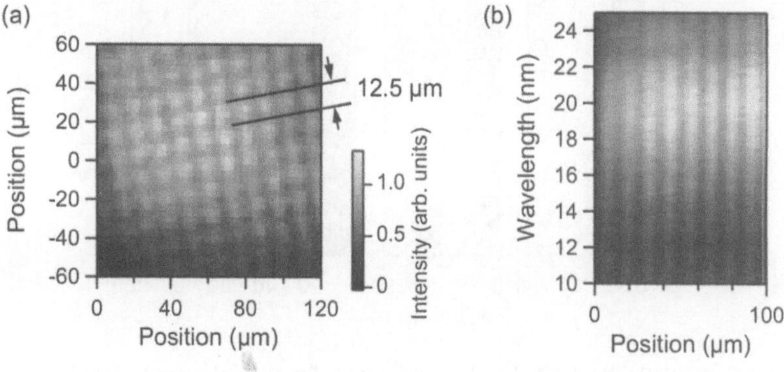


FIGURE 3. (a) Backlit image of a 2000-mesh grid. (b) Spatially resolved spectrum of the backlit image of the 200-mesh.

resolved with one-dimensional resolution using the transmission grating and a slit. Figure 3b shows the spectrally resolved backlit image of the fine grid. The wavelength was calibrated by measuring the absorption edges of aluminum and boron filters. Spatially resolved spectra were clearly obtained. The resolution of the spectrometer was also estimated to be $\Delta\lambda \sim 0.57\,\text{nm}$ (energy resolution of $\Delta E \sim 2.8\,\text{eV}$) from the line emission of Al IX $2p^6$–$2p^5 3s$ at wavelength $\lambda = 16\,\text{nm}$.

# 3 Observation of an Expanding Ablation Plume

Using the XAS system, we measured the aluminum ablation process induced by femtosecond laser irradiation. The pump laser pulses had a pulse width of 120 fs and a main pulse energy of approximately 10 mJ. The pulses were focused normally onto a 40 μm-thick aluminum foil sample. The diameter of the pump pulse estimated from the X-ray emitted region was 180 μm with a corresponding intensity of $3 \times 10^{14}\,\text{W/cm}^2$. Figure 4 shows typical spatially resolved images of the absorbance spectra of the laser-induced ablation plume of aluminum at delay times of (a) −0.5, (b) 0.5, and (c) 5 ns. The target was irradiated with pump laser pulses from the right at a position of 0 μm, and the longitudinal expansion of the generated aluminum plasma and ablating particles was probed perpendicularly by the focused X-rays. The absorbance is defined as $\mu_d = \ln(I_0/I)$, where $I_0$ is a probe spectrum without the pump pulse and $I$ is an absorption spectrum with the pump pulse. The absorbance reflects the density distribution of the ablated particles, and the spatial evolution of the absorbance was observed in the time-resolved spectra. Because of the prepulse irradiation, a little absorption was observed at a delay time of −0.5 ns, when the probe X-ray pulse arrived at the sample before the main pulse. In such a case, plasma or very low density ablated particles would

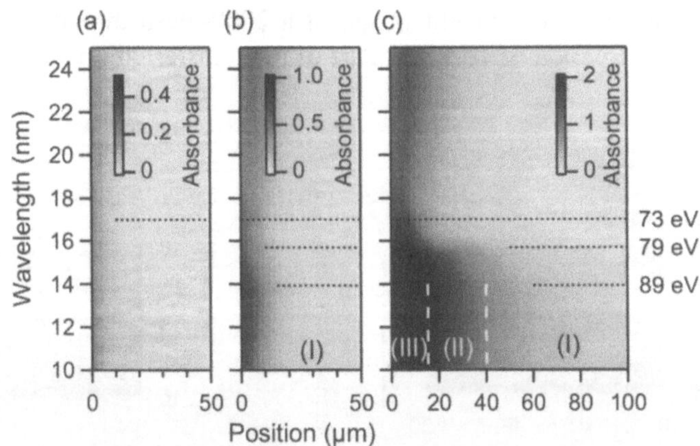


FIGURE 4. Spatially resolved absorbance spectra of the aluminum plasma at probe delay times of (a) −0.5, (b) 0.1, and (c) 5 ns.

FIGURE 5. Normalized absorbance spectra of the reference filter and the aluminum ablation plume for regions I, II, and III at a delay time of 5 ns.

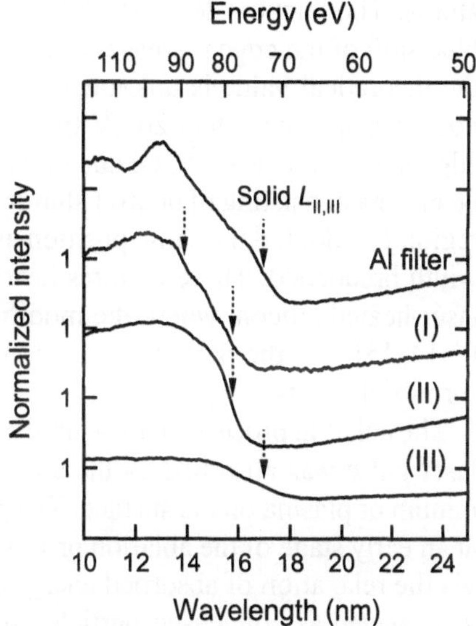


expand from the sample surface. After the main pulse irradiation, increments of absorbance were observed with increasing probe delay time, which indicates that the amount of ablated material increases for more than several nanoseconds. The ablation plume expanded more than 40 μm at a delay time of 5 ns with a corresponding velocity of $10^4$ m/s. This value is consistent with the measured plasma expansion speed induced by femtosecond laser irradiation [11].

In the spectra, the $L_{II,III}$ absorption edge of a solid aluminum is shown as a dotted line at a wavelength of 17 nm ($E \sim 73$ eV) [12]. The $L_{II,III}$ edge of the ablated aluminum clearly shifted toward shorter wavelengths. To investigate this shift, we divided the plume into three regions according to the modification of the $L_{II,III}$ edge as shown in Fig. 4. Region I appears in the early stages of ablation and is far from the target surface when the delay is large. Region II follows region I. Region III occurs late in the ablation process. Figure 5 shows the normalized intensity versus energy of these regions and a reference spectrum for a 0.1-μm-thick aluminum filter. In spectrum I, the $L_{II,III}$ edge has shifted to about 79 eV ($\lambda \sim 15.8$ nm) and a new edge has appeared at 89 eV ($\lambda \sim 13.9$ nm). In spectrum II, there is only a shift to 79 eV. In spectrum III, the edge is at the same position as it is for solid aluminum with a broadening of the edge width.

The edge provides a measure of the energy of the transition of a core electron to free or continuum levels. The edge shift can be understood as the difference between continuum levels. For aluminum in the solid or condensed state, the edge corresponds to transitions to the Fermi level. In contrast, the continuum level of aluminum in the atomic state, for example in a vapor, changes to the vacuum level, which is the ionization potential. This difference in potential appears in regions I and II. The calculated value of the edge of an isolated aluminum atom is 82 eV ($\lambda \sim 15.1$ nm) [13]. This energy agrees fairly well with the measured value of 79 eV. Region II is considered to consist largely of neutral aluminum atoms. The vacuum level is different for ionic aluminum. The amount of the blue shift of the edge increases as the ionization state increases [4, 14]. Although the theoretical value is unknown for aluminum ions, the edge shift of $Si^+$ was observed approximately 20 eV from the edge of solid-state silicon [14]. So, the edge in region I seems to be due to $Al^+$ ions. Therefore, region I is considered to be plasma consisting of neutral aluminum atoms with low ionization. The edge in region III adopted the same position as the reference aluminum filter as the edge width broadened. These features have been observed in several experiments on laser-heated silicon, where the modification of the edge was caused by melting [6, 14, 15]. So, the particles in region III are probably molten aluminum in the form of droplets.

The ablation process we measured can be understood as follows: First, the main laser pulse was absorbed by the aluminum foil and the preformed gaseous aluminum or plasma on the surface. The plasma excited by the main pulse expanded at an early stage of the ablation process. Subsequently, the sample was vaporized via the relaxation of absorbed energy into it. Finally, some of the molten sample was ejected as condensed particles such as droplets. This scheme is supported by a previously reported molecular dynamics simulation [16]. This study clearly shows that our space- and time-resolved XAS system is very suitable for observing the time evolution of expanding ablated materials consisting of various states of matter.

## 4 Summary

We developed a system for obtaining space-resolved measurements using time-resolved X-ray absorption spectroscopy in order to investigate the ablation process of photoexcited materials. The system consists of a femtosecond-laser-induced plasma X-ray source and a K-B microscope with critical illumination. The temporal and spatial resolution was 23 ps and 12.5 μm, respectively. We succeeded in obtaining snapshots of the expanding ablation plume of aluminum as spatially resolved absorption spectra. They showed a shift and modification in the photoabsorption edge, indicating the emission of droplets and vaporized aluminum following the expansion of the plasma. These results demonstrate that this system would be a useful tool for the study of photoexcited material dynamics with spatial evolution, such as laser ablation.

**Acknowledgments.** This work was supported by a grant-in-aid from the Ministry of Education, Culture, Sports, Science and Technology of Japan (Grant No. 16032219).

## References

1. F. Qian, R. K. Singh, S. K. Dutta, P. P. Pronko: Appl. Phys. Lett. **67**, 3120 (1995).
2. T. E. Glover: J. Opt. Soc. Am. B. **20**, 125 (2003).

3. H. Nakano, Y. Goto, P. Lu, T. Nishikawa, N. Uesugi: Appl. Phys. Lett. **75**, 2350 (1999).
4. J. Workman, M. Nantel, A. Maksimchuk, D. Umstadler: Appl. Phys. Lett. **70**, 312 (1997).
5. P. Audebert, P. Renaudin, S. Bastiani-Ceccotti, et al.: Phys. Rev. Lett. **94**, 025004 (2005).
6. K. Oguri, Y. Okano, T. Nishikawa, H. Nakano: Appl. Phys. Lett. **87**, 011503 (2005).
7. P. Kirkpatrick, A. V. Baez: J. Opt. Soc. Am., **38**, 766 (1948).
8. H. Nakano, T. Nishikawa, H. Ahn, N. Uesugi: Appl. Phys. Lett. **69**, 2992 (1996).
9. Y. Okano, K. Oguri, T. Nishikawa, H. Nakano: Proc. Soc. Photo-Opt. Instrum. Eng. **5714**, 215 (2005).
10. B. L. Henke, J. P. Knauer, K. Premaratne: J. Appl. Phys **52**, 1509 (1981).
11. Y. Okano, H. Kishimura, Y. Hironaka, K. G. Nakamura, K. Kondo: Appl. Surf. Sci. **197/198**, 281 (2002).
12. J. A. Bearden, A. F. Burr: Rev. Mod. Phys. **39**, 125 (1967).
13. P. Indelicato, S. Boucard, E. Lindroth: Eur. Phys. J. D **3**, 29 (1998).
14. T. Ohyanagi, A. Miyashita, K. Murakami, O. Yoda: Jpn. J. Appl. Phys. **33**, 2586 (1994).
15. S. L. Johnson, P. A. Heimann, A. M. Lindenberg, et al.: Phys. Rev. Lett. **91**, 157403 (2003).
16. D. Perez, L. J. Lewis: Phys. Rev. Lett. **89**, 255504 (2002).

# Observation of Ultrafast Bond-length Expansion at the Initial Stage of Laser Ablation by Picosecond Time-resolved EXAFS

Katsuya Oguri, Yasuaki Okano, Tadashi Nishikawa, and Hidetoshi Nakano

NTT Basic Research Laboratories, NTT Corporation, 3-1 Morinosato Wakamiya, Atsugi, Kanagawa 243-0198, Japan
**oguri@ nttbrl.jp**

We have demonstrated a time-resolved extended X-ray absorption fine structure (EXAFS) technique by using a femtosecond laser produced plasma soft X-ray source. By applying this technique to the measurement of the initial stage of the laser ablation in Si foil, we were able to observe a slight shortening of the EXAFS oscillation period. This result suggests that the Si–Si atomic bond length expands as a result of the solid–liquid phase transition in Si. The realization of this technique is the first step toward understanding atomic structural dynamics during a chemical reaction.

## 1 Introduction

The X-ray absorption fine structure (XAFS) technique is widely accepted as a useful tool for obtaining information about the local structure of materials including crystal, amorphous, liquid, and gas phases in physics, chemistry, and biology [1]. The XAFS, which appears near and above X-ray absorption edge in X-ray absorption spectra, arises from the interference between the outgoing photoelectron wave and the wave scattered back at neighboring atoms. Therefore, the XAFS spectrum reflects the modification of the absorption spectrum of an isolated atom by the existence of surrounding atoms. In general, XAFS is classified into two parts according to the energy region: The X-ray absorption near edge structure (XANES), which appears within about 50 eV from the X-ray absorption edge and the extended X-ray absorption fine structure (EXAFS), which appears in the higher energy region well above the X-ray absorption edge.

The former provides the arrangement and symmetry of the surrounding atoms, and the latter provides the bond length and coordination number. Recently, the combination of the XAFS technique and the time-resolved X-ray probe technique has attracted considerable attention as a powerful tool for understanding ultrafast atomic structural dynamics during chemical reactions [2,3]. This was stimulated by the development of various ultrashort X-ray-pulse sources with picosecond and femtosecond pulse widths such as laser-based X-ray sources and synchrotron radiation sources.

Here, we present an ultrafast XAFS technique that employs a soft X-ray pulse emitted from Ta plasma produced by a femtosecond laser [4]. We used this technique to measure the time evolution of the L-edge XAFS, especially EXAFS, in amorphous Si foil during ultrafast laser melting. This is because only a few recent studies in the nanosecond region have demonstrated the time-resolved EXAFS technique experimentally [5–7], while several studies on the time-resolved XAFS technique in the millisecond to picosecond region have been published since the mid 1980s [8]. In the present study, we focus on the ultrafast solid–liquid phase transition induced by the irradiation of a femtosecond laser pulse. This is because the ultrafast mechanism of the laser melting in semiconductors such as Si, GaAs, and C has been of scientific interest as well as of engineering importance in laser processing. This has been investigated with conventional ultrafast optical techniques that measure the change in reflectivity [9], and recently developed time-resolved X-ray diffraction techniques that detect the loss of the long-range order of their crystal structure [10]. However, these techniques cannot provide such local structure information as the change in the atomic bond length, which is the most basic parameters during the solid–liquid phase transition. Although two pioneering pieces of work applied the time-resolved XAFS technique to the study of laser melting in Si, the data quality and the temporal and the spectral resolution were relatively poor [11,12]. The purpose of the present study is to obtain direct structural information such as the bond length during laser melting and to demonstrate the feasibility of our time-resolved EXAFS technique.

## 2 Experimental Setup

Figure 1 shows a schematic illustration of our experimental setup, which is similar to a previously reported configuration [13]. For our laser-pump and X-ray-probe measurement system, we used a chirped pulse amplification based Ti:sapphire laser system that can generate up to 100 mJ with a central wavelength of 790 nm at a 10-Hz repetition rate. Figure 2a shows the inside of the vacuum chamber for plasma X-ray generation. The laser pulse was focused perpendicularly on the surface of Ta tape up to a peak intensity of about $2 \times 10^{16}$ W/cm$^2$. The 25 μm thick Ta tape was moved to provide a fresh surface for each laser shot by a tape drive apparatus (TOYOTA MACS). The X-ray pulse, which was emitted

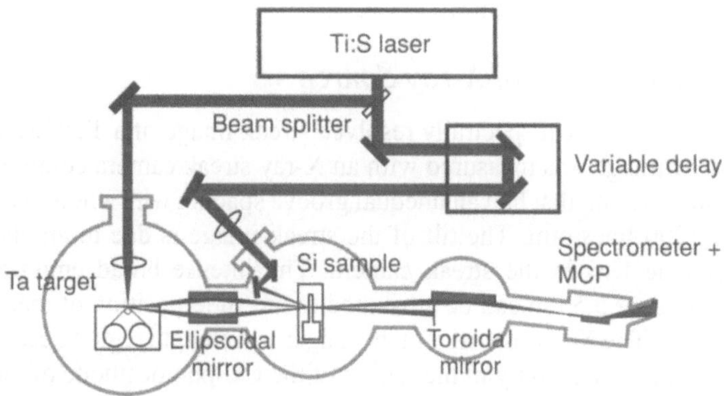


FIGURE 1. Schematic illustration of the time-resolved XAFS system.

perpendicular to the Ta tape normal, was focused onto a Si sample by using a Ni-coated ellipsoidal mirror. The sample consisted of amorphous Si foil deposited on stainless steel mesh that was 120-nm thick and 30 × 20 mm in size. We scanned the sample at 1 mm intervals to refresh the damaged region for each laser shot (Fig. 2b). Therefore, we obtained about 600 shots for a single piece of Si foil. The typical peak intensity of the laser pulse for sample excitation was about $5 \times 10^{12}$ W/cm$^2$ (0.5 J/cm$^2$). The X-ray pulse and the laser pulse overlapped on the sample at an angle of about 7°. After the X-ray pulse had passed through the sample, it was again focused onto the entrance slit of a flat-field grazing incidence spectrograph that had an unequal groove spacing with a nominal grating spacing of 2400 lines/mm. The spectrally resolved soft X-ray was detected with a microchannel plate combined with a cooled charge-coupled device.

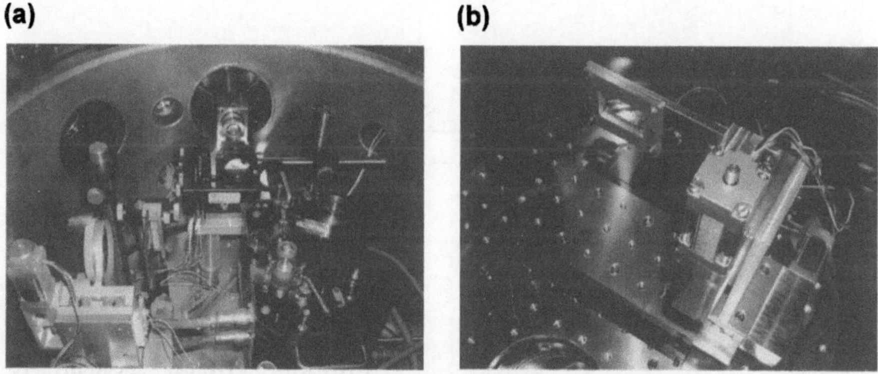


FIGURE 2. Photographs of (a) the inside of the plasma X-ray source chamber and (b) the sample scanning system.

# 3 Results and Discussion

## *3.1 Characterization of X-ray Source*

Figure 3a shows a typical spectrally resolved streak image of a Ta plasma X-ray emission. The image was measured with an X-ray streak camera combined with a transmission grating that had an unequal groove spacing with a nominal grating spacing of 1200 lines/mm. The tilt of the streak image is due to an aberration of the magnetic lens in the streak camera. The intense broad emission band that peaks at about 5.5 nm can be attributed to the superposition of many $4d$–$4f$ transitions [14]. The X-ray intensity in the longer wavelength region became weak because of the low sensitivity in the region of the CsI photocathode of the streak camera. Thus, the X-ray emission from a high-Z material such as Ta provides a broadband and smooth spectrum without any intense line emission. This feature of the spectrum enabled us to use the Ta plasma emission as an X-ray absorption probe. We also plotted X-ray temporal shapes at various wavelength as shown in Fig. 3b. The steep rise and slow decay shown in the temporal shapes are peculiar to the thermal emission of plasma. The pulse duration becomes slightly shorter as the wavelength decreases from 3 to 12 nm. This might be due to the short lifetime of the higher energy states of Ta ions that corresponds to the shorter wavelength X-ray emission. The full width at half maximum of the X-ray pulse that was averaged in the 3–12 nm region was typically 7 ps, which limits the temporal resolution of our time-resolved XAFS system.

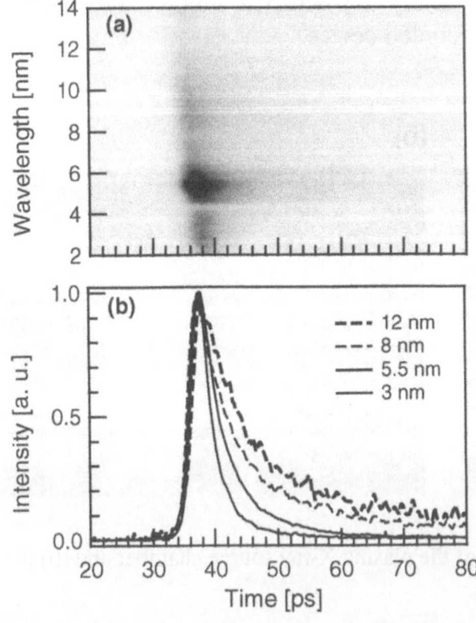


FIGURE 3. Typical spectrally resolved streak image of the soft X-ray emission from Ta plasma (a) and the temporal shapes at various wavelengths (b).

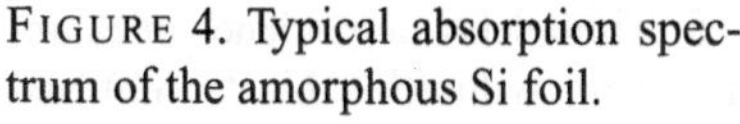

FIGURE 4. Typical absorption spectrum of the amorphous Si foil.

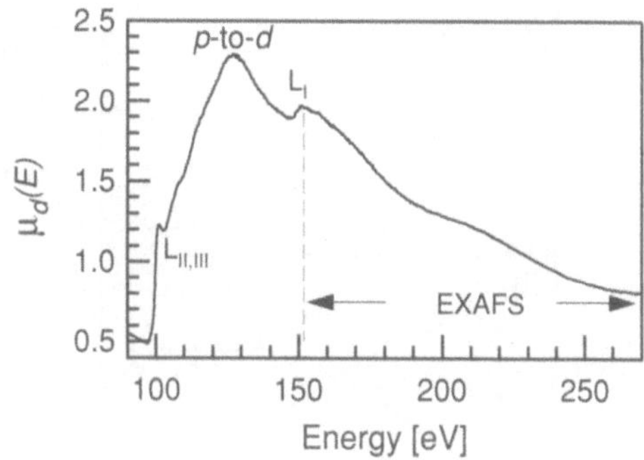


## 3.2 Time-resolved Measurement of Si–L Edge XAFS

Figure 4 shows an example of the absorption spectrum of the Si sample. The absorption coefficient, $\mu_d(E) = \ln[I_0(E)/I(E)]$, where $E$, $I_0(E)$, and $I(E)$ represent the photon energy and the soft X-ray spectra obtained without and with the Si sample. This data consists of 1200 shots. We observed the $L_{\text{II,III}}$ edge at 99 eV, the $L_{\text{I}}$ edge at 150 eV. The intense broad absorption band observed near 125 eV can be attributed to the quasibound state formed above the ionization potential, which results from the centrifugal barrier effects of the 2*p*–3*d* transition [15]. These absorption features agree well with previous X-ray absorption measurements for the $L$-edge of Si [16]. In the 150–270 eV region, the absorption spectrum decreases with oscillation, which corresponds to the EXAFS of the Si $L$-edge. We obtained the nearest neighbor Si–Si bond length of the sample from this EXAFS.

Figure 5 shows the absorption spectra for various time delays between the soft X-ray and the laser pulses. When the X-ray pulse passed through the Si sample 330 ps before the laser pulse arrived, it is natural that there was hardly any difference between the absorption spectra with and without laser irradiation (Fig. 5a). When the X-ray pulse almost overlapped the laser pulse in the time domain, the absorption spectrum exhibited a significant decrease in absorption at the $L_{\text{II,III}}$ edge and a slight shift of the $L_{\text{I}}$ edge toward a lower energy (Fig. 5b). A further decrease in the absorption at the $L_{\text{II,III}}$ edge and a larger shift of the $L_{\text{I}}$ edge were observed for a time delay of 1670 ps (Fig. 5c). Since the absorption feature near the edges partially reflects the density of states of the unoccupied band, the changes near the $L_{\text{II,III}}$ and the $L_{\text{I}}$ edges are most likely caused by the transition in the electronic structure from the semiconductor to a metallic state due to a solid–liquid phase transition [11, 12]. The decrease in the absorption at the $L_{\text{II,III}}$ edge also depended on the energy density of the laser pulse. Figure 6 shows the differential absorption at the $L_{\text{II,III}}$ edge. From this figure, the threshold energy density of the emergence of the $L_{\text{II,III}}$ absorption change can be estimated to be 0.05–0.1 J/cm$^2$. This value agrees well with the damage threshold of Si foil [17].

Figure 7 shows the time evolution of the EXAFS spectra obtained from the absorption spectra shown in Fig. 5. We extracted the EXAFS spectra from these

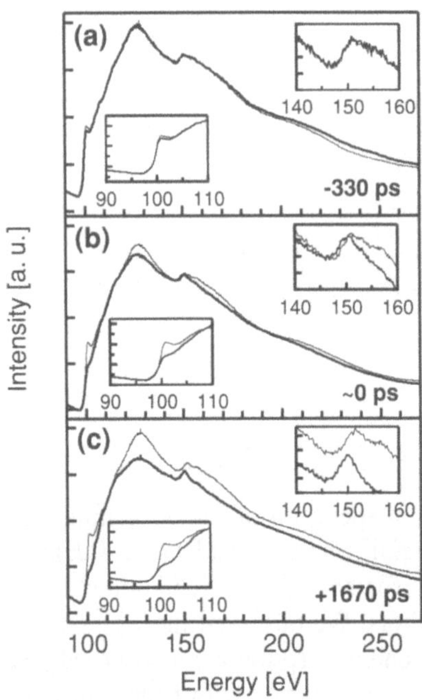


FIGURE 5. Time delay dependence of Si absorption spectra with (thin black line) and without (thick gray line) laser irradiation. The inset shows the expansion plots near the $L_{\mathrm{II,III}}$ and the $L_{\mathrm{I}}$ edges, respectively.

absorption spectra by subtracting the background based on the Victoreen formula and plotting the subtracted oscillation as a function of electron wave number. These EXAFS spectra reveal a clear one and half cycle oscillation, whose period is related to the nearest neighbor Si–Si bond length. We can estimate the bond length of the unpumped amorphous Si to be 2.32 Å by Fourier transformation. The good agreement between the obtained bond length and the previously reported value (2.35 Å) for amorphous Si confirmed the validity of our measurement for the static state [18]. At a time delay of ∼0 ps, we can clearly see that the oscillation amplitude decreased with a small peak shift toward a lower wave number compared with that of the unpumped sample (Fig. 7b). The decrease in the oscillation probably results from a mixture of thermal damping and the structural

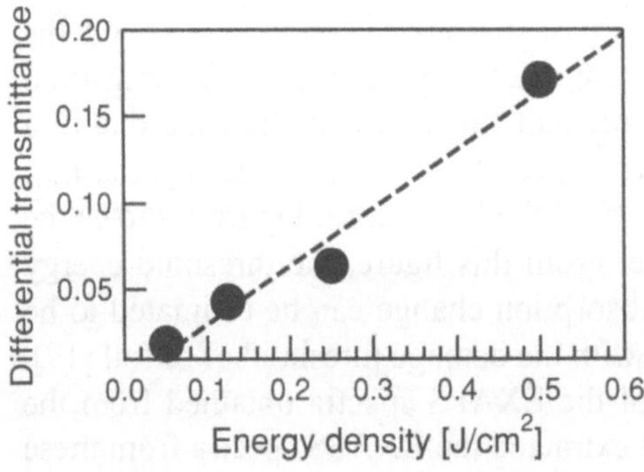


FIGURE 6. The dependence of the edge change on the energy density of the laser pulse. The differential absorption is defined as $-(I(E) - I_0(E))/I_0(E)$.

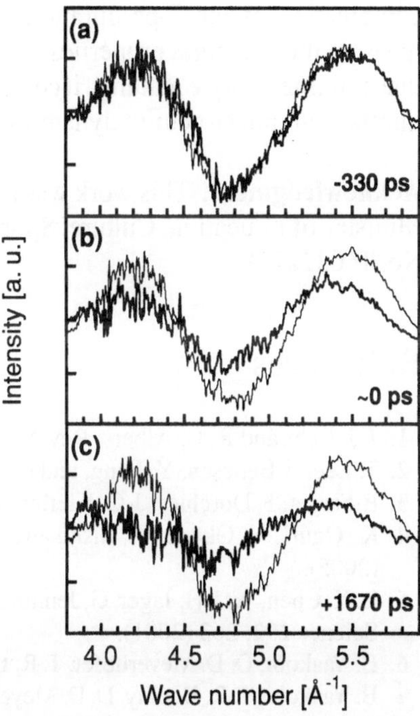


FIGURE 7. Time evolution of EXAFS spectrum of the Si *L* edge.

disordering of the laser-melted Si. The small peak shift indicates that the Si–Si bond length expands slightly due to the solid–liquid phase transition and the bond length was estimated to be 2.43 Å, which is slightly shorter than the value of 2.45–2.50 Å for liquid Si at 1700–1800 K [19]. Since the melting point of amorphous Si was estimated to be about 1480 K, which is about 200 K below the crystalline value, the sample may melt at a relatively low temperature [20], thus exhibiting a slightly shorter Si–Si bond length. In addition, the bond length expansion of $0.15 \pm 0.07$ Å, which is almost the same as the observed value, was predicted by molecular dynamics simulations for laser-melted Si [12]. At a time delay of 1670 ps, the EXAFS amplitude became very weak (Fig. 7c), thus indicating further disordering of the local structure of the Si due to evaporation. Finally, long after the laser excitation, the Si sample in the laser irradiation area was evaporated completely. Thus, since the EXAFS amplitude rapidly decreases with time during the laser ablation process, probing with a sub-10-ps soft X-ray pulse is very effective for detecting the liquid phase that appears transiently at the initial stage of the laser ablation.

## 4 Summary

We have demonstrated a time-resolved XAFS measurement technique that uses an ultrashort soft X-ray pulse emitted from laser-produced plasma by measuring the ultrafast melting of amorphous Si foil induced by a 100-fs laser pulse. The combination of the 7-ps duration and the broadband spectrum enabled us to measure the structural properties of the transient liquid phase. Our results suggest that as this technique matures, it could become a very useful tool for understanding ultrafast atomic structural dynamics in physics and chemistry.

**Acknowledgment.** This work was partially supported by a grant-in-aid from the Ministry of Education, Culture, Sports, Science, and Technology of Japan (Grant No. 16032219).

## References

1. J. J. Rehr and R. C. Albers: Rev. Mod. Phys. **72**, 621 (2000).
2. T. Lee, F. Benesch, Y. Jiang, and C. G. Rose-Petruck: Chem. Phys. **299**, 233 (2004).
3. P. Forget, F. Dorchies, J-C. Kieffer, and O. Peyrusse: Chem. Phys. **299**, 259 (2004).
4. K. Oguri, Y. Okano, T. Nishikawa, and H. Nakano: Appl. Phys. Lett. **87**, 011503 (2005).
5. L. X. Chen, W. J. H. Jager, G. Jennings, D. J. Gosztola, A. Munkholm, and J. P. Hessler: Science **292**, 262 (2001).
6. B. Yaakobi, D. D. Meyerhofer, T. R. Boehly, et al.: Phys. Rev. Lett. **92**, 095504 (2004).
7. B. Yaakobi, T. R. Boehly, D. D. Meyerhofer, et al.: Phys. Rev. Lett. **95**, 075501 (2005).
8. C. Bressler and M. Chergui, Chem. Rev. **104**, 1781 (2004).
9. K. Sokolowski-Tinten, J. Bialkowski, A. Cavalleri, et al.: Phys. Rev. Lett. **81**, 224 (1998).
10. A. Rousse, C. Rischel, S. Fourmaux, et al.: Nature **410** (2001).
11. K. Murakami, H. C. Gerritsen, H. van Brug, F. Bijkerk, F. W. Saris, and M. J. van der Wiel: Phys. Rev. Lett. **56**, 655 (1986).
12. S. L. Johnson, P. A. Heimann, A. M. Lindenberg, et al.: Phys. Rev. Lett. **91**, 157403 (2003).
13. H. Nakano, Y. Goto, P. Lu, T. Nishikawa, and N. Uesugi: Appl. Phys. Lett. **75**, 2350 (1999).
14. P. K. Carroll and G. O'Sullivan: Phys. Rev. A **25**, 275 (1982).
15. T. Sagawa, Y. Iguchi, M. Sasanuma, et al.: J. Phys. Soc. Japan **21**, 2602 (1966).
16. C. Gähwiller and F. C. Brown: Phys. Rev. B **2**, 1918 (1970).
17. C. V. Shank, R. Yen, and C. Hirlimann: Phys. Rev. Lett. **50**, 454 (1983).
18. K. Laaziri, S. Kycia, S. Roorda, et al.: Phys. Rev. B **60**, 13520 (1999).
19. Y. Waseda K. Shinoda, K. Sugiyama, S. Takeda, K. Terashima, and J. M. Toguri: Jpn. J. Appl. Phys. **34**, 4124 (1995).
20. M. O. Thompson, G. J. Galvin, J. W. Mayer, et al.: Phys. Rev. Lett. **52**, 2360 (1984).

# Part VII

# Poster Session

# The Ehrenfest Theorem and Quantitative Predictions of HHG based on the Three-step Model

Ariel Gordon and Franz X. Kärtner

Department of Electrical Engineering and Computer Science, and Research Laboratory of Electronics, Massachusetts Institute of Technology, 77 Massachusetts Avenue, Cambridge, MA 02139, USA

## 1 Introduction

A primary challenge in high-harmonic generation (HHG) experiments is the optimization of the photon yield. To address this challenge, quantitative theoretical modeling of the single-atom response as well as propagation effects is needed. It has been shown that the three-step model (TSM) [1] gives an excellent qualitative description of the single-atom response to an intense laser pulse [2]. The structure of the spectrum and in particular the cutoff frequency are predicted correctly. However, so far there was only poor agreement between the photon yield predicted by the TSM and by direct numerical solutions of the Schrödinger equation (NSE). In fact, there is a discrepancy of about an order of magnitude in the radiation amplitude, which translates to a two orders of magnitude discrepancy in the photon yield. Moreover, the discrepancy is not constant over the spectrum. In other words, the TSM spectrum is also distorted in its shape.

Often HHG experiments are modeled using the TSM single-atom response [3]. The reason is that solving the combined Maxwell and Schrödinger equations simultaneously is still a formidable task even for modern computers. As a result, modern HHG literature does not offer, to our knowledge, quantitative theoretical predictions on HHG photon yields (unless parameter fitting is employed). Therefore, it is difficult to know whether the experimental photon yields are limited by the underlying physics or by the technical implementation.

We show that the TSM agrees with the NSE quantitatively up to a factor of about 1.5 in the photon yield if dipole acceleration is evaluated using the Ehrenfest theorem rather than by using time derivates of the TSM wave function. Our finding is important for quantitative modeling of HHG and for the tomographic imaging of molecular orbitals [4].

First we show that a large error occurs when the dipole acceleration is computed by direct time differentiation of the TSM dipole moment. This is not very surprising, since the quantum mechanical analysis of the TSM, in its original derivation [1] and ever since, is carefully aimed at calculating the dipole moment of the "active" electron. However what is actually required for calculation of the HHG radiation is the dipole acceleration [2].

In order to avoid the differentiation, we propose to employ the Ehrenfest theorem in the TSM as has been already used in the NSE to avoid numerical errors. Using it, however, in the TSM is not to avoid numerical errors but rather to describe the correct physics. We start with the approximate wave function delivered by the TSM, but compute the expectation value of the dipole acceleration under the exact dynamics during the recollision, including the Coulomb potential, using the Ehrenfest theorem.

## 2 Derivation

The TSM starts with the ansatz [1]

$$|\psi(t)\rangle = |0\rangle + |\varphi(t)\rangle, \tag{1}$$

where $|\psi(t)\rangle$ is the electron wavefunction, $|0\rangle$ is the ground state, and $|\varphi(t)\rangle$ represents the continuum part, satisfying

$$i|\dot{\varphi}(t)\rangle = \left[\frac{\mathbf{p}^2}{2} + I_\mathrm{p} - E(t)x\right]|\varphi(t)\rangle - E(t)x|0\rangle. \tag{2}$$

Atomic units ($\hbar = m = e = 1$) are used, $I_\mathrm{p}$ is the ionization potential, and $E(t)$ is the laser field. Depletion of the ground state has been neglected in Eq. (1) to make the discussion simpler. The standard TSM now associates HHG with the interference term in the dipole moment $\langle\psi(t)|x|\psi(t)\rangle$:

$$x(t) = \langle\varphi(t)|x|0\rangle + \langle 0|x|\varphi(t)\rangle. \tag{3}$$

In order to compute Eq. (3), a momentum completeness relation is inserted. That is what gives rise to the "recombination amplitude":

$$a_\mathrm{rec}(\mathbf{v}) = \langle 0|x|\mathbf{v}\rangle, \tag{4}$$

where $\mathbf{v}$ is the velocity of the electron upon return to the parent ion. The dipole moment is then differentiated twice in time to arrive at the dipole acceleration.

Alternatively, however, one can use the Ehrenfest theorem:

$$\ddot{x}(t) = -\left\langle\psi(t)\left|\frac{\partial V(\mathbf{r})}{\partial x}\right|\psi(t)\right\rangle + E(t), \tag{5}$$

where $V(\mathbf{r})$ is the atomic potential. The last term can be dropped since it does not contain HHG components. Using the ansatz (1), the interference term now reads

$$\ddot{x}(t) = -\left\langle 0\left|\frac{\partial V(\mathbf{r})}{\partial x}\right|\varphi(t)\right\rangle - \left\langle\varphi(t)\left|\frac{\partial V(\mathbf{r})}{\partial x}\right|0\right\rangle. \tag{6}$$

By the same reasoning, the "recombination amplitude" will be now replaced by

$$a_{\text{rec}}(\mathbf{v}) = -\left\langle 0 \left| \frac{\partial V(\mathbf{r})}{\partial x} \right| \mathbf{v} \right\rangle. \tag{7}$$

Therefore, if Eq. (4) in the TSM is replaced by Eq. (7), the TSM automatically gives the dipole acceleration rather than the dipole moment.

It is instructive to compare the dipole acceleration obtained by direct differentiation of Eq. (3) to the use of the Ehrenfest theorem. Using Eq. (2), it is easy to show that the leading term in the original TSM gives

$$\ddot{x}(t) = \left\langle 0 \left| \left[ \frac{\mathbf{p}^2}{2} + I_{\text{p}} \right] \right| \varphi(t) \right\rangle + c.c., \tag{8}$$

which corresponds to the "recombination amplitude"

$$a_{\text{rec}}(\mathbf{v}) = -\left\langle 0 \left| \left[ \frac{\mathbf{p}^2}{2} + I_{\text{p}} \right]^2 \right| \mathbf{v} \right\rangle, \tag{9}$$

instead of Eq. (7). Figure 1 compares the magnitude square of recombination amplitudes accoriding to Eqs. (9) and (7) for hydrogen. Indeed, the traditional TSM exceeds the one using the Ehrenfest theorem by one to two orders of magnitude and strong distortions of the spectrum occur.

Figure 2 compares the HHG spectrum obtained form the TSM using the Ehrenfest theorem with the one obtained from the NSE. Quantitative agreement is achieved over a wide range of wavelengths and field intensities, especially at

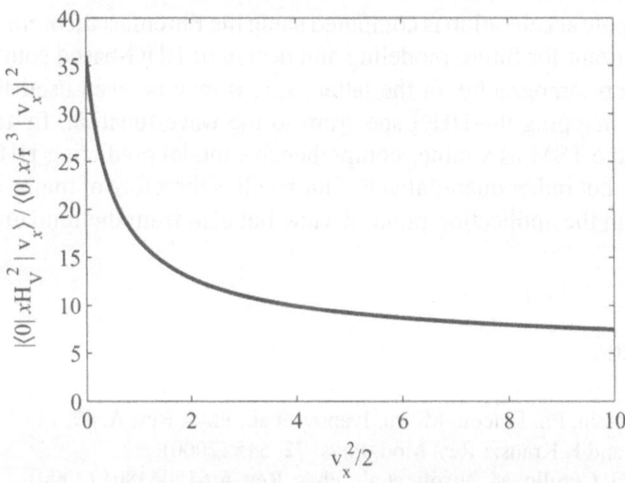


FIGURE 1. HHG spectrum of a hydrogen atom in a single-cycle sinusoidal "pulse" whose amplitude is 0.18 arb. units. The TSM spectrum was computed with the Ehrenfest theorem as explained above. It quantitatively agrees with the NSE spectrum.

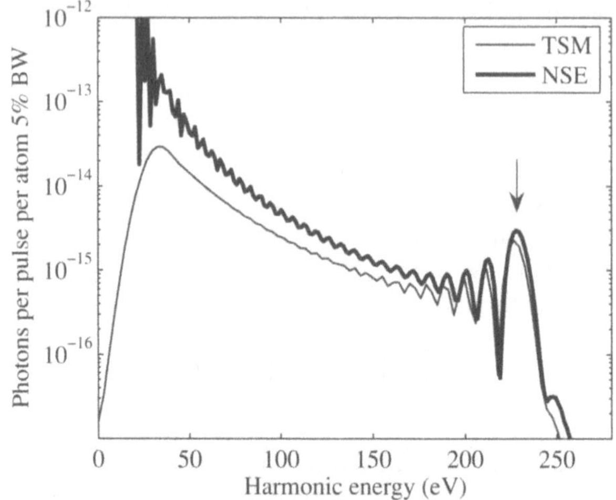


FIGURE 2. HHG spectrum of a hydrogen atom in a single cycle sinusoidal "pulse" whose amplitude is 0.18 arb. units. The TSM spectrum was computed with the Ehrenfest theorem as explained above. It quantitatively agrees with the NSE spectrum.

the cutoff region. The particular example of Fig. 2 is of a hydrogen atom in a single-cycle sinusoidal "pulse" with an amplitude of $E0 = 0.18$ arb. units.

## 3 Discussion

We have shown that the TSM can quantitatively describe HHG if used properly, i.e., if the dipole acceleration is computed using the Ehrenfest theorem. This result is very important for future modeling and design of HHG-based sources and for wave function tomography. In the latter case, it may be even used for defining a transform mapping the HHG spectrum to the wave function. In addition, we established the TSM as a rather comprehensive model predicting HHG not only qualitatively but rather quantitatively. Our result is therefore of major importance not only from the application point of view but also from the fundamental point of view.

## References

1. M. Lewenstein, Ph. Balcou, M. Yu. Ivanov, et al.: Phys. Rev. A. **49**, 2117 (1994).
2. T. Brabec and F. Krausz: Rev. Mod. Phys. **72**, 545 (2000).
3. E. Priori, G. Cerullo, M. Nisoli, et al.: Phys. Rev. A **61**, 063801 (2000).
4. J. Itatani, J. Levesque, D. Zeidleret, et al.: Nature **432**, 867 (2004).

# Modeling of Ni-like Molybdenum X-ray Laser

Sertan Kurnali,[1] Arif Demir,[2,3] and Elif Kacar[2]

[1] Zonguldak Karaelmas University Department of Physics 67100 Zonguldak, Turkey
**sertankurnali@yahoo.com**

[2] University of Kocaeli Department of Physics 41380 Kocaeli, Turkey
**arifd@kou.edu.tr, elifk@kou.edu.tr**

[3] University of Kocaeli Laser Technologies Research and Application Center 41380 Kocaeli, Turkey
**arifd68@yahoo.com**

**Summary.** Detailed simulations of Ni-like Mo X-ray laser are undertaken using the EHYBRID code. The atomic data are obtained using the Cowan code. The optimization calculations are performed in terms of intensity of background and time separation between the background and the short pulse. The optimum value is obtained for the conditions of a Nd:glass laser with 1.2 ps pulse at $3.5 \times 10^{14}$ W $\cdot$ cm$^{-2}$ irradiance pumping a plasma preformed by a 280-ps duration pulse at $1.2 \times 10^{12}$ W $\cdot$ cm$^{-2}$ with peak to peak pulse separation set at 360 ps. X-ray resonance lines between 25 Å and 37 Å emitted from Mo plasmas have been simulated.

## 1 Introduction

Collisional Ni-like Mo x-ray lasers with pumping from optical lasers have been developed using a wide range of elements with saturated output [1–6]. Ni-like X-ray lasers are modeled by various authors[1–7]. Significant reductions in the wavelengths for saturated output have been produced recently, because the necessary driving laser energy for collisional x-ray lasers is greatly reduced by using prepulse and multipulse irradiation. The initial low-intensity pulse produces Ni-like ions over a large plasma volume, and the high-intensity second pulse creates the population inversions. The large volume of the gain region with low electron-density gradients created in this way ensures that the X-ray laser beam propagates and amplifies without refracting out of the gain volume. One of the efficient ways of producing lasing at shorter wavelength uses Ni-like ions, in which the pump intensity can be reduced significantly because the quantum efficiency of the Ni-like systems is much higher than that of the Ne-like systems.

## 2 Method of Simulation

The 1.5-dimension hydrodynamic with atomic physics code EHYBRID simulates laser interaction with a solid target as used to create X-ray laser media. The code calculates electron temperature, electron density, the fractional population of each ionic stage from the neutral to bare nuclei, and using a collisional radiative treatment the populations of the excited levels of Ni- and Co-like ions at each time step for 98 spatial cells [8]. The input parameters for EHYBRID are the FWHM of the pulse duration, the power of the laser pulse, and the length, width, and thickness of the target.

Laser pulses of 1.054-μm wavelength focused on 100-μm wide and 1-cm long slab targets have been simulated. Molybdenum has been assumed pumped with a double pulse configuration in which a 1.2-ps short pulse is superimposed on a 280-ps long background pulse. Peak intensities assumed are $1.2 \times 10^{12}\ \mathrm{W \cdot cm^{-2}}$ and $3.5 \times 10^{14}\ \mathrm{W \cdot cm^{-2}}$ for the background and the short pulse, respectively. The effects of varying the time delay between background pulse and short pulse have also been studied. The peak-to-peak time between the short pulse and background pulse has been varied between 160 ps and 560 ps.

Resonance lines are often optically thick as they have strong oscillator strengths. Line intensities are calculated in a postprocessor to EHYBRID using the simulated $N_i$ populations densities and the $T$ escape factor. An approximate escape factor based on Holstein function [9] for Doppler-broadened lines is used in the simulation. The intensity $I_{\mathrm{tot}}$ of spectral lines at a particular time is calculated using

$$I_{\mathrm{tot}} = \sum N_i \frac{hc}{\lambda_0} T A_{ij} dV, \tag{1}$$

where $N_i$ is the upper state population for a given transition, $A_{ij}$ is the radiative transition probability for the transition, $h$ is Planck's constant, $c$ is the vacuum speed of light, $\lambda_0$ is the spectral wavelength, Vi is the volume of each cell, and the summation is over the EHYBRID cells. The absorption oscillator strengths and excited level energy and statistical weights for all levels are obtained from the Cowan code [10]. 128 Ni-like $1s^2 2s^2 2p^6 3s^2 3p^6 3d^{10} \longrightarrow 1s^2 2s^2 2p^6\, 3s^2 3p^6 3d^9$ nl and 230 Co-like $1s^2 2s^2 2p^6 3s^2 3p^6 3d^9 \longrightarrow 1s^2 2s^2 2p^6 3s^2 3p^6 3d^8$ nl resonance line intensities have been calculated for the simulation of the spectral emission from Mo plasma.

## 3 Results and Discussions

In collisional X-ray laser schemes, gain depends strongly on the electron temperature since the upper-lasing level is excited by collisional excitation. To produce X-ray lasing from Ni-like ions, it is desirable to ionize to the Ni-like ion ground state during the background pulse. The short pulse then creates the population

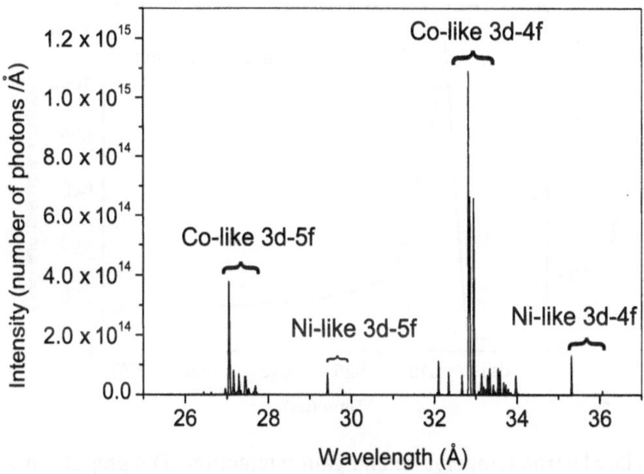


FIGURE 1. The simulated time-integrated Ni-like and Co-like resonance line spectrum. Long pulse duration is 280 ps pulse at $1.2 \times 10^{12}$ W · cm$^{-2}$. Short pulse duration is 1.2 ps pulse at $3.5 \times 10^{14}$ W · cm$^{-2}$.

inversion. Figure 1 shows resonance lines between 25 and 37 Å emitted from Mo plasma. Long pulse duration is 280 ps pulse at $1.2 \times 10^{12}$ W · cm$^{-2}$ and short pulse duration is 1.2 ps pulse at $3.5 \times 10^{14}$ W · cm$^{-2}$ in the calculation.

Figure 2 shows the variation of gain coefficient as a function of peak-to-peak time difference between the background and short pulse. The prepulse has an intensity of $1.2 \times 10^{12}$ W · cm$^{-2}$, and the main pulse has an intensity of $3.5 \times 10^{14}$ W · cm$^{-2}$ on the Mo slab target with 1-cm length, 100-μm width, and 21-μm thickness.

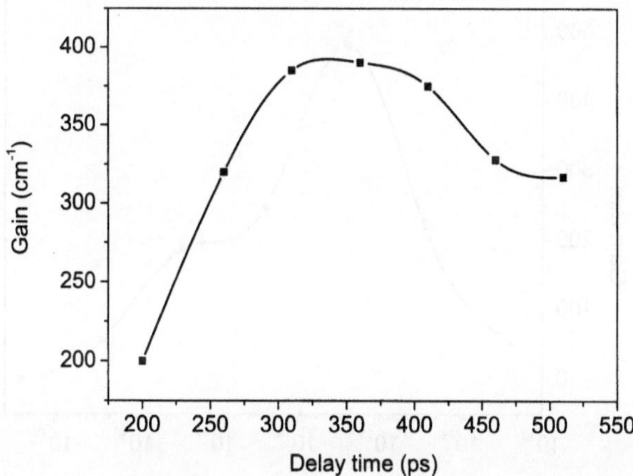


FIGURE 2. The effect of peak-to-peak delay times between pulses. The prepulse has an intensity of $1.2 \times 10^{12}$ W · cm$^{-2}$, and the main pulse has an intensity of $3.5 \times 10^{14}$ W · cm$^{-2}$.

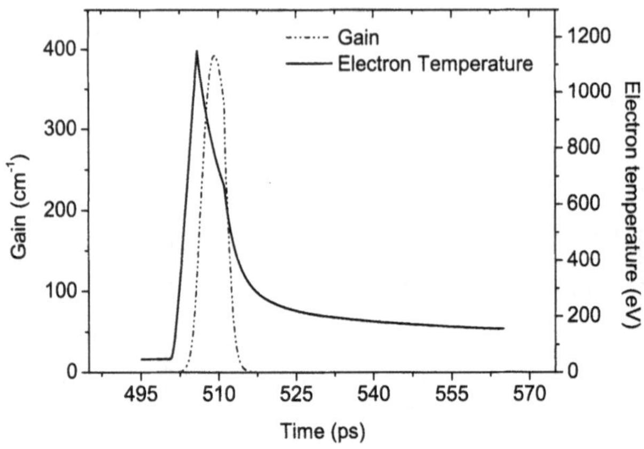


FIGURE 3. The electron temperature and gain interaction. The prepulse has an intensity of $1.2 \times 10^{12}$ W · cm$^{-2}$, and the main pulse has an intensity of $3.5 \times 10^{14}$ W · cm$^{-2}$.

In Fig. 3, the gain and electron temperature interaction in one cell is shown for plasma occurred by Nd:glass pumping laser pulses separated by 360 ps with the intensity value of $1.2 \times 10^{12}$ W · cm$^{-2}$ for prepulse with the duration of 280 ps and $3.5 \times 10^{14}$ W · cm$^{-2}$ for main pulse with the duration of 1.2 ps.

The effect of short pulse intensity on the gain is shown in Fig. 4. The peak gain is obtained at the intensity of $2.5 \times 10^{15}$ W · cm$^{-2}$. When the short pulse intensities increased, the plasma is over ionized and therefore the gain values decrease.

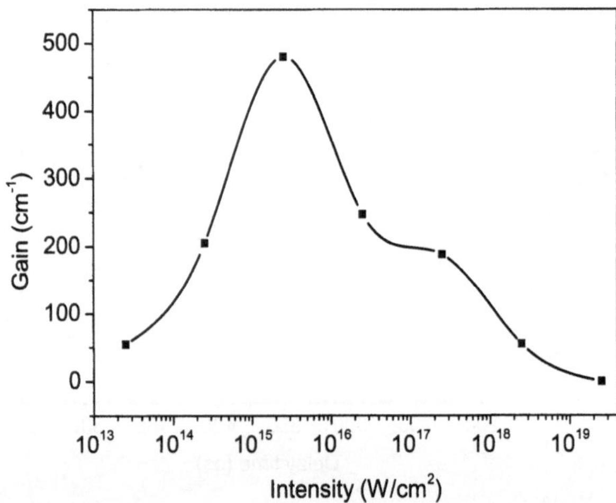


FIGURE 4. The effect of short pulse intensity on the gain. The prepulse intensity kept constant at $1.2 \times 10^{12}$ W · cm$^{-2}$, and the delay time kept constant at 340 ps.

# 4 Conclusions

The Ni-like and Co-like Mo resonance lines between 25 and 37 Å wavelength range have been simulated. The maximum gain ($\sim$400 cm$^{-1}$) is obtained about an intensity $\sim 2.5 \times 10^{15}$ W $\cdot$ cm$^{-2}$. Effect of peak-to-peak laser pulse delays on gain has been obtained. The optimum delay time is approximately 360 ps. FWHM gain is approximately 5.3 ps for the $3.5 \times 10^{14}$ W $\cdot$ cm$^{-2}$ short pulse laser simulation.

**Acknowledgments.** This work is supported by DPT under the contract 2004-K120710.

# References

1. J. Nielsen: J. Opt. Soc. Am. B **14**, 1511 (1997).
2. T. Ozaki, H. Nakano: J. Opt. Soc. Am. B **20**, 402 (2003).
3. T. Ozaki, R. A. Ganeev, A. Ishizawa, et al.: Phys. Rev. Let. **89**, 25, 253902 (2002).
4. T. Ozaki, H. Nakano, and H. Kuroda: Phys. Rev. E **66**, 047402 (2002).
5. J. Nilsen, J. Dunn, and A. L. Osterheld: Phys. Rev. A **60**, 4, R2677–R2680 (1999).
6. M. W. D. Mansfield, N. J. Peacock, C. C. Smith, et al.: J. Phys. B: Atom. Molec. Phys. **11**, 9 (1978).
7. Y. J. Li and J. Zhang: Phys. Rev. E **63**, 036410 (2001).
8. G. J. Pert: J. Fluid Mech. **131**, 401 (1983).
9. T. Holstein: Phys. Rev. **72**, 1212 (1947).
10. R. D. Cowan: J. Opt. Soc. Am. **58**, 808 (1968).

# Enhancement and Spatial Characteristics of K$\alpha$ X-ray Emission from High-contrast Relativistic fs Laser Plasmas

L. M. Chen,[1] M. Kando,[1] H. Kotaki,[1] K. Nakajima,[1,2] S. V. Bulanov,[1,3] T. Tajima,[1] M. H. Xu,[4] Y. T. Li,[4] Q. L. Dong,[4] and J. Zhang[4]

[1] Japan Atomic Energy Agency, 8-1 Umemida, Kizu, Kyoto 619-0215, Japan
chen.liming@jaea.go.jp
[2] High Energy Accelerator Research Organization, 1-1 Oho, Tsukuba, Japan
[3] Moscow Institute of Physics and Technology, Institutskij Preulok 9, Dolgoprudny, 141700 Moscow Region, Russia
[4] Institute of Physics and China Academy of Sciences, Beijing 100080, China

We studied the hard X-ray emission and the K$\alpha$ X-ray conversion efficiency ($\eta_K$) produced by 70 fs, 400 nm high-contrast femtosecond laser pulse focused on Cu solid target. At an intensity $I = 2 \times 10^{18}\ \mathrm{W \cdot cm^{-2}}$, the Cu $\eta_K$ can reach $1 \times 10^{-4}$. Compared with the Cu $\eta_K$ obtained by using a 800-nm low-contrast laser pulse, the second harmonic exhibits a higher X-ray flux by a factor of 2. We show that the use of high-contrast laser pulses may be an effective method to optimize the K$\alpha$ X-ray emission via the vacuum heating mechanism. In addition, the laser focusing was varied widely to give a range of intensities from $10^{15}$–$10^{18}\ \mathrm{W \cdot cm^{-2}}$ by target offset from best laser focus. Two individual emission peaks are obtained, one is at best focal spot and the other is at larger target offset corresponding to $\sim 10^{15}\ \mathrm{W \cdot cm^{-2}}$. Each X-ray emission peak is corresponding to different energy absorption mechanism.

## 1 Introduction

The availability of intense femtosecond laser pulses [1] recently opened a new laser–solid interaction regime in which intense laser pulses are deposited into a solid faster than the hydrodynamic expansion of the target surface. Hot electrons generated via collective absorption mechanisms such as resonant absorption (RA) [2] or vacuum heating (VH) [3] penetrate into the solid target to produce hard

X-rays via K-shell ionization and bremsstrahlung [4]. This new kind of intense and ultrafast hard X-ray source has a number of interesting applications for medical imaging techniques [5] because of its small X-ray emission size, its compactness, and the shortness of its pulse duration.

The control and optimization of the hard X-ray emission produced by high-intensity laser–solid interaction request an understanding of several mechanisms: the laser energy absorption, the hot electron generation, and the X-ray conversion. Several groups have already reported X-ray emission experiments relying on subpicosecond laser systems [6–12]. Previous works [6, 7] used hundreds of femtosecond laser pulse produced by $CO_2$ or Nd laser systems. Plasma density gradient steepened by ponderomotive force and satisfy the optimal conditions for RA, which is the main heating mechanism at this regime. Recently, it was shown that the use of shorter laser pulse durations less than 100 fs involves new X-ray emission processes. Eder et al. reported observing a maximum in K$\alpha$ emission when the target was placed away from best focus [8] and qualitatively explained it with the reabsorption of produced photons inside the target. Based on optimal scale length for RA, Reich et al. theoretically presented a scaling law to estimate the optimal laser intensity and predicted a reduction of the hard X-ray yield if the laser intensity is higher [9]. Zhidkov et al. studied prepulse effects with a low-contrast fundamental 42 fs laser [10]. They observed a decrease of the laser energy absorption for shorter pulse duration with constant laser energy, which was also proved by Schnürer et al. in experiment [11], and they reported the critical influence of the plasma gradient for the hard X-ray emission via the resonant process. All theses publications proved that there is a limitation for hard X-ray enhancement with laser intensity based on RA when tens of fs, low-contrast laser were used. In this paper, we show that this limit does not exist in the case of high-contrast femtosecond laser pulse. Cu $\eta_K$ exhibits a high flux when we work with a high-contrast laser pulse at 400 nm. Hot electrons should be accelerated by the laser pulse via VH mechanism and generate a lower electron temperature. Two individual emission peaks are obtained by displacement of target from best focal spot.

## 2 Experimental Setup

The experiments are realized with the high-intensity IOP (China) Ti:sapphire laser system. The laser delivers a maximum output energy of 600 mJ after compression with a pulse duration of 60 fs and a repetition rate of 10 Hz. The laser pulse energy can be changed conveniently by turning off some of the four laser pumps. After compression, the laser contrasts we measured with a third-order autocorrelator are $1 \times 10^5$ and $1 \times 10^4$ for nanosecond and picosecond pedestal, respectively. A KDP crystal (2-mm thick) is used to get the 400 nm second harmonic pulse. The double-frequency conversion efficiency of the KDP crystal is about 35% at 200 GW $\cdot$ cm$^{-2}$ intensity. It results in a pulse contrast ratio compared to the picosecond pedestal higher than $10^8$, as shown in Fig. 1. The

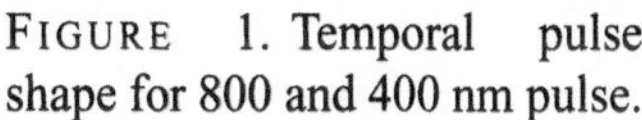

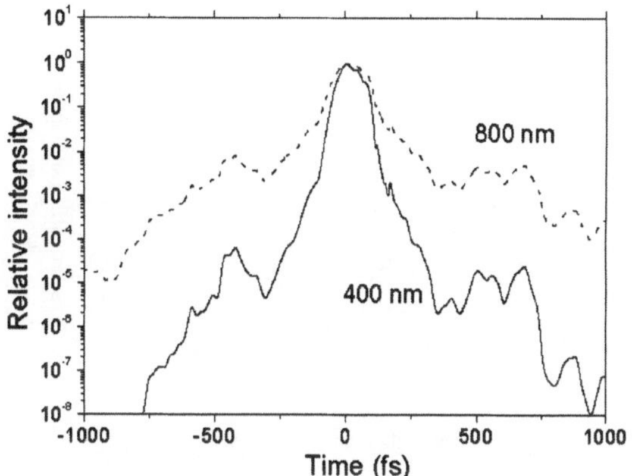


FIGURE 1. Temporal pulse shape for 800 and 400 nm pulse.

800-nm pulse shape is probed by SHG crosscorrelator, and the 400-nm pulse shape is estimated from $I(400)$–$I^2(800)$. Finally the P-polarized laser pulse is obliquely incident on the target with an average $2 \times 10^{18}$ W · cm$^{-2}$ intensity. Cu target was used in the experiments. The measurement of the X-ray spectrum and the determination of the $\eta_K$ are made with NaI (Ti) scintillators followed by a photomultiplier (Philips) [6]. An X-ray CCD camera (Princeton) is used as a dispersionless spectrometer for Cu Kα photon detection. The X-ray emission size is measured with a knife-edge imaging technique. The image magnification factor is equal to 50. The edge-spread function (ESF) obtained with this technique is fitted with a Fermi function $F(x) = a + b/[\exp(x - c)/d + 1]$. The differentiation of the Fermi function gives the line-spread function, which can be well fitted by a Gaussian distribution function. For 800-nm wavelength laser pulses, a diameter larger than 50 μm FWHM was inferred from the line spread function. It indicates that the lateral transport in our experiment condition is high. The X-ray emission size generated by 400 nm laser is 15 ± 5 μm. This result is much smaller than in the case of 800 nm laser pulse.

## 3 Experimental Results and Discussion

Figure 2a represents the Cu $\eta_K$ dependence as a function of the laser intensity with a 70-fs pulse duration. $\eta_K$ is higher for an increasing laser intensity and reaches higher than $1 \times 10^{-4}$ at $I = 2 \times 10^{18}$ W · cm$^{-2}$. This value is higher than the case of 800 nm theoretically predicted by Reith et al. [9] at same laser intensity, because its optimal intensity is $10^{16}$ W · cm$^{-2}$ (shown in dashed line). Fitting the experimental results with an exponential function, we find that the scaling law exponent concerning the intensity dependence of $\eta_K$ is $\gamma \approx 4.0$. Figure 2b shows the spectra measured on a Cu target with an X-ray CCD camera with the following parameters: 70 fs, 240 mJ, $2 \times 10^{18}$ W · cm$^{-2}$ at 800 nm (solid line) and 400 nm (dotted line) wavelength. We observe that the Kα yield at 400 nm is

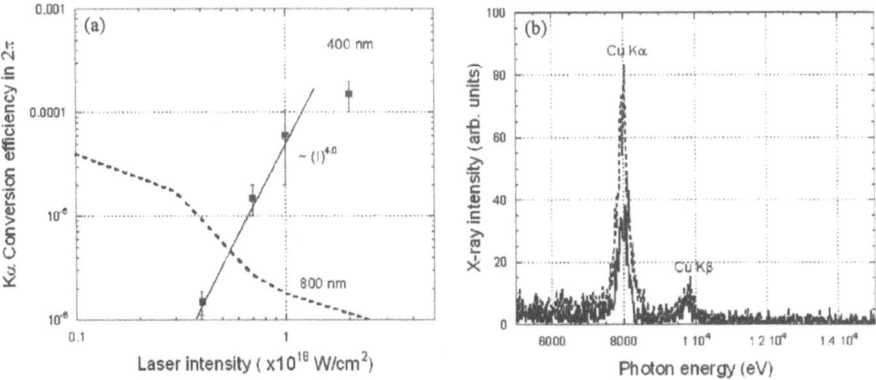


FIGURE 2. (a) Cu K$\alpha$ X-ray conversion efficiency as a function of the laser intensity for a 70-fs laser pulse duration and two different wavelengths: 400 nm (solid line) and 800 nm (dashed line). (b) Cu hard X-ray spectra measured with a CCD camera produced by 800 nm (solid line) and 400 nm (dotted line) laser pulse at $2 \times 10^{18}$ W $\cdot$ cm$^{-2}$ intensity.

higher than at 800 nm by a factor of 2. In addition, X-ray yield in case of 800-nm laser irradiation is also much higher than theoretical prediction at same laser intensity. The other mechanism except for resonant absorption was stimulated that was proved in Ref. [13]. The hard X-ray spectrum, shown in Fig. 3, is measured with a 400-nm laser incident on a Cu bulk target at intensity $2 \times 10^{18}$ W $\cdot$ cm$^{-2}$. The hot electron temperature achieved is about 16 keV fitting with Maxwellian distribution. The temperature obtained is lower than the one at 800 nm at the same laser intensity (~25 keV). We have to notice the spectrum show structure with energy cutoff at electron energy about 30 keV.

$\eta_K$ is controlled by the efficiency of two factors: the conversion of the laser energy into hot electrons ($\eta_{LH}$) and the conversion of hot electrons into K$\alpha$ photons ($\eta_{HK}$): $\eta_K = \eta_{LH}\ \eta_{HK}$. According to theoretical calculations, $\eta_{HK}$ will be optimized when $K_{Th}/E_K$ is about 5 in the case of an Cu target [14]. According

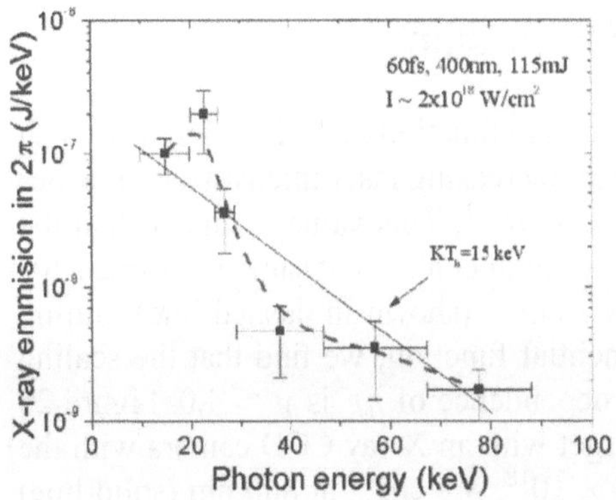


FIGURE 3. Hard X-ray spectrum measured with PMT at $2 \times 10^{18}$ W $\cdot$ cm$^{-2}$ intensity with a 70-fs pulse duration and 400 nm. The hot electron temperature obtained is 16 keV.

to this fact, we believe that $\eta_{HK}$ has the same behavior in our experimental condition at 800 and 400 nm. It means that for a 800-nm laser pulse $\eta_{HK}$ is higher than that at 400 nm. Therefore, a higher $\eta_K$ for a 400-nm laser pulse corresponds to a higher $\eta_{LH}$, which means a greater laser energy absorption by hot electrons.

According to Freshel's equations prediction and Price's experimental results [15], inverse bremsstrahlung (IB) decreases for an increasing intensity, whereas in our results the X-ray emission increases as a function of the laser intensity. Therefore, it implies that an additional absorption mechanism is stimulated. If we consider that RA is the main additional mechanism to generate hot electrons [2], a 800-nm laser pulse should be more effective than a 400 nm one ($L < 0.1\lambda$) because it presents a larger pedestal which induces a longer $L$ and greatly stimulate plasma waves. However, our measurement does not agree with this assumption. A possibility is that RA involves a slow, many-cycle buildup of a resonant plasma wave and electric field at the critical surface that eventually breaks. The necessary plasma condition for nonlinear wave-breaking regime is that L must be greatly superior to the electron quiver amplitude. In our steep gradients, the buildup time is less than a cycle and the critical surface field enhancement should be negligible [16] for 70-fs pulses. Therefore, another energy absorption mechanism should be stimulated instead of RA.

Simulation shows VH dominates RA for steep density gradient [3, 17]. VH means that a P-polarized light pulse is obliquely incident on an atomically abrupt metal surface in order to be strongly absorbed by pulling electrons into vacuum during an optical cycle, then returning to the surface with approximately the quiver velocity [3]. In our laser condition with intensity of $1 \times 10^{18}\ \mathrm{W \cdot cm^{-2}}$, the plasma scale length is less than $0.1\lambda$ according to Ref. [17] in similar conditions. It agrees with the necessary condition to stimulate VH: $X_{\mathrm{osc}} \geq L$. It also satisfies the optimal condition for VH: $V_{\mathrm{osc}}/c \geq 3.1(L/\lambda)^2$ [17] in which $V_{\mathrm{osc}} = eE/me\omega$ is the electron quiver velocity in the laser field that is governed by the quiver energy $E_q = mc^2[(1 + 2U_p/mc^2)^{1/2} - 1]$, where $U_p\ (\mathrm{eV}) = 9.3 \times 10^{-14} I\lambda^2$ is the ponderomotive potential. The most important evidence is the hot electron temperature we detected, i.e., 16 keV, which exactly fits the scaling law for VH: $KT_h - E_q = 15$ keV. This temperature suppresses the possibility of J $\times$ B heating [18] because the increase of electrons energy due to the ponderomotive potential is $E_p = (\gamma - 1)mc^2$ ~30 keV in this intensity. It also does not fit with RA scaling laws which gives ~33 keV. The X-ray emission size we measured with a knife-edge imaging technique [6] confirmed our electron temperature measurement. The X-ray emission size generated by 400 nm laser is $15 \pm 5$ μm. This value is much smaller than in the case of 800 nm laser pulse (>80 μm). So, we conclude that VH is stimulated and may be the main absorption mechanism in our experiment. Some other competing linear mechanisms [19–21] such as anomalous skin effect [19] should be weak in our experiments because the skin depth, where $v_{th} = (2kT/m)^{1/2}$, is much thicker than the required collisionless skin transit depth ($l_s \sim 100$ Å $\ll v_{th}/v$) needed in these mechanisms.

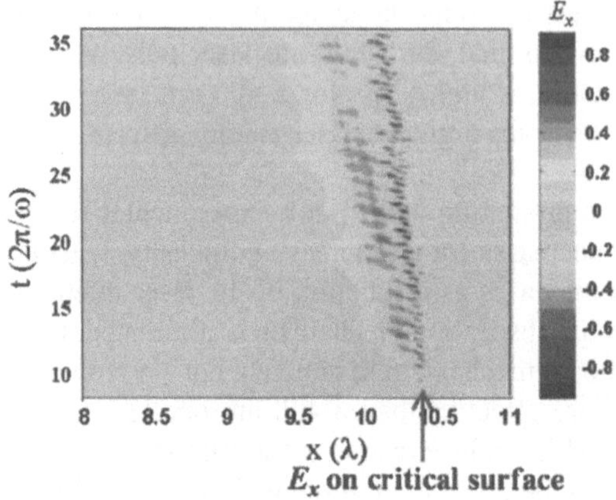


FIGURE 4. (Color) Temporal dependence of the amplitude of electrical field $E_x$. The critical surface start from $x = 10.3$.

Simulations using a 1D fully electromagnetic LPIC++ code have been performed, where an electromagnetic wave is launched obliquely onto an overdense plasma located on the right-hand side. We used the following simulation parameters: $n_e/n_c = 20$, $T_e = 100$ eV, $T_e/T_i = 3$–5 and mass ratio $m_i/Zm_e = 1836$. The initial scale length is $L/\lambda = 0.05$. A square-sine profile for the incident laser pulse is used. Figure 4 shows the temporal dependence of the amplitudes of electric field in the normal direction close to the solid surface. It clearly shows that a group of electrons is pulled out into vacuum at each optical circle and return it to target surface. The electric field at the target surface, which polarity is changed periodically, reflected this "pull–push" procedure in each laser half period [3,17]. Electrons will absorb laser energy continuously in this way. It demonstrates that the VH is greatly stimulated.

Experiment and simulation [17] proved that, with a slight surface expansion of the scale length, the optical field would pull more electrons into vacuum and thus be more strongly absorbed, as long as $L$ does not significantly exceed $x_{osc}$. In our experiments, when the laser intensity increases, a larger pulse pedestal generates more preplasma and higher laser quiver amplitude as well. Both of the parameters increase the energy absorbed. That is why the exponent parameter for $\eta_K$ dependence as a function of laser intensity is high.

A very interesting phenomenon is observed when we offset target from the best focal spot, as shown in Fig. 5. The laser focusing was varied widely to give a range of intensities from $10^{15}$ to $10^{18}$ W $\cdot$ cm$^{-2}$ by target offset from best laser focus. Two individual emission peaks are obtained, one is at the best focal spot position with laser intensity $10^{18}$ W $\cdot$ cm$^{-2}$ and the other is at larger target offset corresponding to $\sim 10^{15}$ W $\cdot$ cm$^{-2}$. Each X-ray emission peak is corresponding to different energy absorption mechanism. The first peak corresponds to Vacuum heating mechanism as discussed above. The second peak, as predicted by D. F. Price et al. [15], is believed to be the contribution of IB mechanism.

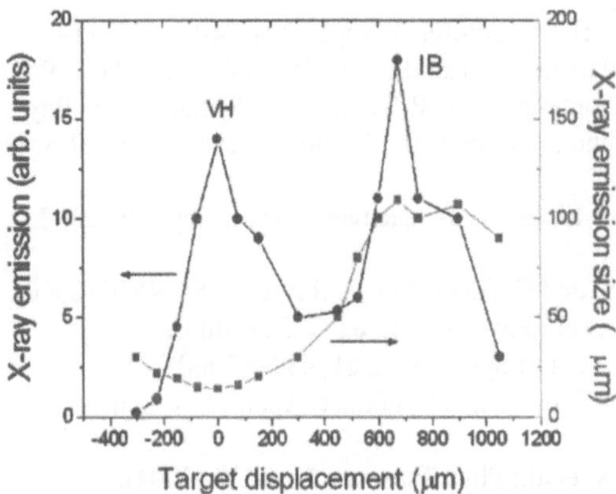


FIGURE 5. X-ray emission source size and emission intensity vs target offset from best focus.

# 4 Summary

In summary, Cu $\eta_K$ produced by a high-contrast laser pulse at 400 nm with intensity $I = 2 \times 10^{18}$ W · cm$^{-2}$ reaches $1 \times 10^{-4}$. While $\eta_K$ is much higher with high-contrast laser pulse at 400 nm, compared to low-contrast laser pulse at 800 nm. It implies an effective method for hard X-ray enhancement in fs–plasma regime: RA maybe noneffective with femtosecond laser dense plasma interactions, whereas high-contrast laser is more efficient for hard X-ray generation via vacuum heating. In addition, the laser focusing was varied widely to give a range of intensities from $10^{15}$ to $10^{18}$ W · cm$^{-2}$ by target offset from best laser focus. Two individual emission peaks are obtained, one is at the best focal spot and the other is at larger target offset corresponding to $\sim 10^{15}$ W · cm$^{-2}$. Each X-ray emission peak is corresponding to different energy absorption mechanism.

# References

1. M. D. Perry and G. Mourou: Science **264**, 917(1994); D. Strickland and G. Mourou: Opt. Commun. **56**, 219 (1985).
2. W. L. Kruer: The Physics of Laser Plasma Interactions. New York: Addison-Wesley (1988).
3. F. Brunel: Phys. Rev. Lett. **59**, 52 (1987); Phys. Fluids. **31**, 2714 (1998).
4. H. Chen, B. Soom, B. Yaakobi, S. Uchida, and D. D. Meyerhofer: Phy. Rev. Lett. **70**, 3431 (1993); T. Feurer, W. Theobald, R. Sauerbrey, et al.: Phys. Rev. E. **56**, 4608 (1997).
5. A. Krol, A. Ikhlef, J. C. Kieffer et al.: Med. Phys. **24**(5), 725 (1997); J. C. Kieffer, A. Krol, Z. Jiang, C. Chamberlain, E. Scalzetti, and Z. Ichalalene: Appl. Phys. B 74, 75 (2002).
6. J. Yu, Z. Jiang, J. C. Kieffer, and A. Krol: Phys. Plasmas. 6(4), 1318 (1999).

7. B. Soom, H. Chen, Y. Fisher, and D. D. Meyerhofer: J. Appl. Phys. **74**(9), 5372 (1993); M. Schnürer, P. V. Nickles, M. P. Kalachnikov, et al.: J. Appl. Phys. **80**(10), 5604 (1996); U. Teubner, J. Bergmann, B. van Wonterghem, F. P. Schäfer, and R. Sauerbrey: Phys. Rev. Lett. **70**, 794 (1993); U. Teubner, I. Uschmann, P. Gibbon, et al.: Phys. Rev. E **54**, 4167 (1996).
8. D. C. Eder, G. Pretzler, E. Fill, K. Eidmann, A. saemann: Appl. Phys. B, **70**, 211 (2000).
9. Ch. Reich, P. Gibbon, I. Uschmann, and E. Forster: Phys. Rev. Lett, **84**, 4846 (2000).
10. A. Zhidkov, A. Sasaki, T. Utsumi, et al.: Phys. Rev. E, **62**, 7232 (2000).
11. M. Schnürer, R. Nolte, A. Rousse, et al.: Phys. Rev. E, **61**, 4394 (2000).
12. U. Andiel, Eidmann, K. Witte, I. Uschmann, and E. Förster: Appl. Phys. Lett. **80**, 198 (2002).
13. L. M. Chen, P. Forget, S. Fourmaux, et al.: Phys. Plasmas, **11**, 4439 (2004).
14. D. Salzmann, Ch. Reich, I. Uschmann, E. Forster: Phys. Rev. E, **65**, 036402 (2002).
15. D. F. Price, R. M. More, R. S. Walling, et al.: Phys. Rev. Lett. **75**, 252 (1995).
16. J. Albritton and P. Koch: Phys. Fluids. **18**, 1136 (1975); P. J. Catto and R. M. More: ibid. **20**, 704 (1977).
17. P. Gibbon and A. R. Bell: Phys. Rev. Lett. **68**, 1535 (1992); L. M. Chen, J. Zhang, Q. L. Dong, et al.: Phys. Plasmas. **8**, 2925 (2001); M. K. Grimes, A. R. Rundquist, Y. S. Lee, M. C. Downer: Phys. Rev. Lett. **82**, 4010 (1999).
18. S. C. Wilks, W. L. Kruer, M. Tabak, and A. B. Langdon: Phys. Rev. Lett, **69**, 1383 (1992).
19. E. S. Weibel: Phys. Fluids. **10**, 741 (1967); A. A. Andreev, et al.: Phys. Rev. E **58**, 2424 (1998).
20. P. J. Catto and R. M. More: Phys. Fluids. **20**, 704 (1977); T. -Y. Brian Yang, W. L. Kruer, R. M. More, et al.: Phys. Plasmas. **2**, 3146 (1995).
21. T. -Y. Brian Yang, W. L. Kruer, A. B. Langdon, et al.: Phys. Plasmas. **3**, 2702 (1996).

# Optical-field Ionization by Femtosecond Laser Pulses of Time-Dependent Polarization

Takamasa Ichino, Kenichi L. Ishikawa, and Masaharu Nakazawa

Department of Quantum Engineering and Systems Science, School of Engineering, University of Tokyo, Hongo 7-3-1, Bunkyo-ku, Tokyo 113-8656, Japan
itino@sophie.q.u-tokyo.ac.jp

**Summary.** We theoretically study the optical-field ionization of a hydrogen atom by an intense femtosecond laser of time-dependent polarization, based on direct solution of the time-dependent Schrödinger equation. In particular, we consider linearly polarized pulses in the broad sense whose plane of polarization rotates in time. The ionization rate depends on the rotation frequency of the plane of polarization, even though the temporal profile of field strength remains unchanged.

## 1 Introduction

Quantum control is guiding quantum-mechanical processes of atoms or molecules toward a desired outcome with specifically designed light fields [1,2]. The main experimental tool for achieving this goal has been spectral phase shaping of femtosecond laser pulses [3]. Various implementations have been reported in recent years. They were about chemical reactions [4], molecular population transfer [5], atomic multiphoton absorption [6], or single Raman spectroscopy [7]. In all of these experiments, spectral-temporal intensity and phase characteristics of laser fields are manipulated and exploited, in many cases using automated feedback learning loops [8, 9]. But such pulse shaping accesses only the scalar properties of ultrashort laser pulses, while the electric field is vectorial quantity.

Recently it has become possible to generate and control a time-dependent polarization (TDP) pulse in which the polarization changes as a function of time within a single femtosecond pulse [10]. It allows us to access the vectorial quantity of laser pulse. The TDP pulses have many potential applications. For example, multiphoton ionization processes in aligned $I_2$ molecules have been actively controlled by the pulse shaping system [11], or the TDP states have been suggested for the generation [12, 13] and characterization [14] of attosecond light pulse.

Basis of these interesting phenomena is ionization of atoms or molecules. In high-intensity laser field, the atomic potential is distorted by the electric field and ionization occurs through tunneling or directly [15]. This ionization process is called optical-field ionization (OFI). Most existing research of OFI treated linear, circular, or elliptical polarization, constant in time [16–19]. However, by the use of the TDP pulses, new types of atomic and molecular phenomena may emerge and more flexible control of quantum processes may become possible.

In this work, we study interaction between the TDP pulses and a hydrogen atom. We develop a numerical code to solve the time-dependent Schrödinger equation (TDSE) for a hydrogen atom in the TDP pulse and, in particular, we study OFI of a hydrogen atom in the linearly polarized pulse in the broad sense whose plane of polarization rotates. We find that the OFI rate depends on the rotation frequency of the plane of polarization, even though the temporal profile of field strength remains unchanged.

## 2 Model

The electric field of laser pulse of the TDP propagating in the $y$-direction has the $x$-component $E_x(t)$ and the $z$-component $E_z(t)$. To study the interaction of a hydrogen atom with the TDP laser pulse, we numerically solve the one-electron TDSE,

$$i\frac{\partial \psi(\mathbf{r},t)}{\partial t} = \left(-\frac{1}{2}\nabla^2 - \frac{1}{r} + zE_z(t) + xE_x(t)\right)\psi(\mathbf{r},t), \tag{1}$$

in the length gauge. We expand the wave function $\psi(\mathbf{r},t)$ in spherical harmonics,

$$\psi(\mathbf{r},t) = \sum_{l=0}^{L}\sum_{m=-l}^{l} \Phi_l^m(r,t) Y_l^m(\theta,\phi), \tag{2}$$

and then discretize the radial wave function $\Phi_l^m(r,t)$ on a spatial grid $r_j = (j-\frac{1}{2})\Delta r$, where $\Delta r$ denotes the mesh size. We now insert this representation into Eq. (1) and then defining $g_{lm}^j = r_j \Phi(r_j,t)$, we derive the following coupled equations for time evolution:

$$\begin{aligned} i\frac{\partial}{\partial t} g_{lm}^j =& -\frac{2}{(\Delta r)^2}\{c_j g_{lm}^{j+1} + c_{j-1} g_{lm}^{j-1} - 2d_j g_{lm}^j\} + \left\{2\frac{l(l+1)}{r_j^2} - \frac{1}{r_j}\right\} g_{lm}^j \\ &+ \frac{1}{2}E_x(t) r \sin\theta\cos\phi\{b_l^{-m} g_{l+1m-1}^j - b_{l-1}^{m-1} g_{l-1m-1}^j \\ &- b_l^m g_{l+1m+1}^j + b_{l-1}^{-m-1} g_{l-1m+1}^j\} \\ &+ E_z(t) r\cos\theta\{a_l^m g_{l+1m}^j + a_{l-1}^m g_{l-1m}^j\}, \end{aligned} \tag{3}$$

where the coefficients are

$$a_l^m = \sqrt{\frac{(l-m+1)(l+m+1)}{(2l+1)(2l+3)}}, \; b_l^m = \sqrt{\frac{(l+m+2)(l+m+1)}{(2l+1)(2l+3)}},$$
$$c_j = \frac{j^2}{j^2-1/4}, \; d_j = \frac{j^2-j+1/2}{j^2-j+1/4}. \tag{4}$$

To prevent reflection of the wave function from the grid boundary, after each time step the wave function is multiplied by a $\cos^{1/8}$ mask function [16] that varies from 1 to 0 over a width of 2/9 of the maximum radius at the outer-radial boundary. We integrate Eq. (3) with a method elaborated based on the finite difference method. Ionization is evaluated as the number of electrons absorbed by the mask function. Typically we use $\Delta r = 0.25$ arb. units and include the values of $l$ up to 40.

# 3 Results and Discussions

## 3.1 Ionization Rate

We consider the ionization of a hydrogen atom in a TDP pulse whose electric field is assumed to be given by

$$E_z(t) = E_0 \cos \Omega t \sin \omega t, \; E_x(t) = E_0 \sin \Omega t \sin \omega t, \tag{5}$$

which represents a linearly polarized pulses whose plane of polarization rotates with a frequency $\Omega$. It should be noted that the temporal profile of the field strength ($\sqrt{E_z(t)^2 + E_x(t)^2} = E_0 |\sin \omega t|$ ) is independent of $\Omega$ and the same as for purely linear polarization. Figure 1 shows laser field strength as a function of time. The wavelength of laser pulse is 800 nm, or angular frequency $\omega$ is 2.35 rad/fs, and laser intensity is $10^{14}$ W $\cdot$ cm$^{-2}$. The pulse shape is trapezoidal, with linear turn on, with a duration of one-fourth of laser period ($T = 2.67$ fs), and the total duration of laser pulse is $8T$. It should also be noted that this pulse is not circularly polarized. Since in the quasistatic approximation instantaneous ionization rate would depend only on field strength, we would expect that the OFI rate is also independent of $\Omega$. Figure 2 shows the temporal evolution of ionization by the TDP pulse for different values of $\Omega/\omega$. Rotation frequency of plane of polarization $\Omega$ alters in $0 \leq \Omega \leq \omega$. We calculate ionization rate as a time-averaged gradient of each curve. Figure 3 shows the calculated OFI rate as a function of $\Omega/\omega$. Unexpectedly, this figure clearly shows that the OFI rate depends on $\Omega$ and oscillates near $\Omega/\omega = 0$. This result cannot be explained by the above discussion and suggests that ionization in high-intensity laser pulses depends also on the history of electric field direction, not only on that of of field strength.

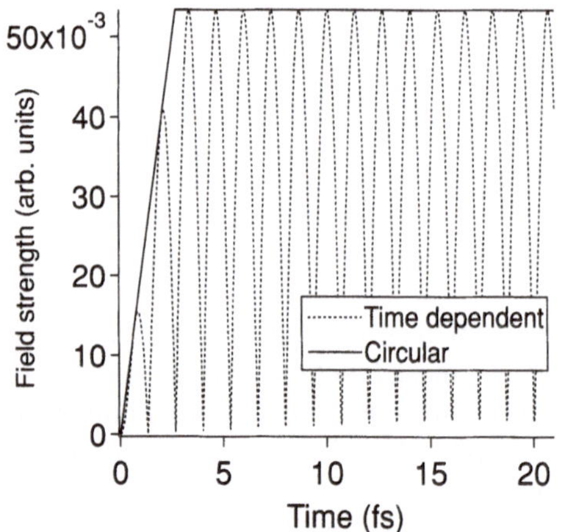


FIGURE 1. Laser field strength as a function of time by the TDP pulse considered in the present study (Eq. (5)) and circular polarized pulse with a peak intensity of $10^{14}\mathrm{W}\cdot\mathrm{cm}^{-2}$ and a wavelength of 800 nm.

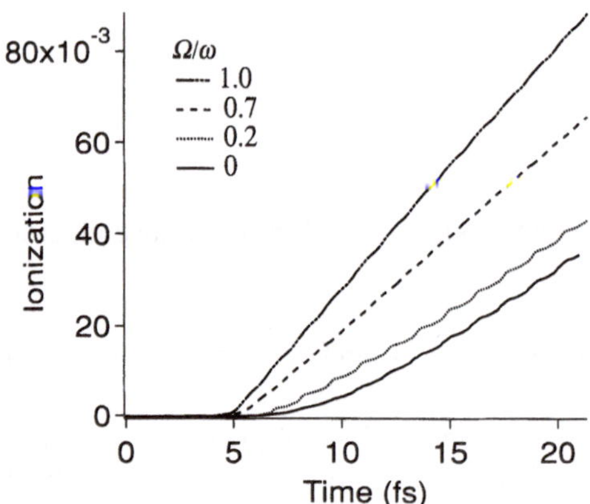


FIGURE 2. Ionization as a function of time by the TDP pulse Eq. (5) with a peak intensity of $10^{14}\mathrm{W}\cdot\mathrm{cm}^{-2}$ and a wavelength of 800 nm for several values of $\Omega/\omega$.

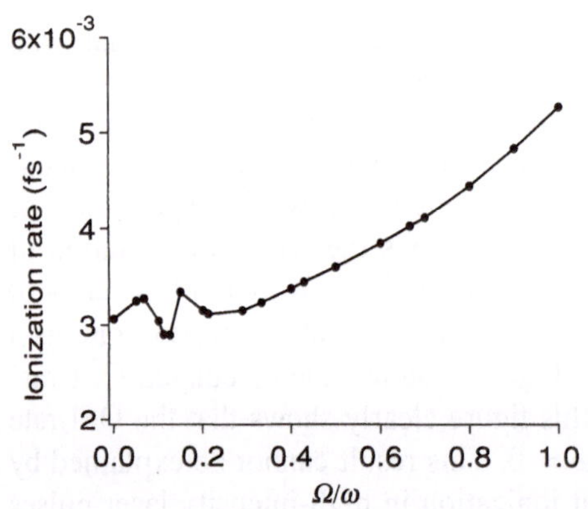


FIGURE 3. Ionization rate as a function of $\Omega/\omega$ by the TDP pulse Eq. (5) with a peak intensity of $10^{14}\ \mathrm{W}\cdot\mathrm{cm}^{-2}$ and a wavelength of 800 nm.

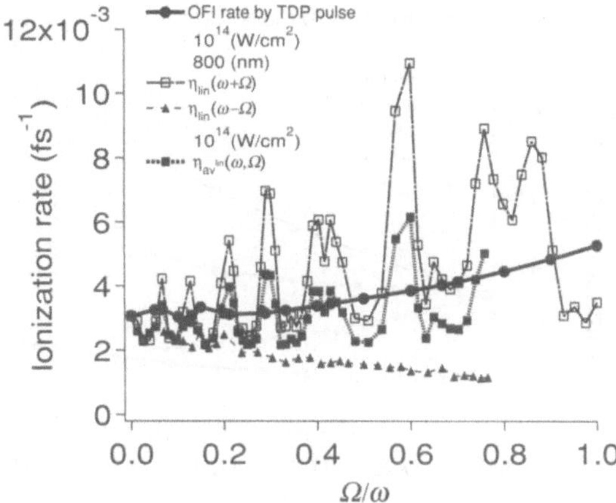


FIGURE 4. Comparison of ionization rate as a function of $\Omega/\omega$ between the TDP pulse and linearly polarized pulse with a peak intensity of $10^{14}$ W $\cdot$ cm$^{-2}$ and a wavelength of 800 nm.

## *3.2 Comparison With Linear and Circular Polarization*

In order to investigate the origin of the dependence of the OFI rate on $\Omega$, we rewrite $E_z$ and $E_x$ in Eq. (5) as follows:

$$E_z(t) = \frac{1}{2} f_z\{\sin(\omega + \Omega)t + \sin(\omega - \Omega)t\} \tag{6}$$

$$E_x(t) = \frac{1}{2} f_x\{-\cos(\omega + \Omega)t + \cos(\omega - \Omega)t\}. \tag{7}$$

This equation indicates that the pulse is composed of two frequency components, $\omega + \Omega$ and $\omega - \Omega$. Here we define $\eta_{\text{lin}}(\omega)$ as a ionization rate of H by linear polarization pulse of frequency $\omega$. We also define

$$\eta_{\text{av}}^{\text{lin}}(\omega, \Omega) = \frac{1}{2}\{\eta_{\text{lin}}(\omega + \Omega) + \eta_{\text{lin}}(\omega + \Omega)\}. \tag{8}$$

Figure 4 shows $\eta_{\text{lin}}(\omega + \Omega)$, $\eta_{\text{lin}}(\omega + \Omega)$, $\eta_{\text{av}}^{\text{lin}}(\omega, \Omega)$, and OFI rate by the TDP pulse as a function of $\Omega/\omega$. The figure shows that the OFI rate by the TDP pulse and linear polarization pulse are of comparable magnitude and have a similar general trend: the rate increases as a function of $\Omega/\omega$. However the ionization rate by the linear polarization pulse exhibits peak structure due to multiphoton resonance.

The laser electric field of the TDP pulse as Eqs. (6) and (7) can also be viewed as combined circular polarization pulses, i.e., right-handed circularly polarized pulse of frequency $\omega + \Omega$ and left-handed circularly polarized pulse of frequency $\omega - \Omega$. So we define $\eta_{\text{cir}}(\omega)$ as ionization rate of H by circular polarization pulse of frequency $\omega$. And we define as linear polarized pulse

$$\eta_{\text{av}}^{\text{cir}}(\omega, \Omega) = \frac{1}{2}\{\eta_{\text{cir}}(\omega + \Omega) + \eta_{\text{cir}}(\omega + \Omega)\}. \tag{9}$$

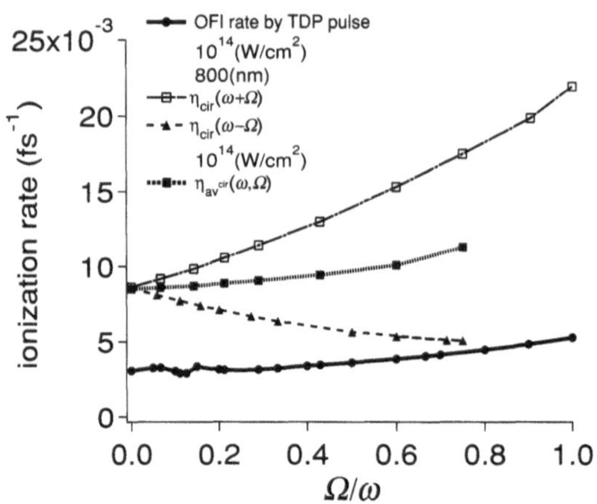


FIGURE 5. Comparison of ionization rate as a function of $\Omega/\omega$ between the TDP pulse and circularly polarized pulse with a peak intensity of $10^{14}$ $\mathrm{W} \cdot \mathrm{cm}^{-2}$ and a wavelength of 800 nm.

Figure 5 shows $\eta_{\mathrm{cir}}(\omega + \Omega)$, $\eta_{\mathrm{cir}}(\omega + \Omega)$, $\eta_{\mathrm{av}}^{\mathrm{cir}}(\omega, \Omega)$, and OFI rate by the TDP pulse as a function of $\Omega/\omega$. The figure shows that the OFI rates with the TDP pulse and circular polarization pulse are both increasing as a function of $\Omega/\omega$. And unlike linear polarization pulse, ionization rate by circular polarization pulse has no peaks by multiphoton resonance. However it is much higher ionization rate than the TDP pulse, since the field strength of circularly polarized pulse is constant in time, while that of the TDP and linearly polarized pulse is oscillating, as is shown in Fig. 1.

From previous comparisons, in magnitude of OFI rate, the TDP pulse has nature of linear polarization pulse whose electric field is oscillating. But in view of the absence of peaks by multiphoton resonance, the TDP pulse has nature of circular polarization pulse. Thus we may say that the TDP pulse has natures of both linearly and circularly polarized pulses. Figure 3 shows small oscillation near $\Omega/\omega = 0$. It is because rotation frequency is so small that the pulse is regarded as nearly linear polarization pulse and that multiphoton resonance takes place.

# 4 Conclusions

We have studied OFI of a hydrogen atom in the TDP pulse by developing a numerical code to solve TDSE. We have calculated OFI rate of H in laser pulses whose plane of polarization rotates with a frequency $\Omega$ and found that OFI rate depends on $\Omega$ even though the temporal profile of field strength remains unchanged.

Our analysis suggests that even in the OFI regime, atoms feel not only the field strength but also the frequency of each of the $x$- and $z$-components of the laser electric field or photon energy. The TDP pulse has features of both linear and circular polarization: linear-polarization-like feature is comparable absolute value of ionization rate, and circular-polarization-like feature is ionization rate with no peaks by multiphoton resonance.

**Acknowledgments.** K. L. Ishikawa sincerely acknowledges financial support by Ministry of Education, Culture, Sports, Science and Technology of Japan Grant 15035203 and 15740252, and also by a grant from Research Foundation for Opto-Science and Technology.

## References

1. T. Brixner and G. Gerber: Chem. Phys. Chem. **4**, 418 (2003).
2. T. Brixner, G. Krampert, T. Pfeifer, et al.: Phys. Rev. Lett. **92**, 208301 (2004).
3. A. M. Weiner: Rev. Sci. Instrum. **71**, 1929 (2000).
4. R. J. Levis, G. M. Menkir, and H. Rabitz: Science **292**, 709 (2001).
5. T. Brixner, N. H. Damrauer, P. Niklaus, and G. Gerber: Nature ( London) **414**, 57 (2001).
6. D. Meshulach and Y. Silberberg: Nature (London) **396**, 239 (1998).
7. N. Dudovich, D. Oron, and Y. Silberberg: Nature (London) **418**, 512 (2002).
8. R. S. Judson and H. Rabitz: Phys. Rev. Let. **68**, 1500 (1992).
9. D. Yelin, D. Meshulach, and Y. Silberberg: Opt. Lett. **22**, 1793 (1997).
10. T. Suzuki, S. Minemoto, and H. Sakai: Appl. Opt. **43**, 6047 (2004).
11. T. Suzuki, S. Minemoto, T. Kanai, and H. Sakai: Phys. Rev. Lett. **92**, 133005 (2004).
12. P. B. Corkum, N. H. Burnett, and M. Y. Ivanov: Opt. Lett. **19**, 1870 (1994).
13. M. Y. Ivanov, P. B. Corkum, T. Zuo, and A. D. Bandrauk: Phys. Rev. Lett. **74**, 2933 (1995).
14. E. Constant, V. D. Taranukhin, A. Stolow, and P. B. Corkum: Phys. Rev. A **56**, 3870 (1997).
15. C. H. Keitel, P. L. Knight, and M. Protopapas: Rep. Prog. Phys. **60**, 389 (1996).
16. K. C. Kulander, K. J. Schafer, and J. L. Krause: Time-dependent studies of multiphoton processes. In: Atoms in Intense Laser Fields. M. Gavrila, ed. Academic New York, pp. 247–300 (1992).
17. L. V. Keldysh: JETP **47**, 1945 (1964); Sov. Phys. JETP **20**, 1307 (1965).
18. M. V. Ammosov, N. B. Delone, and V. P. Krainov: Sov. Phys. JETP **64**, 1191 (1986).
19. A. M. Perelomov, V. S. Popov, and M. V. Tarent'ev: Sov. Phys. JETP **23**, 924 (1965).

# Transient Optical Susceptibility Induced by Nonperturbative Rotational Wave Packets

Omid Masihzadeh,[1] Mark Baertschy,[2] Klaus Hartinger,[1] and Randy A. Bartels[1]

[1] Department of Electrical and Computer Engineering
Colorado State University, Fort Collins, CO 80523-1320, USA
Randy.Bartels@colorstate.edu
[2] Department of Physics, University of Colorado at Denver, Denver, CO 80217-3364, USA

## 1 Introduction

Rotational wave packets formed in a gas of linear molecules can produce a time-dependent index of refraction that can be used as a phase modulator for an ultrafast pulse [1]. A collection of anisotropic molecules can be made to align along the polarization direction of an intense, linearly polarized laser pulse [1–4]. Although the molecules will quickly go out of alignment, the periodic rephasing of the wave packet causes the molecules to come back into alignment at regular intervals. During these so-called rotational revivals, the wave packet evolves through states where the molecules are both aligned and antialigned (i.e., perpendicular to) the polarization of the laser pulse.

In a pump-probe experiment, e.g., [1], the pump pulse is used to form the rotational wave packet. The pump-probe delay is then adjusted so that the probe pulse encounters a particular portion of a revival feature. The time-varying molecular alignment as discussed in [3] produces a transient refractive index in the gas along the eigenpolarization directions of the aligned molecules [1]. The detailed temporal structure of the temporal phase modulation of the probe pulse imposed by the index transient is governed by the detailed structure of the rotational wave packet created by the pump pulse. In order to increase our control over modulation of the probe pulse, we need to understand how changes to the pump pulse can affect the character of the rotational wave packet leading to control over the strength, time scale, and shape of the revival features.

We have investigated the transition of molecular alignment with a weak, perturbative pulse to regimes of strong, nonperturbative alignment with highly energetic, femtosecond laser pulses. In particular, we are studying how highly energetic alignment pulses can affect the angular momentum composition and dynamics of the rotational wave packet. We then analyze the modification of the linear susceptibility tensor under conditions of nonperturbative rotational alignment.

## 2 Theory

Interaction of a short, linearly polarized pump pulse with a gas of anisotropic molecules forms a rotational wave packet that can be interpreted as the orientational probability of the gas as

$$G(\theta,\phi,t)=\sum_{J_0}P(J_0)\sum_{M_0=-J_0}^{J_0}|\psi_{J_0}^{M_0}(\theta,\phi,t)|^2, \tag{1}$$

where $P(J_0)$ is the (thermal) distribution of molecules with total angular momentum quantum number $J_0$ prior to interaction with the laser pulse. Here

$$\psi_{M_0}^{J_0}(\theta,\phi,t)=\sum_J\sum_M c_{M_0\,M}^{J_0\,J}Y_J^M(\theta,\phi)e^{\imath E_J t/\hbar} \tag{2}$$

is the wave function that describes a molecule that was initially in the rotational state specified by quantum numbers $J_0$ and $M_0$ and the expansion coefficients, $c_{M_0\,M}^{J_0\,J}$, are found through numerical integration of Schrödinger's equation as described in [1].

When first formed, the wave packet corresponds to a partial molecular alignment of the gas along the direction of the polarization of the pulse. This alignment is transient, but as the wave packet evolves the molecules will periodically come back into alignment. This time-dependent orientational probability due to the rotational wave packet leads to fluctuations in the optical properties of the gas. These fluctuations are given by the induced polarization density averaged over the molecular ensemble. The corresponding macroscopic polarization density is written as

$$\mathbf{P}=N\langle\langle\mu_I\rangle\rangle\equiv\bar{\bar{\chi}}\cdot\mathbf{E}, \tag{3}$$

where $\bar{\bar{\chi}}$ defines the macroscopic susceptibility tensor of the gas, $N$ is the molecular density of the gas, $\mu_I=\alpha_{IJ}E_J$ is the linear approximation of the dipole moment induced by the applied electric field $E_{\mathrm{A}}$, and $\alpha_{IJ}$ is the linear polarizability tensor given in the principle coordinate frame (indices *[I, J]*) of the molecule.

The induced polarization density given in Eq. (3) defines the macroscopic linear susceptibility tensor that is attributable to the orientational average of the molecular polarizability and is given by

$$\chi_{ij}=N\langle\langle\alpha_{IJ}\rangle\rangle, \tag{4}$$

where the orientational average is defined by

$$\langle\langle\alpha_{IJ}\rangle\rangle=\alpha_{IJ}\int\int_{4\pi}a_{iI}a_{jJ}G(\theta,\phi,t)d\Omega, \tag{5}$$

and $a_{qQ}$ are the direction cosines between the molecular (indices I, J) and lab (indices i, j) frames.

The polarizability tensor of a linear molecule is diagonal in its principle frame and can be written as $\bar{\bar{\alpha}} = \{\alpha_{XX}, \alpha_{YY}, \alpha_{ZZ}\} = \{\alpha_{\perp}, \alpha_{\perp}, \alpha_{\parallel}\}$, where $\alpha_{\parallel}$ is the polarizability along and $\alpha_{\perp}$ is the polarizability perpendicular to the molecular axis. We restrict our discussion to linear molecules in a dilute gas aligned by a linearly polarized femtosecond laser. Under these conditions, it is known that $M$ is conserved in the alignment pulse interaction and only transitions of the total angular momentum, $J$, meeting the selection rules $\Delta J = \pm 2, 0$ are allowed. The conservation of $M$ produces an azimuthally symmetric orientational probability distribution, i.e., $G(\theta, \phi, t) \to G(\theta, t)$.

With the expression above, we can relate the molecular alignment created by the interaction of the molecules and the alignment pulse as given by Eq. (1) to the macroscopic linear optical susceptibility tensor of the gas, $\bar{\bar{\chi}}$. At the thermal equilibrium, the gas molecules are randomly oriented, i.e., $G(\theta, \phi) = \text{constant}$, and the optical response of the gas is isotropic. The isotropic susceptibility tensor is given by

$$\bar{\bar{\chi}}_{\text{iso}} = \frac{N}{3\epsilon_0}\left(\alpha_{\parallel} + 2\alpha_{\perp}\right)\mathbf{I}, \tag{6}$$

where $\mathbf{I}$ is the $3 \times 3$ identity matrix. The linear susceptibility is proportional to the average polarizability along the principle directions of the molecule.

A weak, i.e., perturbative, alignment pulse induces a time-varying birefringence given by $\Delta\chi_{\text{tr}}(t) = \frac{3N}{2\epsilon_0}\Delta\alpha\langle\langle\cos^2\theta(t)\rangle\rangle$, where $\Delta\alpha \equiv \alpha_{\parallel} - \alpha_{\perp}$ is the polarizability anisotropy of the molecule. The transient birefringence is proportional to the transient molecular alignment which is expressed as [1, 3, 4]

$$\langle\langle\cos^2\theta(t)\rangle\rangle = \sum_{J_0} P(J_0) \sum_{M_0=-J_0}^{J_0} \left\langle \psi_{J_0}^{M_0} \left|\cos^2\theta\right| \psi_{J_0}^{M_0} \right\rangle. \tag{7}$$

The linear optical tensor for the transient optical susceptibility that gives rise to this birefringence is diagonal, where the diagonal components of the transient susceptibility tensor, $\bar{\bar{\chi}}_{\text{tr}}$, are given by

$$\bar{\bar{\chi}}_{\text{tr}} = \left\{\chi_{\text{tr}}(t)_{xx}, \chi_{\text{tr}}(t)_{yy}, \chi_{\text{tr}}(t)_{zz}\right\} = \frac{1}{3}\Delta\chi_{\text{tr}}(t)\left\{-1, -1, 2\right\}. \tag{8}$$

The coordinates $\{x, y, z\}$ are those defined by identifying the direction of the linear alignment polarization pulse as $z$. The structure of the transient susceptibility tensor is determined by the symmetry of $G(\theta, \phi)$, where the azimuthal symmetry of $G$ is reflected in the equality $\chi_{\text{tr}}(t)_{xx} = \chi_{\text{tr}}(t)_{yy}$.

Figure 1a shows a typical transient alignment cosine for a perturbative transient alignment of $CO_2$ molecules. In Fig. 1a, we plot the transient alignment, $\langle\langle\cos^2\theta(t)\rangle\rangle$, due to a 20-fs Gaussian pulse with a peak intensity of $I_0 = 3 \times 10^{13}\ \text{W} \cdot \text{cm}^{-2}$ that is interacting with a dilute gas of $CO_2$ molecules. Figure 1b shows the corresponding angular density matrix. For the perturbative pulse

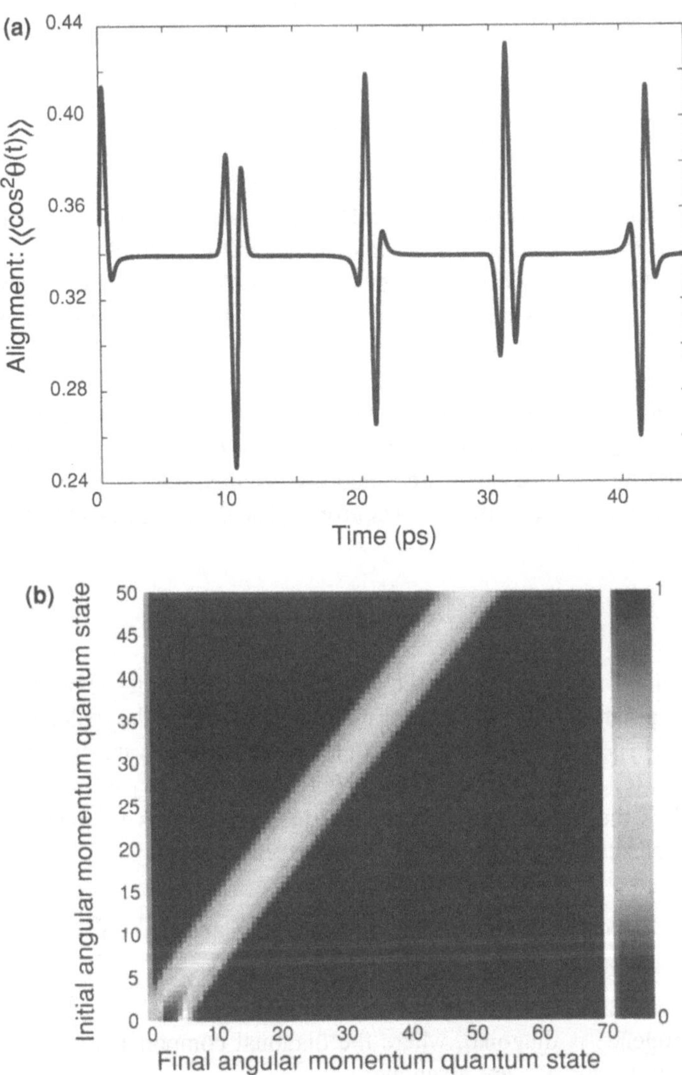


FIGURE 1. (a) Transient alignment, i.e., $\langle\langle\cos^2\theta(t)\rangle\rangle$, for a 20-fs alignment pulse with a perturbative energy. Note that the maximum alignment is relatively small and the average alignment value is close to the $\frac{1}{3}$ expected for no orientational preference. (b) Density matrix of the same pulse as in (a). The perturbative nature of the alignment is indicated by the diagonal nature of the density matrix.

the density matrix is dominated by the diagonal component with weak (i.e., perturbative) off-diagonal components that correspond to the angular coherences which give rise to the transient molecular alignment. Moreover, we see that the average value of the molecular alignment, $\overline{\langle\langle\cos^2\theta(t)\rangle\rangle}$, remains essentially at $\frac{1}{3}$, increasing by only 0.63% to 0.3354. This implies that the birefringence in the gas is dominated by the transient contribution as reported by [4].

Transient alignment of molecules with a linearly polarized ultrafast laser pulse requires peak pulse intensities of order $\sim 10^{13}$ W $\cdot$ cm$^{-2}$, see for example [1]. Increasing the peak intensity of the alignement pulse leads to an appreciable rise in the probability of dissociation of the molecule. As a result, we study the increase in alignment pulse duration while keeping the peak pulse intensity fixed at $I_0 = 3 \times 10^{13}$ W $\cdot$ cm$^{-2}$ to probe the transition from perturbative to nonperturbative molecular alignment. A representative example of a nonperturbative alignment calculation for $CO_2$ is shown in Fig. 2. The pulse duration for this calculation has been increased to 200 fs. The nonperturbative nature of the molecular alignment simulated by the pulse is illustrated by both the increase of the average

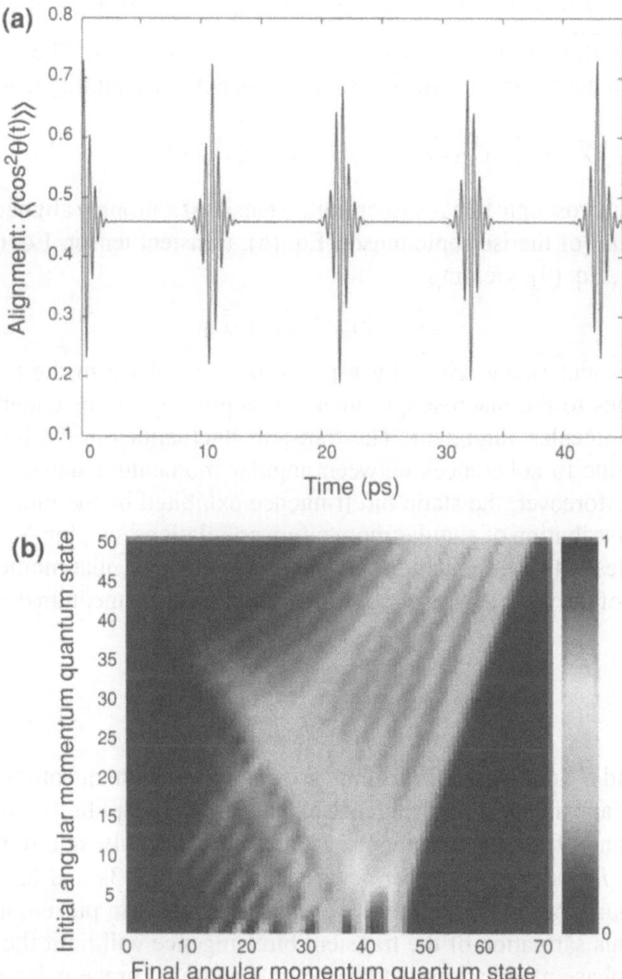


FIGURE 2. (a) Transient alignment, i.e., $\langle\langle \cos^2 \theta(t) \rangle\rangle$, for a 200-fs alignment pulse with a nonperturbative energy. (b) Density matrix of the same pulse as in (a). The nonperturbative nature of the alignment is indicated by a depletion of the density matrix diagonal.

alignment, Fig. 2a, to well above $\frac{1}{3}$ and the complex structure in the angular density matrix, Fig. 2b. Indeed we also see a significant increase in the speed of the susceptibility transient that correlates with an increased excitation of high angular momentum coherences as displayed in the angular density matrix. The nonperturbative transient alignment exhibits faster temporal structure than for perturbative alignment as a result of the excitation of a broader distribution of angular coherences. The fast temporal structure leads to larger spectral changes in a probe pulse propagating in the coherence. The second feature of the nonperturbative transient alignment cosine displayed in Fig. 2a is that the average alignment is substantially higher than $\frac{1}{3}$. This increased average alignment leads to a macroscopic birefringence in the molecular gas that persists in between revivals. This static birefringence may be expressed as $\Delta\chi_{\mathrm{st}} = \frac{3N}{2\epsilon_0}\Delta\alpha\left(\overline{\langle\langle\cos^2\theta(t)\rangle\rangle} - \frac{1}{3}\right)$, where $\overline{\langle\langle\cos^2\theta(t)\rangle\rangle}$ is the time-averaged molecular alignment. The static susceptibility tensor has the same structure as the transient susceptibility tensor, namely,

$$\bar{\bar{\chi}}_{\mathrm{st}} = \left\{\chi_{\mathrm{st},\,xx}, \chi_{\mathrm{st},\,yy}, \chi_{\mathrm{st},\,zz}\right\} = \frac{1}{3}\Delta\chi_{\mathrm{st}}\left\{-1, -1, 2\right\}. \tag{9}$$

The full macroscopic linear susceptibility tensor of a strongly aligned molecular gas is the sum of the isotropic tensor, Eq. (6), transient tensor, Eq. (8), and the static tensor, Eq. (9), yielding

$$\bar{\bar{\chi}} = \bar{\bar{\chi}}_{\mathrm{iso}} + \bar{\bar{\chi}}_{\mathrm{tr}} + \bar{\bar{\chi}}_{\mathrm{st}}. \tag{10}$$

The alignment cosine given by Eq. (7) directly relates to the transient and static changes to the macroscopic linear susceptibility tensor under conditions of strong molecular alignment. The transient fluctuations in the linear susceptibility are due to coherences between angular momentum states separated by $\Delta J = \pm 2$. Moreover, the static birefringence exhibited by the molecular gas is due to the distribution of angular momentum populations, i.e., for $\Delta J = 0$. Since the timescales of these fluctuations are governed by the angular momentum spectral content of the wave packet, we examine the angular momentum density matrix defined as

$$\mathcal{P}(J, J_0) = \sum_{M_0=-J_0}^{J_0}\sum_{M=-J}^{J}\left|c_{M_0\,M}^{J_0\,J}\right|^2, \tag{11}$$

where $J_0$ and $J$ are the initial and final angular momentum quantum numbers.

The static and transient birefringence of the optical susceptibility tensor as given by Eqs. (8) and (9) for alignment pulses with increasing pulse duration and a peak intensity of $I_0 = 3\times10^{13}\ \mathrm{W\cdot cm^{-2}}$ are shown in Figs. 3a and b, respectively. The maximum transient alignment steadily increases with pulse duration up to ~125 fs. This saturation of the transient birefringence will limit the strength of a transient phase modulation that can be imposed on a probe pulse propagating in molecules transiently aligned by a single Gaussian alignment pulse. Figure 3b shows that the calculated static birefringence increases steadily up to the 250-fs pulse durations considered in these calculations.

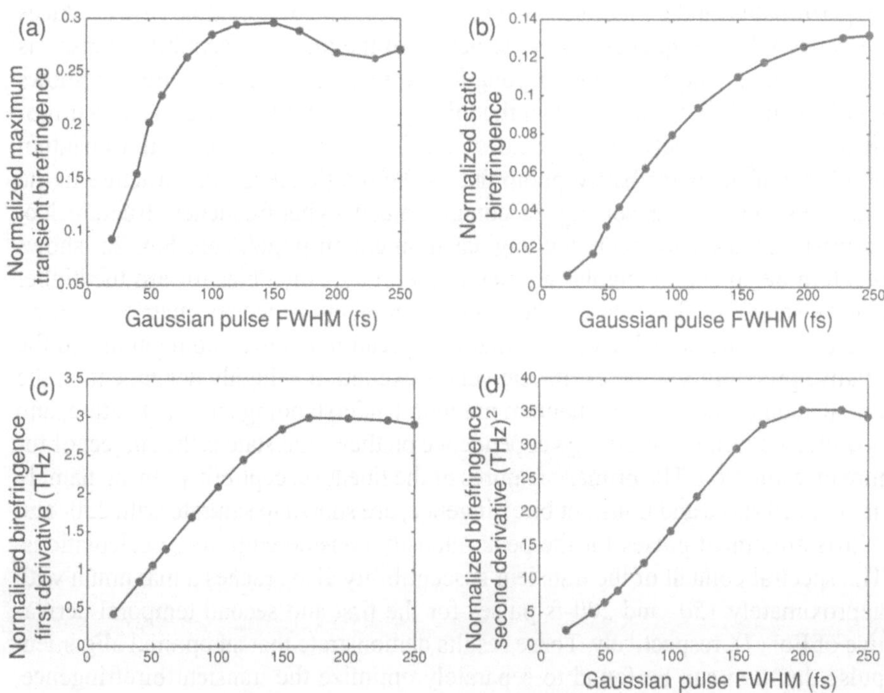


FIGURE 3. (a) Normalized static birefringence, $\frac{2\epsilon_0}{3N\Delta\alpha}\Delta\chi_{st}$, as the alignment pulse energy is increased. (b) Normalized maximum transient birefringence, $\frac{2\epsilon_0}{3N\Delta\alpha}(|\Delta\chi_{tr}(t)|_{max}\,\Delta\chi_{st})$, as the alignment pulse energy is increased. (c,d) Normalized, first derivative (shift in the frequency) and second derivative (chirp) of the transients.

A comparison of the transient alignment cosine, i.e., Eq. (7), for the perturbative pulse shown in Fig. 1a and the nonperturbative pulse illustrated in Fig. 2b show an increase in the oscillation frequency of the rotational revivals. This increased spectral content can be quantified by considering the temporal derivatives of the alignment cosine as shown in Figs. 3c and d. These data demonstrate an increase in the frequency of the oscillation of the transient optical susceptibility. The first-order derivative leads to spectral shifts, whereas the second-order temporal derivative is responsible for spectral broadening of a probe pulse propagating in the transient optical medium. Thus, the increase in oscillation frequency found in our calculations should produce larger frequency shifts in experiments where a probe pulse propagates in the transient optical medium, e.g., [1].

## 3 Summary

With our numerical simulations, we are able to probe, in great detail, the differences in the transient optical susceptibility due to rotational alignment induced by perturbative and nonperturbative laser pulses. The perturbative case produces a relatively low-frequency revival structure in the linear susceptibility tensor as seen in Fig. 1a. The corresponding angular momentum population, Fig. 1b, shows relatively modest deviations from the initial state with the spread in angular momentum being largely independent of the initial angular momentum quantum number. For nonperturbative pump pulses the rotational revival structure in the linear susceptibility tensor, Fig. 2a, contains much higher frequency structure. Not surprisingly, the corresponding angular momentum population, Fig. 2b, shows much more spread in angular momentum with a distinct bias toward the higher energy states. In these cases, the angular momentum density matrices exhibit interesting structure which shows that the spread in angular momentum and the relative probability of de-excitation versus excitation is highly dependent on the initial angular momentum quantum number. Understanding the exact nature, and potential consequences, of this dependence on the initial state is the subject of future investigation. The primary features of the linear susceptibility tensor, namely an induced static and transient birefringence, are shown to saturate with 250- and 125-fs alignment pulses for the peak intensity considered in these calculations. The spectral content of the transient susceptibility also reaches a maximum with approximately 150- and 200-fs pulses for the first and second temporal derivative of Eq. (7), respectively. These results demonstrate that an optimal alignment pulse duration can be found to separately optimize the transient birefringence, frequency shifting, and spectral broadening of a probe pulse.

**Acknowledgments.** The authors gratefully acknowledge support for this work from the National Science Foundation CAREER Award ECS-0348068, the Office of Naval Research Young Investigator Award, the Beckman Young Investigator Award. One author (RAB) gratefully acknowledges generous support from a Sloan Research Fellowship.

## References

1. R. A. Bartels, T. C. Weinacht, N. Wagner, et al.: Phys. Rev. Lett. **88**, 013903 (2002).
2. B. Friedrich and D. Herschbach: J. Phys. Chem. **99**, 15686 (1995).
3. H. Stapelfeldt and T. Seideman: Rev. Mod. Phys. **75**, 543 (2003).
4. B. H. Lin, J. P. Heritage, T. K. Gustafson, R. Y. Chiao, and J. P. McTague.: Phys. Rev. A **13**, 813 (1976).

# THz Wave Generation and Detection System using ~1000 nm Yb-doped Fiber Laser

Atsushi Syouji,[1] Shingo Saito,[1] Kiyomi Sakai,[1] Masaya Nagai,[2] Koichiro Tanaka,[2] Hideyuki Ohtake,[3] Toshiaki Bessho,[3] Toshiharu Sugiura,[3] Tomoya Hirosumi,[3] and Makoto Yoshida[3]

[1] KARC, National Institute of Information and Communications Technology, Nishi Ku, Kobe, Hyogo 651-2492, Japan
[2] Department of Physics, Kyoto University, Kyoto 606-8502, Japan
[3] Aisin Seiki Co., Ltd., Kojiritsuki, Hitotsugi-cho, Kariya 448-0003, Japan
`a2syouji@po.nict.go.jp`

**Summary.** The THz generation and detection system using ~1000 nm Yb-doped ultrashort pulse fiber laser has been constructed. We compared three binary zinc blende semiconductors CdTe, GaAs, and ZnTe as the nonlinear optical crystals for THz radiation. The phase matching condition of THz generation is satisfied in CdTe at the pumping wavelength of ~1000 nm. At the condition, the intensity of THz wave becomes very high and its spectrum spreads widely. This behavior agrees with our simulation based on phase-matching condition.

## 1 Introduction

In resent years, THz-TDS (Terahertz Time-Domain Spectroscopy) measurement is closely watched, because the femtosecond ultrashort pulse laser is easy to use. THz-TDS has the advantage that we can get not only the spectral strength but also the phase information measuring the THz electric field. This means that we are able to obtain the real and imaginary part of the dielectric constant. Therefore, this way of measurement is very important for analysis of various materials. THz wave is generated by optical rectification using ultrashort pulse laser and nonlinear optical crystal. Above all, the phase-matching condition is very important for THz generation and detection. Nagai et al. [1] calculated the phase-matching condition at 2 THz for some semiconductor, and they conclude that CdTe and GaP semiconductor are optimum for nonlinear crystal pumping laser wavelength ~1000 nm as Yb-doped fiber laser. So, we measured the THz spectra and total power from CdTe crystal pumping under several wavelengths condition using optical parametric amplifier (OPA) system. On the other hand, we simulated THz generation intensity from CdTe for various wavelengths. Furthermore, we constructed THz generation and detection system with Yb-doped fiber laser. And we compare several crystals CdTe, GaAs, and ZnTe as the THz wave generator.

## 2 Experiment

The light source is Ti:S regenerative amplifier (regen) and OPA. Output beam from regen system with a wavelength of 800 nm, a repetition rate of 200 kHz, a pulse duration of 50 fs, and a output power of 900 mW is divided into two beams for pumping the OPA system and for detecting of THz wave form. The output beam from OPA laser was reduced by neutral density filter to 3 mW, and the beam was focused on 1-mm thick <110> CdTe crystal. The wave form of THz radiation is detected by an EO sampling method using 1-mm thick ZnTe crystal and 800-nm sampling pulse. The modulated sampling pulse is divided by Wolston prism and detected by balanced photo-diode detector. A 343-Hz optical chopping and lock-in detection were used in order to improve the sensitivity. Further experiment was done with Yb-doped fiber laser, IMRA Femtolite D-200. The output beam from the laser with a wavelength of 1045 nm, a repetition rate of 43 MHz, a pulse duration of 200 fs, and a output power of 200m W is divided into two beams. One beam pumps zinc-blende semiconductor CdTe, GaAs, or ZnTe and the other is used for EO sampling of THz wave form. Here, we used CdTe as the EO crystal.

## 3 Simulation

We simulated THz wave generation and detection by nonlinear optical crystal based on phase-matching condition. THz wave is generated by differential wave generation of two frequencies included in pump laser pulse. So, we calculated the polarization vibration spectra of THz region by overlap integral between pump laser and THz wave. The obtained vibration spectrum is converted to intensity of THz radiation multiplying the factor of square polarization frequency. Therefore, the efficiency of THz spectra is calculated in Eq. (1). Here, $\omega_T$, $\lambda_T$, and $n_T$ is frequency, wavelength, and refractive index of THz wave, and $\omega_{1,2}$, $\lambda_{1,2}$ and $n_{1,2}$ is frequency, wavelength, and refractive index of pump laser, respectively. The index 1 and 2 indicate two different frequencies included in pump laser pulse and the distance conform to THz wave. $d$ is thickness of sample and $F(\lambda_0, \lambda_1, \lambda_2)$ is product of electric field intensity at wavelength $\lambda_1$ and $\lambda_2$ of pump laser with center wavelength $\lambda_0$. The parameter of refractive index of near and far IR region of CdTe, ZnTe, and GaAs are referred [2–7], respectively.

$$I^G(\omega_T) = \int d\omega_1 \int_0^d dz\, \omega_T^2 \sum_{\omega_2=\omega_1\pm\omega_T} F(\lambda_0, \lambda_1, \lambda_2) e^{2\pi i \left(\left|\frac{n_1}{\lambda_1}-\frac{n_2}{\lambda_2}\right|-\frac{n_T}{\lambda_T}\right)z}. \qquad (1)$$

On the other hand, THz wave form by EO sampling is measured as the rotation of polarization of sampling pulse. The rotation is in proportional to the equivalence of the group velocity of the sampling laser and the phase velocity of THz wave. Therefore, the sensitivity of THz wave by the EO sampling is derived from Eq. (2). Here, $V_g$ is the group velocity of NIR and $V_p$ is the phase velocity of THz wave. The total THz detection intensity is derived multiplying THz efficiency by Eq. (1) and EO sensitivity by Eq. (2).

$$I^D(\omega_T) = \int_0^d dz\, e^{2\pi i \frac{V_p - V_g}{V_g \lambda_T} z}. \tag{2}$$

# 4 Results

Figure 1 is the wavelength dependent of THz radiation intensity from CdTe crystal. The solid line of the figure is experimental results pumped by several wavelengths, and the broken line is simulated one. The simulated result is consistent with the experimental one, which provides evidence that phase-matching condition is the major process for enhancement of THz radiation. Apparently, the phase-matching condition of CdTe crystal for THz wave is satisfied under 1000 nm pumping condition. In accordance with above conclusion, we constructed ~1000 nm pumping and detecting system of THz wave using Yb-doped fiber laser. Figure 2 is the experimental result and simulated one. Because, the coherent length in CdTe is enhanced at 1000 nm, the generated THz wave intensity from CdTe is larger than other two crystals. This fact should be same as EO crystal. Therefore, CdTe crystal is suitable for the THz radiation and detection system using Yb-doped fiber laser. The deviations of experimental spectrum and simulated one in the case of CdTe generation may come from the chirping of pump pulse, because the dispersion of refractive index of CdTe at ~1000 nm is larger than those of other crystals.

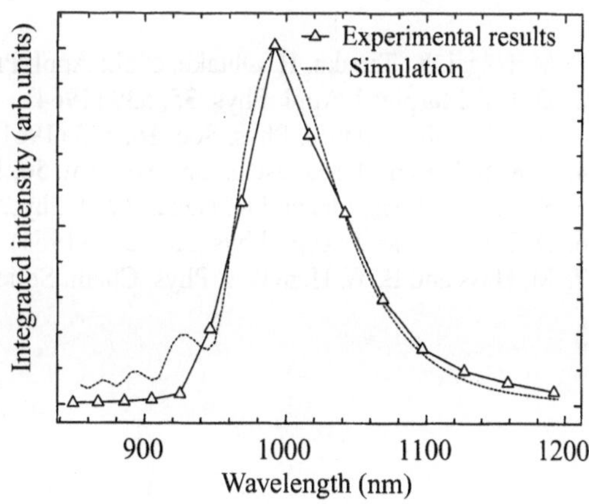


FIGURE 1. Intensities of THz radiation from CdTe crystal for several pumping wavelengths. These intensities are integrated from 0 to 4 THz of Fourier amplitude spectrum calculated from the THz wave form. The solid line shows experimental results and the broken line is calculation. Here, THz wave is detected by an EO sampling method using ZnTe crystal and 800-nm sampling pulse.

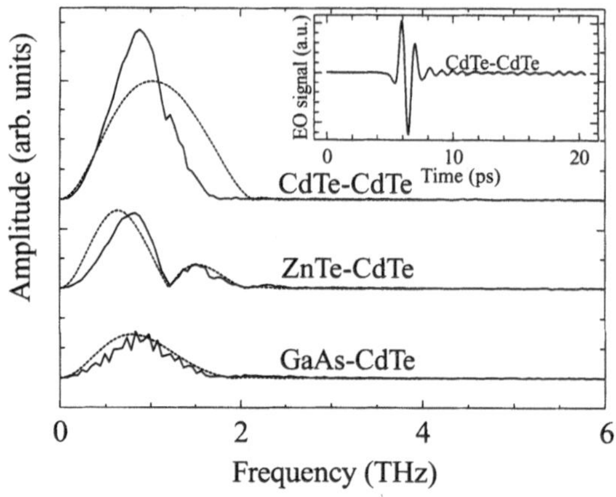


FIGURE 2. THz spectra generated from three kinds of zinc-blende crystals GaAs, ZnTe, and CdTe pumped by Yb-doped fiber laser. The solid line shows experimental result and the broken line is the calculation. These THz wave is detected by an EO crystal of CdTe. Here, the description of, e.g., "ZnTe-CdTe" in figure indicates that ZnTe is used as THz radiation and CdTe is as detection. The insert figure is THz wave form generated from CdTe.

## 5 Conclusions

In this report, we measured THz radiation intensity from CdTe for several pumping wavelengths. The phase-matching condition of CdTe crystal for THz wave is suitable under 1000-nm pumping condition; and therefore the radiation intensity is higher than other wavelength. The fact is explained by our simulation. And we constructed the THz generation and detection system using ~1000 nm Yb-doped fiber laser. In the case of using CdTe as nonlinear optical crystal for THz generation, the THz intensity is stronger than other semiconductors such as ZnTe and GaAs. It is shown that CdTe is the one of the most suitable EO crystal in binary semiconductors for THz sensing system using Yb-doped fiber laser.

**Acknowledgement.** This work was financially supported by the Ministry of Public Management, Home Affairs, Posts, and Telecommunications, Japan.

## References

1. M. Nagai, K. Tanaka, H. Ohtake, et al.: Appl. Phys. Lett. 85, 3974 (2004).
2. D. T. F. Marple: J. Appl. Phys. 35, 539 (1964).
3. A. Mitsuishi: Japan. J. Phys. Soc. 16, 533 (1961).
4. T. R. Sliker and J. M. Jost: J. Opt. Soc Am. 56, 130 (1966).
5. S. Narita, H. Harada, and K. Nagasaka: J. Phys. Soc. Japan 22, 1176 (1967).
6. D. T. F. Marple: J. Appl. Phys. 35, 1241 (1964).
7. M. Hass and B. W. Henvis: J. Phys. Chem. Solids 23, 1099 (1962).

# Cavity-enhanced Optical Parametric Chirped-pulse Amplification

F. Ö. Ilday and F. X. Kärtner

Department of Electrical Engineering and Computer Science, and Research Laboratory of Electronics, Massachusetts Institute of Technology, 77 Massachusetts Avenue, Cambridge, MA 02139, USA

The generation of few-cycle (sub-10-fs) pulses is important for optical frequency metrology, attosecond physics, and extreme nonlinear optics. There is much motivation to develop simple, robust, and inexpensive techniques for the amplification of these few-cycle pulses, especially to peak and average power levels exceeding the current limits of the Ti:sapphire laser technology. For instance, high-harmonic generation would benefit greatly from the availability of higher average power, without compromising on the peak power.

Optical parametric chirped-pulse amplification (OPCPA) is a promising method to amplify few-cycle pulses [1]. Large amplification bandwidths are achieved through noncollinear phase matching, and amplification is energy-scalable with larger beam sizes. OPCPA can be thought of as the utilization of an "engineered" gain medium, where the traditional energy states of an amplifying medium are replaced by virtual states arising from the coupling of the pump and signal beams through the nonlinear polarization of a quadratic crystal. Without the storage of the pump light, synchronous pumping is necessary.

Meanwhile, coherent addition of femtosecond pulses in an enhancement cavity has been proposed [2] and demonstrated for picosecond [3, 4] and femtosecond pulses [5]. Enhancement of optical fields in a resonator is a well-established and commonly used technique, including femtosecond pulse addition for efficient second-harmonic generation. However, this method gets increasingly difficult for shorter pulses as the intracavity dispersion distorts the mode comb, preventing full enhancement. At high powers, self-phase modulation within the passive cavity emerges as a challenging limitation [6].

Here, we propose a scheme, which can be regarded conceptually as a combination of the OPCPA and enhancement cavity techniques (Fig. 1): a pump beam is coherently added up in a high-finesse external cavity, which contains a nonlinear crystal, phase-matched for parametric amplification of a signal pulse. The cavity mirrors are chosen to be transparent to the signal wavelength. The (stretched) signal pulses are synchronized and triggered to enter the cavity and overlap spatially

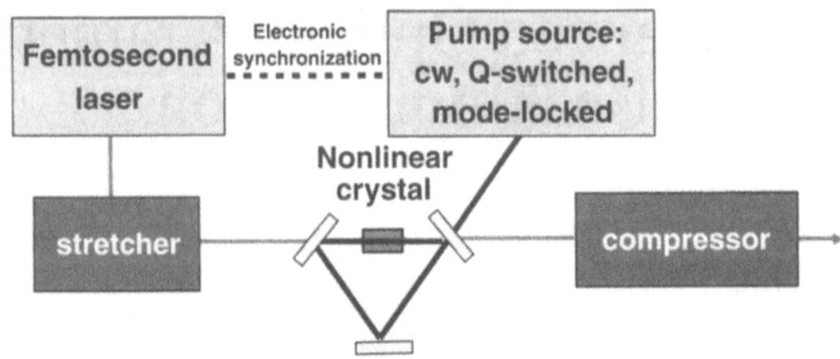


FIGURE 1. Schematic of the experimental setup for cavity-enhanced OPCPA.

and temporally with the pump beam once the cavity is fully loaded, undergoing parametric amplification. In this way, signal pulses with a bandwidth corresponding to 5 fs in the time domain can be amplified with proper phase matching. Thereby, the pump energy is extracted out of the cavity by way of conversion to the signal wavelength. Conceptually, the cavity-enhanced parametric amplification can be viewed as a regular amplifier, where the enhancement cavity stores the pump light in contrast to storage by population inversion. In general, the pump source can be cw, Q-switched, or a pulse train from a mode-locked laser. Here, we focus on the case of a pulsed pump source.

Several major advantages of this scheme can be identified. The pump light can be in the picosecond range or longer, and thus dispersion ceases to be a limitation to the enhancement cavity. With increasing finesse of the cavity, the peak power and pulse energy required from the pump source is reduced, opening up the use of laser sources which excel in high average power but are limited in peak power, such as fiber lasers. Unloading of the cavity is through a virtually instantaneous optical process, removing any restrictions on the cavity size due to finite switching speed. In fact, the cavity may consist entirely of a nonlinear crystal with appropriate dielectric coatings. Importantly, the high-finesse cavity will act as a very effective spatial mode cleaner, ensuring a high beam quality, at the expense of a small amount of power loss.

The dynamics of the cavity-enhanced OPCPA can be understood in two largely independent steps, permitting a two-step quantitative analysis. During loading of the enhancement cavity, no light at the signal wavelength is present and the nonlinear crystal is simply a passive element. Once the cavity is fully loaded, the signal pulse is triggered to interact with the pump beam within the crystal. Since the signal pulse passes through the cavity only once, the parametric amplification is not influenced by the presence of the enhancement cavity and therefore is not discussed further here. After unloading of the cavity, the entire sequence is repeated.

In practice, the achievable enhancement factor will depend on, among others, the roundtrip loss of the cavity. Enhancement factors of 30–100 appear to be readily achievable, with a potential for factors of up to 1000. However, any phase variations over the temporal profile of the pulse will cause its frequency comb to shift with respect to the cavity modes, limiting the enhancement factor to much less than the theoretical finesse of the cavity. Thus, even if all of the practical limitations are addressed, the stored energy will be ultimately limited by the Kerr nonlinearity occurring in the crystal.

We propose the use of nonlinearity management to overcome limitations of the Kerr nonlinearity at high peak powers. Several material systems have been identified to provide negative (self-defocusing), large (up to $10^{-13}\text{cm}^2/\text{W}$) Kerr nonlinearity, including semiconductors for $h\nu > 0.7E_{\text{gap}}$. In particular, cascaded quadratic nonlinear processes give access to self-defocusing nonlinearity of tunable magnitude and have been used in soliton mode locking of lasers [6]. A major technical advantage of using picosecond or longer pump pulses is that group-velocity mismatch is not a limiting factor for the cascade process and dispersion of even the longest crystals becomes unimportant. An overall compensation of $n_2$ can be, to first order, complete. Energy loss due to residual second-harmonic generation can be kept well below 1%.

In order to test the validity of these expectations, we numerically simulated the loading of a cavity with a pulse train from a high-power, mode-locked laser. Dispersion and nonlinearity of the cavity elements are taken into account. Pulse propagation is described by the nonlinear Schrödinger equation. The cavity comprises of a 2 mm-long periodically poled lithium niobate (PPLN) crystal for parametric amplification and dispersion-compensating mirrors. For nonlinearity compensation, another PPLN crystal of equal length is added for the cascade process. For computational speed, we approximate the cascade process with an effective negative $n_2$. The benefits of nonlinearity compensation are illustrated in Fig. 2a. In these calculations, the pulse duration is fixed at 10 ps. For efficient

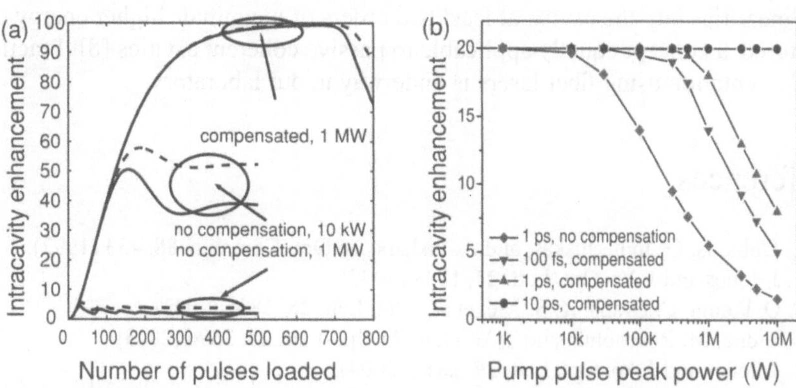


FIGURE 2. (a) Results of numerical simulations showing the loading of the cavity with and without nonlinearity compensation. Solid (red) lines depict intracavity energy enhancement and dashed (blue) lines depict intracavity peak power enhancement. (b) Numerical simulations depicting intracavity buildup factors as a function of pump pulse peak power with and without compensation of the nonlinearity. The lines connecting the dots are only to guide the eye.

operation, the transmittance of the coupling mirror is set equal to the remaining losses of the cavity, resulting in a cavity finesse of $100\pi$. Without compensation, about 50 of the 10 kW pump pulses can be added up. With nonlinearity compensation, even pulses with 100 times higher peak power (1 MW) can attain the theoretical limit of 100 times enhancement in the cavity. As a control experiment, the use of 1 MW pulses without compensation is shown, leading to a buildup of only about 5.

However, for shorter pulses, the buildup process becomes complicated due to the interplay of nonlinearity and dispersion: perfect compensation of nonlinearity is not possible anymore since these effects do not commute, a scenario previously analyzed for femtosecond fiber lasers [7]. In Fig. 2b, we plot the cavity buildup factor achieved as a function of peak power of the pump pulses, for pulse durations of 100 fs, 1 ps, and 10 ps and employing nonlinearity compensation. Also plotted is the uncompensated case for 1 ps pulses (the pulse duration was found to be largely immaterial in the absence of nonlinearity compensation). In all cases, compensation improves the loading process dramatically at high powers, but the best results are obtained using the longest pulses.

In conclusion, we propose a new technique, cavity-enhanced OPCPA, which is a conceptual combination of parametric amplification and coherent pulse addition with an enhancement cavity, as a practical route to high average and peak power sources of femtosecond optical pulses. The role of the cavity is analogous to the storage of pump light in an excited atomic state in regular amplification. The use of ps pump pulses eliminates dispersion-related problems during the cavity loading, while the chirped signal pulse can have a broad bandwidth, permitting amplification of down to few-cycle pulses with proper phase matching. In order to reach extremely high energies, nonlinearity management in the enhancement cavity can be utilized. Numerical simulations show that by introducing self-defocusing nonlinearities into the cavity, at least two orders of magnitude higher energy can be stored, a concept equally applicable to passive coherent cavities [8]. Practical implementation using fiber lasers is underway in our laboratory.

## References


1. A. Dubietis, G. Jonusauskas, and A. Piskarskas: Opt. Commun. **88**, 433 (1992).
2. R. J. Jones and J. Ye: Opt. Lett. **27**, 1848 (2002).
3. E. O. Potma, C. Evans, X. S. Xie, et al.: Opt. Lett. **28**, 1835 (2003).
4. Y. Vidne, M. Rosenbluh, and T. W. Hänsch: Opt. Lett. **28**, 2396 (2003).
5. R. J. Jones and J. Ye: Opt. Lett. **29**, 2812 (2004).
6. F. Wise, L. Qian, and X. Liu: J. Nonlin. Opt. Phys. Mat. **11**, 317 (2002).
7. F. Ö. Ilday and F. W. Wise: J. Opt. Soc. Am. B **19**, 470 (2002).
8. K. D. Moll, R. J. Jones, and J. Ye: Opt. Express. **13**, 1672 (2005).

# Measuring Spatiotemporal Distortions with GRENOUILLE

Xun Gu, Ziyang Wang, Selçuk Aktürk, and Rick Trebino

School of Physics, Georgia Institute of Technology, Atlanta, Ga 30332-0430, USA
**akturk@socrates.physics.gatech.edu**

## 1 Introduction

Researchers often implicitly assume in their ultrafast studies that their laser beams have separable spatial and temporal dependencies. This underlying assumption greatly simplifies the theoretical treatment of ultrashort laser pulses, but is increasingly difficult to maintain as ever shorter pulse lengths and ever greater bandwidths are routinely generated in ultrafast laboratories. With a large frequency bandwidth, any imperfect operations, such as the stretching, compression, and amplification of ultrashort laser pulses, tend to introduce spatiotemporal distortions, which can be detrimental to the application of these pulses.

In other cases, researchers have begun to utilize controlled spatiotemporal couplings in an ultrafast laser beam for some special applications, as in noncollinear OPA phase matching [1] and pulse shaping [2, 3]. Precise knowledge of the couplings is imperative for the success of these experiments.

Whatever the purpose, in order to accurately control the spatiotemporal couplings in an ultrashort laser pulse, it is very important to have a simple, sensitive, and reliable technique to measure them.

## 2 GRENOUILLE with Spatiotemporal Distortions

The pulse intensity and phase characterization device, frequency-resolved optical gating (FROG [4]), measures the nonlinear–optical spectra generated by the interaction of the pulse and its replica at a series of delays, which constitute a 2D FROG trace. Through an iterative computer algorithm, the temporal intensity and phase of the pulse field are retrieved by searching for a solution that produces a theoretical FROG trace closest to the measured one. Because a 2D FROG trace is completely determined by but contains many more data points than the 1D pulse intensity and phase, the redundant information has important implications.

For example, when the spatiotemporal separability assumption fails, the FROG trace will be distorted in a way that is dictated by the type and amount of spatiotemporal distortion in the beam. If we understand the relationship between the distortion experienced by the FROG trace and the spatiotemporal distortion in the input beam, it is conceivable to develop an advanced FROG algorithm to measure not only the temporal (spectral) intensity and phase of the pulse but also the spatiotemporal distortions.

In this work, we will focus on the measurement of spatiotemporal distortions using a special, extremely simple type of FROG: the grating-eliminated no-nonsense observation of ultrafast incident laser-light E-fields (GRENOUILLE) [5]. In previous research, we have shown that a GRENOUILLE trace can exhibit distinctive features with two of the most common spatiotemporal distortions: spatial chirp and pulse-front tilt [6, 7]. With spatial chirp, the GRENOUILLE trace undergoes a shear in the frequency direction, which destroys the usual delay symmetry of an SHG-FROG trace. With pulse-front tilt, the GRENOUILLE trace experiences a shift in the delay direction, which likewise destroys the delay symmetry and therefore may also be easily detected. The amounts of the frequency shear and the delay offset can then be used to measure the values of frequency gradient and pulse-front tilt in the input beam.

Although these results are valid for many pulses and have been experimentally verified, we later discovered that the GRENOUILLE trace can suffer other types of distortions from spatial chirp, if the pulse has a complicated spectrum or spectral phase. The discrepancy stems from the recently observed fact that there are two types of the spatial chirp: "spatial dispersion" and "frequency gradient." The former is more common, but GRENOUILLE only straightforwardly measures the latter. In order to measure spatiotemporal distortions in the widest range of cases, we need a more rigorous theory and a more sophisticated retrieval algorithm. In this work, we derive a rigorous theory of GRENOUILLE applicable to an arbitrary 1D input spatiotemporal electric field, discuss the effects of first-order spatiotemporal distortions to a GRENOUILLE trace, and demonstrate two algorithms to retrieve spatiotemporal distortions from a GRENOUILLE trace.

To this end, we have successfully developed a general model of GRENOUILLE for arbitrary spatiotemporal input beams. Analyzing each element of the device, we derive

$$\begin{aligned} G(x,\omega) &= \left|\int E^{+}(x,t)\,E^{-}(x,t)\exp(-\mathrm{i}\omega t)\,\mathrm{d}t\right|^{2} \\ &= \left|\int E(x-L\theta,t-\theta x/c)\,E(x+L\theta,t+\theta x/c)\exp(-\mathrm{i}\omega t)\,\mathrm{d}t\right|^{2}, \end{aligned} \tag{1}$$

where $\theta$ is the angle that the Fresnel biprism crosses the two beams at, and $L$ is the separation between the Fresnel biprism and the SHG crystal.

Equation (1) is the main result of our rigorous GRENOUILLE theory. It may be applied to arbitrary input 1D spatiotemporal fields. No assumptions haven been made here about the pulse temporal profile, spatial profile, or the complexity of the spatiotemporal distortions.

# 3 Retrieving Spatiotemporal Distortions from a GRENOUILLE Trace

The GRENOUILLE trace given by Eq. (1) has a different mathematical form than a traditional FROG trace. Therefore, it requires some thoughts to develop a suitable algorithm to extract the spatiotemporal distortions and pulse temporal/spectral profile from the measured trace, while maintaining an appropriate balance between speed and accuracy. Two algorithms have been developed for this purpose, which are described as follows.

## *3.1 Perturbative FROG Algorithm*

If the spatiotemporal distortions in the input beam are minor, the measured GRENOUILLE trace will only be slightly different from a standard SHG-FROG trace. In this case, the well-established FROG pulse retrieval code can be utilized to obtain an estimate of the intensity and phase of the pulse (which would be the "local" pulse at the center of the beam, with a frequency bandwidth reduced from that of the whole beam due to spatial chirp). This retrieved local pulse is then used in the retrieval of the spatiotemporal distortion parameters, which will be a simple one- or two-parameter optimization process.

The perturbative recipe consists of the following two steps: (1) running the standard FROG algorithm on the perturbed trace to obtain an estimated pulse intensity and phase profile; (2) running an optimization procedure to obtain the spatiotemporal distortion parameters, using the pulse intensity and phase estimation from the first step and the distorted GRENOUILLE trace as the inputs.

The standard FROG pulse retrieval algorithm is remarkably reliable even with noise and systematic errors. Its exceptional reliability can be fundamentally attributed to the high redundancy of information in a FROG trace. As long as the spatiotemporal distortions are small, the perturbation to the GRENOUILLE trace will not greatly affect the ability of the algorithm to produce a decent estimation of the pulse temporal/spectral profile.

The advantage of this perturbative approach is that it requires very little modification to the existing FROG program, which is very fast, reliable, and robust. The appended spatiotemporal distortion retrieval step takes only seconds to complete. We have implemented the perturbative approach in MatLab, using the standard Levenberg-Marquardt nonlinear least-squares minimization routine in the optimization toolbox for the optimization of the spatial dispersion parameter $\zeta$. The effect of the angular dispersion parameter $\beta$ is only a pulse-front tilt, which corresponds to a shift of the trace in the delay axis. Therefore, we measure the $\beta$ parameter using the trace center position. We have performed numerous tests of this perturbative algorithm and verified that it works very well with a large range of practical pulses, when the GRENOUILLE trace resembles a standard SHG-FROG trace reasonably well. One example of the test of the algorithm is shown in Fig. 1.

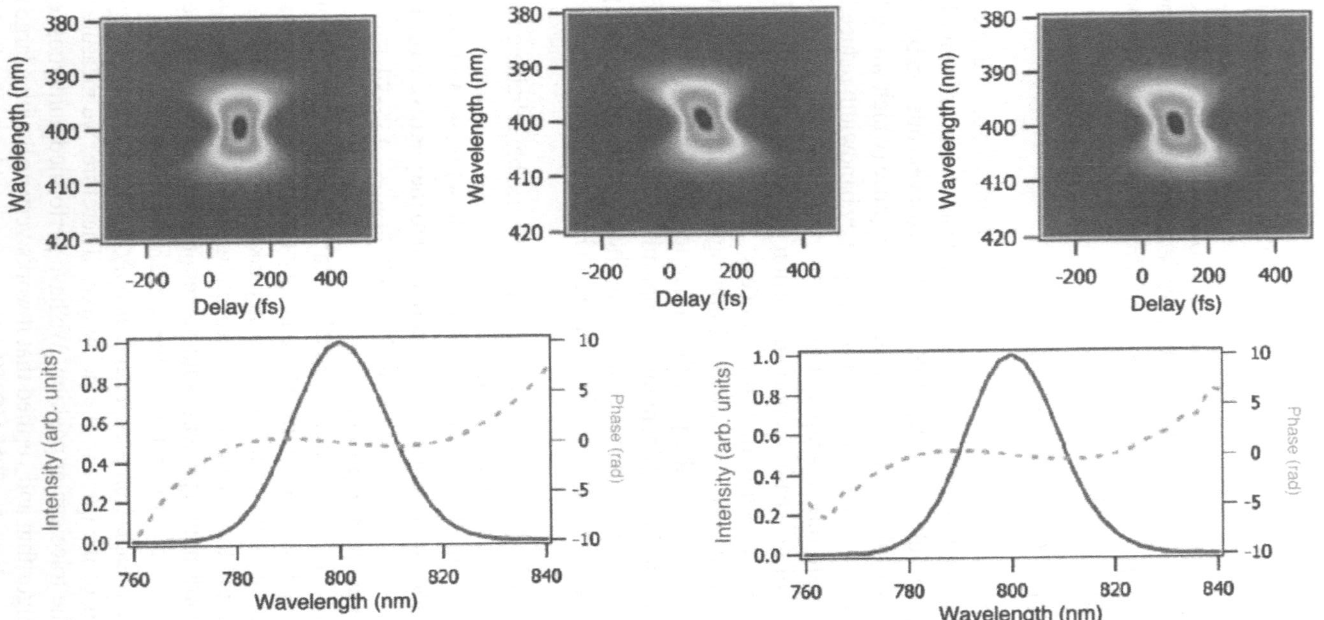


FIGURE 1. Top row: GRENOUILLE traces of the test pulse, with no spatiotemporal distortions (left), in the presence of spatial chirp (middle), and the algorithm-reconstructed trace (right). Bottom row: spectral intensity and phase of the test pulse (left), and the algorithm-retrieved spectral intensity and phase (right). The imposed spatial dispersion value is $\zeta = 30$ mm.fs/rad; the algorithm retrieves $\zeta = 29.6$ mm.fs/rad. The imposed angular dispersion value is $\beta = 2.03 \times 10^{-3}$ fs; the algorith retrieves $\beta = 2.04 \times 10^{-3}$ fs.

It is evident that the complexity of the pulse form will begin to affect the accuracy of the spatial dispersion measurement. This is largely due to the fact that the local pulse retrieved from the perturbed GRENOUILLE trace using the standard SHG-FROG algorithm is only approximate in nature. This is increasingly a problem when the spatiotemporal distortion of the beam becomes fairly large, which results in a distortion in the GRENOUILLE trace that significantly alters the shape of the trace. In such cases, the first-order perturbative algorithm ceases to be effective, and we will have to employ a more general global-optimization approach which will be discussed below.

However, for small spatiotemporal distortions, which are by far the majority of experimental cases, the advantages of the perturbative algorithm are clear: it is very simple and fast. The perturbative spatiotemporal distortion algorithm can be an easy extension to the existing FROG programs, as it involves only appending the current FROG retrieval code by one fast optimization step. We hope to see this method of measuring spatiotemporal distortions widely adopted in ultrafast laser laboratories.

## *3.2 Rigorous Pulse and Spatiotemporal Distortion Retrieval Algorithm*

To further improve the accuracy and applicable range of the technique, we would require an entirely new rigorous algorithm that globally optimizes the pulse profile and the spatiotemporal distortions simultaneously. Although the efficient regular FROG code cannot be directly applied in the process, it turns out that the powerful generalized projection scheme used in the regular FROG code can still be utilized in the design of the new algorithm, which renders similar strength and robustness to the new algorithm.

We have implemented this rigorous iterative algorithm, using a full frequency-domain expression of the GRENOUILLE trace, equivalent to Eq. (1). In this algorithm, each iteration consists of a pulse retrieval step, a spatial dispersion $\zeta$ retrieval step, and a magnitude replacement step which enforces the measured data constraint, the same as in a regular FROG algorithm. The angular dispersion retrieval is conducted at the very end of the entire algorithm, using the delay center position of the trace just as in the perturbative algorithm. A schematic of the algorithm is shown in Fig. 2.

The intraiteration pulse retrieval step uses a modified Newtonís method, in which a direction of search is calculated using the firstorder (gradient) and secondorder (Hessian matrix) of the error function with respect to the pulse profile points, and an optimal step size is calculated from a polynomial. After obtaining the new pulse profile, the spatial dispersion $\zeta$ is optimized using Brentís method, a parabolic interpolation method. Then the measured trace intensity is imposed onto the data in the magnitude replacement step, and the entire process is run over again, until convergence is reached.

It should be pointed out that it is not necessary for the in-iteration parameter optimization steps to be truly complete, because they will be repeated many times

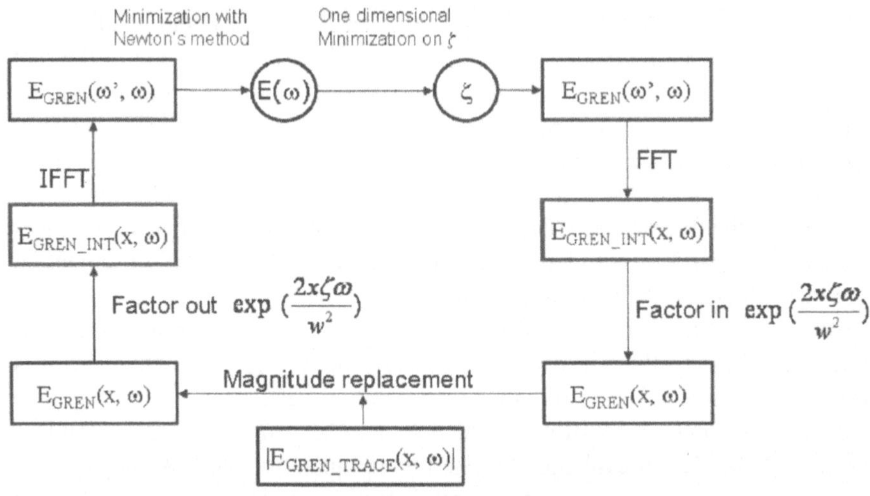


FIGURE 2. Flowchart of the rigorous GRENOUILLE retrieval algorithm.

as iterations continue on. Therefore, the goal is to design an optimization strategy for these parameters to strike a proper balance between speed and effectiveness.

We performed a large number of numerical simulations in order to test the performance of this rigorous algorithm. Overall, the performance and robustness

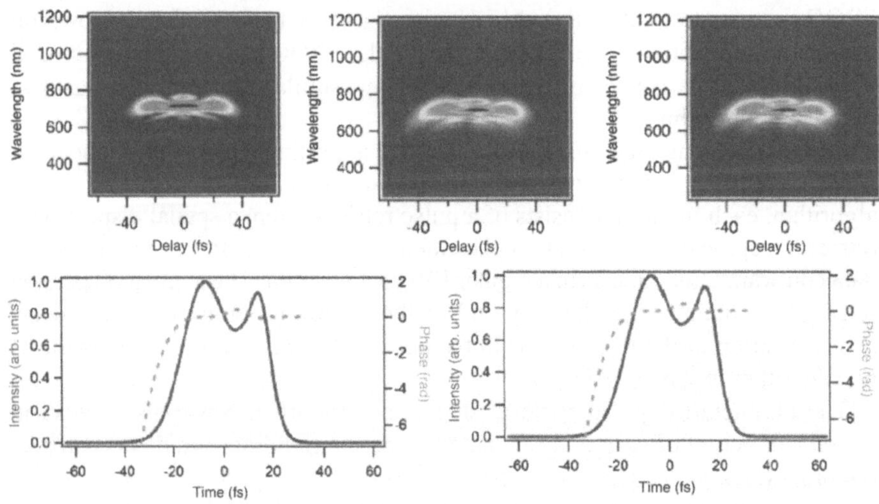


FIGURE 3. Top row: GRENOUILLE traces of the test pulse, with no spatiotemporal distortions (left), in the presence of spatial chirp (middle), and the algorithm-reconstructed trace (right). Bottom row: temporal intensity and phase of the test pulse (left), and the algorithm-retrieved temporal intensity and phase (right). The imposed spatial dispersion value is $\zeta = 7.65$ mm fs/rad; the algorithm retrieves an identical value. No angular dispersion is imposed in this test.

of the algorithm are quite satisfactory. Although the speed of the algorithm is much slower than the perturbative algorithm, taking 5–10 minutes to converge on a typical 64 × 64 test trace, it should still be acceptable for scientists who do not need immediate feedback and who require more reliable and accurate information about the spatiotemporal distortions in their beams. Despite the slowness, the accuracy of the results is impressive, with even some highly distorted GRENOUILLE traces. Figure 3 shows a numerical example of a test pulse.

## 4 Conclusions

In conclusion, we have developed two algorithms to retrieve spatiotemporal distortions from distorted GRENOUILLE traces. The first one is fast and works in the case of relatively simple pulses, and small distortions. The second one is more rigorous, therefore slower, but very sensitive even in the case of complicated pulses and severe distortions. Both algorithms are tested with theoretical pulses with spatiotemporal distortions and are shown to work well in most cases. These developments in the characterization of spatiotemporal distortions will provide a much needed tool for ultrafast scientists.

## References

1. A. V. Smith: Group-velocity-matched three-wave mixing in birefringent crystals. Opt Lett. 26(10):719–721 (2001).
2. J. P. Heritage, A. M. Weiner, and R. N. Thurston: Picosecond pulse shaping by spectral phase and amplitude manipulation. Opt Lett. 10(12):609–611 (1985).
3. M. M. Wefers and K. A. Nelson: Analysis of programmable ultrashort waveform generation using liquid-crystal spatial light modulators. J Opt Soc Am B. 12(7):1343–1362 (1995).
4. D. J. Kane and R. Trebino. Characterization of arbitrary femtosecond pulses using frequency resolved optical gating. IEEE J Quant Electron. 29(2):571–579 (1993).
5. P. O'Shea, M. Kimmel, X. Gu, and R. Trebino: Highly simplified device for ultrashort-pulse measurement. Opt Lett. 26(12):932–934 (2001).
6. S. Akturk, M. Kimmel, P. O'Shea, and R. Trebino: Measuring pulse-front tilt in ultrashort pulses using grenouille. Opt Express. 11:491–501 (2003).
7. S. Akturk, M. Kimmel, P. O'Shea, and R. Trebino: Measuring spatial chirp in ultrashort pulses using single-shot frequency-resolved optical gating. Opt Express. 11:68–78 (2003).

# First-order Spatiotemporal Distortions of Gaussian Pulses and Beams

Selçuk Aktürk, Xun Gu, Pablo Gabolde, and Rick Trebino

School of Physics, Georgia Institute of Technology, Atlanta, GA 30332-0430 USA
**akturk@socrates.physics.gatech.edu**

## 1 Introduction

Ultrashort laser pulses are usually expressed in terms of the temporal and spectral dependencies of their electric field. In this way, it is implicitly assumed that the spatial dependencies of the pulses are completely separable from time/frequency dependencies. However, this assumption fails to be valid very easily as the generation and manipulation of ultrashort pulses require introduction of couplings between space and time and/or frequency (spatiotemporal couplings). Moreover, disregarding the spatial dependencies of ultrashort pulses also greatly limits the potential applications. For these reasons, the spatiotemporal couplings have received much attention from researchers in recent years [1–3]. However, most of this work focuses on a particular effect or two, disregarding possible relations between different effects. All of the subtleties encountered in the past and the broad range of possible applications suggest that ultrafast optics researchers will benefit extensively from a detailed study of spatiotemporal couplings. This study will also provide a much more intuitive understanding of the behavior of ultrashort pulses.

Indeed, the study of spatiotemporal distortions is proving to be more subtle than previously thought. Various distortions are not independent because the pulse fields expressed in different domains are related by Fourier transforms. This fact was exploited to derive a relation between angular dispersion (AGD) and pulse-front tilt (PFT) [4], with the result that it is widely believed that AGD and PFT are equivalent phenomena. This observation is, however, not general, and, as only recently shown in [1], PFT is also generated by simultaneous spatial and temporal chirp. In another work, Gu et al. [5] showed that even defining spatial chirp (SPC) is not a trivial matter, and two possible, physically distinct, approaches are possible. All of the subtleties encountered in the past and the broad range of possible applications suggest that ultrafast optics researchers would benefit extensively from a detailed study of spatiotemporal couplings. Such a study would also provide a much more intuitive understanding of the behavior of ultrashort pulses.

In this work, we investigate all possible first-order coupling effects, providing a general theory, and we derive analytical expressions that not only properly describe them but also yield explicit relations between different effects, some of which have not been previously considered. We consider beams that are Gaussian in the transverse spatial ($x$) domain and hence also in its Fourier conjugate domain ($k$). We also consider pulses that are Gaussian in time ($t$) and hence also in frequency ($\omega$). While we use Gaussian profiles for convenience in analytical work, our approach is general and can be applied to arbitrary profiles using computational methods.

Previously identified spatiotemporal couplings include PFT tilt (in the $x - t$ domain), spatial dispersion (SPD) (in the $x - \omega$ domain), AGD (in the $k - \omega$ domain), and a less well-known distortion called "time vs. angle" (TVA) (in the $k - t$ domain). These distortions occur in the real component of the coupling and hence can be called amplitude couplings. Here we identify four new spatiotemporal distortions, which occur in the imaginary part of the coupling in the relevant domains and hence are spatiotemporal phase distortions. They are "wave-front rotation" (WFR), "wave-front-tilt dispersion" (WFD), "angular frequency chirp" (AFC), and "angular temporal chirp" (ATC), respectively in the same domains above. While there are eight spatiotemporal distortions in all, only two are independent. Each spatiotemporal distortion can then be written as the weighted sum of two distortions in a neighboring domain.

## 2 The Pulse Electric Field Expressions in Different Domains

We start by expressing the electric field of an ultrashort pulse (with Gaussian profiles) in $x - t$ domain:

$$E(x,t) \propto \exp\left\{\widetilde{Q}_{xx}x^2 + 2\widetilde{Q}_{xt}xt - \widetilde{Q}_{tt}t^2\right\} \tag{1}$$

where the complex $\widetilde{Q}_{xx}$ and $\widetilde{Q}_{tt}$ coefficients are given by

$$\widetilde{Q}_{xx} = -i\frac{\pi}{\lambda R(z)} - \frac{1}{w^2(z)} \tag{2}$$

$$\widetilde{Q}_{tt} = -i\beta + \frac{1}{\tau^2}. \tag{3}$$

When expressed in this form, the pulse can also be easily propagated through linear optical elements, by using the matrix formed by the $Q$ elements (after inversion and multiplication by a constant) in the $4 \times 4$ matrix formalism proposed by Kostenbauder [6].

The cross term in Eq. 1 yields the spatiotemporal couplings. The real part of $\widetilde{Q}_{xt}$, the coupling term, yields the PFT. The imaginary part of this coupling term

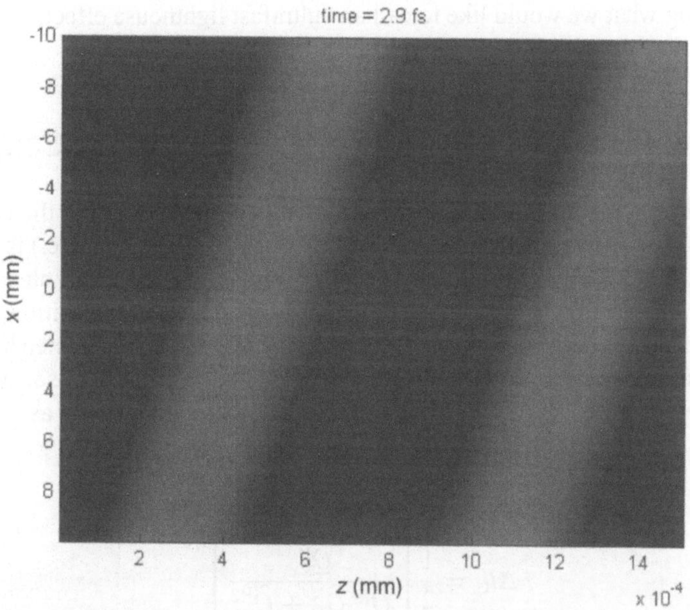


FIGURE 1. The effect of the WFR. The figure shows the wave fronts in $(x, t)$ domain at an instant of time.

yields an effect that has never been considered before and which we identify as WFR, as it corresponds to tilt of phase fronts evolving in time. This effect is displayed in Fig. 1, where the tilted phase fronts are evident at the instant of time given.

We then Fourier transform Eq. (1) to $x - \omega$, $k - \omega$, and $k - t$ domains. In $x - \omega$ domain, we get

$$E(x, \omega) \propto \exp\left\{R_{xx}x^2 + 2R_{x\omega}x\omega - R_{\omega\omega}\omega^2\right\}, \tag{4}$$

where the real part of $R_{xt}$ yields the SPC and the imaginary part of this coupling term yields WFD, which causes the wavefront to be tilted by an amount that depends on the frequency.

Proceeding to $k - \omega$ domain:

$$E(k, \omega) \propto \exp\left\{S_{kk}k^2 + 2S_{k\omega}k\omega - S_{\omega\omega}\omega^2\right\}. \tag{5}$$

The real off-diagonal parameter $S_{k\omega}$ is the well-known AGD. The imaginary part is a new quantity and it indicates the presence of what we will call ASC because it involves a variation in the frequency with angle.

Lastly, in $k - t$ domain:

$$E(k, t) \propto \exp\left\{P_{kk}k^2 + 2P_{kt}kt - P_{tt}t^2\right\}. \tag{6}$$

The real off-diagonal parameter $P_{kt}$ is the TVA. The presence of TVA indicates that the propagation direction of the pulse changes as the pulse proceeds in time, generating what we would like to call an "ultrafast lighthouse effect." The imaginary part of the coupling term is ATC.

## 3 Defining the Widths and Normalization of the Couplings

For a pulse that does not have spatiotemporal couplings, defining the pulse and widths (e.g., beam spot size, pulse width, frequency bandwidth, etc.) is straightforward. When there is coupling, however, the situation is more complex. In $x - t$ domain, for example, one can either find the pulse width at each position, yielding the "local pulse width" (see Fig. 2). Alternatively, the pulse width, when integrated over position yields the "global pulse width." With the root-mean-square definition, these two widths can be written in terms of the $Q$ parameters as

$$\Delta t_{\mathrm{L}} = \left[\frac{1}{4\widetilde{Q}_{tt}^{\mathrm{R}}}\right]^{1/2} \tag{7}$$

$$\Delta t_{\mathrm{G}} = \frac{1}{2}\left[\frac{\widetilde{Q}_{xx}^{\mathrm{R}}}{\widetilde{Q}_{xx}^{\mathrm{R}}\widetilde{Q}_{tt}^{\mathrm{R}} + \widetilde{Q}_{xt}^{\mathrm{R2}}}\right]^{1/2}, \tag{8}$$

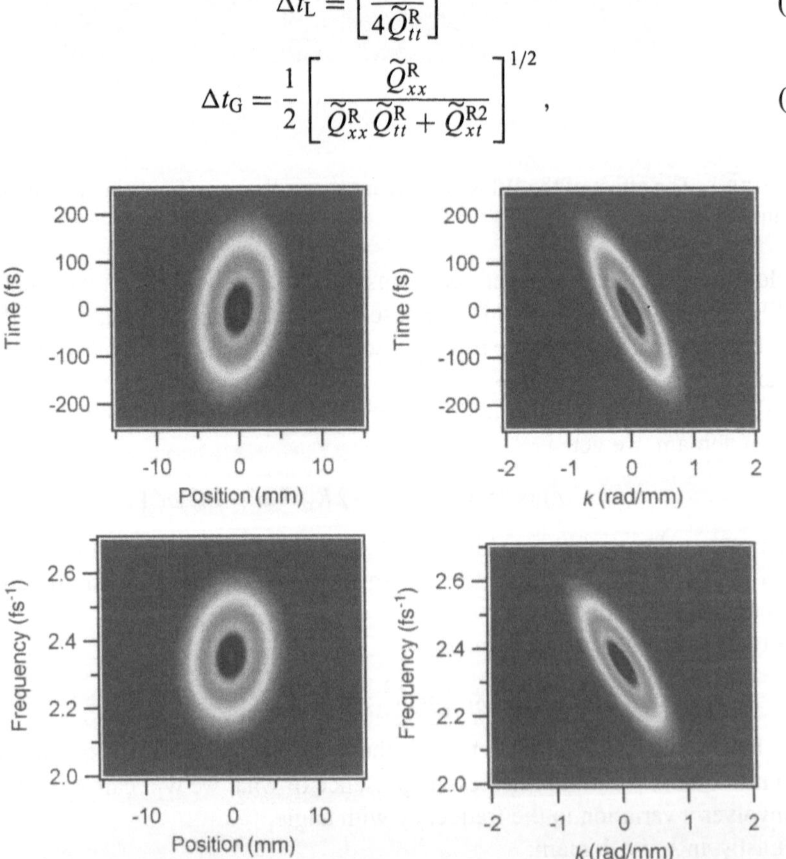


FIGURE 2. Intensity profile of a pulse expressed in different domains. This pulse simultaneously has all four spatiotemporal couplings, PFT, FCH, AGD, and TVA, as can be seen from the tilted images.

TABLE 1. The correlation coefficients of the pulse field shown in Fig. 2

| Coupling | $\rho_{xt}$ | $\rho_{kt}$ | $\rho_{x\omega}$ | $\rho_{k\omega}$ |
|---|---|---|---|---|
| Value | 0.22 | −0.81 | 0.14 | −0.82 |

where subscripts L and G denote local and global and superscript R denotes the real part. Equations (7) and (8) reduce to the expected values when the cross term (PFT) is zero.

We have identified the real parts of the cross terms in each domain as equivalent to the spatiotemporal couplings. However, the dimensions of the various cross terms as they are usually considered are unintuitive and do not correspond to logical definitions of spatiotemporal distortions. Thus, for experimental convenience and better intuition, normalization is helpful. As an example, we can normalize PFT with respect to pulse width and get arrival time of maximum intensity as a function of position:

$$\frac{\partial t_0}{\partial x} = \frac{\widetilde{Q}^{R}_{xt}}{\widetilde{Q}^{R}_{tt}}. \tag{9}$$

Or, we can normalize with respect to beam width and get the position of maximum intensity as a function of time, analogous to Eq. (9).

There is yet another approach for normalization based on correlation coefficients. The correlation coefficient is defined as

$$\rho_{xt} \equiv \frac{\int_{-\infty}^{\infty}\int_{-\infty}^{\infty} xtI(x,t)dxdt}{\Delta x_G \Delta t_G}. \tag{10}$$

Then, in terms of the $Q$ coefficients, we find

$$\rho_{xt} = \frac{\widetilde{Q}^{R}_{xt}}{\sqrt{-\widetilde{Q}^{R}_{xx}\widetilde{Q}^{R}_{tt}}}. \tag{11}$$

Defining the coupling in terms of these correlation coefficients has an advantage that the coefficient $\rho$ always lies between 1 and $-1$. Therefore, this quantity can easily show whether the coupling is "large" or "small." Table 1 shows the correlation coefficients for the couplings of the pulse displayed in Fig. 2. Note that the correlation coefficients are also unitless.

## 4 Relations Between the Spatiotemporal Couplings

To first order, there are four distinct intensity spatiotemporal couplings: PFT, SPC, AGD, and TVA, and there are four imaginary phase spatiotemporal couplings: WFR, WFD, ASC, and ATC. Obviously, they are not all independent; if one has all six (three real and three imaginary) coefficients in any single domain, say

TABLE 2. The relations of spatiotemporal distortions in all four domains

| $x - t$ | $x - \omega$ | $k - \omega$ | $kt$ |
|---|---|---|---|
| $\widetilde{Q}_{xt}$ | $\frac{i}{2}\frac{R_{x\omega}}{R_{\omega\omega}}$ | $\frac{1}{4}\frac{S_{k\omega}}{S_{kk}S_{\omega\omega}+S_{k\omega}^2}$ | $-\frac{i}{2}\frac{P_{kt}}{P_{kk}}$ |
| $-\frac{i}{2}\frac{\widetilde{Q}_{xt}}{\widetilde{Q}_{tt}}$ | $R_{x\omega}$ | $-\frac{i}{2}\frac{S_{k\omega}}{S_{kk}}$ | $\frac{1}{4}\frac{P_{kt}}{P_{kk}P_{tt}+P_{kt}^2}$ |
| $\frac{1}{4}\frac{\widetilde{Q}_{xt}}{\widetilde{Q}_{xx}\widetilde{Q}_{tt}+\widetilde{Q}_{xt}^2}$ | $\frac{i}{2}\frac{R_{x\omega}}{R_{xx}}$ | $S_{k\omega}$ | $-\frac{i}{2}\frac{P_{kt}}{P_{tt}}$ |
| $\frac{i}{2}\frac{\widetilde{Q}_{xt}}{\widetilde{Q}_{xx}}$ | $\frac{1}{4}\frac{R_{x\omega}}{R_{xx}R_{\omega\omega}+R_{x\omega}^2}$ | $\frac{i}{2}\frac{S_{k\omega}}{S_{\omega\omega}}$ | $P_{kt}$ |

$x - t$, then the field in all other domains can be calculated by taking Fourier transforms. Thus, for Gaussian pulse/beams, since four parameters are diagonal components that describe the pulse width, spot size, etc., there are only two independent spatiotemporal couplings. The other six can be computed from these two (and the relevant diagonal components, such as the pulse width, temporal chirp, spot size, and wavefront radius of curvature). The explicit relations between the spatiotemporal couplings are summarized in Table 2 Each row of the table shows the cross terms in one domain and its equivalent representation in the others.

While Table 2 presents the complete listing of the relations between the first-order spatiotemporal couplings, these expressions are somewhat unintuitive, as they are expressed entirely in terms of the $P$, $Q$, $R$, and $S$ matrices. Most of these equations, however, can easily be written in a more compact an intuitive manner. Skipping the details:

$$\mathrm{PFT} = -2\mathrm{WFD} + 2\mathrm{FCH} \times \mathrm{FRG} \tag{12}$$

$$\mathrm{SPD} = 2\mathrm{ASC} - 2\mathrm{AGD} \times \mathrm{APC} \tag{13}$$

$$\mathrm{AGD} = -2\mathrm{ATC} + 2\mathrm{TVA} \times \mathrm{TCH} \tag{14}$$

$$\mathrm{TVA} = -2\mathrm{WFR} + 2\mathrm{PFT} \times \mathrm{WFC}. \tag{15}$$

# 5 Constructing the Electric Field in Different Domains

Up to this point, we defined a total of eight phenomena resulting from couplings. Because of the Fourier transforms involved, all of these couplings depend on each other. In fact, each spatiotemporal distortion can then be written as the weighted sum of two distortions in a neighboring domain. We derive analytical expressions that relate all these distortions to each other, yielding a complete picture of spatiotemporal couplings. For example, the pulse-front tilt is the weighted sum of the angular dispersion and spatial/temporal chirps.

Another very important point to be mentioned here is that Eq. (1) involves only six independent parameters. They are the spatial beam parameters (beam width and radius of curvature), the temporal pulse parameters (pulse width and chirp), and PFT and WFR. Since the field in all other domains can be found via Fourier transforms, this set of six parameters completely determines all of the eight couplings described above.

Obviously, the set of six parameters can be chosen in each of the other domains as well. However, more interestingly, it is also possible to choose the real parts of the couplings term and instead of the imaginary part, to use the real part of the coupling in the adjacent domain. For example, spatial beam parameters, the temporal pulse parameters and PFT and FCH also completely determine the pulse in all domains. In other words, provided that spatial and temporal parameters are known, all of the eight couplings can be derived from two. This is demonstrated in Fig. 2. Starting form the six parameters mentioned above, we constructed the electric fields of an ultrashort pulse in all four domains. While the figures only show the intensity profiles, the phase terms are also calculated.

## 6 Conclusion

In conclusion, we demonstrated a definitive approach in determining and describing the spatiotemporal couplings in ultrashort pulses and have identified several previously unidentified but common distortions. Our approach completely determines the explicit relations between various spatiotemporal couplings. It can also be generalized to arbitrary profiles by using computational analysis instead of the analytical approach described here. We hope that our calculations and definitions will provide a useful tool for ultrafastoptics researchers to better understand the distortions that commonly occur (intentionally or not) in ultrashort pulses.

## References


1. S. Akturk, X. Gu, E. Zeek, and R. Trebino: Opt. Exp. **12**(19), 4399 (2004).
2. S. Akturk, M. Kimmel, P. O'Shea, and R. Trebino: Opt. Exp. **11**, 491 (2003).
3. S. Akturk, M. Kimmel, P. O'Shea, and R. Trebino: Opt. Exp. **11**, 68 (2003).
4. O.E. Martinez: Opt. Comm. **59**(3), 229 (1986).
5. X. Gu, S. Akturk, and R. Trebino: Opt. Comm. **242**, 599 (2004).
6. A.G. Kostenbauder: IEEE J. Quant. Electron. **26**(6), 1148 (1990).

# Characterization of Femtosecond Optical Pulses with Wavelet Transform of Spectral Shearing Interferogram

Yuqiang Deng,[1] Zubin Wu,[1] Shiying Cao,[1] Lu Chai,[1]
Ching-yue Wang,[1] and Zhigang Zhang[1,2]

[1] Ultrafast Laser Laboratory, School of Precision Instrument and Optoelectronics Engineering, Tianjin University, Tianjin 300072, China
yuqdeng@yahoo.com

[2] Institute of Quantum Electronics, School of Electronics Engineering and Computer Science, Peking University, Beijing 100871, China
zhgzhang@pku.edu.cn

**Summary.** We report the recent progress in the application of wavelet-transform technique in the characterization of femtosecond optical pulses. The technique eliminates the uncertainty from filter with Fourier-transform technique, and calculations from a structured spectrum demonstrate that it is a more accurate technique for spectral phase retrieval. This technique will be helpful in ultrashort optical pulses generation and measurement.

## 1 Introduction

Femtosecond optical pulses have shown their robust applications in various fields. The improvement of phase measurement is helpful in compressing optical pulses. There are two most widely used methods to measure phases of ultrashort optical pulses: frequency resolved optical gating (FROG) [1] and spectral phase interferometry for direct electric-field reconstruction (SPIDER) [2].

The advantage of FROG is that it results in a two-dimensional trace [1, 3, 4], one axis of which responds to time and the other responds to frequency. The two-dimensional topography shows full information of the optical pulse, and the phase of pulse can be identified from the trace. However, the phase retrieval algorithm of FROG is iterative, which is time consuming. Therefore it is not suitable for real-time measurements.

However, with SPIDER [2, 5], which is based on spectral interference theory, the spectral phase can be retrieved in short time and therefore it is suitable for real-time measurements and is used as on-line feedback compensation loop [6, 7].

We have introduced wavelet transform (WT) [8] into the phase extraction from SPIDER signals [9, 10]. Wavelet transform can produce a “SPIDER trace,” from which spectral phase can be judged apparently. In this chapter, we will make a brief introduction of WT concept to femtosecond laser pulse characterization and demonstrate that WT is an accurate technique for structured spectral phase retrieval. With this technique, the uncertainty introduced from filter with Fourier-transform (FT) technique [11] is eliminated.

## 2 Mathematics Description of Wavelet Transform

The basis of WT is a function limited in both time domain and frequency domain. It can shift in time domain (or frequency domain alternatively) to obtain the frequency information (or time information) of the signal at local time position (or local frequency position) by dilation. A series of wavelets generated from an appointed wavelet by dilations and translations called daughter wavelets. Accordingly, the appointed wavelet is called mother wavelet [8].

Wavelet transform is the inner scalar between signal and daughter wavelet functions. It converts one-dimensional time (or frequency) signal into two-dimensional time–frequency amplitude topography [9, 10], which displays the time and frequency information of the signal in an apparent time–frequency plane. On the other hand, it produces a phase topography simultaneously. One could first find the ridgeline of the amplitude topography and then read out the phase along the ridgeline [12] in the phase topography, which eliminates the filter and the second FT in conventional SPIDER algorithm.

The WT $W(t, \omega)$ of a spectral shearing interferogram (SSI) [2, 5] $f(\omega')$ can take the following form [8, 13]:

$$W(t, \omega) = \frac{t}{2\pi} \int_{-\infty}^{+\infty} f(\omega') \psi^* \left[ \frac{(\omega' - \omega) \cdot t}{2\pi} \right] d\omega'. \tag{1}$$

We choose the best time–frequency resolution wavelet, Gabor wavelet [14, 15], as the mother wavelet. The expression of the Gabor wavelet is

$$\psi(\omega') = \exp\left(-\omega'^2/2\sigma^2 + i2\pi\omega'\right)/(\sigma^2\pi)^{1/4}, \tag{2}$$

where $\sigma = (2 \ln 2)^{-1/2}$ is a constant, indicating the width of the wavelet.

## 3 Experimental Examples

We reconstruct the spectral phase for a practical example. Figure 1a is a measured interferogram from a mirror dispersion controlled femtosecond Ti:sapphire laser. The central wavelength was measured as 848 nm with a bandwidth of 21.6 THz. The spectral interferogram was obtained with a homemade SPIDER instrument with the frequency shear $\Omega = 24.5 \times 10^{12}$ rad/s and the pulses separation $\tau = 1.15$ ps. Following conventional procedure of SPIDER [2,5], we can derive different phase

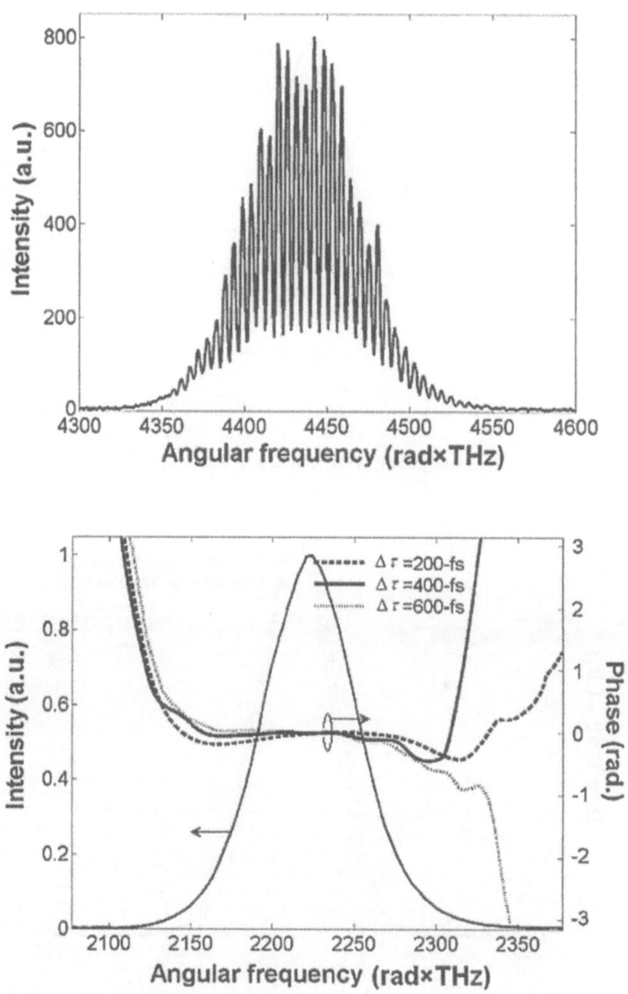


FIGURE 1. (a) Measured spectral interferogram. (b) Spectrum and reconstructed phases with FT techniques through different widths of filter.

results from different widths of filter windows (see phases in Fig. 1b). The figure demonstrates that the uncertainty resulted from the filter widths: Different width of filter windows contains different quantities of noise, which results different retrieved phases.

Since WT technique directly reads the phase from the transformed interference by detecting the maximum WT of the interferogram, we can retrieve only one phase, which is most accurate because it essentially eliminates the phase noise from the filter windows. Following the procedure, we obtained spectral phase (see Fig. 2a). For comparison, the retrieved phase with a conventional FT technique by carefully selected window size (400fs) is also shown in Fig. 2a. The reconstructed pulses with the two spectral phases as well as transform-limited pulse are shown

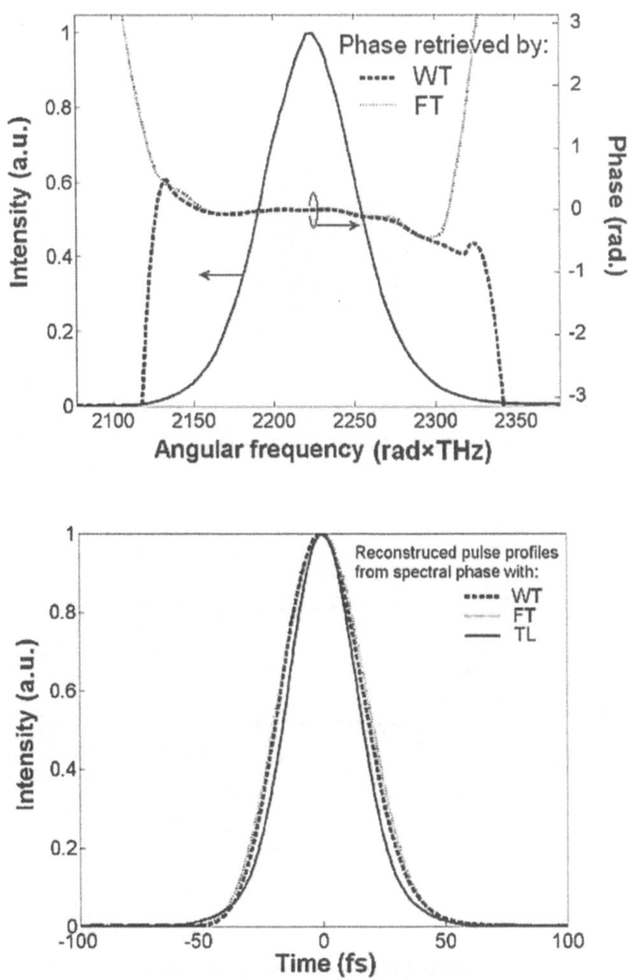


FIGURE 2. (a) Spectrum and reconstructed phase with WT and FT techniques. (b) Reconstructed pulse profiles in comparison with the transform-limited pulses, TL, transform-limited pulse.

in Fig. 2b. The phase obtained with WT technique fairly agrees with the phase with FT technique with carefully selected window size.

# 4 Error Analysis

We take a practical spectrum from our homemade chirped pulse amplifier (CPA) system as an example. The transform-limited pulse width is 16.03 fs. The following two assumed phases are applied to the pulse respectively, which are shown in Fig. 3(a):

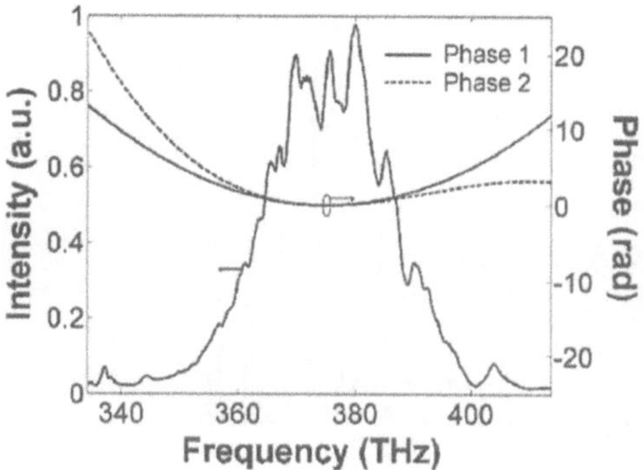


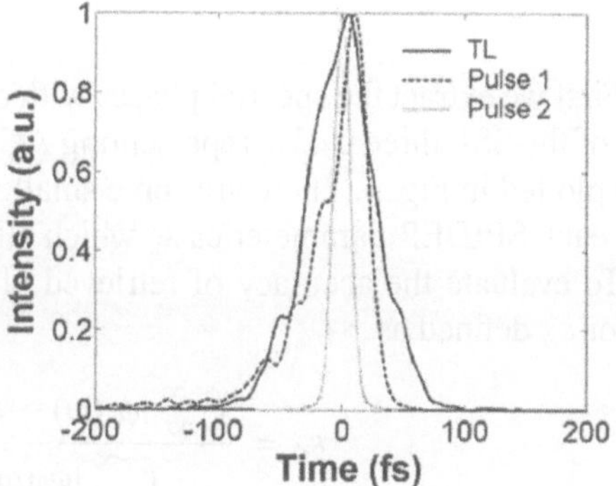


FIGURE 3. Spectrum from CPA, simulated spectral phases, and pulse profiles. (a) Spectrum and assumed spectral phases. (b) Assumed pulse profiles. TL, transform-limited pulse.

Phase 1: $\phi(\omega) = 200fs^2 \times (\omega - \omega_0)^2$
Phase 2: $\phi(\omega) = 200fs^2 \times (\omega - \omega_0)^2 - 600fs^3 \times (\omega - \omega_0)^3$

where $\omega_0$ is the central angular frequency. The pulses will be stretched to 60.53 fs and 41.16 fs in time domain, respectively, which are shown in Fig. 3(b).

Figure 3a shows the spectrum and the two assumed phases and Fig. 3b shows the intensity profiles of the reconstructed pulses. Then we build SSI with selected SPIDER parameters. We assume pulse separations $\tau$ from 0.2 to 1.2 ps, at a step of 0.2 ps. The pulse separation $\tau$ and corresponding spectral shearing $\Omega$ are listed in Table 1.

TABLE 1. Assumed SPIDER parameters

| $\tau$(ps) | $\Omega(10^{12}$rad/s) |
|---|---|
| 0.2 | 6.28 |
| 0.4 | 12.56 |
| 0.6 | 18.84 |
| 0.8 | 25.12 |
| 1.0 | 31.40 |
| 1.2 | 37.68 |

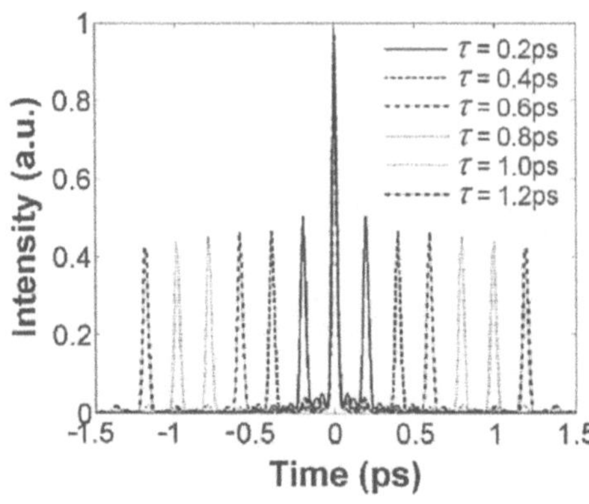


FIGURE 4. Fourier transform of the interferograms in time domain for the six pulse separations.

First we extract the spectral phases with conventional FT algorithm [11]. After FT of the SSI, three peaks, representing AC and DC components in time domain, are plotted in Fig. 4. There are some small waves between the AC and DC peaks for each SPIDER parameter case, which alleviative as the separation increases.

To evaluate the accuracy of retrieved phase, we introduce the relative phase error $\varepsilon_\phi$ defined as

$$\varepsilon_\phi = \frac{\int_{-\infty}^{+\infty} |\phi_r(\omega) - \phi_0(\omega)| I(\omega) d\omega}{\int_{-\infty}^{+\infty} |\phi_0(\omega)| I(\omega) d\omega}, \tag{3}$$

where $\phi_r(\omega)$ is the retrieved phase, $\phi_0(\omega)$ is the given phase, and $I(\omega)$ is the intensity of the spectrum.

The extracted relative phase errors for the six time delays with different filter widths of a rectangular shape are shown in Fig. 5.

Then we extract the spectral phases with WT algorithm. Wavelet transform converts the SSI into a two-dimensional time–frequency plane, and the phase is extracted form the ridge of phase topography; therefore, the resulted phase is unique for each pulse separation. The retrieved relative phase errors with WT algorithm for the six time delays are also shown in Fig. 5.

It can be observed from Fig. 5 that the phase error is much smaller with WT than with FT algorithm for a large range of filter width and for a large range of pulse separation.

## 5 Conclusions

In conclusion, we have applied the WT technique in characterization of femtosecond laser pulses. We demonstrated that with WT technique we could obtain very much similar phase results to that using FT, with less phase errors and with no window choosing procedure. This technique has potential applications in on-line automatic feedback control of the phase.

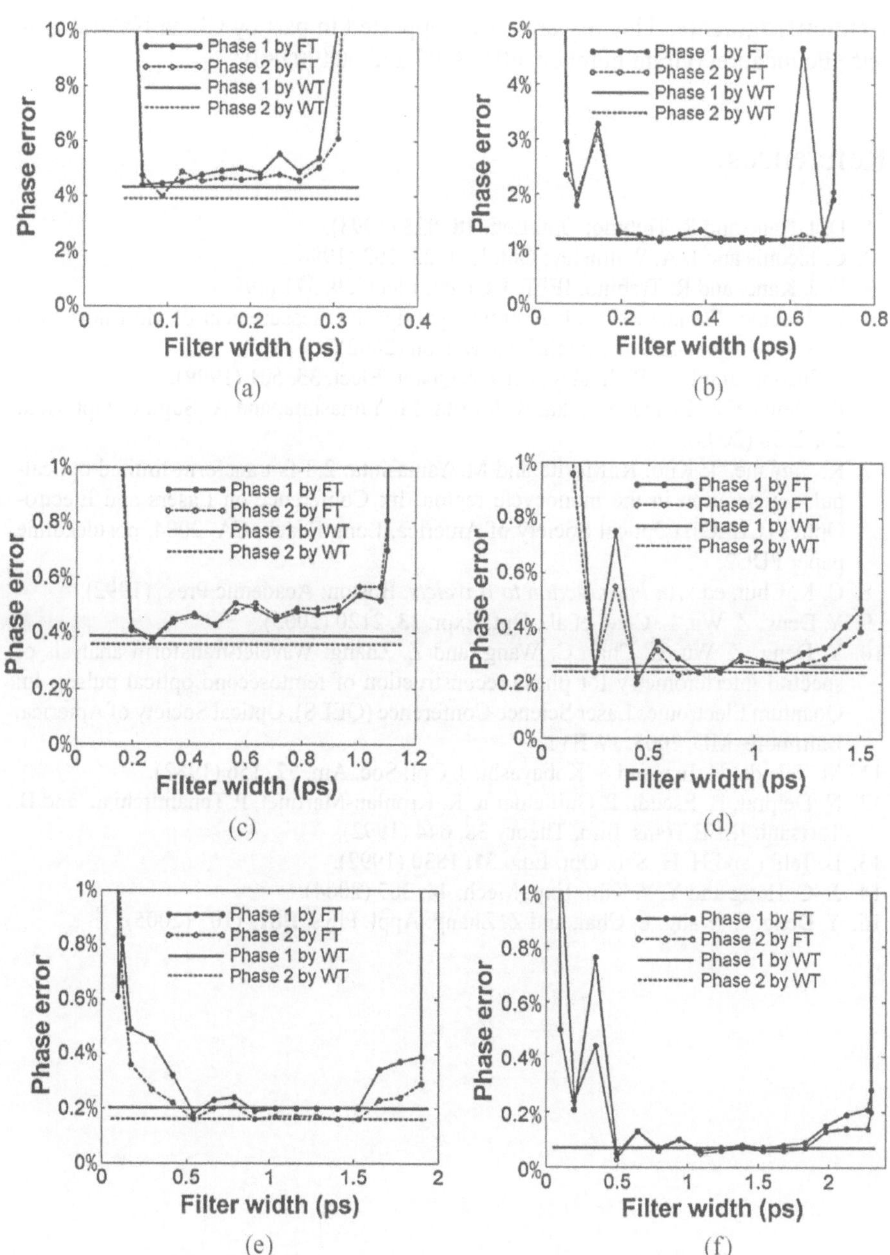


FIGURE 5. Retrieved relative phase errors. Pulse separation (a) $\tau = 0.2$ ps, (b) $\tau = 0.4$ ps, (c) $\tau = 0.6$ ps, (d) $\tau = 0.8$ ps, (e) $\tau = 1.0$ ps, and (f) $\tau = 1.2$ ps.

**Acknowledgments.** This research was supported in part by China Natural Science Foundation (Grant numbers 60178007 and 60490280).

## References

1. D. J. Kane and R. Trebino: Opt. Lett. **18**, 823 (1993).
2. C. Iaconis and I. A. Walmsley: Opt. Lett. **23**, 792 (1998).
3. D. J. Kane, and R. Trebino: IEEE J. Quant. Elect. **29**, 571 (1993).
4. R. Trebino: Frequency-resolved optical gating: the measurement of ultrashort laser pulses, Kluwer academic publishers, Boston (2002).
5. C. Iaconis and I. A. Walmsley: IEEE J. Quant. Elect. **35**, 501 (1999).
6. K. Yamane, Z. Zhang, K. Oka, R. Morita, M. Yamashita, and A. Suguro: Opt. Lett. **28**, 2258 (2003).
7. K. Yamane, T. Kito, R. Morita and M. Yamashita: 2.8-fs transform-limited optical-pulse generation in the monocycle region. In: Conference on Lasers and Electro-Optics (CLEO), (Optical Society of America, Long Beach, CA, 2004, postdeadline paper PDC2.
8. C. K. Chui, ed.: *An Introduction to Wavelets*. Boston: Academic Press (1992).
9. Y. Deng, Z. Wu, L. Chai, et al.: Opt. Expr. **13**, 2120 (2005).
10. Y. Deng, Z. Wu, L. Chai, C. Wang, and Z. Zhang: Wavelet-transform analysis of spectral interferometry for phase reconstruction of femtosecond optical pulses. In: Quantum Electronics Laser Science Conference (QELS), Optical Society of America, Baltimore, MD, 2005, JWB11.
11. M. Takeda, H. Ina, and S. Kobayashi: J. Opt. Soc. Am. **72**, 156 (1982).
12. N. Delprat, B. Escudi, P. Guillemai n, R. Kronlan-Martinet, P. Tchamitchian, and B. Torrsani: IEEE Trans. Info. Theory **38**, 644 (1992).
13. B. Telfer and H. H. Szu: Opt. Eng. **31**, 1830 (1992).
14. J. -C. Hong and Y. Y. Kim: Exp. Mech. **44**, 387 (2004).
15. Y. Deng, C. Wang, L. Chai, and Z. Zhang: Appl. Phys. **B81**, 1107 (2005).

# Water Vapor Absorption Spectroscopy in Terahertz Range Using Wavelet-transform Analysis

Yuqiang Deng,[1] Qirong Xing,[1] Liying Lang,[1] Shuxin Li,[1] Lu Chai,[1] Ching-yue Wang,[1] and Zhigang Zhang[1,2]

[1] Ultrafast Laser Laboratory, School of Precision Instrument and Optoelectronics Engineering, Tianjin University, Tianjin 300072, China
yuqdeng@yahoo.com

[2] Institute of Quantum Electronics, School of Electronics Engineering and Computer Science, Peking University, Beijing 100871, China
zhgzhang@pku.edu.cn

**Summary.** We introduce wavelet transform into terahertz time-domain spectroscopy, and report the analysis of experimental results of water vapor absorption in terahertz range. The extracted absorption lines are in excellent agreement with those previously reported. Wavelet transform in terahertz time-domain spectroscopy can be considered as a novel method for the analysis of the time-domain properties of absorption spectra of terahertz radiation.

**Key words:** Ultrafast lasers, wavelet transform, terahertz radiation, optical spectrum analysis

## 1 Introduction

Coherent terahertz (THz) wave is the electromagnetic radiation from 0.1 to several tens terahertz in frequency domain [1]. In recent years, the progress of femtosecond optical pulses technique has promoted the studies of THz wave. Based on optical rectification theory [2], it is a widely used technique using femtosecond optical pulses as pump source, to stimulate nonlinear crystal for THz wave generation. The stimulated THz wave, associated with photoemission of THz detection, constitutes terahertz time-domain spectroscopy (THz–TDS) [3].

THz–TDS is the phase coherent electromagnetic radiate detection technique in nature. It can detect electric field of reciprocity between THz wave and matters, i.e., detect the amplitude and phase of the electromagnetic wave simultaneously. Therefore, it can accurately measure the dispersion and absorption characterizations of the matters. The technique has been used in the spectroscopy of gases [4], liquids [5], dielectrics [6], semiconductors [7], superconductors [8], and even in biology specimens [9].

Conventional THz–TDS is based on Fourier transform (FT) technique [3]. Recently, we have applied wavelet transform (WT) [10] in the phase extraction from spectral phase interferometry for direct electric-field reconstruction (SPIDER) signals [11, 12]. In this chapter, we introduce WT into THz–TDS, and show our experimental water vapor absorption lines of THz and compare with the data obtained using other techniques. The results show that WT analysis of the absorption spectrum delivers accurate absorption lines in THz range. Because the results are shown in a time–frequency domain, it gives a more intuitive image on when the absorption happens at which frequency.

Combining WT technique with THz–TDS, we hope there births a new spectroscopy: the wavelet transform terahertz time-domain spectroscopy (WT–THz–TDS).

## 2 Mathematics Description of Wavelet Transform

The WT is a powerful tool for signals analysis in time-frequency domain; the WT of a temporal signal $f(t)$ transforms the signal information on an apparent time vs. frequency plane $W(\tau, \nu)$, on which frequency component on each time point can be identified.

The basis of WT is a function limited in both time domain and frequency domain. It can shift in time domain to obtain the frequency information of the signal at local time position by dilation. A series of wavelets generated from an appointed wavelet by dilations and translations called daughter wavelets, accordingly the appointed wavelet are called mother wavelet.

WT is the inner scalar between signal and daughter wavelet functions. It converts one-dimensional time signal into two-dimensional time-frequency amplitude topography, which displays the time and frequency information of the signal in an apparent time-frequency plane. Wavelet-transform $W(\tau, \nu)$ of a time-domain function $f(t)$ can be expressed in the following form [10, 13]:

$$W(\tau, \nu) = \frac{t}{2\pi} \int_{-\infty}^{+\infty} f(t) \cdot \psi^* \left[ \frac{(t-\tau) \cdot \nu}{2\pi} \right] dt. \tag{1}$$

The mother wavelet we choose is Gabor wavelet (also called Gaussian wavelet) [14, 15], because it has the smallest Heisenberg box, i.e. the best time-frequency resolution [14]. Gabor wavelet is expressed as

$$\phi(t) = \exp\left(-t^2/2\sigma^2 + i2\pi \cdot t\right)/(\sigma^2\pi)^{1/4}, \tag{2}$$

here $\sigma = (2\ln 2)^{-1/2}$ is a constant, showing the width of wavelet.

## 3 Terahertz Radiations Spectra Analysis

In the conventional THz–TDS [3], the THz signal is usually taken in time domain and the spectrum is obtained by using FT; therefore, it gives only an averaged spectrum and the temporal evolution is hidden. WT has the potential of time-frequency analysis and should be able to be used in the time-frequency analysis of THz radiation signals.

Figures 1a and b are temporal THz radiation signals for air and water vapor absorption, respectively, measured in our laboratory [16].

Figure 1 shows WT on temporal THz radiation signals. Time vs. frequency information is shown in Fig. 2 in an apparent two-dimensional plane. Figures 2a and b are the time-frequency topographies for Figs. 1a and b given by WT technique, respectively, which clearly show the frequency components of THz signals at each time-domain position.

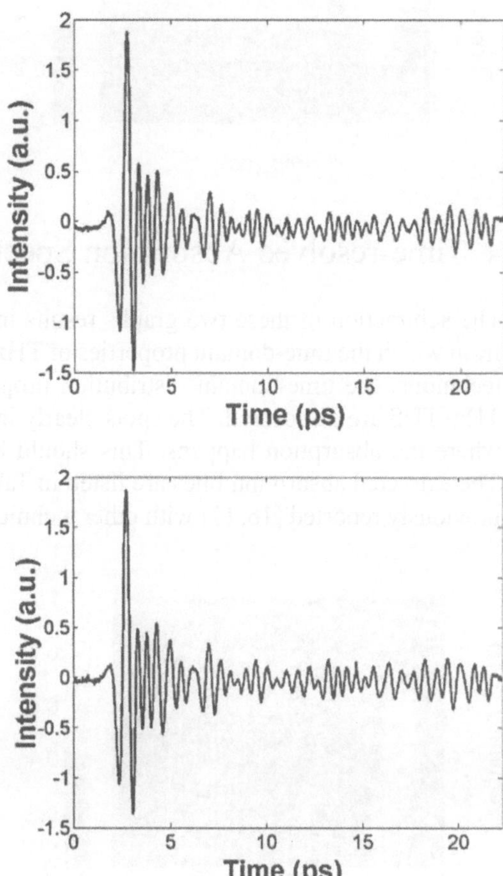


FIGURE 1. Terahertz radiation signals. (a) Terahertz radiation in air. (b) Terahertz radiation in water vapor.

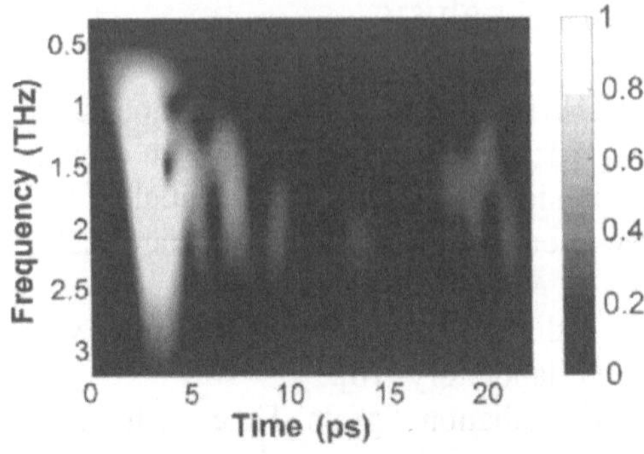


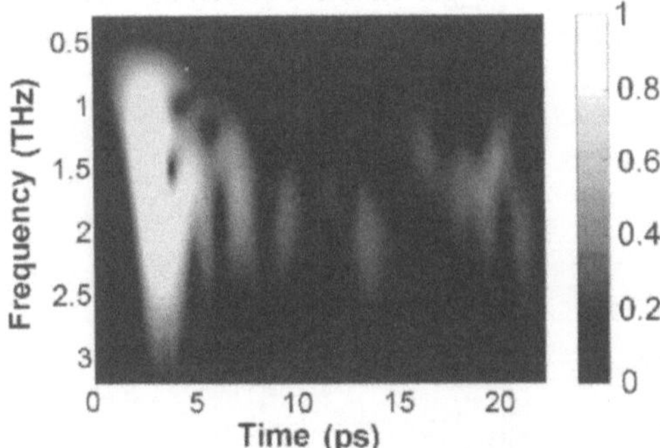


FIGURE 2. Time–frequency distributions of THz signals. (a) Air; (b) water vapor.

# 4 Time-resolved Absorption Spectra of Water Vapor

The subtraction of these two graphs results in pure absorption signals (Fig. 3), from which the time-domain properties of THz–TDS can be identified. With WT technique, the time-domain distribution properties hidden in the conventional THz–TDS are discovered. The spots clearly indicate the time and the frequency where the absorption happens. This should be useful in molecular dynamics. The extracted absorption lines are listed in Table 1 and are compared with those previously reported [16, 17] with other techniques. The agreement is excellent.

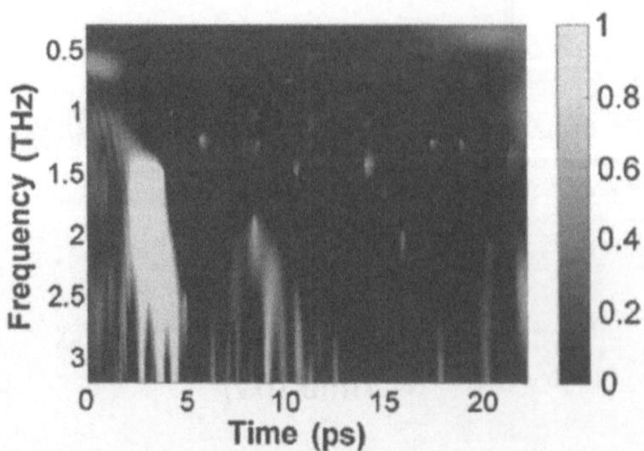


FIGURE 3. Time-resolved water vapor absorption spectroscopy in THz range.

TABLE 1. Comparisons of the absorption lines between using WT with previously reported

| Spectrum line | Data 1 | Data 2 | Data 3 |
|---|---|---|---|
| 1 | 1.099 | 1.094 | 1.097 |
| 2 | 1.116 | 1.126 | 1.125 |
| 3 | 1.163 | 1.167 | 1.162 |
| 4 | 1.209 | 1.201 | 1.209 |
| 5 | 1.231 | 1.233 | 1.231 |
| 6 | 1.411 | 1.413 | 1.410 |
| 7 | 1.603 | 1.605 | 1.605 |
| 8 | 1.671 | 1.669 | 1.670 |
| 9 | 1.718 | 1.721 | 1.720 |
| 10 | 1.798 | 1.793 | 1.798 |
| 11 | 1.870 | 1.868 | 1.867 |
| 12 | 1.920 | 1.928 | 1.917 |
| 13 | 2.041 | 2.045 | 2.042 |
| 14 | 2.166 | 2.165 | 2.165 |
| 15 | 2.199 | 2.199 | 2.198 |
| 16 | 2.257 | 2.263 | 2.256 |

Data 1: data reported in [17]; data 2: data reported in [16] using FT; data 3: data obtained using WT.

## 5 Discussions

There are some differences between WT–THz–TDS and the conventional THz–TDS:

1. The conventional THz–TDS based on FT technique transforms the whole signal from time domain into frequency domain. Therefore, time-resolved absorption spectroscopy is hidden. However, WT is a two-dimensional time-frequency analysis. Time-resolved absorption spectroscopy is clearly shown through the dilations and translations of the daughter wavelets in the time-frequency plane.
2. Based on Heisenberg uncertainty theory, the accuracy of absorption spectrum line analyzed by conventional THz–TDS depends on the length of the data. There is a need to add a great amount of zeros to lengthen the data for improving the frequency resolution [18]. However, the multiresolution analysis of wavelet has excellent local characterizations in both time domain and frequency domain. Therefore WT–THz–TDS can focus on any fine frequency details by adopting gradually fine steps in frequency domain. Therefore, the unnecessary expansion of the data size is eliminated.

## 6 Conclusions

In conclusion, we have applied the WT technique in the absorption spectra analysis of water vapor in THz range. The extracted spectral lines are in excellent agreement with previously reported using other techniques. In addition, it also shows the absorption in time domain, which can be considered as a kind of time-resolved absorption spectroscopy. This technique could have potential applications in molecular dynamics.

**Acknowledgments.** This research was supported in part by China Natural Science Foundation (Grant numbers 60278002, 60178007, and 60490280).

## References

1. D. Dragoman, M. Dragoman: Prog. Quant. Elect. **28**, 1 (2004).
2. A. Rice, Y. Jin, X.-F. Ma, et al.: Appl. Phys. Lett. **64**, 1324 (1994).
3. P. R. Smith, D. H. Auston, and M. C. Nuss: IEEE J. Quant. Elect. **24**, 255 (1998).
4. H. Harde, S. R. Keiding, and D. Grischkowsky: Phys. Rev. Lett. **66**, 1834 (1991).
5. B. L. Yu, F. Zeng, Q. Xing, and R R. Alfano: Appl. Phys. Lett. **82**, 4633 (2003).
6. D. Grischkowsky, S. R. Keiding, M. van Exter, and Ch. Fatinger: J. Opt. Soc. Am. **B7**, 2006 (1990).
7. W. Zhang, A. Azad, and D. Grischkowsky: Appl. Phys. Lett. **82**, 2841 (2003).
8. Martin C. Nuss, P. M. Mankiewich, M. L. O'Malley, E. H. Westerwick, and Peter B. Littlewood: Phys. Rev. Lett. **66**, 3305 (1991).
9. B. Yu, F. Zeng, Y. Yang, et al.: Biophys. J. **86**, 1649 (2004).
10. C. K. Chui (ed.): *An Introduction to Wavelets*. Boston: Academic Press (1992).
11. Y. Deng, Z. Wu, L. Chai, et al.: Opt. Exp. **13**, 2120 (2005).
12. Y. Deng, Z. Wu, L. Chai, C. Wang, and Z. Zhang: Wavelet-transform analysis of spectral interferometry for phase reconstruction of femtosecond optical pulses. In: Quantum Electronics Laser Science Conference (QELS). Optical Society of America, Baltimore, MD, 2005, JWB11.
13. B. Telfer and H. H. Szu: Opt. Eng. **31**, 1830 (1992).
14. J.-C. Hong and Y. Y. Kim: Exp. Mech. **44**, 387 (2004).
15. Y. Deng, C. Wang, L. Chai, and Z. Zhang: Appl. Phys. **B81**, 1107 (2005).
16. L. Lang, Q. Xing, S. Li, F. Mao, L. Chai, and Q.Wang: Chin. Opt. Lett. **2**, 677 (2004).
17. R. A. Cheville and D. Grischkownsky: J. Opt. Soc. Am. **B.16**, 317 (1999).
18. P. R. Griffith: *Chemical Infrared Fourier Transform Spectroscopy*. New York: Wiley (1975).

# Some Newly Developed Crystals for Measurement of Ultrafast Laser Pulses by Second Harmonic Generation

Pathik Kumbhakar,[1] Sunanda Chatterjee,[1] and Takayoshi Kobayashi[2]

[1] Department of Physics, National Institute of Technology, Durgapur (Deemed University), Durgapur 713209, India
**nitdgpkumbhakar@yahoo.com**
[2] Department of Physics, Graduate School of Sciences, University of Tokyo, 7-3-1 Hongo, Bunkyo-ku, Tokyo 113-0033, Japan

**Abstract:** Second harmonic generation in nonlinear optical crystal is a well-established technique for the characterization of ultrafast laser pulses. However, only selected nonlinear crystalline materials can be used due to their limitations in transparency cut-off and limited phase-matching range. It is observed that some newly discovered crystals, such as CLBO, KABO, LB4, and KBBF, have suitable characteristics for use in ultrafast nonlinear optical applications. We have calculated several linear and nonlinear optical parameters including phase-matching angle, angular and spectral acceptance bandwidths, walk-off angle, nonlinear coupling coefficient, and group-velocity mismatch for the generation of vacuum ultraviolet laser radiation in these crystals by type-I and type-II second harmonic generation techniques.

## 1 Introduction

The nonlinear optical (NLO) frequency conversion technique continues to be the best choice for the generation of tunable coherent ultraviolet (UV) radiation, and also second harmonic generation (SHG) in a NLO crystal is a simple and widely used technique for the characterization of broadband ultrafast laser pulses generated via either noncollinear optical parametric amplification (NOPA) or by some other methods [1–4]. In view of transparency cut-off near the UV range and limited phase-matching (PM) range, most NLO materials are unsuitable for use in conversion to UV and thus their use for characterization of broadband UV–visible pulses becomes limited. Ultrafast optical pulses can provide high conversion efficiency due to their high peak power, but in this case the additional limitation is the temporal walk-off caused by different group velocities of the fundamental and the second harmonic radiations. This walk-off sets a limit on the interaction length and consequently thin samples should be used and this further limits the conversion efficiency. The special performances of borate-based NLO crystals are their capability to broaden the range of harmonic generation even in VUV spectral region in some newly discovered crystals [5]. Here we have considered investigating numerically the NLO properties of some potential newly discovered borate crystals including BBO, CLBO, KABO, KBBF, and LB4 for their capability to generate SHG wavelength in the deep-UV region including the VUV region. The optical transparency ranges of BBO, CLBO, KABO, KBBF, and LB4 crystals are 188–3500, 175–2750, 180–3600, 155–3780, and 160–3500 nm, respectively [5]. We have calculated some basic NLO properties of these crystals including (PM) angles, acceptance angle, acceptance wavelength, and walk-off angle. We have also calculated the values of group-velocity mismatch (GVM) for the generation of UV and VUV laser radiation by type-I and type-II SHG in these crystals. It is very much important to know the value of GVM for NLO interactions with ultrafast laser pulses.

## 2 NLO Parameters of Some Newly Discovered Crystals for Frequency Conversion with Ultrashort Laser Pulses

The simplest method for obtaining the shorter wavelength in UV and deep-UV region is the SHG in a NLO media where a frequency doubled ($2\omega$) optical radiation is generated, when two optical radiations of frequency $\omega$, having same or orthogonal polarizations, incident in an anisotropic crystal. Energy transfer from the input to the generated radiation becomes optimum when the phase velocities of the interacting radiations become same, i.e., at the PM condition [6]. The wave dispersion results in a GVM in the creation of harmonic wave due to the two different group velocities of the fundamental and harmonic waves and for efficient conversion both group velocity and phase velocity matching are required to achieve simultaneously. In practical situations, except in some special cases, both the conditions are not satisfied simultaneously, since the group velocity indices exhibit stronger dispersion than the refractive indices. Group velocity dispersion (GVD) in a nonlinear medium produces pulse stretching. However, it has been pointed out [7] that pulse broadening due to crystal bulk dispersion is negligibly small compared to the GVM. This means that the appropriate crystal thickness should mostly be determined from the PM condition. Therefore, efficient SHG with ultrashort laser pulses generally requires a small GVM. Due to GVM, the fundamental and harmonic pulses propagate with different velocities inside the crystal and this may lead to temporal walk-off between the interacting radiations. This effect is detrimental to achieve large gain in SHG process as it reduces the effective interaction length of the crystal. For maximizing the conversion efficiency a reasonable large size crystal is required, but too large size is useless because only a small portion of the crystal becomes effective due to the presence

TABLE 1. Some important NLO properties of some newly developed crystals for type-I SHG of 800 nm

| Crystal name | $\theta_{pm}$ (deg) | $d_{eff}$ (pm/V) | $\rho$ (mrad) | L$\Delta\theta$ (cm mrad) | L$\Delta\lambda$ (cm nm) | $L_{eff}$ (mm) |
|---|---|---|---|---|---|---|
| BBO | 29.0 | 2.09 | 67.8 | 0.31 | 0.24 | 0.51 |
| CLBO | 36.2 | 0.5 | 34.7 | 0.68 | 0.44 | 0.94 |
| KABO | 35.5 | 0.37 | 47.9 | 0.48 | 0.33 | 0.69 |
| KBBF | 24.0 | 0.41 | 41.2 | 0.58 | 0.66 | 1.38 |
| LB4 | 36.0 | 0.1 | 36.0 | 0.61 | 0.40 | 0.84 |

of spatial as well as temporal walk-off effect. The effective interaction length, $L_{eff}$, is calculated using the following relation:

$$L_{eff} = pw/GVM, \tag{1}$$

where $pw$ = temporal width of the fundamental pulse. For type-I (o + o − e) interaction GVM = $GVM_{13} = 1/(u_1^{-1} - u_3^{-1})$, $u_1$ and $u_3$ are group velocities of the o-polarized input ($\lambda_1$) and the e-polarized generated ($\lambda_3$) waves, respectively. For type-II (o + e − e) interaction GVM = $GVM_{i3} = 1/(u_i^{-1} - u_3^{-1})$, $u_i$ $(i = 1, 2)$ and $u_3$ are group velocities of the o (or e)-polarized input ($\lambda_i$) and the e-polarized generated ($\lambda_3$) waves, respectively. Several NLO properties viz. for generation 400 nm laser radiation by type-I SHG in BBO, CLBO, KABO, LB4, and KBBF crystals from Ti:sapphire fundamental 800 nm laser radiation have been presented in Table 1. The methods of calculation of PM angle ($\theta_{pm}$), acceptance angle ($\Delta\theta$), acceptance wavelength ($\Delta\lambda$), and walk-off angle ($\rho$) for type-I and type-II SHG are given in [6]. We have also found a simple method to obtain SHG cut-off wavelength for both type-I and type-II processes and that can be evaluated by solving Eqs. (2) and (3), respectively, by using any mathematical software or computer programming:

$$n_{F1}^{o} + n_{F2}^{o} - 2n_{SH}^{e} = 0, \tag{2}$$

$$n_{F1}^{e} + n_{F2}^{o} - 2n_{SH}^{e} = 0, \tag{3}$$

where $n_{SH}^{o}(n_{SH}^{e})$ and $n_{F}^{o}(n_{F}^{e})$ ordinary (extraordinary) are refractive indices at SH and fundamental wavelengths, respectively. These indices are calculated using the available Sellmeier dispersion relations [5, 6] of the respective crystals.

Figures 1 and 2 demonstrate the PM and GVM characteristics of five potential NLO crystals considered for investigation in this work. Solid curves correspond to PM and dotted curves show GVM. For type-II interaction only $GVM_{13}$ have been plotted. From the Figs. 1 and 2 it is clearly observed that it is always type-I interaction that permits the generation of shortest VUV wavelengths by SHG. It may also be noted that type-II interaction is not allowed in LB4 crystal by its crystal symmetry.

Figure 1 shows that only KBBF crystal has the capability of generation of VUV wavelength even below 200 nm, being the type-I SHG cut-off at 162 nm. This is the shortest wavelength that can be generated till date in any crystal by SHG [5].

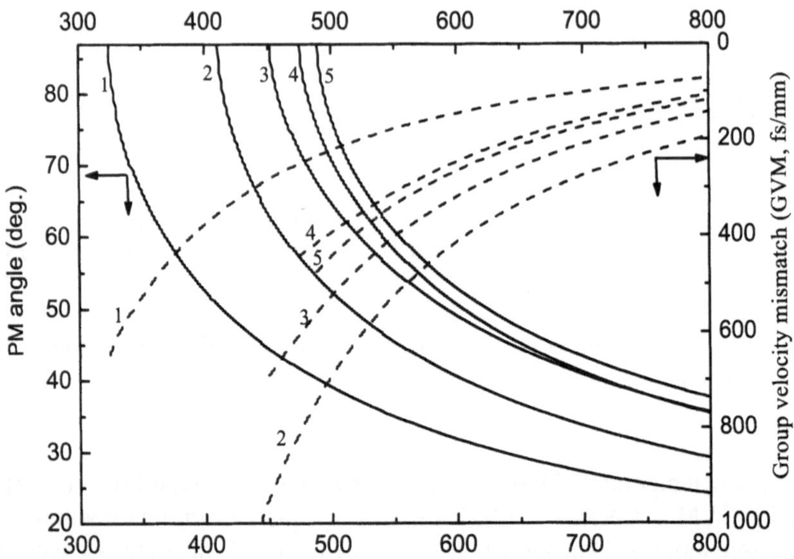


FIGURE 1. Type-I PM (solid) and GVM (dotted) characteristics. Curves marked 1, 2, 3, 4, and 5 correspond to KBBF (162), BBO (204.8), KABO (225.5), CLBO (236.8), and LB4 (243.8) crystal, respectively. Bracketed data show the type-I SHG cut-off wavelength (in nm) of the respective crystal.

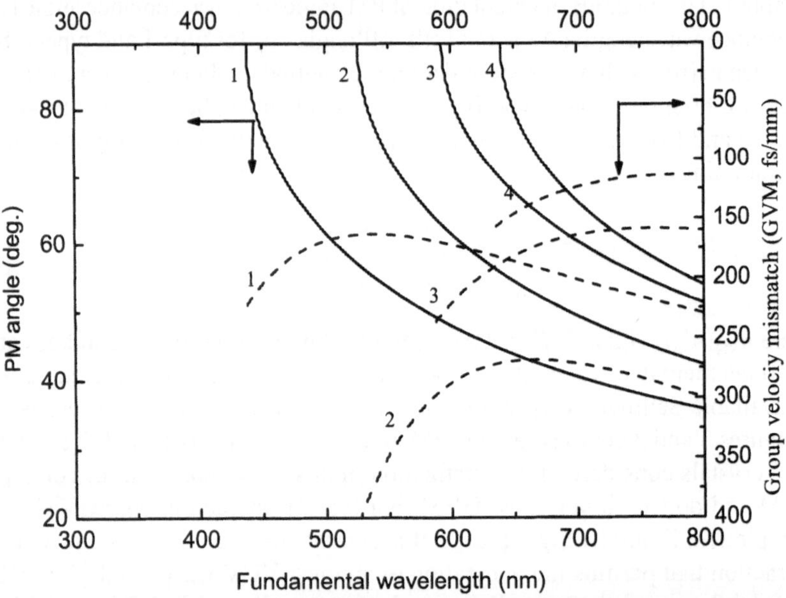


FIGURE 2. Type-II PM (solid) and GVM (dotted) characteristics. Curves marked 1, 2, 3, and 4 correspond to KBBF (224.0), BBO (262.7), KABO (294.6), and CLBO (318.0), respectively. Bracketed data show the type-I SHG cut-off wavelength (in nm) of the respective crystal. In LB4 type-II is not allowed.

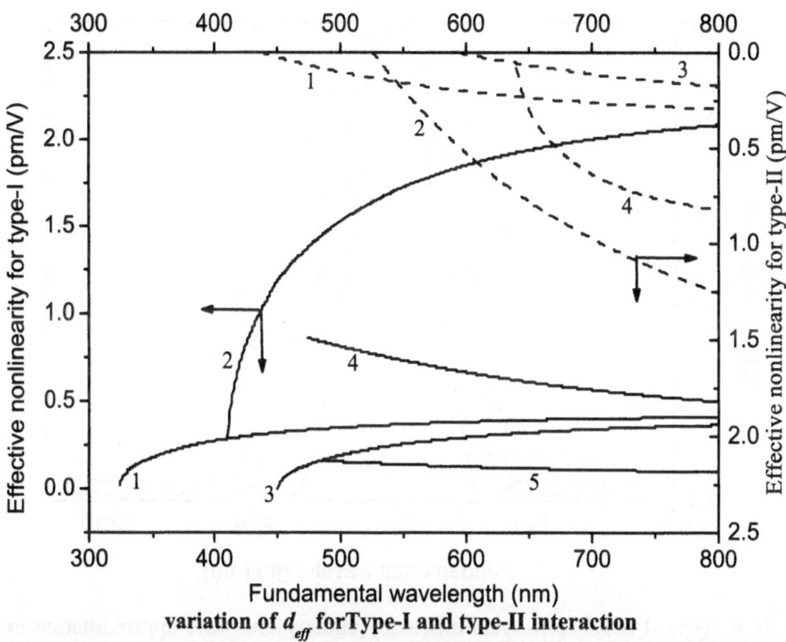


FIGURE 3. Variation of effective nonlinearity with fundamental wavelength for type-I (solid) and type-II (dotted) interaction, respectively. Curves marked 1, 2, 3, 4, and 5 correspond to KBBF, BBO, KABO, CLBO, and LB4, respectively.

Reported [5] values of the laser damage threshold (LDT) of KBBF crystals are as high as 4000 GW/cm$^2$ at 532 nm with 7-ns pulse width and 10 Hz and above 1000 GW/cm$^2$ at 400 nm with 50-fs pulse width and 1 KHz. However, the problem with this crystal is that its growth is heavily limited by the strong layering tendency [5]. BBO crystal is next to generating SHG wavelength near 200 nm and this also has the highest value of NLO coefficient for both type-I and type-II SHG as can be seen in Fig. 3. However, growth of good quality large size BBO crystal is not easy and it is also hygroscopic in nature. CLBO crystal has lowest GVM for type-II phase-matching interactions and moderate $d_{\mathrm{eff}}$ for both type-I and type-II SHG as is shown in Fig. 3. Growth of CLBO crystal is although comparatively easier, it is highly hygroscopic in nature and requires specific atmospheric conditions for its best performances. KABO and LB4 crystals are nonhygroscopic in nature and particularly LB4 crystal is easily grownable to good quality large size. However, the disadvantage of KABO and LB4 crystals are that the value of $d_{\mathrm{eff}}$ is very small and it is lowest for LB4 as is evident from Fig. 3. This can be overcome at least partially by their high-value LDT employing tight focusing geometries.

LDT of LB4 is about 40 GW/cm$^2$ at 1064 nm, which is ~3 times that of BBO (13.5 GW/cm$^2$) and ~1.5 times that of CLBO (26 GW/cm$^2$). Figure 4 shows the calculated approximate type-I SHG efficiency, considering the crystal length ($L$) = 100 μm in each case. The solid, dotted, dashed-dot-dot, dashed, and dashed-dot curves are for BBO, KABO, LB4, CLBO, and KBBF crystals, respectively.

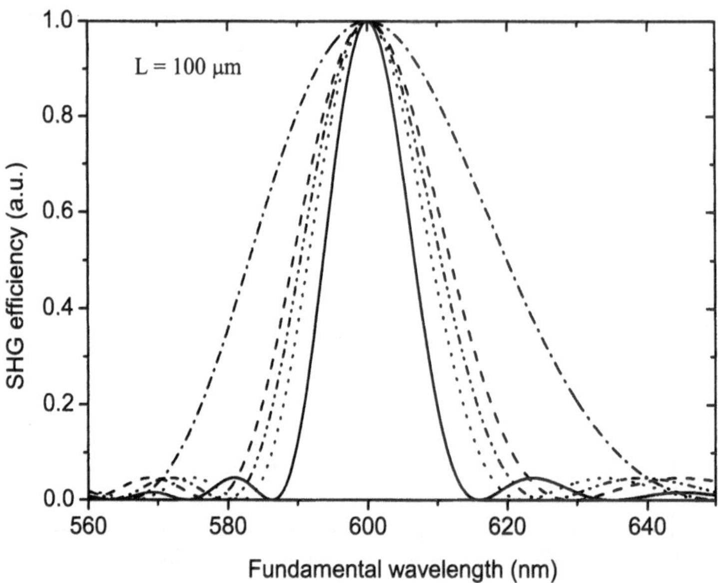


FIGURE 4. Type -I SHG conversion efficiency under plane-wave approximation in different crystals, each of 100 μm length. The solid, dotted, dashed-dot-dot, dashed, and dashed-dot curves are BBO (13), KABO (17), LB4 (21), CLBO (23), and KBBF (37) crystals, respectively. Bracketed data show FWHM (in nm) of the respective curves.

The SHG efficiency have been calculated using infinite plane wave approximation and it is given by $\mathrm{Sinc}^2(\Delta kL/2)$ function, where $\Delta k = k^e_{\mathrm{SH}}(\theta_{\mathrm{pm}}) - k^o_{\mathrm{F}}$, $k_j$ ($j$ = F and SH) is the wave vector corresponding to $\lambda_j$ radiation. It can be seen from Fig. 4 that the full widths of the curves at half maximum level of SHG efficiency are 13, 17, 21, 23, and 37 nm, for BBO, KABO, LB4, CLBO, and KBBF crystals, respectively, centered at 600 nm. During the calculations the refractive indices were calculated using the available Sellmeier dispersion relations [5, 6] of the respective crystals.

## 3 Summary

In conclusion, we have presented several NLO parameters for type-I and type-II PM and GVM characteristics of various NLO crystals including some recently discovered crystals whose potentialities for use in femtosecond experiments are yet to be verified conclusively. The widely used BBO crystal has the highest angular and wavelength sensitivity and large spatial walk-off angle. These properties may have detrimental effects in some NLO experiments. Another crystal CLBO is highly hygroscopic, which limits its practical application although it has several other advantageous characteristics. Comparative newly discovered LB4 and KABO have good mechanical properties and high value of LDT [5]. However, these suffer due to low NLO coefficient. A recently discovered crystal KBBF has moderate walk-off angle, and wider spectral and angular bandwidths. It is found that the value of type-I SHG spectral bandwidth for KBBF crystal is 37 nm and that of BBO, KABO, LB4, and CLBO crystals are 13, 17, 21, and 23 nm, respectively, for generation of second harmonic of fundamental radiation centered at 600 nm. LDT of KBBF crystal is 4000 GW/cm$^2$ at 532 nm with 7-ns pulse and 10 Hz rep. rate and >1000 GW/cm$^2$ at 400 nm with 50-fs pulse and 1 KHz rep. rate [5]. These properties of KBBF may be very useful in different ultrafast NLO applications viz. for characterization of high power ultrashort laser pulses with ultra-wide bandwidth.

**Acknowledgements.** Kumbhakar is grateful to NIT, Durgapur, for providing the partial financial support and computational facilities.

## References

1. A. Shirakawa and T. Kobayashi: IEICE Trans. Electron **E81-C**, 246 (1998).
2. G. Cerullo and S. De Silvestri: Rev. Sci. Instrum. **74**, 1 (2003).
3. A. Baltuska and T. Kobayashi: Appl. Phys. B. Lasers Opt. **B75**, 427 (2002).
4. S. Adachi, P. Kumbhakar, and T. Kobayashi: Opt. Lett. **29**, 1150 (2004).
5. C. Chen, Z. Lin, and Z. Wang: Appl. Phys. B. Lasers Opt. **B80**, 1 (2005).
6. V. G. Dmitriev, G. G. Gurzadyan, and D. N. Nikogosyan: *Handbook of Nonlinear Optical Crystals*, 3rd edn. Berlin: Springer, pp. 3 (1999).
7. A. M. Weiner: IEEE J. Quantum Electron **QE-19**, 1276 (1983).

# Part VIII

# High-field Physics and Applications II

# Intensity Scalings of Attosecond Pulse Generation by the Relativistic-irradiance Laser Pulses

Alexander S. Pirozhkov,[1,2] Sergei V. Bulanov,[1,3,4]
Timur Zh. Esirkepov,[1,4] Akito Sagisaka,[1] Toshiki Tajima,[1]
and Hiroyuki Daido[1]

[1] Advanced Photon Research Center, Japan Atomic Energy Research Institute, 8-1 Umemidai, Kizu-cho, Soraku-gun, Kyoto 619-0215, Japan
pirozh@apr.jaeri.go.jp
[2] Division of Optics, P. N. Lebedev Physical Institute of the Russian Academy of Sciences, 53 Leninskiy prospekt, 119991 Moscow, Russia
[3] A. M. Prokhorov Institute of General Physics of the Russian Academy of Sciences, 38 Vavilov Street, 119991 Moscow, Russia
[4] Moscow Institute of Physics and Technology, 9 Institutskiy pereulok, 141700 Dolgoprudny, Moscow Region, Russia

**Summary.** We present the theoretical comparative analysis of different attosecond pulse generation techniques in which the relativistic-irradiance driver pulses are used. In particular, we concentrate on the intensity scalings of the attosecond pulse duration, wavelength, and conversion efficiency. We also discuss the optimum conditions and the major implementation challenges.

**Key words:** attosecond pulses, relativistic laser–plasma interaction

## 1 Introduction

Recent progress in attosecond physics relies on the high-order harmonic generation in gases [1]. Advantages of this process are the possibility to use the present day laser technology and the availability of the appropriate theoretical description [2, 3]. However, due to the ionization of a gas medium, the technique is not easily scalable to high driver pulse energies. Furthermore, the conversion efficiency is low [4, 5]. For these reasons, the attosecond pulse characterization and applications are challenging.

Other kinds of nonlinearities are the ones existing in the relativistic laser–plasma and charged particle interactions [6–13]. Several theoretical proposals concerning attosecond pulse generation employing relativistic nonlinearities can be found in literature [14–20]. In this paper, we perform the theoretical analysis of two of these proposed techniques from the point of view of the duration, wavelength, and conversion efficiency scalings. We also address the requirements on a driving laser as well as other implementation issues.

All theoretical models of the attosecond pulse generation considered here have several parameters. One parameter common for all these models is the dimensionless electric field amplitude $a_0 = eE_0/(mc\omega_0)$ and, so the peak intensity of the laser pulse is $I_0 = a_0^2 \times 1.37 \times 10^{18}$ W/cm$^2$ $\mu$m$^2/\lambda_0^2$. Here $\omega_0$, $\lambda_0$, and $E_0$ are the laser frequency, wavelength, and electric field amplitude, $e$ and $m$ are the electron charge and mass, and $c$ is the velocity of light in vacuum. Other parameters are specific for each model. In order to compare different models, it is necessary to derive the scaling laws on the same parameter, i.e. on $a_0$. To do so, we consider each model in an optimum regime, described below. In the optimum regimes, the other model parameters are expressed in terms of $a_0$, thus providing an opportunity for direct comparison.

## 2 Flying Mirror Model

The first technique we consider is the reflection of the source laser from the "flying mirror" [14] (Fig. 1). In this technique, two laser pulses, namely driver (subscript 0) and source (subscript s), are focused simultaneously into the underdense plasma. Relativistic-irradiance driver pulse produces a wake wave when the phase velocity equals the group velocity of the driver pulse, which is close to $c$ in the underdense plasma. The corresponding gamma-factor is $\gamma_{ph} \approx (1 + a_0^2/2)^{1/4}\omega_0/\omega_{pe}$, where $\omega_{pe} = (4\pi ne^2/m)^{1/2}$ is the Langmuir frequency and $n$ is the plasma density. Near the threshold of the wave-breaking, the electron density in the wake wave has sharp spikes containing approximately

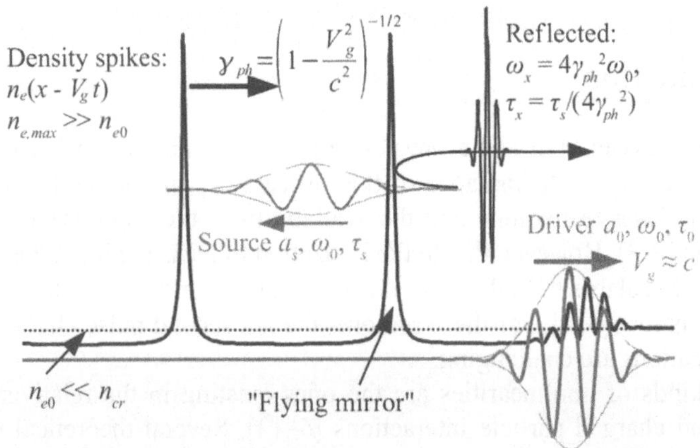


FIGURE 1. "Flying mirror" interaction geometry.

half of the electrons. Peak electron density in these spikes can be much larger than the original plasma density. The spikes act as a partially transparent mirrors moving with the velocity close to $c$ (the "flying mirrors"). Counter-propagating source pulse with duration $\tau_s$ is partially reflected from these spikes, and its frequency and duration become $\omega_x \approx 4\gamma_{ph}^2\omega_0$ and $\tau_x \approx \tau_0/(4\gamma_{ph}^2)$, respectively (we assume that $\omega_s = \omega_0$). The reflected pulse has the form of a first main pulse with several weaker postpulses. In the limit of small $a_s$, the number of reflected photons is $N_x \approx N_s/(2\gamma_{ph}^3)$ and so the energy reflectivity is $R \approx 2/\gamma_{ph}$. Conversion efficiency is

$$\eta = \frac{Ra_s^2\tau_s}{a_0^2\tau_0 + a_s^2\tau_s}. \tag{1}$$

In the optimum regime, the driver pulse duration is determined by the resonance wake wave excitation condition: $\tau_0 \approx \lambda_p/(4c)$ [21], where $\lambda_p = 4(2\gamma_{ph})^{1/2}\, c/\omega_{pe}$ and the amplitude by the wave breaking threshold $a_0 = (2\gamma_{ph})^{1/2}$ [21]. Reflected pulse energy $16^{-1}(\pi \ln 2)^{-1/2} Ra_s^2 c\tau_s/a_0^2(mc\omega_0/e)^2$ comes mainly from the energy of the flying mirror. Reflected pulse energy should be much smaller than the energy of the electron density spike $mc^2\gamma_{ph}n\lambda_p/2 = (2\pi)^{-1}a_0(1 + a_0^2/2)^{1/4}c/\omega_0(mc\omega_0/e)^2$. Assuming that the reflected pulse takes 10% of the mirror's energy, the maximum source amplitude is $a_s \approx 0.24a_0^{9/4}/(\omega_0\tau_s)^{1/2}$ and the conversion efficiency is $\eta \approx 0.5/a_0^3$ (Table 1). For very high driver pulse intensity, the permissible source amplitude can approach or even exceed 1 and so the flying mirror reflectivity will be decreased. This strongly nonlinear regime requires further investigation.

The advantage of the attosecond pulse generation using the flying mirror is an $a_0^{-4}$ scaling of the duration and wavelength (see Fig. 2a and Table 1), which allows generating attosecond pulses in the "water window" spectral region using driver pulses with $a_0 \approx 4$ (Fig. 3). Another advantage comes from the fact that

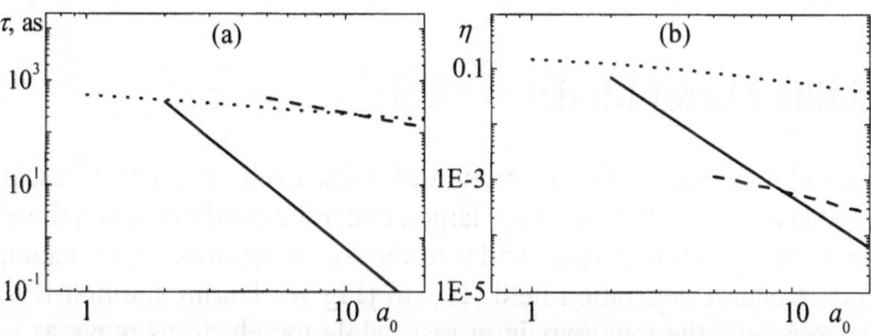


FIGURE 2. Attosecond pulse duration (a) and conversion efficiency (b) scalings with the driver pulse amplitude for flying mirror (solid lines), sliding mirror with spectral filter $[a_0\omega_0, 2a_0\omega_0]$ (dashed lines), and Sliding mirror without spectral filtering (dotted lines). ($\lambda_0 = 800$ nm; for flying mirror, $\tau_0 = \lambda_p/(4c)$, $\tau_s = 15/\omega_0 = 6.3$ fs; for sliding mirror, $\tau_0 = 15/\omega_0 = 6.3$ fs).

TABLE 1. Scalings for the attosecond pulse duration, efficiency, and wavelength ($\lambda_0 = 800$ nm; for "Flying mirror", $\tau_0 = \lambda_p/(4c)$; for "Sliding mirror", $\tau_0 = 15/\omega_0 = 6.3$ fs)

| | $\tau_x$ | $\eta$ | $\lambda_x$ |
|---|---|---|---|
| Flying mirror | $\tau_s/a_0^4$ | $\approx 0.5/a_0^3 \approx 0.5(\tau_x/\tau_s)^{3/4}$ | $\lambda_0/a_0^4$ |
| Sliding mirror $[a_0\omega_0, 2a_0\omega_0]$ | $\approx 2000/a_0^{0.9}$ as | $\approx 0.008/a_0^{1.1} \approx 10^{-6}(\tau_x/\text{as})^{1.2}$ | $\approx 0.7\lambda_0/a_0$ |
| Sliding mirror (no filter) | $\approx 540/a_0^{0.4}$ as | $\approx 0.2/a_0^{0.5} \approx 5 \times 10^{-5}(\tau_x/\text{as})^{1.34}$ | $\sim \lambda_0/4 - \lambda_0$ |

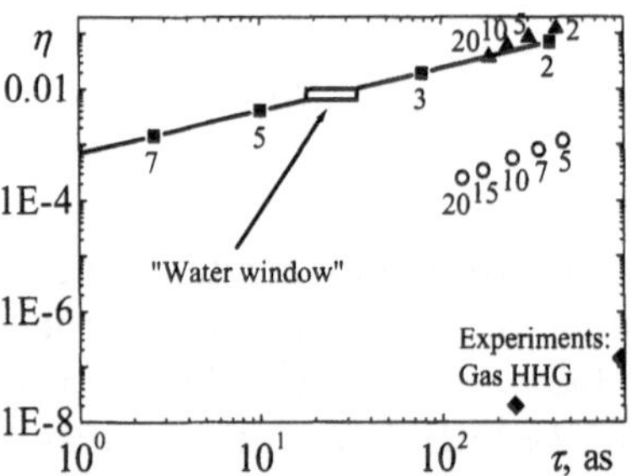


FIGURE 3. Conversion efficiency versus attosecond pulse duration for flying mirror (squares), sliding mirror with spectral filter $[a_0\omega_0, 2a_0\omega_0]$ (*circles*), and sliding mirror without spectral filtering (triangles). Numbers near the symbols denote the driver laser amplitude $a_0$. ($\lambda_0 = 800$ nm; for flying mirror, $\tau_0 = \lambda_p/(4c)$, $\tau_s = 15/\omega_0 = 6.3$ fs; for sliding mirror, $\tau_0 = 15/\omega_0 = 6.3$ fs). Diamonds represent the experimental achievements using high-order harmonics generated in gases [4,5].

the reflected pulse is a compressed replica of the source pulse:

$$E_x(t) = 2(2\gamma_{ph})^{1/2} E_s(4\gamma_{ph}^2 t), \tag{2}$$

which allows generating attosecond pulses with the prescribed dependence of the electric field $E_x(t)$ by shaping source pulses in the visible or near-infrared spectral regions. Finally, a gas target is automatically renewed before each laser shot.

## 3 Sliding Mirror Model

The second technique is the interaction of a few cycle relativistic irradiance p-polarized laser pulse with a thin foil target, under the condition when the electron motion in the direction perpendicular to the foil is suppressed, for example, by the electric charge separation field [10, 16] (Fig. 4). During the first few cycles of the laser field, the ions remain at rest, while the electrons move as a whole mainly in the foil plane (the electrons "slide" along the foil). As a result of the laser–"sliding mirror" interaction, part of the laser energy is transmitted through the foil and the remaining part is reflected. Both reflected and transmitted pulses contain harmonics of the incident laser radiation. After spectral filtering, these harmonics form isolated attosecond pulses [15, 16] like in the case of harmonics

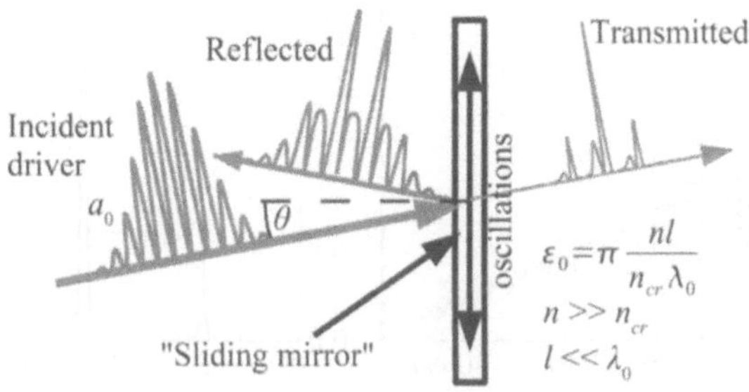


FIGURE 4. "Sliding mirror" interaction geometry.

generated in gases for few-cycle driver pulses [4,22]. Furthermore, under appropriate conditions even without spectral filtering transmitted pulses have the form of high-energy attosecond pulse trains with the spectrum consisting of few low-order harmonics. Relative satellite pulse intensity is typically 0.2–0.4 and so these pulse trains can be considered as low-contrast isolated pulses. For both reflected and transmitted attosecond pulse generation, the carrier-envelope phase control of the driver laser is essential. The most important model parameters are the pulse amplitude $a_0$, the normalized foil density $\varepsilon_0 = \pi nl/(n_{cr}\lambda_0)$, the incidence angle $\theta$, and the carrier-envelope phase $\psi$; $l$ is the foil thickness and $n_{cr} = m\omega_0^2/(4\pi e^2)$ is the critical density.

Attosecond pulses can be generated in different spectral regions using the corresponding spectral filters [16]. For the reflected attosecond pulse, we use the spectral filter of $[a_0\omega_0, 2a_0\omega_0]$. To find the optimum parameter values, we use the merit function

$$MF_r = I_{\max}[1 - (5I_2/I_{\max})^2]/a_0^2, \tag{3}$$

where $I_{\max}$ and $I_2$ are the main attosecond pulse and satellite pulse peak intensities. Thus, there is a trade-off between the conversion efficiency and the attosecond pulse contrast. We find that the optimum values of parameters are $\varepsilon_0 = 1.9 + 0.75a_0$, $\theta = 40 - 50°$, and $\psi = 0.5 - 1$ rad. The resulting duration and efficiency scalings are listed in Table 1; the resulting contrast ratio is $I_2/I_{\max} \approx 0.1$.

To find the optimum parameter values for the transmitted attosecond pulse, we use the merit function

$$MF_t = I_{\max}/\tau_{\text{eff}}^2, \tag{4}$$

where $\tau_{\text{eff}} = \int I(t)dt/I_{\max}$ is the effective attosecond pulse duration. Thus, there is a trade-off between the conversion efficiency and the pulse duration. The optimum parameter values can be fitted as $\varepsilon_0 = 0.86a_0$, $\theta = 50°/a_0^{0.3}$, and $\psi = 1.7/a_0^{0.55}$ rad. The resulting duration and efficiency scalings are listed in Table 1, the attosecond pulse contrast ratio changes from 0.5 at $a_0 = 1$ to 0.2 at $a_0 = 20$.

In both cases, with or without spectral filter (see Figs. 5 and 6), the optimum condition is $\varepsilon_0 \approx a_0$. If $\varepsilon_0$ is smaller, the number of electrons is too small and so the efficiency is low. If $\varepsilon_0$ is larger, the laser electric field is not strong enough

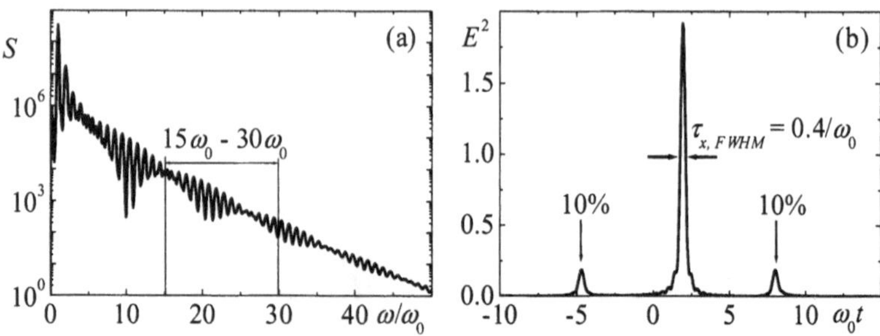


FIGURE 5. Attosecond pulse generation by the sliding mirror using spectral filtering. $a_0 = 15$, $\varepsilon_0 = 13.2$, $\theta = 46°$, $\psi = 0.47$ rad, $\tau_0 = 15/\omega_0$, spectral filter is $[15\omega_0, 30\omega_0]$. (a) Spectrum of the reflected pulse. (b) Isolated attosecond pulse in the time domain after spectral filtering; the duration is $\tau_x = 0.4/\omega_0$ (170 as for $\lambda_0 = 800$ nm), and the conversion efficiency is $\eta = 3.4 \times 10^{-4}$.

to move the electrons at the relativistic velocity and so in the limit $v \ll c$ the nonlinear effect is small and efficiency is low.

The advantage of the attosecond pulse generation using the sliding mirror without spectral filtering is high conversion efficiency, which is typically several percent (Fig. 2b). Furthermore, in both variants of the sliding mirror laser interacts with a small volume of matter. For this reason, even for the solid-density plasma, laser pulse is able to accelerate all electrons up to the relativistic velocity. Thus, the sliding mirror provides simultaneously high stability inherent to all solid-state techniques and efficient nonlinear interaction even for moderate laser amplitudes. Note that in the case of a bulk target (the oscillating mirror model [8, 9]), the plasma density should be much smaller than the solid-state density: the optimum condition is $n = n_{cr} a_0/2$ [18, 19]. Experimental implementation of the sliding mirror requires high driver pulse contrast, which can be achieved using either plasma mirrors [23] or frequency doubling [24] and the careful control of the pre-formed plasma (see [25] and paper by Sagisaka et al. in this volume). To reach the optimum conditions, high-energy few-cycle driver pulses are necessary. The condition $a_0 = \varepsilon_0$ is equivalent to $I_0 = 1.1 \times 10^{19}$ W/cm$^2 \times (n/10^{24}\text{cm}^3)^2 (l/\text{nm})^2$. For a 10-nm fully ionized carbon foil, $\varepsilon_0 \approx 14$ and the corresponding intensity is $I_0 \approx 4 \times 10^{20}$W/cm$^2$.

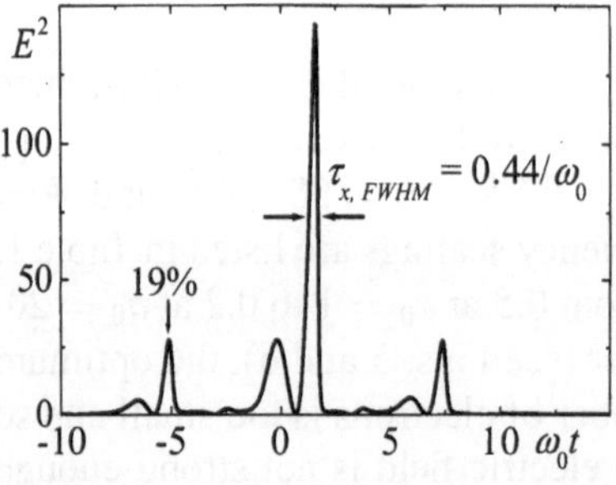


FIGURE 6. Attosecond pulse train generation by the sliding mirror in transmission without spectral filtering. $a_0 = 16$, $\varepsilon_0 = 13.6$, $\theta = 13°$, $\psi = 0.14$ rad, $\tau_0 = 15/\omega_0$. Duration is $\tau_x = 0.44/\omega_0$ (190 as for $\lambda_0 = 800$ nm), the conversion efficiency into the main peak is $\eta = 0.03$.

## 4 Conclusion

Theoretical models predict that the attosecond pulse generation with the use of relativistic nonlinearities is several orders of magnitude more efficient than with the use of harmonics generated in gases. In particular, the sliding mirror model predicts the generation of low-frequency attosecond pulse trains with the conversion efficiency of several percent. Short-wavelength attosecond pulses can be generated using the flying mirror technique, e.g., 30-as pulses in the water window spectral region can be generated with the conversion efficiency of the order of $10^{-3}$.

**Acknowledgments.** We gratefully acknowledge the discussions with M. Zepf, E. N. Ragozin, J. Nees, N. M. Naumova, and G. Mourou. The work was supported by the Japan Society for the Promotion of Science (JSPS P-03543), the Japan Atomic Energy Research Institute, and the Japanese Ministry of Education, Culture, Sports, Science and Technology.

## References

1. P. Agostini and L. F. DiMauro: Rep. Prog. Phys. **67**, 813 (2004).
2. P. B. Corkum: Phys. Rev. Lett. **71**, 1994 (1993).
3. M. Lewenstein et al.: Phys. Rev. A **49**, 2117 (1994).
4. R. Kienberger et al.: Nature (London) **427**, 817 (2004).
5. T. Sekikawa, A. Kosuge, T. Kanai, and S. Watanabe: Nature (London) **432**, 605 (2004).
6. A. I. Akhiezer and R. V. Polovin: Sov. Phys. JETP **3**, 696 (1956).
7. E. S. Sarachik and G. T. Schappert: Phys. Rev. D **1**, 2738 (1970).
8. S. V. Bulanov, N. M. Naumova, and F. Pegoraro: Phys. Plasmas **1**, 745 (1994).
9. R. Lichters, J. Meyer-ter-Vehn, and A. Pukhov: Phys. Plasmas **3**, 3425 (1996).
10. V. A. Vshivkov, N. M. Naumova, F. Pegoraro, and S. V. Bulanov: Phys. Plasmas **5**, 2727 (1998).
11. Y. Ueshima, Y, Kishimoto, A, Sasaki, and T. Tajima: Las. Part. Beams **17**, 45 (1999).
12. D. Li et al.: Nucl. Instr. Meth. Phys. Res. A **528**, 516 (2004).
13. T. Nakano et al.: Phys. Rev. Lett. **71**, 012002 (2003).
14. S. V. Bulanov, T. Esirkepov, and T. Tajima: Phys. Rev. Lett. **91**, 085001 (2003).
15. A. S. Pirozhkov, H. Daido, and S. V. Bulanov: Attosecond pulse generation during the laser pulse reflection at the plasma-vacuum interface. In: *Ultrafast Phenomena XIV*, Springer Ser. Chem. Phys. Vol. 79, T. Kobayashi, T. Okada, T. Kobayashi, K. A. Nelson, and S. De Silvestri (eds). Berlin: Springer, p. 216 (2004).
16. A. S. Pirozhkov et al.: Phys. Lett. A (2005) doi: 10.1016/j.physleta.2005.09.024.
17. N. M. Naumova et al.: Phys. Rev. Lett. **91**, 063902 (2004).
18. J. Nees et al.: J. Mod. Opt. **52**, 305 (2005).
19. N. M. Naumova, J. Nees, and G. A. Mourou: Phys. Plas **12**, 056707 (2005).
20. A. V. Isanin, S. S. Bulanov, F. F. Kamenets, and F. Pegoraro: Phys. Rev. E **71**, 036404 (2005); Phys. Lett. A **337**, 107 (2005).

21. S. V. Bulanov, V. I. Kirsanov, and A. S. Sakharov: JETP Lett. **53**, 565 (1991).
22. A. Baltuska et al.: Nature (London) **421**, 611 (2003).
23. B. Dromey et al.: Rev. Sci. Instr. **75**, 645 (2004).
24. U. Teubner et al.: Phys. Rev. Lett. **92**, 185001 (2005).
25. A. Sagisaka et al.: Appl. Phys. B **78**, 919 (2004).

# Observation of Thin Foil Preformed Plasmas with a Relativistic-intensity Ultra-short Pulse Laser by Means of Two-color Interferometer

A. Sagisaka,[1] H. Daido,[1] K. Ogura,[1] S. Orimo,[1] Y. Hayashi,[1] M. Nishiuchi,[1] M. Mori,[1] A. Yogo,[1] M. Kado,[1] T. Morita,[1] M. Yamagiwa,[1] A. Fukumi,[1,2] Z. Li,[1,2] Y. Oishi,[3] T. Nayuki,[3] T. Fujii,[3] K. Nemoto,[3] S. Nakamura,[1,4] A. Noda,[4] S. V. Bulanov,[1,5,6] T. Zh. Esirkepov,[1,6] A. S. Pirozhkov,[1,7] and T. Utsumi[8]

[1] Advanced Photon Research Center, Kansai Research Establishment, Japan Atomic Energy Research Institute, 8-1 Umemidai, Kizu, Kyoto 619-0215, Japan
**sagisaka.akito@jaea.go.jp**
[2] National Institute of Radiological Sciences, 4-9-1, Anagawa, Inage, Chiba 263-8555, Japan
[3] Central Research Institute of Electric Power Industry, 2-6-1 Nagasaka, Yokosuka, Kanagawa 240-0196, Japan
[4] Institute for Chemical Research, Kyoto University, Gokasho, Uji, Kyoto 611-0011, Japan
[5] A. M. Prokhorov Institute of General Physics of the Russian Academy of Sciences, 38 Vavilov Street, 119991 Moscow, Russia
[6] Moscow Institute of Physics and Technology, 9 Institutskiy pereulok, 141700 Dolgoprudny, Moscow Region, Russia
[7] Division of Optics, P. N. Lebedev Physical Institute of the Russian Academy of Sciences, 53 Leninskiy prospekt, 119991 Moscow, Russia
[8] Electronics and Computer Science, Faculty of Science and Engineering, Tokyo University of Science, Yamaguchi, Daigaku-dori 1-1-1, Onoda city, Yamaguchi, 756-0884, Japan

Intense laser driven thin foil target acts as a double layered coulomb system that produces collimated ions as well as high-order harmonics, hard X-ray, and so on. From this point of view, we observed the electron density distribution of preformed plasmas of a tantalum foil target with two-color probe beams as an interferometer. The preformed plasma at the front side of the target was generated by a prepulse with a high-intensity Ti:sapphire laser pulse. On the other hand, the preformed plasma at the rear side was not observed. In our experimental condition, high-energy electrons and protons were observed. The characterization of preformed plasmas will contribute to describe the high-intensity laser–thin foil interactions.

# 1 Introduction

High-intensity laser–thin foil interactions produce collimated particles, incoherent X-ray, and coherent soft x-ray [1–5]. For example, Teubner et al. have measured the harmonic emission from the rear side of a thin foil target [3]. Pirozhkov et al. reported the analytical theory of attosecond pulse generation in the relativistic intensity laser–thin foil interaction [4,5]. Matsukado et al. reported on the production of high-energy protons from a tantalum (Ta) plasma produced by a high-intensity Ti:sapphire laser [6]. To interpret the experimental results, they proposed an ion acceleration mechanism in which ions are accelerated in a preformed underdense plasma slab created by a prepulse. In all these cases [1–6], the preformed plasma plays an important role for interaction with the high-density thin foil plasma. Therefore, it is necessary to understand the preformed plasma condition experimentally for analyzing the high-intensity laser–matter interactions. In this experiment, we measured the profiles of preformed plasmas produced by a high-intensity Ti:sapphire laser on a metal foil target in conjunction with the detection of X-ray, electrons, and ions.

# 2 Experimental Setup

We used a Ti:sapphire laser system at Central Research Institute of Electric Power Industry [7]. The central wavelength was 800 nm and the pulse duration was 70 fs [full width at half maximum (FWHM)]. The intensity ratio of a prepulse at 5 ns before the main pulse versus that of the main pulse was measured to be $\sim 10^{-5}$ by a photodiode detector. Figure 1 shows the schematic of the experimental setup. A p-polarized laser beam was transported into a target chamber and was divided into main and probe beams by a beam splitter. The main beam was focused by an off-axis parabolic mirror with a focal length of $f = 179$ mm and an incident angle of 45°. The focal point of the attenuated laser beam was imaged with a magnification of $\sim 30$ with an achromatic lens and detected by a charge-coupled

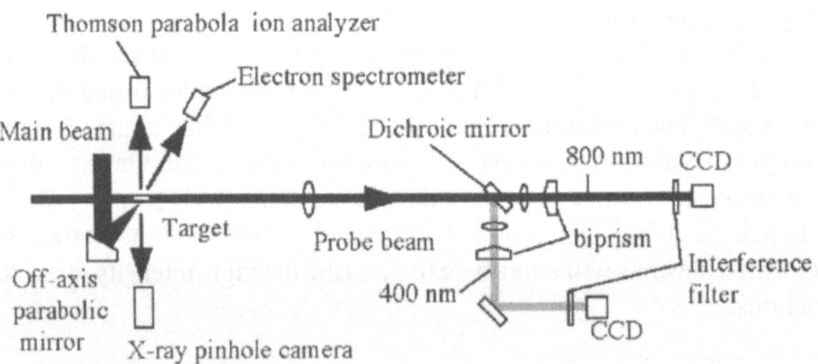


FIGURE 1. Schematic of the experimental setup.

device (CCD) camera. The spot size was 7 μm [vertical (FWHM)] × 8 μm [horizontal (FWHM)], containing ~28% of the energy estimated from the profile of the focusing pattern. In this condition the estimated peak intensity was up to $2.7 \times 10^{18}$ W · cm$^{-2}$ with an energy of 250 mJ at the target surface. A tape target driver provided a fresh surface of the 3-μm thick Ta target with a width of 5 mm [8]. On the other hand, the probe beam was frequency-doubled with a 2-mm thick KDP crystal (type-I). The pulse duration of the second harmonic pulse was estimated to be ~200 fs because of an effect of group-velocity dispersion. We used both fundamental (800 nm) and second harmonic (400 nm) pulses as an optical probe beam. The two-color probe beams passed through the optical delay line. A linear translation stage was used to vary the delay between the main and probe beams. Interference fringes were produced using a Fresnel biprism with an apex angle of 176° and fringe separation was adjusted by the position of the biprism [9]. The plasma image was magnified by a factor of ~10 and detected by the CCD camera. The intensity of the probe beam was attenuated by neutral density filters so as to fall within the dynamic range of the camera. We placed the narrowband interference filters, with the bandpass width of 10 nm for each wavelength, in front of the CCD cameras to reject unwanted emission from the plasma.

## 3 Experimental Results

Figure 2 shows the interferograms and electron density distributions of the preformed plasmas on a 3-μm thick Ta foil target for different shots by using the fundamental (Fig 2a) and second harmonic (Fig 2c) probe beams. These were taken at 50 ps before the main laser pulse arrived. The arrow and dashed lines show the direction of the laser pulse propagation and the target positions corresponding to the center of the figures. In fact, the visibility of the fringes dropped to zero near the target surface because of the target shadow. The fringe shifts could be determined from ~30 μm with respect to the target position toward the laser in the front side. In the rear side of the target, no fringe shifts were observed within the region ~30 μm away from the target position. In this case, the intensity of the main pulse was $2.7 \times 10^{18}$ W · cm$^{-2}$ and that of the prepulse at 5 ns before the main pulse was ~$3 \times 10^{13}$ W · cm$^{-2}$. The emissions at the center of the preformed plasma were observed for both wavelengths. It might come from the scattering of the main pulse for Fig. 2a and the second harmonic emission of main pulse for Fig. 2c at the preformed plasma. A phase shift was determined from the fringe shift in the interferograms for each Figs. 2a and c. Next, the electron density distributions were obtained by using the Abel inversion of the phase shift distributions with the techniques of smoothing and interpolation. Figs. 2b and d show the contour plots of the electron density distribution obtained from the interferogram of Figs. 2a and e, respectively. The contour levels are labeled in units of $10^{18}$ cm$^{-3}$ and separated by $5 \times 10^{18}$ cm$^{-3}$. In Figs. 2b and d, shaded areas denote the region of the interferogram where the fringe shift could not be determined. The highest densities were ~$2 \times 10^{19}$ cm$^{-3}$ for Fig. 2b

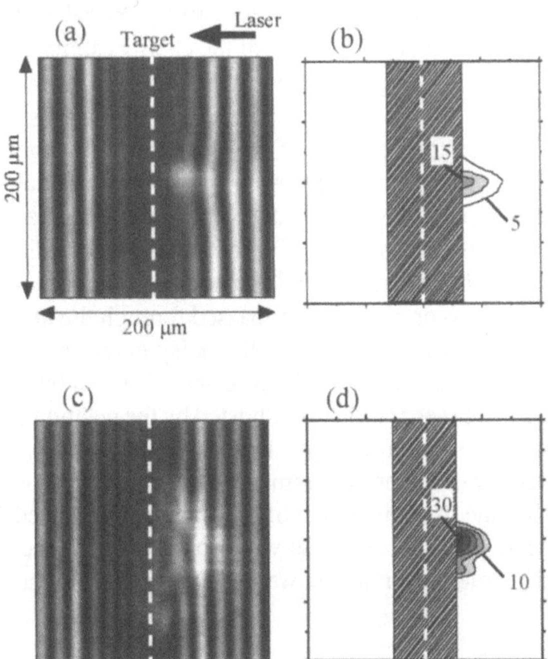


FIGURE 2. (a) Interferogram of the preformed plasma by using a probe beam with fundamental wavelength. (b) Electron density distribution obtained from (a). (c) Interferogram of the preformed plasma by using second harmonic probe beam. (d) Electron density distribution obtained from (c). All figures have same scales. The contour levels of (b) and (d) are labeled in units of $10^{18}$ cm$^{-3}$ and separated by $5 \times 10^{18}$ cm$^{-3}$.

and $\sim 3 \times 10^{19}$ cm$^{-3}$ for Fig. 2d. In this measurement we could not observe the high-density region ($>10^{20}$ cm$^{-3}$). A plasma expansion length was ~60 μm. If we assume that the preformed plasma was created by the prepulse at 5 ns before the main pulse, we could estimate the plasma expansion velocity to be $\sim 10^6$ cm/s. It corresponded to the thermal electron temperature of ~60 eV.

We measured the X-ray, electrons, and ions generated from the main laser pulse interactions [10]. The X-ray image of the plasma was observed by using the pinhole camera with the 0.8-micron thick aluminum filter for each shot. This measurement indicated that the spatial stability of the target was less than a few tens of micron. High-energy electrons and protons were clearly observed by using the Thomson parabola ion energy analyzer and the magnetic electron energy spectrometer. The electrons with energies up to ~2 MeV and the protons with maximum energy of ~1 MeV were measured with the Ta foil target.

## 4 Conclusion

We observed the preformed plasma of a 3-μm thick Ta foil target with two-color interferometer at the front and rear sides simultaneously. On the rear side, no fringe shift was observed within the region ~30 μm away from the original surface. The characterization of preformed plasmas will significantly contribute to describe the relativistic-intensity laser–thin foil interactions for generation of high-energy particles as well as X-ray.

**Acknowledgments.** This work was partly supported by the Ministry of Education, Culture, Sports, Science and Technology of Japan (Advanced Compact Accelerator Development project). We acknowledge the support of Dr. T. Kimura and Dr. A. Nagashima of the JAERI, Advanced Photon Research Center.

## References

1. T. Esirkepov, M. Borghesi, S. V. Bulanov, G. Mourou, and T. Tajima: Phys. Rev. Lett. **92**, 175003 (2004).
2. P. Audebert, R. Shepherd, K. B. Fournier, et al.: Phys. Rev. Lett. **89**, 265001 (2002).
3. U. Teubner, K. Eidmann, U. Wagner, et al.: Phys. Rev. Lett. **92**, 185001 (2004).
4. A. S. Pirozhkov, H. Daido, and S. V. Bulanov: In: *Ultrafast Phenomena XIV* (Springer Ser. Chem. Phys. Vol. 79). Berlin: Springer, p. 216 (2004).
5. A. S. Pirozhkov, S. V. Bulanov, T. Zh. Esirkepov, M. Mori, A. Sagisaka, and H. Daido: Phys. Lett. A **349**, 256 (2006).
6. K. Matsukado, T. Esirkepov, K. Kinoshita, et al.: Phys. Rev. Lett. **91**, 215001 (2003).
7. T. Fujii, Y. Oishi, T. Nayuki, et al.: Appl. Phys. Lett. **83**, 1524 (2003).
8. T. Nayuki, Y. Oishi, T. Fujii, et al.: Rev. Sci. Instrum. **74**, 3293 (2003).
9. A. Sagisaka, H. Daido, K. Ogura, et al.: Appl. Phys. B **78**, 919 (2004)
10. A. Fukumi, M. Nishiuchi, H. Daido, et al.: Phys. Plasmas **12**, 100701 (2005).

# Part IX

# Terahertz (THz) and Nonlinear Processes I

# Generation of Ultrasmooth Broadband Spectra by Gain-assisted Self-phase Modulation in a Ti:sapphire Laser

H. Crespo,[1] M. V. Tognetti,[1] M. A. Cataluna,[2,3] J. T. Mendonça,[2,4] and A. dos Santos[5]

[1] CLOQ/Departamento de Física, Faculdade de Ciências, Universidade do Porto, R. do Campo Alegre 687, 4169-007 Porto, Portugal
**hcrespo@fc.up.pt**

[2] GoLP/Centro de Fisica de Plasmas, Instituto Superior Técnico, Av. Rovisco Pais 1, 1049-001 Lisboa, Portugal

[3] School of Physics and Astronomy, University of St. Andrews, St. Andrews KY16 9SS, Scotland

[4] Rutherford Appleton Laboratory, CCLRC, Chilton, Didcot, Oxon OX11 0QX, England

[5] Laboratoire d'Optique Appliquée, École Nationale Supérieure de Techniques Avancées, École Polytechnique, F91761 Palaiseau, France

**Summary.** We demonstrate the generation of broadband symmetric spectra centered at 800 nm directly from a standard prism dispersion controlled mode locked Ti:sapphire laser built using standard quarter-wave stacks. To our knowledge, these are the smoothest broadband spectra ever generated from a Ti:sapphire laser. Measurements of the net intracavity dispersion, and a one-dimensional simulation, suggest that spectral generation and shaping result from intracavity gain-assisted self-phase modulation.

## 1 Introduction

Ever since the discovery of Kerr-lens modelocking (KLM) in Ti:sapphire over a decade ago [1,2] much effort has been put into improving ultrashort pulse generation from solid-state laser oscillators. In early systems that employed thin crystals and low-dispersion fused-silica prisms as the main source of negative group-delay dispersion (GDD), it was found that residual third-order dispersion (TOD) was a limiting factor that prevented further reduction in pulse length [3]. Additional compensation of TOD [4, 5] proved somewhat insufficient for broadband sub-10-fs pulse generation, since higher order dispersion would now come into play, with the resulting spectra becoming asymmetric and red-shifted with respect to the maximum of the gain-profile of Ti:sapphire [5]. The invention of chirped mirrors [6] provided engineered dispersion for the first time, which together with thinner (2-mm) highly doped crystals resulted in the direct generation of sub-10-fs pulses centered at 800 nm [7]. Pure mirror-dispersion-controlled (MDC) lasers are very robust and compact, but do not allow for fine tuning of intracavity GDD, as opposed to hybrid systems where both prisms and chirped mirrors are used for broadband high-order-dispersion compensation, setting the present world record of 5-fs sub-two-cycle pulses [8].

State-of-the-art ultrashort pulse generation has relied upon achieving the flattest possible intracavity dispersion over the broadest possible bandwidth. Due to the dispersive characteristics of both standard optical materials and quarter-wave stack dielectric mirrors, it is usually assumed that broadband pulses with durations below 11 fs can be generated only from a standard prism-dispersion-controlled KLM Ti:Sapphire oscillator at the expense of spectral asymmetry (which may compromise pulse quality) and detuning (from 800 nm to around 850 nm), making the pulses less suitable for important applications such as subsequent amplification in Ti:sapphire. However, we have showed that broadband (over 150 nm FHWM) spectra centered at 800 nm could be directly generated from a system comprising standard mirrors and fused-silica prisms [9]. The use of standard mirrors designed for 850 nm operation in a prism-dispersion-controlled laser allows for the simultaneous minimization of GDD and TOD in this region and can also provide a second region of low GDD at lower wavelenghts, which is known to result in a double-peaked spectrum; the corresponding pulses are sufficiently short so as to experience strong self-phase modulation (SPM) within the laser crystal.

Intracavity SPM has recently been explored in more advanced chirped-mirror laser designs to generate ultrabroadband spectra [10, 11] and ultrashort pulses falling outside the gain bandwidth of Ti:sapphire [12]. In this work we show that SPM and net intracavity gain (and losses) will determine the ultimate broadband and ultrasmooth spectral shapes and that a flat GDD is not required for efficient broadband operation.

## 2 Description of the Setup

Our Ti:sapphire oscillator (Fig. 1) comprises a relatively thick (4.5 mm) crystal (Crystal Systems) placed in a standard z-fold subcavity [3] composed of two 10-cm radius mirrors, a pair of fused-silica prisms for intracavity GDD compensation (54 cm interprism separation), commercially available standard ultrafast laser mirrors designed for 850 nm (Spectra-Physics and TecOptics), and a 3.5% output coupler (Spectra-Physics). Modelocked output power is typically 150 mW for a pump power of 4.5 W (all-lines Ar-ion). Pulse repetition rate is 85 MHz and KLM action is started by a shaking cavity mirror. A protected silver mirror was used as the end mirror (HR) in the negative dispersion section of the laser, where the intracavity spectrum is widest. This mirror does not reduce the total modelocked power, even though the CW output decreases by almost 50% when

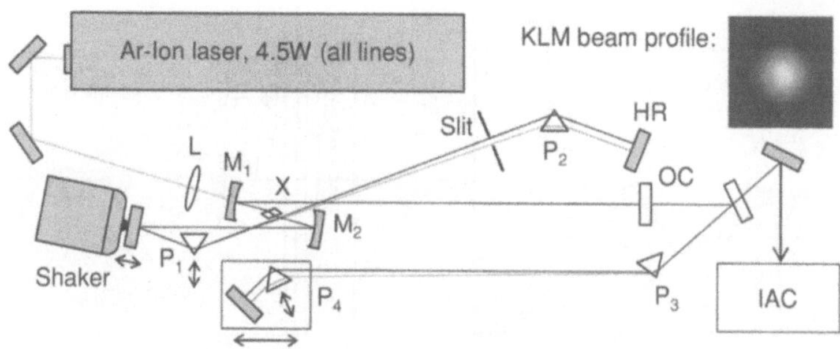


FIGURE 1. Schematic of the Ti:sapphire laser (IAC denotes a home-built low-dispersion interferometric autocorrelator). All mirrors used in the setup are either commercial quarter-wave stacks designed for 850 nm operation or protected silver mirrors; the inset is the measured beam profile.

compared to a dielectric HR. A second prism pair is used for extracavity dispersion compensation. The output beam (see inset of Fig. 1) has a smooth $TEM_{00}$ spatial profile (99.6% Gaussian in both axis).

# 3 Results and Discussion

An adjustable slit placed between the two intracavity prisms is used to induce losses at the edges of the spectrum, with a dramatic effect in the output, as evidenced in Fig. 2. The spectrum obtained without the aperture (dashed line) has a double-peaked structure with a minimum at 800 nm as well as a sharp peak in the blue region. Optimized intracavity aperturing results in an ultrasmooth spectrum peaked at 800 nm (solid line).

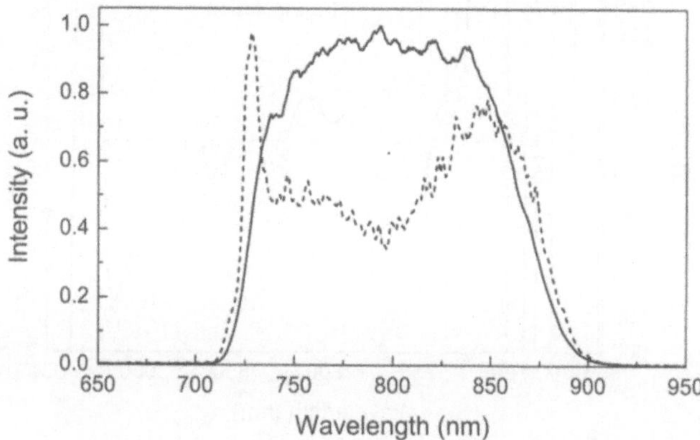


FIGURE 2. Normalized laser output spectra obtained with an optimized intracavity aperture (solid line) and without aperture (dashed line).

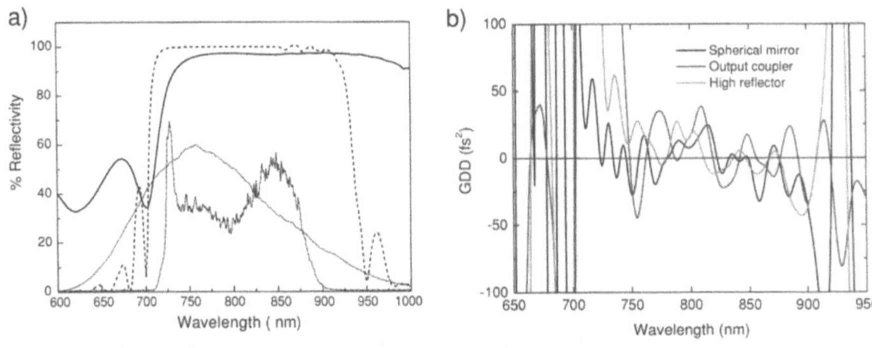


FIGURE 3. (a) Measured reflectivity of the output-coupler (solid line) and combined reflectivity of all high-reflectors (dashed line); a typical laser spectrum (dotted line) and the measured Ti:sapphire fluorescence (dashed-dotted line) are given for comparison purposes. (b) Measured GDD of each dielectric mirror.

In order to better interpret these results, reflectivity and dispersion of all mirrors were measured with an improved white light interferometer [13] (Fig. 3).

The net round-trip GDD of the cavity is shown in Fig. 4, and this accounts for all of the dielectric mirrors, the output-coupler, the laser crystal, the exact prism-pair dispersion, and the air path in the cavity. We see that the total GDD is not flat, although there is a region around 850 nm where GDD is negative and TOD remains practically constant (estimated at below 650 $fs^3$).

Figure 5a shows the measured spectrum for increasing prism insertion (total insertion range is 0.5 mm). The spectrum is initially peaked at 800 nm (60 nm

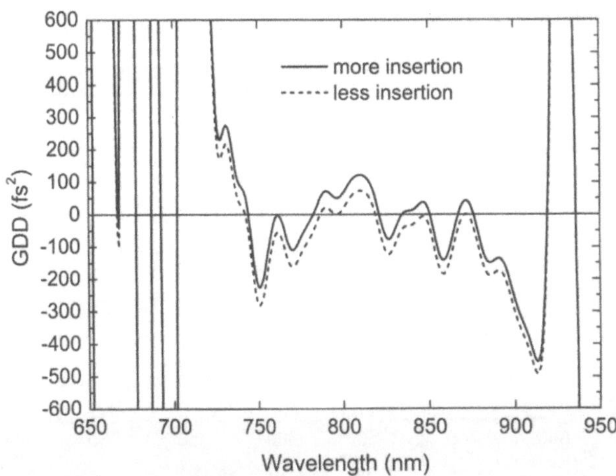


FIGURE 4. Total round-trip intracavity GDD for the prism insertion range used in Fig. 5a (dotted line corresponds to less insertion).

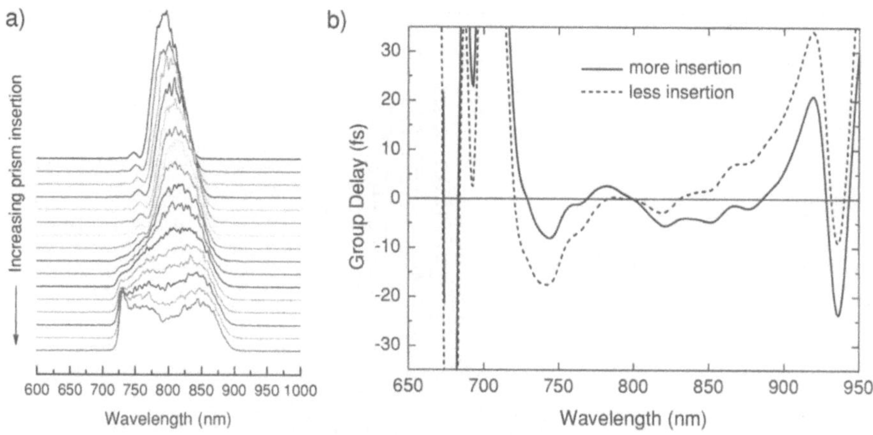


FIGURE 5. (a) Spectral evolution as a function of prism insertion (total insertion range is 0.5 mm). (b) Total intracavity group delay for the same insertion range.

FWHM), with a second small peak at 750 nm. This is in agrement with the total intracavity group delay shown in Fig 5b, where we see that for small insertions there are two regions with negative GD and approximately zero GDD around 750 nm and 800 nm. The small peak at 750 nm has also been observed in an ultralow-threshold system that uses similar 850 nm optics [14]). For small insertions, the generated spectrum has a relatively narrow bandwidth (and correspondingly large duration), which is why linear intracavity dispersion dominates over nonlinear propagation effects.

As the prism insertion is increased, both peaks move to the longer wavelength region. For the main part of the pulse, this is in very good agreement with the evolution of the GDD curve of Fig. 4; the shift observed in the small peak at 750 nm has a somewhat different origin and can be interpreted in the light of the heuristic model described in [15]. From Fig. 5b, we see that for larger prism insertions, two regions of negative group delay appear around 750 nm and 850 nm. In one round trip, these two wavelengths will then be advanced with respect to the central wavelength in the spectrum and will consequently experience SPM induced by the leading edge of the pulse, which results in the observed continuous redshift. This shift can then be seen as a measure of the amount of SPM in the cavity. For larger prism insertions, intracavity GDD can support very short (sub-11-fs) pulses with spectra peaking at 750 nm and 850 nm, where strong SPM and amplification in the crystal result in additional spectral buildup around these two wavelengths. The final spectrum (without intracavity aperturing) is then a double-peaked structure centered at 800 nm (over 150 nm FHWM; 200 nm at −20dB). This was not observed in [14] probably because of the lower intracavity power and insufficient SPM. The spectrum has a third sharp peak at 730 nm (due to coupling between the soliton and the continuum via TOD), which is further enhanced by the roll-off characteristics of the OC (see Fig. 3a). The output coupler

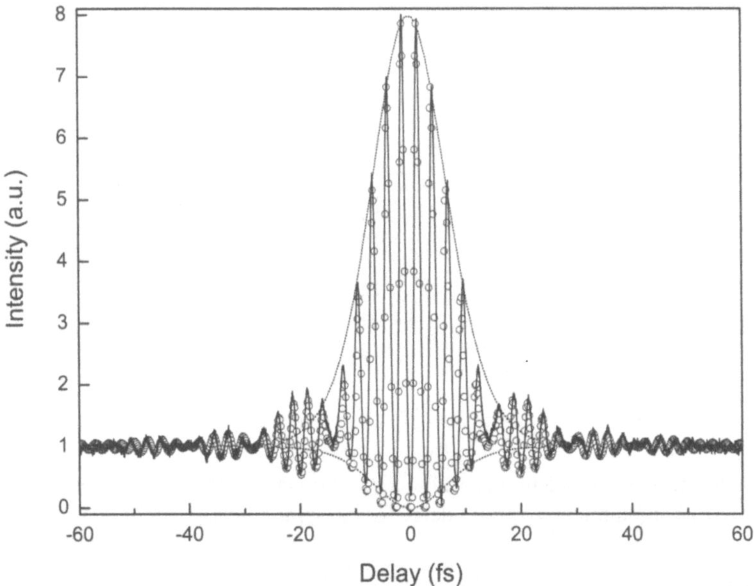


FIGURE 6. Interferometric autocorrelation trace of the shortest measured pulses (black line) and calculation using the Fourier transform of the spectrum assuming a flat spectral phase (open dots); pulse duration is 10.4 fs.

efficiently transmits the bluer part of the spectrum, where intracavity dispersion cannot support stable soliton-like pulses. The final spectrum is then limited on the blue side by HR reflectivity and on the red side by intracavity gain and dispersion. Optimum self-focusing in the crystal corresponds to a compromise between large spectral broadening and pulse profile quality. The resulting spectral shape could be helpful for amplification since it can reduce gain narrowing. Also, and due to the large spectral bandwidtha achieved, temporal stretching as required for chirped-pulse amplification (CPA) can be effectively accomplished by simply passing the pulses through a piece of dispersive glass [16].

The interferometric autocorrelation (IAC) trace of the shortest measured pulses obtained using two-photon absorption in a GaAsP photodiode is shown in Fig. 6, together with the IAC calculated from the measured spectrum (last spectrum of Fig. 5a) assuming a flat spectral phase. The agreement is very good and the retrieved pulse duration is 10.4 fs. The observed side wings are not due to residual extracavity TOD (estimated at less than 250 $fs^3$) but to the flat-topped spectral shape. Assuming a flat spectral phase is in agreement with the fact that, under strong SPM conditions, KLM action tends to smooth out spectral modulations caused by GDD oscillations, as discussed in [8], the measured spectral phase is then affected only by extracavity elements, as shown in [11]. The compressed output pulses from the laser are essentially transform-limited, even in the case of spectra that clearly deviate from a $sech^2$ shape.

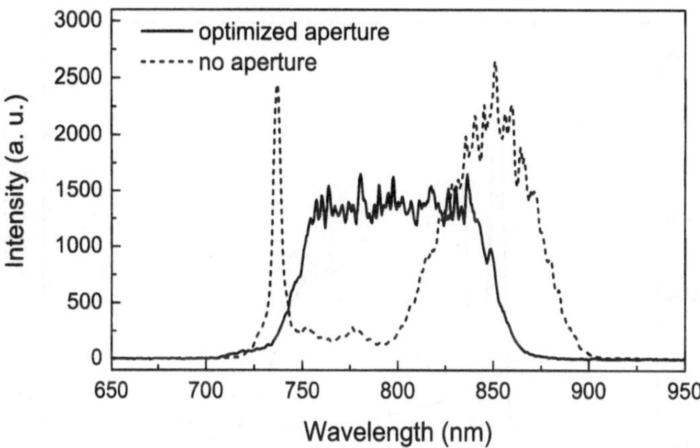


FIGURE 7. Measured laser output spectra for the case of a dielectric HR, obtained without aperture (dashed line) and with an optimized intracavity aperture (solid line).

Using an optimized intracavity aperture (see Fig. 1), an ultrasmooth spectrum with 0.1 dB spectral modulation (Fig. 2) can be obtained, which does not at all reflect the total net GDD of the cavity. If laser action is interrupted, KLM operation is always resumed for the previous spectral shape (with or without hard aperturing). This rounder spectrum is very similar the one expected for a strongly dispersion managed mode-locked laser [17] and could find application in high-resolution optical coherent tomography. This effect is even more noticeable when we replace the protected silver HR with a second dielectric mirror. The total dispersion in this case does not favor generation at 800 nm, resulting in the spectrum shown in Fig. 7 (dashed line). Again, an optimized intracavity aperture can be used to produce a smoothed spectrum centered at 800 nm, although the required higher aperturing will result in a narrower spectrum when compared to the silver mirror case.

It is known that intracavity aperturing corresponds to a frequency dependent gain, which can explain why hard-aperturing at the wings results in enhanced amplification of the central part of the spectrum. To better understand this mechanism, we implemented a detailed one-dimensional numerical simulation based on an iterative procedure that relates the spectral component of the field envelop obtained after the $k$th pass in the cavity with the one obtained in the previous pass [18]. The model includes building up of the laser pulse from noise through the action of active and KLM (modeled by a fast saturable absorber), the measured reflectivity and phase distortion of every optical element, the measured gain bandwidth of the 4.5-mm thick Ti:sapphire crystal, and exact propagation inside the active medium, by solving the nonlinear Schrödinger equation in the presence of saturable gain and KLM using a split-step Fourier method with error-optimized variable grid spacing [19]. For a prism insertion range that is in agreement with the measured value, the evolving pulse always reaches a stable periodic solution after approximately 500–550 iterations. The final spectrum obtained at the

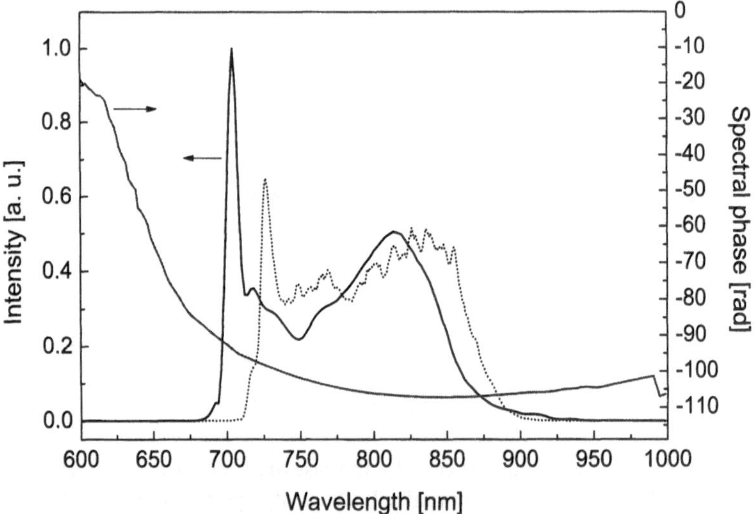


FIGURE 8. Simulated laser spectrum and corresponding spectral phase. The measured spectrum is shown for comparison (dashed line).

output-coupler end of the laser for a typical value of insertion is shown in Fig. 8. We see that this result is in good agreement with the measured spectra. Also, the spectral phase is extremely smooth and basically corresponds to the positive dispersion imposed by the laser crystal (pulses extracted from the OC end of the laser are known not to be transform-limited). Further simulation work aimed at studying the effect of intracavity aperturing is presently under development.

## 4 Conclusions

We demonstrate the generation of broadband symmetric spectra centered at 800 nm directly from a prism-dispersion-controlled KLM Ti:sapphire laser built with standard quarter-wave stack mirrors. Spectral shape can be controlled to a large extent with an intracavity aperture while maintaining 10-fs pulse durations and flat intracavity GDD is not required for efficient broadband operation. Under strong SPM conditions, intracavity losses appear to play a major role in spectral shaping. To our knowledge, these are the smoothest broadband spectra ever generated with a standard Ti:sapphire laser. Measurements of the intracavity dispersion, together with a one-dimensional computer simulation, strongly suggest that spectral generation and shaping result from intracavity gain-assisted SPM. Further study of this regime could lead to more simple, reliable, and efficient methods for ultrashort pulse generation in self-mode-locked laser oscillators.

**Acknowledgments.** The authors gratefully acknowledge support for this work from the Foundation for Science and Technology, Portuguese Ministry of Science and Higher Education Grant No. POCTI/FIS/48709/2002, partially financed by FEDER. The authors wish to thank Carla Carmelo Rosa for the help in the white light interferometer setup and dispersion measurements.

## References

1. D. E. Spence, P. N. Kean, W. Sibbett: Opt. Lett. **16**, 42 (1991).
2. U. Keller, G. W. 't Hooft, W. H. Knox, J. E. Cunningham: Opt. Lett. **16**, 1022 (1991).
3. M. T. Asaki, C.-P. Huang, D. Garvey, J. Zhou, H. C. Kapteyn, and M. M. Murnane: Opt. Lett. **18**, 977 (1993).
4. B. E. Lemoff and C. P. J. Barty: Opt. Lett. **18**, 57 (1993).
5. J. Zhou, G. Taft, C.-P. Huang, M. M. Murnane, H. C. Kapteyn, and I. P. Christov: Opt. Lett. **19**, 1149 (1994).
6. R. Szipöcs, K. Ferencz, Ch. Spielmann, and F. Krausz: Opt. Lett. **19**, 201 (1994).
7. A. Stingl, M. Lenzner, Ch. Spielmann, amd F. Krauzs, R. Szipöcs: Opt. Lett. **20**, 602 (1995).
8. U. Morgner, F. X. Kärtner, S. H. Cho, et al.: Opt. Lett. **24**, 411 (1999).
9. H. Crespo, M. A. Cataluna, A. Guerreiro, J. T. Mendonça: High-order Self-dispersion compensation in a Ti:sapphire laser oscillator. In: *Ultrafast Phenomena XIII* Springer Series in Chemical Physics, Vol. 71. D. R. Miller, M. M. Murnane, N. F. Scherer, A. M. Weiner (eds). Berlin: Springer, p. 161 (2002).
10. R. Ell, U. Morgner, F. X. Kärtner, et al.: Opt. Lett. **26**, 373 (2001).
11. T. Fuji, A. Unterhuber, V. S. Yakovlev, et al.: Appl. Phys. B **77**, 125 (2003).
12. A. Bartels and H. Kurz: Opt. Lett. **27**, 1839 (2002).
13. H. Crespo, A. A. Amorim, M. Miranda, C. C. Rosa, and L. M. Bernardo (in preparation).
14. A. M. Kowalevicz Jr., T. R. Schibli, F. X. Kärtner, and J. G. Fujimoto: Opt. Lett. **27**, 2037 (2002).
15. A. Rundquist, C. Durfee, Z. Chang, et al.: Appl. Phys. B **65**, 161 (1997).
16. S. Sartania, Z. Cheng, M. Lenzner, et al.: Opt. Lett. **22**, 1562 (1997).
17. Y. Chen, F. X. Kärtner, U. Morgner, et al.: JOSA B**16**, 1999 (1999).
18. M. Tognetti and H. Crespo (in preparation).
19. O. V. Sinkin, R. Holzlhner, J. Zweck, and C. R. Menyuk: J. Lightwave Technol. **21**, 61 (2003).

# Part X

# Terahertz (THz) and Nonlinear Processes II

# Terahertz (THz) Pigtail Assembly Utilizing a Lens Duct for Effective Coupling of THz Radiation into Teflon Photonic Crystal Fiber Waveguide

Alex Quema,[1,2] Gilbert Diwa,[1] Elmer Estacio,[1] Romeric Pobre,[2] Glenda Delos Reyes,[1] Carlito Ponseca Jr.,[1] Hidetoshi Murakami,[1] Shingo Ono,[1] and Nobuhiko Sarukura[1]

[1] Institute for Molecular Science (IMS), Myodaiji, Okazaki 444-8585, Japan
alexq@ims.ac.jp
[2] Department of Physics, De La Salle University, 2401 Taft Avenue, Manila 1004, Philippines

**Abstract:** The design of a lens duct to facilitate the launching of terahertz (THz) radiation into Teflon photonic crystal fiber waveguide is presented. The InAs THz emitter, lens duct, and waveguide, collectively called THz pigtail are found to be potential mean of effectively channeling and directing THz radiation.

## 1 Introduction

Since the development of lasers that can produce ultrafast optical pulses several techniques in generating terahertz (THz) radiation have been reported. In particular, InAs under magnetic field is considered to be the practical source since it provides intense THz radiation without the need for any chemical process or microfabrication technique in the sample preparation [1]. With these devices, THz spectroscopy and imaging can be applied to virtually all kinds of samples, including chemicals, polymers, and biomolecules [2–4]. At present, THz spectroscopic techniques use free space propagation and are to some extent difficult to control and guide. This then led to the development of THz waveguides using various materials [5–7]. Recently, we developed a Teflon photonic crystal fiber (PCF), which showed effective wave guiding of THz radiation with strong polarization maintaining property [8]. Nakajima et al. reported the use an MgO hemispherical lens coupler to enhance THz radiation power from InAs surfaces excited by ultrashort laser pulses [9]. Such enhancement was explained by the increase of the transmission efficiency of the THz wave from InAs to free space. Although these waveguides and lens coupler have in part provided a solution to control and guide THz radiation, still the problem of efficiently launching THz wave into the waveguide remains. In most cases, THz radiation is guided into the waveguide using a paraboloidal mirror coupled with a hemispherical lens at the entry point of the waveguide. Using these optical devices, one has to consider the coupling efficiency between air and hemispherical lens and between hemispherical lens and waveguide. More often than not, this trivial problem leads to an enormous amount of signal loss.

In this paper, the design and fabrication of an integrated optics in the THz regime, here referred to as "THz pigtail," is presented. The THz pigtail is an assembly comprising of an InAs emitter, a lens duct, and a Teflon PCF waveguide. The lens duct is designed to facilitate the launching of THz radiation from an InAs emitter into a Teflon PCF waveguide. The fabricated lens duct is found to successfully channel and focus the THz radiation thereby ably transmitting the radiation from emitter to the core of the waveguide.

## 2 Teflon Photonic Crystal Fiber Waveguide

The PCF samples are constructed using Teflon rods as core and Teflon tubes for the periodic cladding. The periodic claddings mounted in the samples have two layers, wherein the first and second layers have 8 and 16 Teflon tubes, respectively. The diameter of the core is 1 mm while the clad has an outer diameter of 0.75 mm and an inner diameter of 0.25 mm. For comparison purposes, a hollow metal tube (inner diameter of 0.25 mm, outer diameter of 5 mm) and a solid Teflon rod whose diameter is the same as that of the PCF is also used. Teflon is chosen for this study because several reports have shown that this material is highly transparent in the THz frequency region [10, 11] and our group found that the absorption coefficient of this material is approximately 0.3 $\text{cm}^{-1}$ at 1 THz. A photograph of the constructed PCFs is shown in Fig. 1a. The upper-right-hand corner inset shows the cross section of the PCF. The cladding layers and the core are fused together without any furnace heating. Holding the first cladding layer and the core together is ordinary plastic tube, which is placed between the first and second cladding layers. Another thin plastic tube is placed on the outer portion to envelop the whole PCF thereby fusing together the core and the first and second layers into one fiber. To test the confinement of the incident THz wave inside and to confirm that the wave propagation is mainly in the core of the constructed PCF, knife-edge measurements at the exit point are conducted. The intensity profile of the beam in one of the PCFs is shown at the lower-left-hand corner inset of Fig. 1a. From the plot shown, it can be seen that the data is well described via Gaussian fitting. This clearly indicates that almost all the beam is confined within a diameter of 1 mm thereby confirming that almost all the THz radiation propagate in the core. Using a wire grid polarizer, it is found that the constructed Teflon PCFs are highly polarization maintaining as compared to the hollow metal tube.

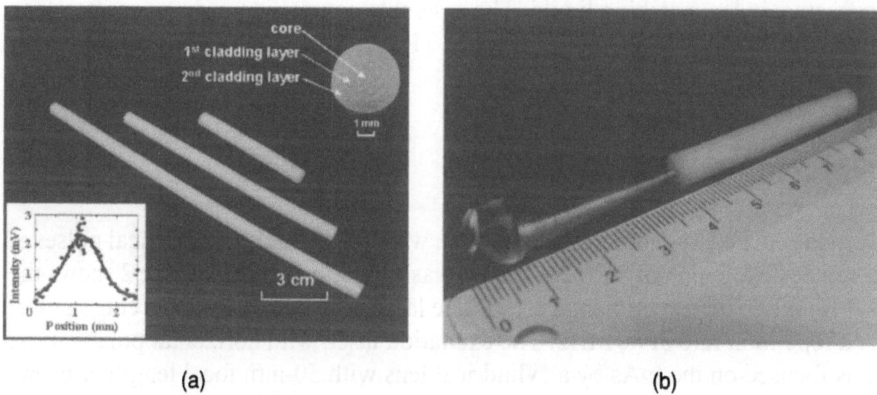


FIGURE 1. (a) Photograph of Teflon PCF waveguide. The lower-left-hand corner inset shows that almost all the THz radiation propagates in the core of the PCF. (b) Photograph of THz pigtail wherein the lens duct is coupled to a 5-cm PCF.

## 3 Terahertz Pigtail

A lens duct is a simple optical device that has found applications in the coupling of pump radiation from a semiconductor laser diode into a solid-state laser gain medium. Basically, this device relies on the combined effects of lensing at the input surface, channeling by means of total internal reflection, and focusing to its apex [12]. With this in mind, a lens duct for THz radiation is designed and fabricated and the material used is polymethylpentane, commonly known as TPX. This material was chosen due to its relatively high transparency in the visible, infrared, and THz frequency regions [13]. In constructing the lens duct, three important aspects are considered. These are an input surface for the excitation laser, a coupling surface for the InAs emitter, and a provision to guide and channel the THz radiation into the core of the Teflon PCF waveguide. The input surface is oriented in such a way that the incidence of the horizontally polarized excitation laser is at Brewster's angle with respect to the input-surface normal. As for the coupling surface, its orientation is adjusted so that the excitation laser is incident at Brewster's angle on the InAs emitter. The generated THz radiation from the InAs emitter is then contained in a provision for channeling and focusing the radiation into the waveguide. The containment provision is shaped into a cone-like structure with its apex formed to accommodate the core of the Teflon PCF waveguide. This conical portion of the lens duct has a conic angle of 8° and a conic length of 30 mm. The emitter used is an undoped slightly $n$-type InAs (100) wafer. Using silicone grease, the InAs is attached to the lens duct. To readily facilitate the entry of THz radiation into the waveguide, a small portion of the Teflon PCF core was allowed to protrude. This overhang portion of the core was inserted into the apex of the lens duct. To aid the coupling of the lens duct and Teflon PCF core, silicone grease was applied on the outer surface of the protruding portion of the waveguide core. Silicone grease was used since it is a good low-loss refractive index matching liquid. The assembly comprising of InAs emitter, lens duct, and Teflon PCF waveguide is called THz pigtail and the photograph is shown in Fig. 1b.

## 4 Experiment

A mode-locked Ti-sapphire laser system, which delivered 100 fs optical pulses at a center wavelength of 800 nm, was used as excitation source. Figure 2 shows the schematic of the experimental setup. The laser provided 300-mW average power at a repetition rate of 80 MHz. The excitation laser, with horizontal polarization, was focused on the InAs by a cylindrical lens with 50-mm focal length at Brewster's angle incidence and the laser spot size on the sample surface was about 1 mm × 3 mm. A 1-T permanent magnet was applied parallel to the surface of the InAs emitter. The THz radiation power from the THz pigtail was measured using a liquid helium cooled Ge-bolometer and its spectrum was obtained by a polarizing Michelson interferometer.

## 5 Results and Discussions

Figure 3a shows the experimentally obtained spectra (normalized values) of the bare InAs emitter (dotted line) and the InAs emitter coupled to the TPX lens duct (open circles), both of which are under a 1-T magnetic field. Knowing the absorption coefficient in THz region of TPX, the transmittance of a 3-cm thick slab of such material is determined and is shown here by the dashed line. Using this numerically determined THz transmittance of the 3-cm thick TPX slab and multiplying it with the THz spectrum of the bare InAs emitter, the spectrum of the 3-cm TPX lens duct is obtained, which is shown here by the solid line. It can be seen that the experimentally obtained and the numerically obtained spectra of the emitter-lens duct assembly are in reasonable agreement. By measuring the output THz radiation power and placing a wire grid polarizer in front of the lens duct, it is found that the constructed lens duct is polarization maintaining.

The THz radiation power as a function of length is shown in Fig. 3b. The plot is divided into two regions: region I (0–3 cm) and region II (3–18 cm). The THz radiation power of the lens duct is in region I while that of the pigtail is in region II. The data points represent the calculated integrated intensities of the spectra obtained from 0.1 to 1.5 THz. These values are found to be comparable to those obtained via direct THz radiation power measurements using the bolometer. The solid line in region II is a linear fit of the data. It can be seen that at the 3-cm mark, the linear fit, and the data point do not coincide. This connotes a certain loss due to coupling at the interface between lens duct and waveguide and is estimated to be ~0.7 dB [15]. Although a rather large amount of THz radiation dissipated upon propagating through the lens duct, which can be attributed to large index mismatch between TPX [11] and InAs [14], this assembly is able to minimize the insertion loss between lens duct and waveguide.

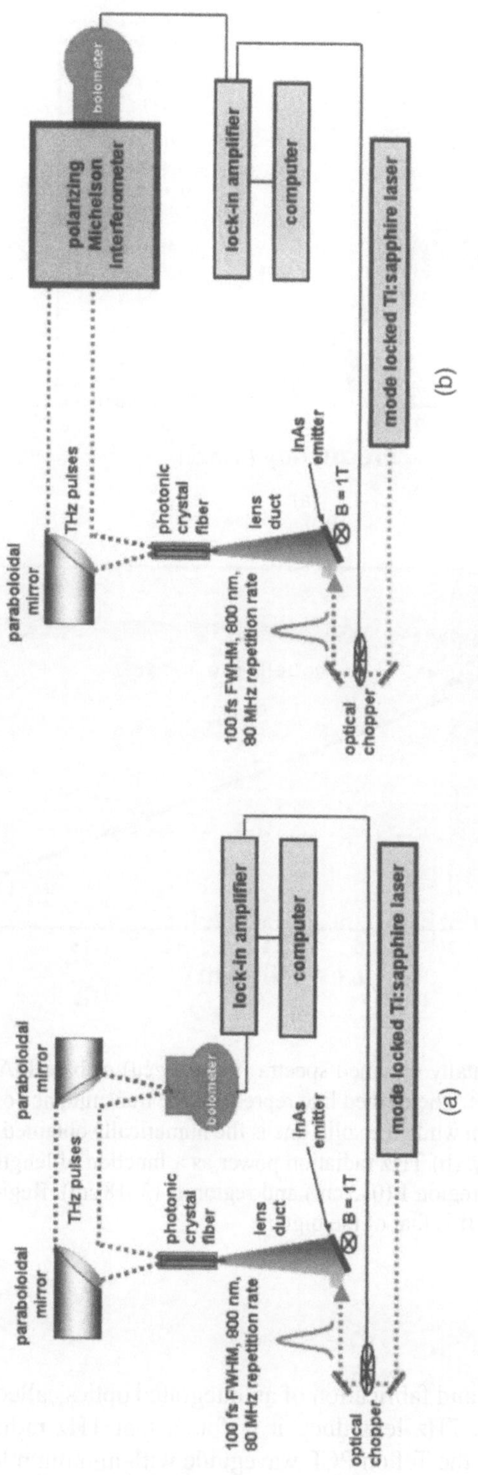


FIGURE 2. Schematic of the experimental setup for (a) THz power measurement and (b) THz radiation spectrum.

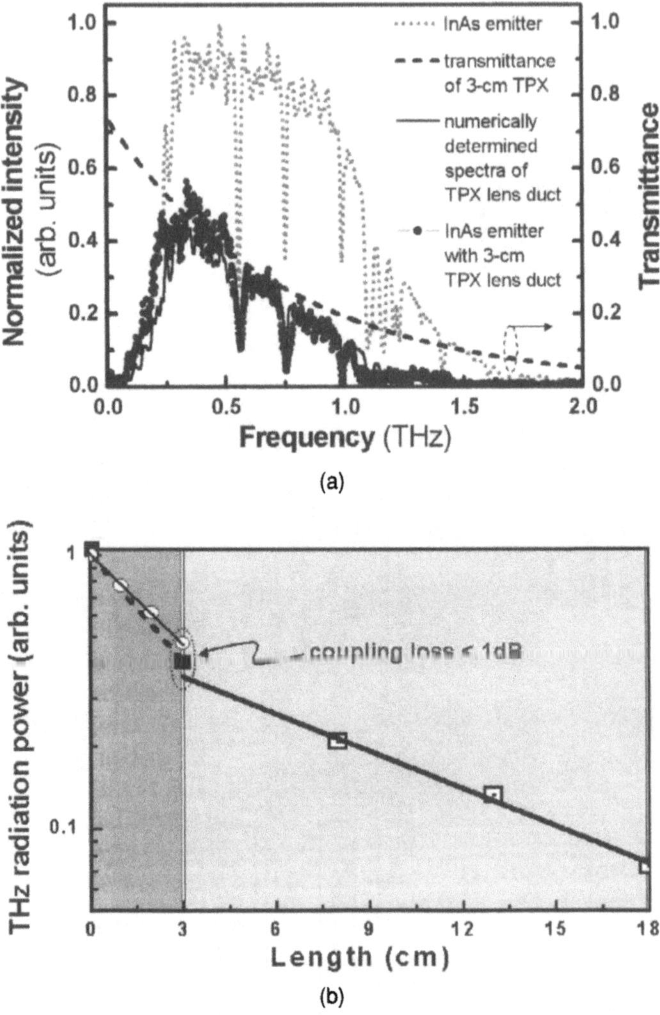


FIGURE 3. (a) Experimentally obtained spectra (normalized) of bare InAs emitter and InAs emitter with lens duct. The dashed line represents the transmittance of a 3-cm thick TPX slab in the THz region while the solid line is the numerically obtained spectra of the emitter-lens duct assembly. (b) THz radiation power as a function of length. This plot is divided into two regions: region I (0–3 cm) and region II (3–18 cm). Region I is that of the lens duct while region II is that of the pigtail.

## 6 Summary

In summary, the design and fabrication of an integrated optics called THz pigtail is discussed. Using the THz lens duct, it is found that THz radiation can be effectively coupled into the Teflon PCF waveguide with minimum loss (<1 dB).

Based on the results obtained here, the THz pigtail can be a means of guiding THz radiation without the need of using expensive optical devices.

## References

1. N. Sarukura, H. Ohtake, S. Izumida, and Z. Liu: J. Appl. Phys. **84**, 654 (1998).
2. K. Kawase, M. Sato, and H. Itoh: Appl. Phys. Lett. **68**, 2483 (1996).
3. Y. Shen, P. Upadhya, E. Linfield, and A. Davies: Appl. Phys. Lett. **82**, 2350 (2003).
4. A. Quema, M. Goto, N. Sarukura, et. al: Appl. Phys. Lett. **85**, 3914 (2004).
5. R. Mendis and D. Grischkowsky: J. Appl. Phys. **88**, 4449 (2000).
6. S. Jamison, R. McGowan, and D. Grischkowsky: Appl. Phys. Lett. **76**, 1987 (2000).
7. H. Han, H. Park, M. Cho, and J. Kim: Appl. Phys. Lett. **80**, 2634 (2002).
8. M. Goto, A. Quema, H. Takahashi, Shingo Ono, and N. Sarukura: Jpn. J. Appl. Phys. **43**, L317 (2004).
9. M. Nakajima, K. Uchida, M. Tani, and M. Hangyo: Appl. Phys. Lett. **85**, 191 (2004).
10. G. Chantry, J. Fleming, and P. Smith: Chem. Phys. Lett. **10**, 473 (1921).
11. J. Birch, J. Dromey, and J. Lesurf: Infrared Phys. **21**, 225 (1981).
12. R. Beach: Appl. Opt. **35**, 2005 (1996).
13. A. Quema, M. Goto, H. Takahashi, S. Ono, N. Sarukura, and N. Yamada: In: The Japan Society of Applied Physics 64th Autumn Meeting Extended Abstracts, Fukuoka University, Fukuoka, Japan, 2003, Session No. 29-ZD, p. 84.
14. E. Palik and R. Holm: In *Handbook of Optical Constants of Solids*. E. Palik (ed). New York: Academic Press, pp. 479 (1985).
15. G. Diwa, A. Quema, E. Estacio, et al.: Appl. Phys. Lett. **87**, 151114-1 (2005).

# Terahertz Radiation from Photoconductive Switch Fabricated from a Zinc Oxide Single Crystal

Hidetoshi Murakami,[1] Elmer Estacio,[1] Alex Quema,[1,5] Glenda Delos Reyes,[1] Shingo Ono,[1,4] Nobuhiko Sarukura,[1,4] Yo Ichikawa,[2] Hiraku Ogino,[3] Akira Yoshikawa,[3] and Tsuguo Fukuda[3]

[1] Institute for Molecular Science (IMS), Myodaiji, Okazaki 444-8585, Japan
murakami@ims.ac.jp
[2] Nagoya Institute of Technology, Showa, Gokiso, Nagoya 466-8555, Japan
[3] Tohoku University, 2-1-1 Katahira, Aoba-ku, Sendai 980-8577, Japan
[4] Department of Photo Science, The Graduate University for Advanced Studies, Hayama, Japan
[5] Department of Physics, De La Salle University, 2401 Taft Avenue, Manila 1004, Philippines

**Abstract:** We present results on the terahertz (THz) radiation from a photoconductive switch fabricated on a zinc oxide (ZnO) single crystal. Due to its high transmittance in the visible, near-infrared, mid-infrared, and THz regions, ZnO may present itself as a viable material for integrated active optics operating in the THz region.

## 1 Introduction

The generation of terahertz (THz) pulses from a photoconductive switch was demonstrated for the first time using radiation-damaged silicon on sapphire crystals [1]. Subsequently, various materials have been used for photoconductive antennas, such as low temperature grown GaAs [2,3]. Simply increasing the applied bias voltage improves the output power from a photoconductive switch since the electric fields acting on the photoexcited carriers are enhanced. However, this THz enhancement method is constrained by the maximum applied bias voltage, which is limited by the dielectric breakdown of the material. For this reason, wide bandgap materials with higher breakdown voltages are potential candidates for photoconductive antennas that operate as high-power THz radiation emitter. Recently, high-power THz radiation emitted from a diamond photoconductive antenna array was demonstrated [4]. The results show that the breakdown electric field of diamond was 2 MV/cm. A diamond photoconductive switch requires a KrF excimer laser as excitation source since its bandgap is located in the deep ultraviolet (UV) region (5.5 eV) [5]. Such excitation source is disadvantageous due to the size, running cost, and stability. In addition, a chemical vapor deposited polycrystalline diamond has a smaller mobility (1–70 $cm^2/V$ s) than single diamond crystal (1800 $cm^2/V$ s) [4]. Unlike diamond, ZnO can be excited by the second or third harmonics of a Ti:sapphire laser. With the recent developments of new nonlinear crystals such as $CsLiB_6O_{10}$, $Li_2B_4O_7$, $BaB_2O_4$, and $LiB_4O_7$ [6–8], this laser system is significantly efficient. Additionally, efficient UV gain media, which include $Ce^{3+}$:$LiCaAlF_6$(Ce:LiCAF) and $Ce^{3+}$:$LiLuF_4$, have been developed [9–12]. All these works demonstrate that a solid-state UV excitation source is indeed well established.

In this paper, we report on the THz generation from a ZnO photoconductive antenna excited by UV pulses. Due to its highly transparent nature in the visible, near-infrared, mid-infrared, and THz frequency regions, this work demonstrates that ZnO is a suitable hybrid optical material for integrated active optics.

## 2 Experiment

In our experiment, the frequency tripling of the Ti:sapphire regenerative amplifier output, which has a 290 nm wavelength, was used as the excitation source. High peak power laser pulses can be easily obtained in the 280–330 nm region using a Ce:LiCAF crystal as a gain medium. Femtosecond pulses with a 1-kHz repetition rate had an average output power of 25 mW with pulse duration of 200 fs. The ZnO single crystals, which were substrates for photoconductive antennas, had dimensions of $10 \times 10 \times 0.5$ mm and the crystallographic $c$-axis of the crystals were oriented parallel to the excitation laser. Two parallel 5-mm long and 1-mm wide coplanar silver strip lines were fabricated onto these substrates. The gap between the transmission lines was 1 mm. The excitation laser was focused on the anode side using a 100-mm focal length cylindrical lens. The laser spot size on the sample surface was about 1 mm × 5 mm. A liquid-helium-cooled germanium bolometer was used to detect the total transmitted THz-radiation power.

## 3 Results and Discussions

ZnO has attracted considerable interests in the development of blue and UV light emitting devices [13,14]. Large-sized and high-quality crystals are essential for substrates in light-emitting devices. Recently, the improved crystal growth techniques have led to the fabrication of substantially large single crystals. In particular, the hydrothermal method has allowed large-sized ZnO single crystals to be grown [15], which allows flexibility when designing optics in the THz region, including focusing optics and lens duct [16].

As shown in Fig. 1, the ZnO crystal has high transmittance in the THz region. The periodic structure is attributed to the interference of the signals, which originate from the front and back surfaces of the ZnO crystal. The solid circles indicate

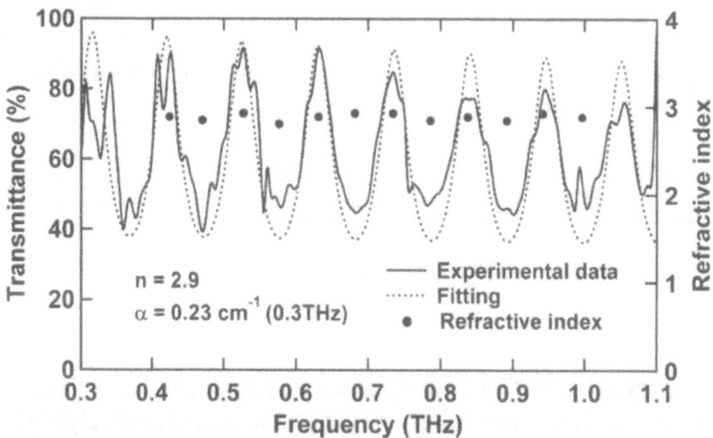


FIGURE 1. Transmission spectrum of a ZnO single crystal in the THz region. The periodic structure is due to the interference from the front and back surfaces of the ZnO crystal.

the refractive index of ZnO calculated by the fringe gap. A strong dispersion of refractive index is not observed in the 0.3–1.1 THz frequency regions. Assuming refractive index $n = 2.9$, the dotted-curve is the least-square fitting solution for the transmittance $T$ expressed as [17]

$$T = \frac{(1 - R_0)^2 e^{-\alpha d}}{(1 - R_0 e^{-\alpha d})^2 + 4R_0 e^{-\alpha d} \sin^2(\frac{2\pi}{\lambda})nd}, \tag{1}$$

where $R_0 = \frac{(n-1)^2+\kappa^2}{(n+1)^2+\kappa^2}$ is the reflectance and $\alpha = \left(\frac{4\pi\kappa}{\lambda}\right)$ is the absorption coefficient. Here, the sample thickness is $d = 0.5$ mm, $\kappa$ is the attenuation coefficient, and $\lambda$ is the wavelength. From this result, $\kappa$ as fitting parameter is evaluated to be 0.0018. The THz radiation used to obtain the transmission spectrum of ZnO was generated from an InAs semiconductor surface irradiated with femtosecond pulses under a magnetic field with a spectrum central frequency of 0.3 THz [18, 19]. The absorption coefficient at 0.3 THz were evaluated to be 0.23 $cm^{-1}$. In addition to its high-transmission characteristic in the THz region, ZnO is also transparent in the visible, near-infrared, and mid-infrared regions as shown in Fig. 2. Unlike silicon and germanium, such high transparency suggests the possibility of using this material for integrated hybrid optics and waveguide. We have already demonstrated the THz pigtail that consists of a THz emitter, lens duct, and photonic crystal fiber [16, 20]. For the lens duct, the material should have a high transparency in the near-infrared and THz regions along with a high refractive index. TPX, a polymer based on poly 4-methyl pentene-1, was used as a material of lens duct. The refractive index of TPX is 1.4 in the THz region [16]. It was found that ZnO is better than TPX, which is reasonable since a material with a higher refractive index is more ideal for lens ducts.

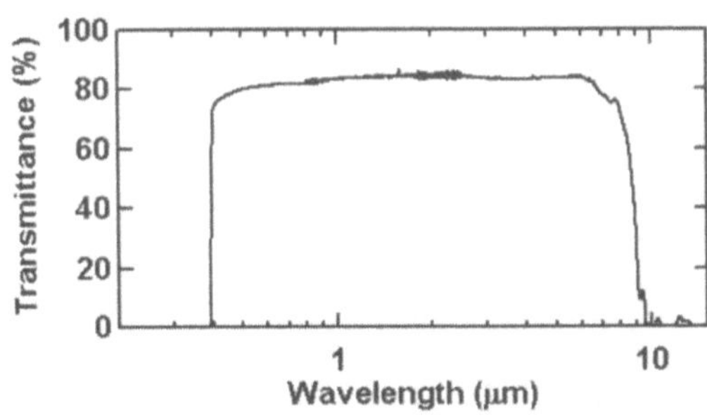


FIGURE 2. Transmission spectrum in the visible, near-infrared, and mid-infrared regions.

The ZnO crystals are of interest as photoconductive material for THz generation due to the ease of fabrication, wide bandgap character, and relatively high mobility and resistivity. In terms of mobility, ZnO (200 $cm^2/V$ s) [21] is one order of magnitude greater than that of polycrystalline diamond (1–70 $cm^2/V$ s) [4]. ZnO also has the capacity to be excited by the second harmonics of a Ti:sapphire laser since its transmission edge is at 390 nm (3.2-eV bandgap). All these qualities make ZnO a suitable candidate for a high power THz emitter.

Figure 3 shows that the THz-radiation power quadratically depends on the electric field. The sample is biased up to 2000 kV/cm and is observed to saturate above 800 V/cm. The saturation is attributed to the increased dark current. It is predicted that the dark current will decrease under high bias voltage and the THz-radiation power will increase as the crystal quality is improved.

Figure 4 is the spectrum of the THz radiation obtained by using a polarizing Michelson interferometer and clearly shows that the spectrum extends up to 1 THz with a peak frequency centered at around 0.35 THz. Clear water vapor absorption lines are observed [22]. The main lines are shown as solid lozenge.

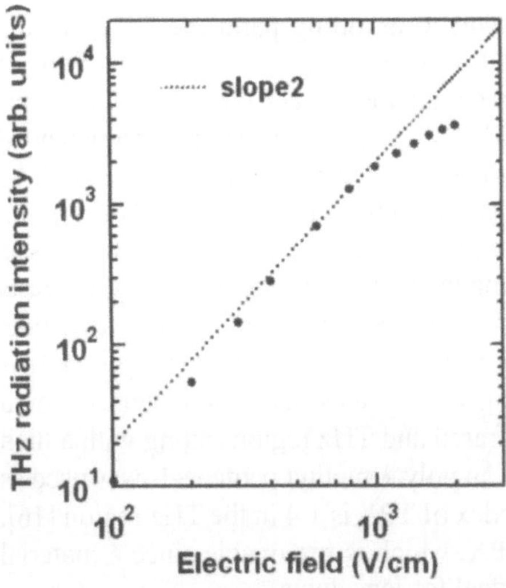


FIGURE 3. THz radiation power dependence on the electric field. Saturation is observed above 800 V/cm.

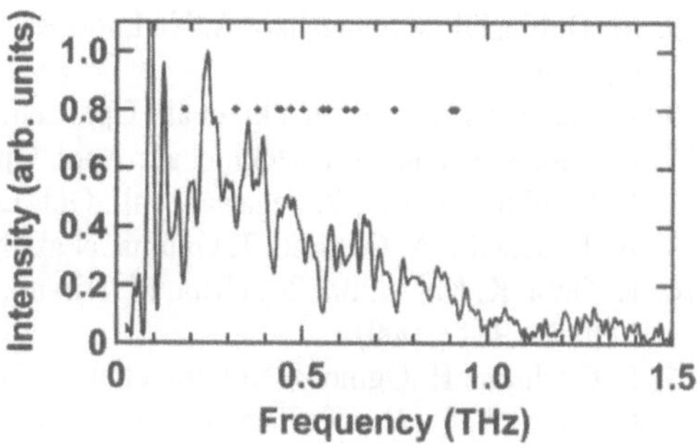


FIGURE 4. THz radiation spectrum from a ZnO photoconductive switch.

## 4 Summary

In conclusion, we have successfully generated THz pulses from a photoconductive antenna fabricated on ZnO single crystals. Using an all-solid-state UV femtosecond laser as the excitation source for ZnO reveals that the THz-radiation power saturates above 800 V/cm and that the spectrum range extends up to 1 THz. Furthermore, the high transparent nature of ZnO is manifested in the near-infrared and THz frequency regions. These observed characteristics could open up the possibility of using ZnO in integrated active optics.

**Acknowledgments.** This research was partially supported by grant-in-aids for Creative Scientific Research, for Young Scientists (B) (16760043), for Creative Scientific Research (17GS0209), for Scientific Research on Priority Areas (16032216), for Creative Scientific Research Collaboration on Electron Correlation (13NP0201), and for Scientific Research (B) (17360029); by Special Coordination Funds for Promoting Science and Aid Technology from the Ministry of Education, Culture, Sports, Science and Technology (MEXT); and a grant-in-aid for JSPS Fellows (16-04077) from Japan Society for the Promotion of Science (JSPS).

## References

1. D. Auston and P. Smith: Appl. Phys. Lett. **43**, 631 (1983).
2. S. Gupta, J. Whitaker, and G. Mourou: IEEE J. Quant. Electron **28**, 2464 (1992).
3. M. Tani, M. Matsuura, and K. Sakai: Appl. Opt. **36**, 7853 (1997).
4. H. Yoneda, K. Tokuyama, K. Ueda, H. Yamamoto, and K. Baba: Appl. Opt. **40**, 6733 (2001).
5. Y. Nabekawa, K. Kondo, N. Sarukura, K. Sajiki, and S. Watanabe: Opt. Lett. **18**, 1922 (1993).
6. Y. Mori, I. Kuroda, S. Nakajima, T. Sasaki, and S. Nakai: Jpn. J. Appl. Phys. **34**, L296 (1995).
7. R. Komatsu, T. Sugawara, K. Sassa, et al.: Appl. Phys. Lett. **70**, 3492 (1997).
8. Y. Suzuki, S. Ono, H. Murakami, et al.: Jpn. J. Appl. Phys. **41**, L823 (2002).

9. M. Dubinskii, V. Semashko, A. Naumov, R. Abdulsabirov, and S. Korableva: J. Mod. Opt. **40**, 1 (1993).
10. Z. Liu, T. Kozeki, Y. Suzuki, et al.: Opt. Lett. **26**, 301 (2001).
11. S. Ono, Y. Suzuki, T. Kozeki, et al.: Appl. Opt. **41**, 7556 (2002).
12. N. Sarukura, Z. Liu, Y. Segawa, et al.: Opt. Lett. **20**, 294 (1995).
13. A. Tsukazaki, A. Ohtomo, T. Onuma, et al.: Nature Mater. **4**, 42 (2005).
14. H. Ohta, K. Kawamura, M. Orita, M. Hirano, N. Sarukura, and H. Hosono: Electron. Lett. **36**, 984 (2000)
15. E. Ohshima, H. Ogino, I. Niikura, et al.: J. Cryst. Growth **260**, 166 (2004).
16. G. Diwa, A. Quema, E. Estacio, et al.: Appl. Phys. Lett. **87**, 151114-1 (2005)
17. M. Born and E. Wolf: *Principles of Optics*. Cambridge: Cambridge University Press, p. 323 (1998).
18. X. C. Zhang, Y. Lin, T. Hewitt, T. Sangsiri, L. Kingsley, and M. Weiner: Appl. Phys. Lett. **62**, 2003 (1993).
19. N. Sarukura, H. Ohtake, S. Izumida, and Z. Liu: J. Appl. Phys. **84**, 654 (1998).
20. M. Goto, A. Quema, H. Takahashi, S. Ono, and N. Sarukura: Jpn. J. Appl. Phys. **43**, L317 (2004).
21. K. Maeda, M. Sato, I. Niikura, and T. Fukuda: Semicond. Sci. Technol. **20**, S49 (2005).
22. H. M. Pickett, R. L. Poynter, and E. A. Cohen: Submillimeter, millimeter, and microwave spectral line Catalog. Jet Propulsion Laboratory, Pasadena, CA. Available at: http://spec. jpl.nasa.gov.

# Action Spectra of GaAs/AlGaAs Multiple Quantum Wells Exhibiting Terahertz Emission Peak at Excitation Energies Below the Bandgap

E. Estacio,[1] A. Quema,[1,3] G. Diwa,[1] G. De los Reyes,[1] H. Murakami,[1] S. Ono,[1] N. Sarukura,[1] A. Somintac,[2] and A. Salvador[2]

[1] Institute for Molecular Science (IMS), Myodaiji, Okazaki 444-8585, Japan
eestacio@ims.ac.jp
[2] National Institute of Physics, University of the Philippines-Diliman, Quezon City 1101, Philippines
[3] Department of Physics, De La Salle University, 2401 Taft Avenue, Manila 1004, Philippines

**Abstract:** We present terahertz (THz) emission of optically pumped 5- and 12.5-nm GaAs/AlGaAs multiple quantum wells (MQWs) even at excitation energies below the bandgap. The effect of the externally applied magnetic field parallel to the MQW plain is to enhance or reduce the THz emission intensity for the up or down direction, respectively. For both MQW samples, the excitation energy corresponding to the peak THz emission is red-shifted with respect to the photoluminescence (PL) and PL excitation peak. This is attributed to an instantaneous bandgap renormalization that occurs at the same time scale as the generation of the THz transients. Moreover, an emission shoulder at ~40 meV below the THz emission peak of the 5-nm MQW was observed. Deep level transient spectroscopy results do not indicate this to be due to electron traps. However, an indistinct LO phonon-related, below-bandgap PL feature was seen at low temperature, which agrees well with the observed THz radiation feature. These results suggest that the THz action spectrum is more sensitive to phonon-mediated processes as compared to more conventional optical spectroscopy techniques.

## 1 Introduction

It has been known that the surface-field terahertz (THz) emission in a semiconductor and in a semiconductor quantum well occurs when a short laser pulse creates electron-hole (e-h) dipoles that are accelerated in opposite directions by the surface electric field. The spatial separation of the e-h pairs and the acceleration of the charges are two accepted mechanisms of THz emission in surface-irradiated semiconductors [1–7]. Other mechanisms include intersubband spontaneous emissions and the excitation of quantum beats in the quantum well subbands [8,9]. In these scenarios, it is always surmised that the energy of the excitation laser pulse is higher than the bandgap of the semiconductor for it to be able to create e-h pairs. In this work, we present results of THz radiation emission from narrow and wide uncoupled GaAs/AlGaAs multiple quantum well (MQW) heterostructures at excitation energies lower than the bandgap. The red-shifted THz emission peak is owed to an instantaneous bandgap renormalization process. Although deep level transient spectroscopy (DLTS) measurements revealed the presence of electron traps in the samples, the calculated trap activation energies failed to corroborate the energetic red shift of the THz peak. Moreover, a THz emission shoulder appearing ~40 meV below the THz peak of the narrow MQW sample is attributed to an LO phonon-assisted emission. Low temperature photoluminescence (PL) spectroscopy affirms this. Even for experiments performed at room temperature, the THz action spectrum appears to be more sensitive to this process than to other conventional optical spectroscopy methods.

## 2 Experiment

The 5-nm and 12.5-nm GaAs/$Al_{0.3}Ga_{0.7}As$ MQW heterostructures were grown via a Riber 32P molecular beam epitaxy (MBE) machine on semiinsulating (carrier density $\sim 10^{15}$ cm$^{-3}$) GaAs (100) substrates. The structure schematics for the MQWs consist of an undoped GaAs buffer layer, followed by 40 periods of 5-nm undoped GaAs QWs with undoped 10-nm AlGaAs barriers, and finally a 10-nm Si-doped ($\sim 10^{17}$ cm$^{-3}$) GaAs cap. For the experiments, the samples were cut into a 2 mm × 2 mm pieces. Initially, PL spectroscopy measurements were performed using an $Ar^+$ laser ($\lambda_e = 488$ nm) as excitation and a 0.5-m spectrometer fitted with a GaAs photomultiplier to disperse and detect the PL signal, respectively. This was done to probe the band structure and to assess the quality of the layers. The THz emission experiments were carried out using a mode-locked, 100-fs-pulsed (86-MHz repetition rate) Ti: sapphire laser as the excitation. The laser was tuned from 1.38 to 1.55 eV (900–800 nm) at 10-meV increments to obtain the THz action spectra. All throughout the experiment, the spectral width, beam-spot diameter, and the optical power of the laser were maintained at 20 meV, ~1 mm, and 160 mW, respectively. A 1-T permanent magnet was used for the externally applied magnetic field oriented parallel to the quantum well plains. The broadband THz radiation was collected, collimated, and focused unto a liquid He-cooled Ge bolometer using paraboloidal mirrors. For selected excitation energies, the time-domain-spectroscopy (TDS) traces for the MQWs were taken. In conjunction with this, room temperature photoluminescence excitation (PLE) measurements using the same optical excitation parameters (i.e., a suitable bandpass filter, and silicon photodiode) was performed. To determine the presence of electron traps, the MQW layers underwent DLTS. The samples were fabricated into arrays of 250-mm-diameter circular mesas, using standard photolithography techniques. The bottom and the top indium contacts were then deposited via thermal evaporation.

Temperature scans were performed at 77 to 370 K. To generate the Arrhenius plots and to ascertain the activation energy of the traps, four frequency scans (10 Hz, 100 Hz, 500 Hz, and 1 KHz) were also done on the sample with a reverse bias of -6 V and a filling pulse of 1 V. Finally, PL spectroscopy at 10 K was performed to verify the optical contribution of these traps.

## 3 Results and Discussions

Shown in Fig. 1a is the action spectrum of the 12.5-nm MQW sample for three different conditions; namely, $B_{up}$, $B_{down}$, and no B as shown in the diagram inset. The magnetic field dependence of the THz action spectrum exhibits an enhancement for the $B_{up}$ case and degradation for the $B_{down}$ case. This is understood in the context of the mechanism of THz radiation in semiconductors. In MQW samples, the main mechanism of THz radiation is the e-h dipole [4]. For low magnetic field strengths (i.e., no significant additional confinement for the carriers is expected) and high excitation fluence, the reorientation of the dipole with respect to the surface plane causes the modification in the efficiency of THz radiation extraction [10].

The observed THz emission peak sits at 10 meV below the calculated bandgap using effective mass approximation methods [11]. This is attributed to an instantaneous bandgap renormalization process and will be discussed later in the text [12]. No excitation fluence dependence studies have been done but at this excitation condition, we expect carrier density-dependent shrinkage of the bandgap to occur (photocarriers are estimated to be of the order of $10^{17}$ cm$^{-3}$ at 1.45-eV excitation) [13]. Interestingly, the room temperature PL and PLE spectra agreed well with the calculated bandgap energy, as shown in Fig. 1b. It must be pointed out that although the PLE spectrum was taken using the same optical excitation conditions as in the THz action spectrum, no bandgap shrinkage was observed.

The THz action spectra for the 5-nm MQW sample are shown in Fig. 2a. Again the excitation energy corresponding to the THz emission peak is redshifted with respect to the PL/PLE data in Fig. 2b. The TDS data for selected excitation energies is also shown in the inset of Fig. 2a. Note that the TDS waveform did not change for excitation energy below and above the bandgap. Although it is not shown here, similar TDS waveforms for different excitation energies above and near the bandgap for the 12.5-nm MQW were also observed. This implies that the THz radiation mechanism did not change for different pump energies. Another interesting result is the appearance of a THz emission shoulder at an excitation energy ~40 meV below the observed THz emission peak in the 5-nm MQW sample.

An instantaneous bandgap renormalization phenomenon is believed to have caused the redshifted THz emission peak for both MQW samples. This discrepancy is explained by considering the difference in the mechanism of luminescence and of THz transient generation. While the THz radiation transients are caused

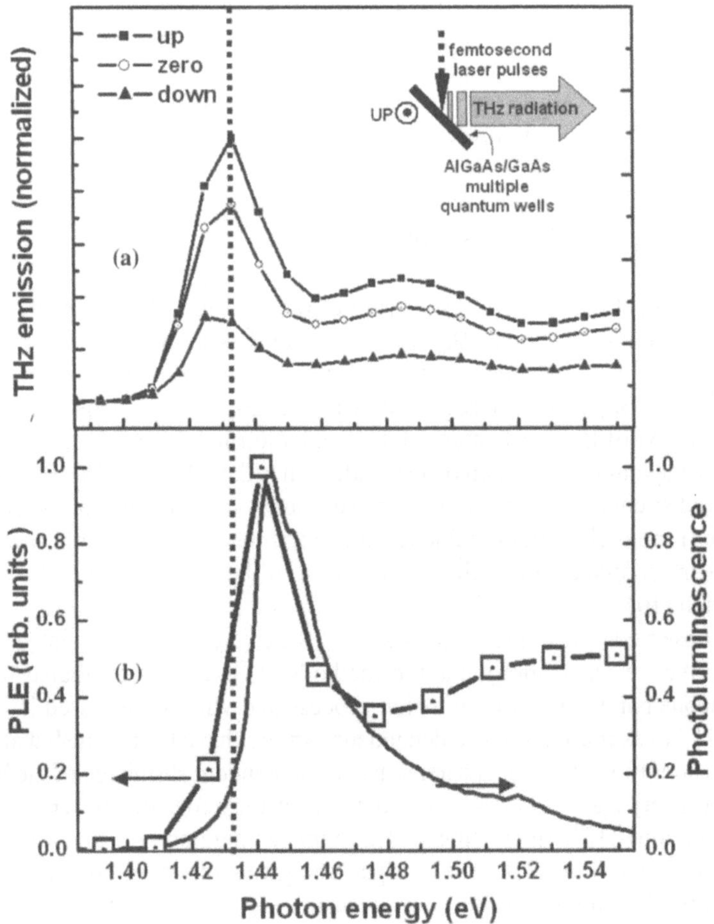


FIGURE 1. **(a)** Terahertz action spectrum of the 12.5-nm MQW sample. The THz emission peak occurs at excitation energy, 10 meV below the bandgap. This is attributed to bandgap renormalization. The inset shows the schematic diagram of the experiment and the orientation of the external magnetic field with respect to the sample. An enhancement or reduction in the THz emission intensity is observed depending on the orientation of the applied magnetic field. **(b)** The corresponding PL and PLE spectra showing good agreement with the expected bandgap value. No bandgap shrinkage was observed for this case.

by the e-h dipole, the luminescence is only observed after the photogenerated electrons have recombined with the holes. This indicates that the two mechanisms occur at different temporal regions. The inset in Fig. 2b shows the time-resolved PL of the MQW sample at room temperature. Clearly, the PL persists, even as long as 1200 ps. The time-integrated PL signal as detected by the Si photodiode did not exhibit the temporal region where the photocarrier density is high. Furthermore,

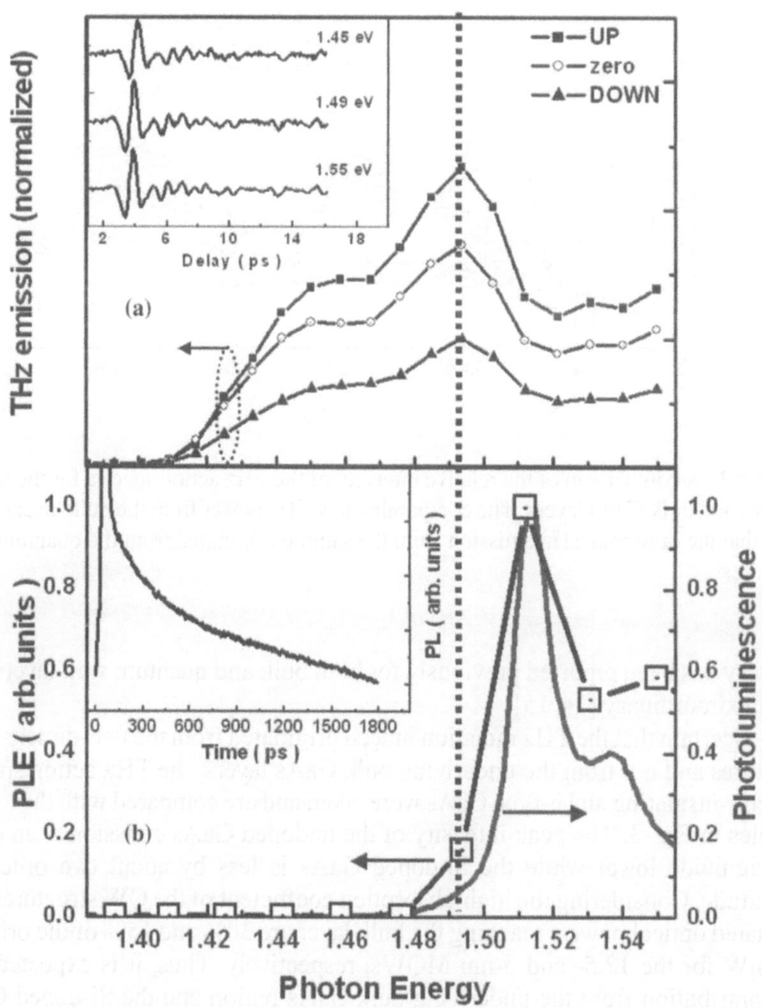


FIGURE 2. **(a)** Terahertz action spectrum of the 5-nm MQW sample. The THz emission peak occurs at excitation energy, 20 meV below the bandgap. This is attributed to bandgap renormalization. The spectral feature at 1.45 eV is owed to an LO phonon-mediated transition. The inset shows the TDS spectra for selected excitation energies. **(b)** The corresponding PL and PLE spectra, again, showing good agreement with the expected bandgap value. The inset shows the time decay of the PL signal.

the PL signal has long decayed before the next optical pulse arrives after ~11 ns (8-MHz repetition rate). This shows that no excess carriers persist in the MQW to cause long-term bandgap renormalization. In contrast to the PL mechanism, the THz transient is caused by the creation of the e-h dipole and disappears after ~9 ps as per the TDS data. This happens during the time when the density of the photogenerated carriers is large. This transient bandgap renormalization and fast

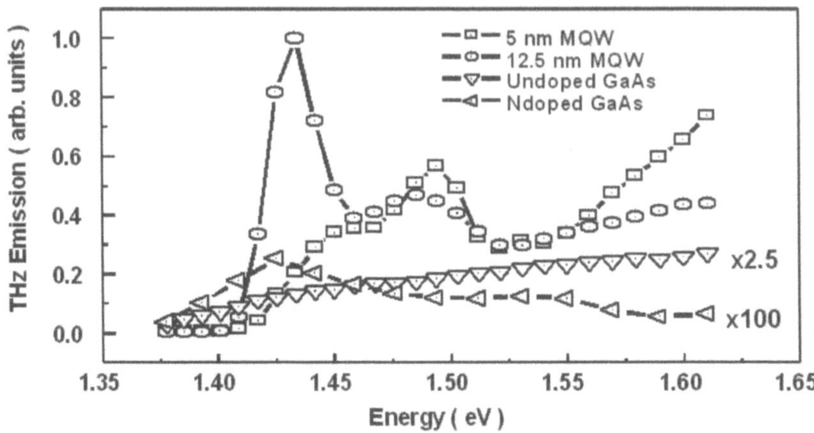


FIGURE 3. A comparison of the relative intensity of the THz action spectra for the MQW samples and bulk GaAs layers. The comparably low THz power from the bulk layers lends proof that the observed THz emission from the sample originated from the quantum well structures.

recovery has been reported previously for both bulk and quantum well layers and is not extraordinary [14, 15].

To ascertain that the THz radiation indeed originated from the two-dimensional structures and not from the underlying bulk GaAs layers, the THz action spectra for semi-insulating and n-type GaAs were taken and are compared with the MQW samples in Fig. 3. The peak intensity of the undoped GaAs emission is an order of magnitude lower while the Si-doped GaAs is less by about two orders of magnitude. Considering the high-absorption coefficient of the QW structures, the estimated optical powers reaching the bulk layer are 30% and 15% of the original 160 mW for the 12.5- and 5-nm MQWs, respectively. Thus, it is expected that the contribution from the underlying bulk GaAs region and the Si-doped GaAs cap should be significantly small for both MQW samples. This strongly suggests that the below-gap THz emission from the MQWs did not originate from the bulk layers underneath the 2D structures.

The presence of deep level traps that may have been incorporated during the growth process was initially surmised to cause the below-bandgap-excited THz emission for both MQW samples. The corresponding Arrhenius plots ($\ln(v/T^2)$ vs $1000/T$) that were derived from DLTS data are shown in Figs. 4a and b. The MQWs exhibited EL5, EL6, EL10, and two other unidentified traps. Two traps are found for the 12.5-nm MQW layer and three for the 5-nm MQW sample. There were two unidentified traps that were detected in addition to the EL5 and EL6 traps that are attributed to Ga vacancies or a complex defect of AsGa [16, 17]. The EL10 trap is a low-energy GaAs trap described by Martin et al. [18]. In addition, the deep levels of a bare GaAs substrate are also studied and it is ascertained that the DLTS signals came from the MBE-grown layers and not from the substrate itself.

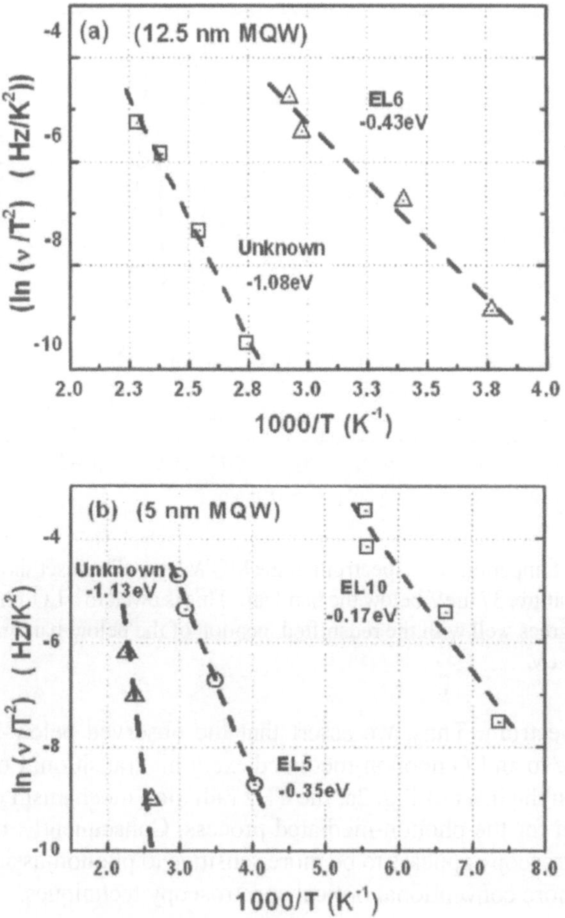


FIGURE 4. Arrhenius plots that were derived from DLTS data of the MQW samples showing five types of traps. The calculated activation energies of the traps are not in the range of the observed below-bandgap excited THz emission for the 5-nm MQW sample.

The activation energy of the traps, however, could not account for the energetic location of the below bandgap THz emission.

To further assess the optical contribution from these traps, PL spectroscopy was performed at 10 K for the 5-nm MQW. Shown in Fig. 5 is the PL peak associated with the first conduction band to heavy-hole band transition. This is in good agreement with the 10-K bandgap of the heterostructure. However, no below-bandgap PL features were observed in the energy region corresponding to the below-bandgap THz emission. A higher resolution PL scan was performed on the low-energy side of the main PL peak and is shown in the inset of Fig. 5. The log plot revealed an indistinct PL feature at an energy, ~37 meV below the excitonic emission peak, which is attributed to an LO- phonon assisted PL emission [19]. This coincides well with the ~40-meV below-bandgap excited feature in the

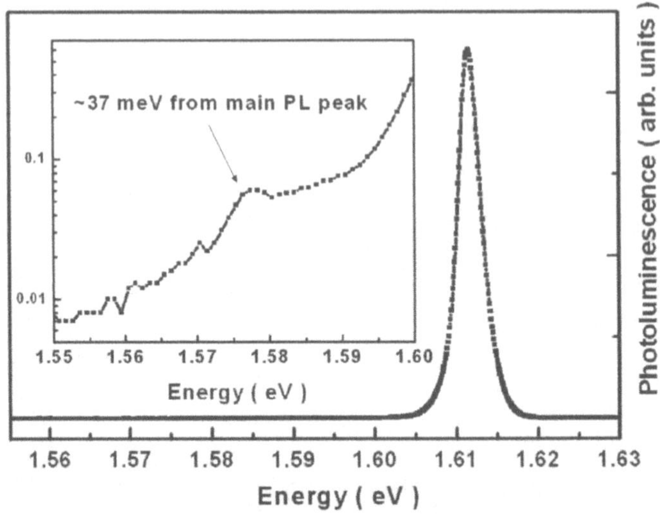


FIGURE 5. Low temperature PL spectrum of the MQW layer. The inset shows an indistinct luminescence feature, 37 meV below the bandgap. This is owed to a LO-phonon-mediated transition and agrees well with the redshifted location of the below-bandgap-excited THz radiation at 1.45 eV.

THz action spectrum. Thus, we assert that the observed below-bandgap THz emission is due to an LO phonon-mediated excitonic transition. Looking back at the TDS data in the inset of Fig. 2a, the THz radiation mechanism did not appear to change even for the phonon-mediated process. Consequently, the data imply that THz spectroscopy appears to be more sensitive to phonon-assisted processes compared to more conventional optical spectroscopy techniques.

## 4 Summary

In summary, the THz action spectrum of 5- and 12.5-nm GaAs/AlGaAs MQW samples exhibited emission even for excitation energies below the bandgap. The THz extraction efficiency is either enhanced or reduced for the $B_{up}$ or $B_{down}$ directions, respectively. Moreover, the pump energy corresponding to the THz emission peak is redshifted by 10 meV for the wide MQW and 20 meV for the narrow MQW as compared to the PL and PLE experiment. This is attributed to an instantaneous bandgap renormalization process that was not sensed by the time-integrated PL and PLE experiments. Furthermore, the appearance of a THz emission feature at 40 meV below the THz peak of the the 5-nm MQW sample is owed to an LO phonon-mediated transition in the MQW structure. This is supported by low-temperature PL spectroscopy. These results suggest that even though the experiments were performed at room temperature, the THz action spectrum of the MQW sample is more sensitive to that of an LO phonon-assisted optical process.

**Acknowledgments.** This research was partially supported by a Grant-in-Aid for JSPS Fellows (16-04077) from Japan Society for the Promotion of Science (JSPS), a Grant-in-Aid for Creative Scientific Research, a Grant-in-Aid for Scientific Research on Priority Areas (16032216), a Grant-in-Aid for Creative Scientific Research Collaboration on Electron Correlation (13NP0201), a Grant-in-Aid for Young Scientists (B) (16760043), and Special Coordination Funds for Promoting Science and Aid Technology from the Ministry of Education, Culture, Sports, Science and Technology (MEXT). A. Salvador thanks DOST-PCASTRD for the research grant. Thanks also to C. Ison, M. Bailon, V. Mag-usara, and J. Mateo for their invaluable contributions to this work.

## References

1. P. Planken, M. Nuss, I. Brener, K. Goosen, M. Luo, S. Chuang, and L. Pfeiffer: Phys. Rev. Lett. **69**, 3800–3803 (1992).
2. J. Cerne, J. Kono, M. Sundaram, A. Gossard, and G. Bauer: Phys. Rev. Lett. **77**, 1131–1134 (1996).
3. N. Sekine, K. Hirakawa, M. Vobebrger, P. Haring Bolivar, and H. Kurz: Phys. Rev. Lett. **64**, 201323-1–201323-4 (2001).
4. B. B. Hu, E. A. de Souza, W. H. Knox, et al: Phys. Rev. Lett. **74**, 1689–1692 (1995).
5. S. L. Chuang, S. Rink, B. I. Greene, P. N. Saeta, and A. F. J. Levi: Phys. Rev. Lett. **68**, 102–105 (1992).
6. T. Liu, K. Huang, C. Pan, S. Ono. H. Ohtake, and N. Sarukura: Jpn. J. Appl. Phys. **40**, L681–L683 (2001).
7. I. Morohashi, K. Komori, S. Wang, T. Hidaka, M. Ogura, and M. Watanabe: Jpn. J. Appl. Phys. **41**, 2710–271340 (2002).
8. A. Maslov and D. Citrin: J. Appl. Phys. **93**, 10131–10133 (2003).
9. H. Liu, C. Song, A. SpringThorpe, and J. Cao: Appl. Phys. Lett. **84**, 4068–4070 (2004).
10. H. Takahashi, A. Quema, R. Yoshioka, S. Ono, and N. Sarukura: App. Phys. Lett. **83**, 1068–1070 (2003).
11. F. Pollak: Properties of III-V quantum wells and superlattices. In: *EMIS Data Reviews Series* Vol. 15. P. Bhattacharya (ed.), England: INSPEC, pp. 232–240 (1996).
12. S. Schmitt-Rink, D. S. Chemla, and D. A. B. Miller: Adv. Phys. **38**, 89–138 (1990).
13. H. Haug and S. Koch: Phys. Rev. A **39**, 1887–1898 (2003).
14. P. W. Juodawlskis and S. E. Ralph: Appl. Phys. Lett. **76**, 1722–1724 (2000).
15. H. Roskos, B. Rieck, A. Seilmeier, W. Kaiser, and G. G. Baumann: Phys. Rev. B **40**, 1396–1399 (1989).
16. H. Shiraki, Y. Tokuda, and K. Sassa: J. Appl. Phys. **84**, 3167–3174 (1998).
17. K. Yokota, H. Kuchii, K. Nakamura, M. Sakaguchi, and Y. Ando: J. Appl. Phys. **88**, 5017–5021 (2000).
18. G. Martin, A. Mittoneau, and A. Mircea: Electron. Lett. **13**, 191–193 (1977).
19. B. V Shanabrook, S. Rudin, T. L. Reinecke, W. Tseng, and P. Newman: Phys Rev. B **41**, 1577–1580 (1990).

# Ultrabroadband Detection of Terahertz Radiation from 0.1 to 100 THz with Photoconductive Antenna

H. Shimosato,[1] M. Ashida,[1] T. Itoh,[1] S. Saito,[2] and K. Sakai[2]

[1] Graduate School of Engineering Science, Osaka University, 1-3 Machikaneyama-cho, Toyonaka, Osaka, 560-8531, Japan
**shimosato@laser.mp.es.osaka-u.ac.jp**

[2] KARC, National Institute of Information and Communications Technology, 588-2 Iwaoka, Nishi-ku, Kobe, Hyogo, 651-2492, Japan

We demonstrate ultrabroadband detection of terahertz electric field from 0.1 up to 100 THz with a dipole-type photoconductive antenna using an ultrashort pulse laser whose pulse width is approximately 10 fs. Ultrabroadband terahertz pulses are generated with a thin GaSe crystal and a low-temperature grown GaAs photoconductive antenna. The sensitivity of the photoconductive antenna is examined in mid-infrared region.

## 1 Introduction

Terahertz time-domain spectroscopy (THz-TDS), which allows us to detect a waveform of electric field by using ultrashort pulse lasers, is a unique method for infrared spectroscopy [1]. There are two detection methods for THz-TDS: electro-optic (EO) sampling and photoconductive (PC) antenna detection. EO sampling utilizes phase change of laser pulses by EO effect (Pockels effect) in a nonlinear optical crystal such as ZnTe or GaSe. On the other hand, the PC antenna detects photocurrent induced by the incident THz electric field gated with laser pulses. Kübler et al. reported the EO sampling detection of the ultrabroadband THz radiation beyond 120 THz [2]. With generation and detection using nonlinear optical crystals, we cannot observe the spectrum in the Restrahlen band due to optical phonons of the crystals (between 5 and 10 THz for typical nonlinear optical crystals). Therefore, PC antennas should be used for both generation and detection for the spectroscopy in this frequency region [3]. Kono et al. reported PC antenna detection up to 60 THz with a ZnTe emitter and a dipole-type PC antenna detector [4]. Recently, we observed THz radiation over 80 THz with a GaSe emitter and a PC antenna detector and suggested the detection sensitivity beyond 100 THz from a model calculation [5]. In the present experiment, we demonstrated ultrabroadband detection from 0.1 up to 100 THz using a dipole-type PC antenna detector in two experiments changing generation methods. We also investigated the sensitivity of the PC antenna for mid-infrared region by calibrating the spectrum of the THz radiation.

## 2 Experiment

We used a mode-locked Ti: sapphire laser whose pulse width was approximately 10 fs at a repetition rate of 78 MHz and center wavelength was 800 nm with 480 mW average power. Figure 1 shows an experimental arrangement and a schematic geometry of PC antennas. Two types of experiments were performed by changing THz generation methods. In the first experiment, a PC antenna with a large aperture composed of two Au electrodes separated by 400 μm fabricated on a low-temperature grown (LT) GaAs substrate was used for generating THz pulses and a 30-μm-long dipole-type antenna with a 5-μm gap at the center fabricated on an LT-GaAs substrate was used for detection. The laser pulse from the Ti: sapphire laser was divided into pump and probe beams with a beam splitter. The pump beam was focused onto the biased gap of the PC antenna emitter with an off-axis parabolic mirror. In this arrangement, the THz radiation emitted backward was

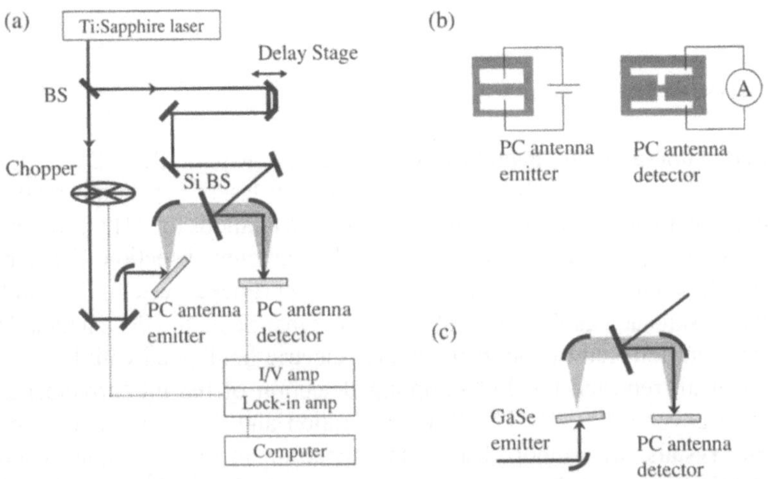


FIGURE 1. (a) Experimental arrangement for the first experiment: PC antenna generation and detection. (b) Schematic geometry of PC antennas. Left: the large aperture PC antenna used for generation. Right: the dipole-type PC antenna used for detection. (c) Modified part of experimental arrangement for the second experiment: generation with a GaSe crystal and detection with the dipole-type PC antenna.

collected to minimize absorption and dispersion in the LT-GaAs substrate [6]. After transmitting a Si beam splitter, the emitted THz radiation was focused onto the PC antenna detector with an off-axis parabolic mirror. On the other hand, after being passed through optical delay line, the probe beam was reflected with the Si beam splitter and focused onto the PC antenna detector. Detected photocurrent was amplified with a current amplifier and fed to a lock-in amplifier referenced to an optical chopper whose modulation frequency was 2 kHz. In the second experiment, experimental arrangement was almost the same as that of first one. Only the part of emitter arrangement was changed as shown in Fig. 1c. Instead of the PC antenna emitter, a z-cut GaSe crystal with thickness of 30 μm was used for generating ultrabroadband THz pulses by optical rectification [2]. We used the same dipole PC antenna as in the first experiment for detection.

# 3 Results and Discussion

Figures 2a and b show a time-domain waveform of the THz electric field and its Fourier transformed power spectrum in the first experiment, emitted and detected with different types of PC antennas. In the frequency domain, except around 8 THz, monotonously decreasing spectral distribution from 0.1 to 25 THz was observed. The spectral features around 8 THz were caused by the reflection of the THz radiation at the boundary between an LT-GaAs substrate and the air. A number of sharp dips seen in Fig. 2b were due to the water vapor absorption in the air. To avoid this absorption, the whole apparatus should be put in a box purged with dry gas. Shen et al. reported the observation of the broadband THz radiation over 15 THz, and it has been the highest frequency so far reported with PC antennas for both generation and detection [3]. The difference between their experimental arrangement and ours is the shape of the PC antenna detector. They used a bow-tie PC antenna as a detector, while we used the dipole-type PC antenna. The sizes of the antennas are also different.

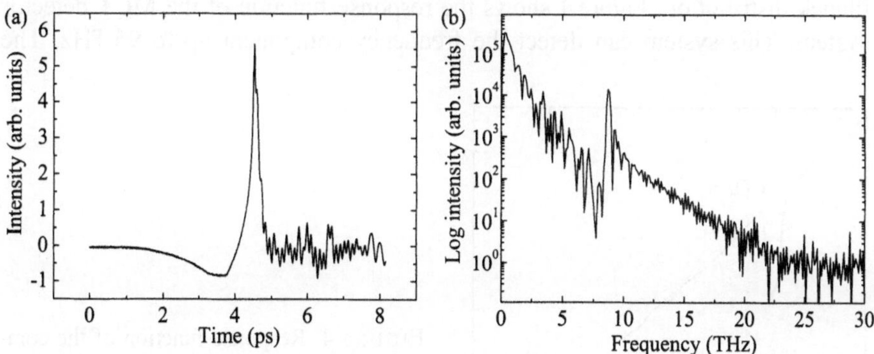


FIGURE 2. (a) Temporal waveform of the THz electric field generated and detected with PC antennas (b) Fourier transformed power spectrum of the THz electric field shown in (a).

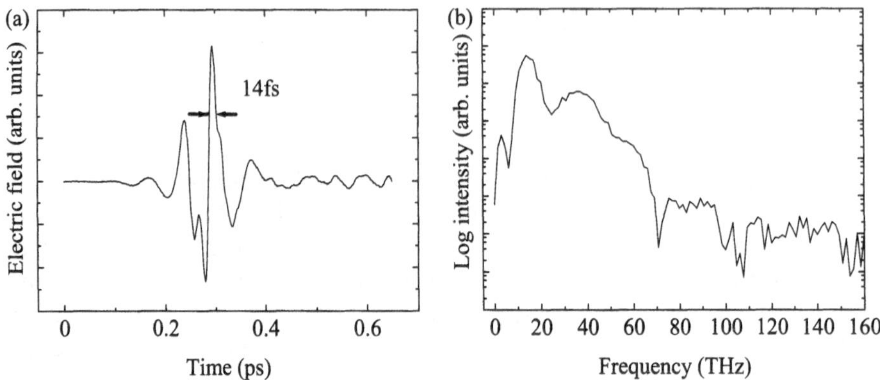


FIGURE 3. (a) Temporal waveform of the THz electric field generated with GaSe crystal and detected with PC antenna (b) Fourier transformed power spectrum of the THz electric field shown in (a).

Figure 3a shows a time-domain waveform of the THz electric field emitted from the GaSe crystal and detected with the PC antenna. This waveform includes an ultrafast frequency component, e.g., the width of the shortest peak is 14 fs as shown in the figure. Figure 3b shows the Fourier transformed power spectrum of the THz electric field in logarithmic scale. Ultrabroadband spectrum of the THz radiation extends up to 100 THz. The spectral peak at 15 THz and the broad dip between 20 and 30 THz originates from the phase-matching condition of the GaSe crystal. The sharp dip around 70 THz is due to the absorption of $CO_2$ and the phase-matching condition. The steep decrease below 10 THz is caused by the phonon absorption of the GaSe crystal. In order to discuss the sensitivity of the PC antenna, we performed the calibration of the spectrum of the THz radiation emitting from the GaSe crystal with a combination of a monochromator and an MCT (HgCdTe) detector. The calibrated spectrum from the GaSe crystal was estimated as follows. First, the response function of this detection system was estimated by measuring a black-body radiation at 500°C and comparing the obtained spectrum with the Planck distribution. Figure 4 shows the response function of the MCT detector system. This system can detect the frequency component up to 95 THz. The

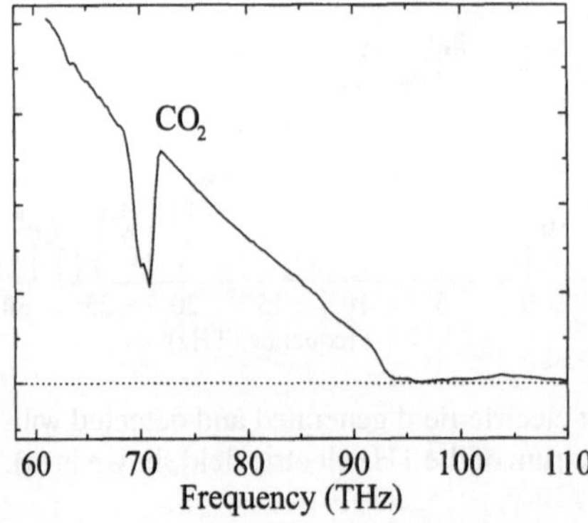


FIGURE 4. Response function of the combination of monochromator and MCT detector. Sharp dip around 70 THz is attributed to the absorption by $CO_2$. The dashed line indicates the noise level.

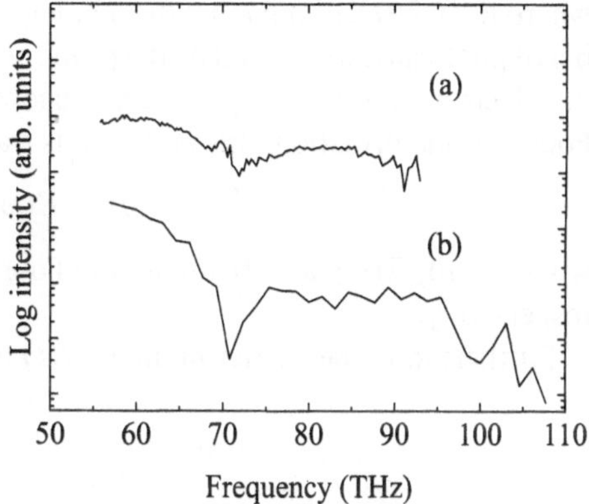


FIGURE 5. (a) Calibrated power spectrum of the THz radiation with a monochromator and an MCT detector. (b) Power spectrum of the THz radiation with the PC antenna extracted from Fig. 3b.

response around 70 THz is distorted by the absorption by $CO_2$ in the air. Using this response function and the spectrum of the THz radiation from the GaSe crystal detected with the MCT detector, we obtained the calibrated spectrum from the GaSe emitter as shown in Fig. 5a. The power spectrum detected with PC antenna extracted from Fig. 3b in the corresponding region is also shown in Fig. 5b. It should be remarked that the calibrated spectrum around 70 THz, which may be distorted by the absorption of $CO_2$, is unreliable. However, the spectral shape with the PC antenna detection is similar to the calibrated one.

With the spectrum of the THz radiation detected with the PC antenna and the calibrated one, we estimated the sensitivity of the PC antenna in the frequency region between 50 and 90 THz. Figure 6 shows the sensitivity obtained from the experiments and estimated from a model calculation as follows. The photocurrent by the incident THz radiation at a time delay $t$ is described by the following equation [7]:

$$J(t) = e\mu \int_{-\infty}^{\infty} E(t')N(t'-t)dt', \tag{1}$$

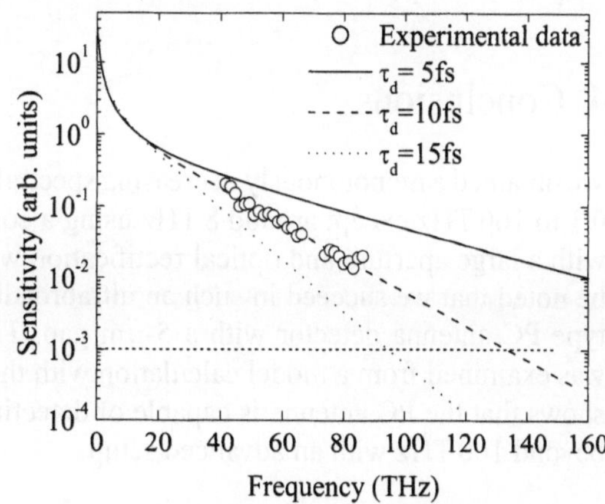


FIGURE 6. Sensitivity of the PC antenna estimated from the experiment and from the model calculation for different probe pulse width $\tau_d$.

where $E(t)$, $N(t)$, $e$, and $\mu$ are the incident electric field of the THz radiation, number of photocarriers excited by the probe beam, the elementary electric charge, and the electron mobility, respectively. According to the convolution theorem of the Fourier transformation, this equation is transformed into the following equation:

$$J(\omega) \propto E(\omega) \cdot N(\omega), \tag{2}$$

where $J(\omega)$, $E(\omega)$, and $N(\omega)$ are the Fourier transforms of $J(t)$, $E(t)$, and $N(t)$, respectively.

Furthermore, the carrier number $N(t)$ is described by the following equation:

$$N(t) \propto \int_{-\infty}^{\infty} G(t')D(t' - t)dt', \tag{3}$$

where $G(t)$ and $D(t)$ are the shape of the probe laser pulse, time response of carriers of the LT-GaAs substrate, respectively. We assume that $G(t) \propto \mathrm{sech}^2\,(1.76t/\tau_d)$ with the pulse width of $\tau_d$. $D(t)$ is estimated from the transient reflection measurement. Equation (3) is transformed into the following equation:

$$N(\omega) \propto G(\omega) \cdot D(\omega), \tag{4}$$

where $G(\omega)$ and $D(\omega)$ are the Fourier transform of $G(t)$ and $D(t)$, respectively. Thus, we obtain the photocurrent from the PC antenna as the following equation:

$$J(\omega) \propto E(\omega) \cdot D(\omega) \cdot G(\omega). \tag{5}$$

The term $G(\omega) \cdot D(\omega)$ in Eq. (5) corresponds to the sensitivity of the PC antenna. Experimental data in Fig. 6 plotted except around 70 THz show the smooth spectral response. The resonant structure due to the shape of the antenna was not observed. The slope of the sensitivity curve from the experiment is similar to that of the calculated one in the case of $\tau_d = 10$ fs. The pulse width is consistent with that of the laser in the present experiment. This result indicates the feasibility of ultrabroadband detection over 130 THz using the PC antenna if broader and stronger THz radiation is obtained, because the horizontal dashed line indicates the noise level in the present experimental setup. Moreover, according to the calculated curve $\tau_d = 5$ fs, the PC antenna has the possibility to detect much higher frequency components, if a shorter probe laser pulse is utilized.

## 4 Conclusions

We obtained a monotonously decreasing spectral response of the PC antenna from 0.1 to 100 THz except around 8 THz using a combination of PC antenna emitter with a large aperture and optical rectification with a thin GaSe crystal. It should be noted that we succeed in such an ultrabroadband detection using one dipole-type PC antenna detector with a 5-μm gap. The sensitivity of the PC antenna was examined from a model calculation with the calibrated spectrum. The result shows that the PC antenna is capable of detecting ultrabroadband THz radiation beyond 100 THz with an advanced setup.

# References


1. B. Ferguson and X-C Zhang: Nat. Mater. **1**, 26 (2002).
2. C. Kübler, R. Huber, S. Tubel, and A. Leitenstorfer: Appl. Phys. Lett. **85**, 3360 (2004).
3. Y. C. Shen, P. C. Upadhya, H. E. Beere, et al.: Appl. Phys. Lett. **85**, 164 (2004).
4. S. Kono, M. Tani, and K. Sakai: IEE Proc. Optoelectron. **149**, 105 (2002).
5. M. Ashida, A. Doi, H. Shimosato, S. Saito, K. Sakai, and T. Itoh: IQEC/CLEO-PR JFH1-2 (2005).
6. Y. C. Shen, P. C. Upadhya, E. H. Linfield, H. E. Beere, and A. G. Davies: Appl. Phys. Lett. **83**, 3117 (2003).
7. S.-G. Park, M. R. Melloch, and A. M. Weiner: Appl. Phys. Lett. **73**, 3184 (1998).

# Improvement of the Photoluminescence Decay Response Characteristics of an Oxide-confined Vertical Cavity Surface Emitting Laser Probed by Femtosecond Laser Pulses

F. D. Recoleto Jr., J. N. Mateo, M. S. Dimamay, E. S. Estacio, A. S. Somintac, and A. A. Salvador

National Institute of Physics, University of the Philippines, Quezon City, Philippines
**florencio.recoleto@up.edu.ph arnels@nip.upd.edu.ph**

**Summary.** Room temperature time-resolved photoluminescence decay measurements have been performed on oxidized and unoxidized vertical cavity surface emitting lasers. The oxidized device showed faster photoluminescence (PL) decay time of 370 ps compared to the 1200 ps of the unoxidized device. Oxidation of the AlAs layers brought a reduction of the aperture area, a shift in the cavity resonance wavelength relative to the unoxidized device, and the enhancement of the standing wave effect inside the microcavity resulting to the increase in the intensity of the optical field. These factors increase the transition probability of electron hole recombination resulting in the fast PL decay time.

## 1 Introduction

Vertical cavity surface emitting lasers (VCSELs) have been developed with unique designs [1]. They have several well-known characteristics that differ with other types of lasers, including small beam divergence, low threshold current, ease of integrability to fiber optics, and testability. Driven by a wide variety of applications in optical communications, such as optical interconnects, high-speed local area networks, and optical signal processing, VCSEL technology has made significant progress [2].

Previous works on time resolved photoluminescence (TRPL) spectroscopy for VCSELs concentrated mainly on the measurements of the carrier lifetime when the quantum wells constituting the active region inside the microcavity are properly arranged to achieve enhancement of the spontaneous emission [3, 4]. In the present state-of-the-art VCSEL design, where the threshold current is lowered by selectively oxidizing AlAs layers imbedded in the VCSEL heterostructure, no optical technique on the TRPL measurements of the carrier lifetime has been reported. Differential carrier lifetime in oxide-confined VCSEL has been reported but it is obtained by electrical technique thru impedance measurements [5]. The method of oxidizing AlAs layers confines the current within the unoxidized region only, thus reducing the aperture size of the laser structure [6, 7].

Decay time measurement is an important tool in semiconductor laser spectroscopy, because it yields information on the speed characteristic of the grown device. In this paper, we present results on the photoluminescence (PL) decay time comparison of an oxidized GaAs/AlGaAs VCSEL with its unoxidized counterpart. Results show that the oxidized VCSEL device has a faster PL decay time than the unoxidized device.

## 2 Methodology

The VCSEL heterostructure was grown by a Riber32P Molecular Beam Epitaxy (MBE) machine. The nominal growth design is a *p–i–n* structure consisting of a 1-μm-thick Si-doped GaAs buffer grown initially on a (100) *n*-GaAs substrate. This was followed by 24 pairs of Si-doped AlAs (692 Å)/$Al_{0.2}Ga_{0.8}As$ (593 Å) quarter-wave distributed Bragg reflector (DBR) pairs, and a half-wave cavity spacer consisting of undoped $Al_{0.2}Ga_{0.8}As$ layers. The active region is a 90 Å GaAs/$Al_{0.2}Ga_{0.8}As$ single quantum well (SQW) structure. A thick quarter-wave *p*-doped AlAs (2077 Å) layer was grown just above the active region (oxidation layer), a *p*-type DBR consisting of 24 pairs of Be-doped AlAs (692 Å)/ $Al_{0.2}Ga_{0.8}As$ (593 Å) quarter-wave pairs and a 1000 Å Be-doped GaAs capping layer.

Device fabrication was carried out via standard UV-photolithography and developing techniques. The wafer surface was processed into arrays of 250-μm-diameter mesas with *p*-ring contacts allowing a 135-μm window. The *p*-type contact is composed of AuZn/Ti/Au with 1000 Å/100 Å/1000 Å thickness. Prior to the *n*-metallization, one batch was then oxidized leaving an inner 30-μ diameter, while the other batch was left unaltered. This was confirmed by taking Scanning Electron Microscope (SEM) micrographs of the cross-section of the device. *N*-type contacts of AuGe/Ti/ Ni, with the same thickness as the *p*-metal alloy, were employed. Figure 1 shows the schematic diagram of the device after undergoing fabrication process.

PL spectra and TRPL measurements were performed using a Ti: sapphire mode-locked 100-fs laser pumped at 795 nm at a repetition rate of 86 MHz. Low group velocity dispersion optics were used to steer the beam at normal incidence to the device. The scattered light, which constitutes the PL emission, were collected and focused unto a spectrometer after which PL images configured in temporal and spatial (wavelength) axis were captured by an Optronis SCMU-ST synchroscan streak camera. All TRPL measurements were done at the detection

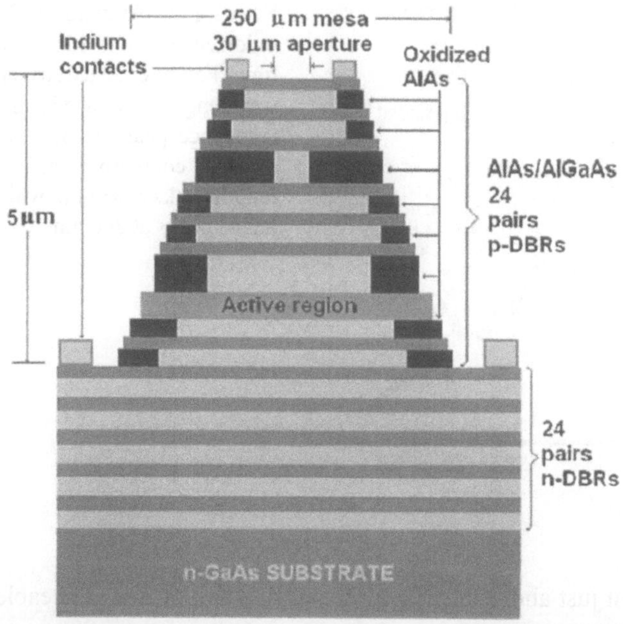


FIGURE 1. Schematic diagram of the VCSEL device after fabrication. Oxidation of the thick AlAs layers reduces the diameter of the aperture to 30 μm. The sloped device edge is characteristic of the fabrication etching process.

wavelength corresponding to the PL spectral peaks. The emission spectra or electroluminescence (EL) for the devices were obtained using a Hamamatsu GaAs photomultiplier and dispersed by a 0.5-m spectrometer fitted with a 1200 gr/mm grating blazed at 750 nm. The EL was taken at room temperature with an input current of 3.1 mA.

## 3 Results and Discussions

Figure 2 shows the normalized PL spectra of the oxidized and unoxidized VCSEL devices. The two peaks correspond to the resonance and off resonance characteristics of the two devices. Off resonance wavelength peaks were observed because the reflective zone of the DBR mirrors ranges from 810 nm to 870 nm. The oxidized VCSEL device has its PL resonance peak which is observed at about 831 nm, while the unoxidized counterpart has its resonance peak at 833 nm. This shows that the oxidized cavity resonance is blue shifted by ~2 nm (20 Å) relative to the unoxidized cavity resonance wavelength. It is well known that reducing the aperture radius causes a blue shift [8]. According to the literature [9], the magnitude of the shift is reduced when the oxide layer in the DBR is inserted away from the active region. For our VCSEL structure, the AlAs layers (oxidation region)

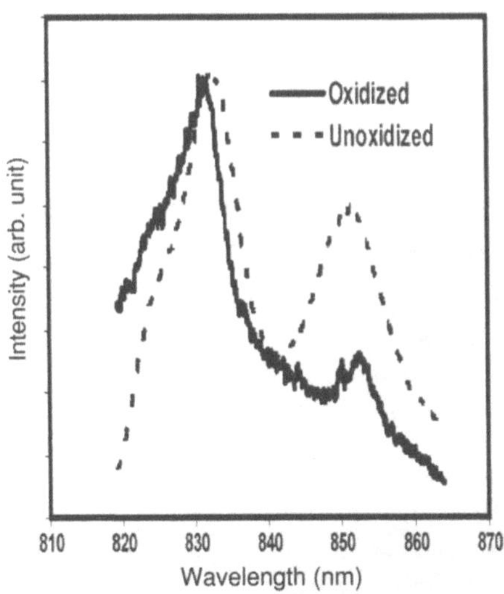


FIGURE 2. Normalized PL spectra for oxidized and unoxidized VCSEL structures. For the oxidized structure, resonance wavelength peaks at 831 nm while the off resonance peaks at 851 nm. For its unoxidized counterpart the resonance wavelength peaks at 833 nm while the off resonance peaks at 846 nm.

were grown just above the active region; thus the shift is noticeable as depicted in the PL spectra.

The PL spectra of the oxidized and unoxidized VCSELs correspond to the EL spectra of the samples as depicted in Fig. 3. As shown, the EL of the oxidized and unoxidized devices has approximately the same peaks as the PL for both the resonance and off-resonance wavelengths. The measured FWHM is ~40 Å at the resonant wavelengths, and this fairly agrees with the FWHM for an idealized RCE/VCSEL device which is ~30 Å according to the literature [10].

The TRPL profiles at the resonance wavelength peaks are shown in Fig. 4. The same was done for the off-resonance peaks. Using an exponential decay fitting,

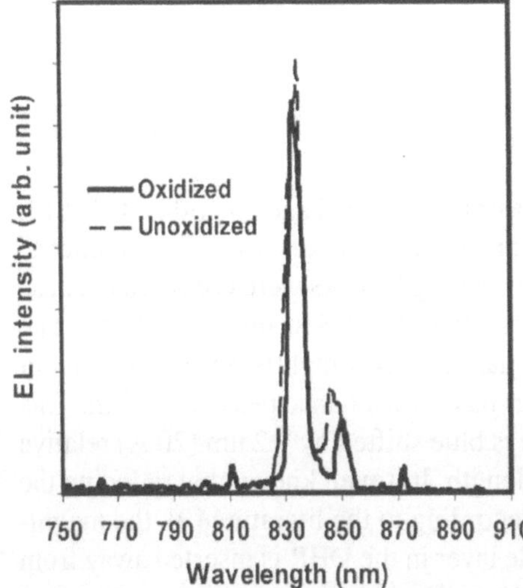


FIGURE 3. Normalized EL spectra of the oxidized and unoxidized devices. For the oxidized device, resonance and off-resonance wavelengths are observed at 831 nm and 851 nm, respectively. EL spectra corresponds approximately to that of the PL spectra of the devices.

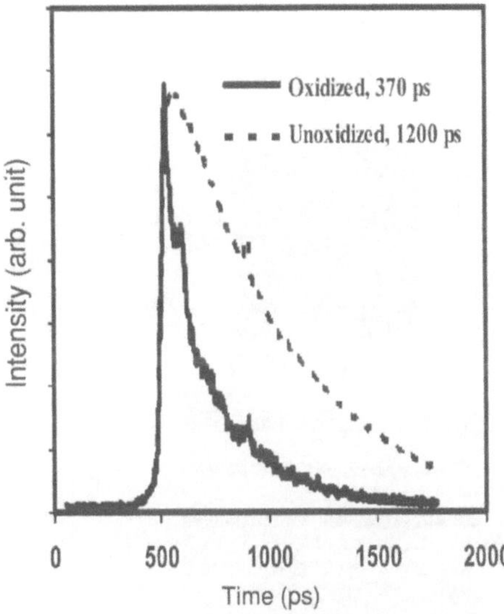


FIGURE 4. Time-resolved photoluminescence of the oxidized and unoxidized VCSEL structures detected at the resonant peaks. The oxidized device showed faster decay time of 370 ps compared to its unoxidized counterpart which is 1200 ps.

the oxidized device showed a faster decay time of about 370 ps, compared to the unoxidized counterpart which is about 1200 ps. The faster PL decay time for the oxidized device can be ascribed to the increase in magnitude and better confinement of the optical field inside the microcavity, since it underwent aperture reduction. The transition probability of electron hall recombination W is increased because it is dependent on the optical-field intensity $|E|^2$. Since the decay time (carrier lifetime) is inversely proportional to the transition probability, a faster decay time will result.

To account for the difference in the optical-field intensity inside the cavity for the two devices, a simulation of the optical-field intensity in the active region for the oxidized and unoxidized VCSEL devices was done. Figure 5 shows the different regions being considered for the two devices. The edges of the devices were approximated to be vertical so that the sloped device edge, which is a characteristic of the fabrication-etching process, is assumed to have negligible effect in the calculation of the optical-field intensity. As shown, the unoxidized device has only one cylindrical region for the top DBR layers (Region 1), while the oxidized device has two regions: one for the unoxidized cylindrical region (Region 1) which corresponds to the aperture and the other for the oxidized cylindrical shell region (Region 2). To derive the expression of the optical fields for the two devices, we considered the models to be cylindrical waveguides and employ an optical fiber type problem [11]. Starting from Maxwell's curl equations in cylindrical coordinates, the differential equation to be satisfied by the fields for the cylindrical geometry is given by

$$\frac{1}{r}\frac{\partial}{\partial r}\left(r\frac{\partial \Psi}{\partial r}\right)+\frac{1}{r^2}\frac{\partial^2 \Psi}{\partial \phi^2}+\beta_c^2\Psi=0,$$

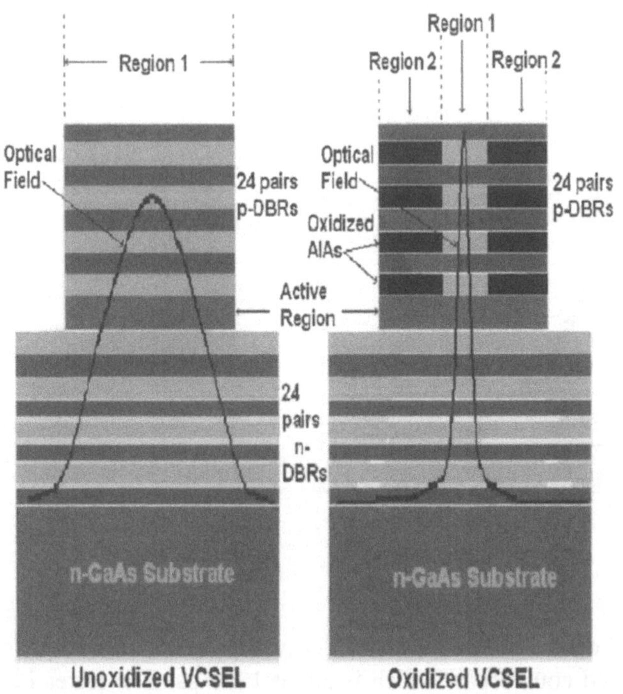


FIGURE 5. Cross-sectioning of the VCSEL models. The unoxidized device is modeled as having only one region labeled as Region 1, while the oxidized device has two regions labeled as Region 1 and Region 2. Sketch of the optical fields are shown for both models to indicate the expected outcome of the simulation.

where $\psi$ stands for $E_z$ in the case of TM waves or $H_z$ in the case of TE waves, r is a variable representing the radius of the cylinder, and $\beta_c = \beta^2 - \beta_z^2 = \omega^2\mu\varepsilon - (2\pi/\lambda_g)^2$. The solution for the $r$-variation in the core and the cladding regions is a superposition of the Bessel functions so that the longitudinal and transverse components are derived. Furthermore, invoking E = 0 at the boundary conditions (i.e., the tangential components of the electric field for the core and cladding regions at the interface are equal—same is true for the normal components) resulted to an expression for the axial component of the optical field as a function of the aperture radius. Figure 6 shows the simulated results of the axial distribution of the optical field, which indicated a much higher optical-field intensity for the oxidized device than its unoxidized counterpart. The optical field inside the cavity for the oxidized device is also squeezed to a smaller region in the aperture indicating a better confinement. These findings can be ascribed to the reduction of the aperture area that increases the optical-field interaction inside the cavity causing stronger diffraction effects.

The fast PL decay time characteristic for the oxidized device can then be attributed to the reduction of the aperture area which increases the intensity of the optical field inside the cavity. This optical field increases the transition probability

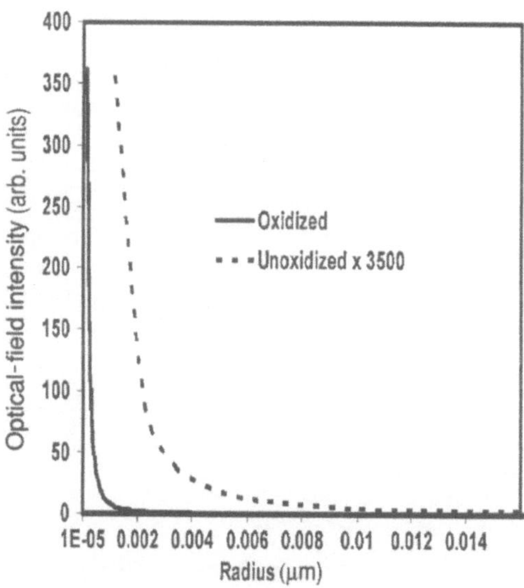


FIGURE 6. A simulation of the optical field intensity for the unoxidized and oxidized VCSEL devices. The optical field intensity for the oxidized device is 3,500 times greater than the unoxidized device and it is also squeezed to a smaller region.

of electron-hole recombination, resulting to a faster decay time characteristic for the oxidized device. This clearly shows that oxidation brings an improvement to the performance of the device ensuing information of the increased speed characteristic for the oxidized VCSEL heterostructure.

## References


1. K. Iga, S. Ishkawa, S. Oukouchi, et al.: IEEE J. Quan. Elec. **6**, 1201–1215 (2000).
2. K. M. B. Sinclair, P. L. Ourley, T. M. Brennan, et al.: Appl. Phys. Lett. **66**, 662–664 (1995).
3. H. Yokohama, K. Nishi, T. Anan, et al.: Appl. Phys. Lett. **57**, 2814–2816 (1990).
4. T. Yamauchi, Y. Arakawa, M. Nishioka: Appl. Phys. Lett. **58**, 21 (1991).
5. G. E. Giudice, D. V. Kuksenkov, H. Temkin, et al.: Appl. Phys. Lett. **74**, 899–901 (1998).
6. J. M. Dallesasse, N. Holonyak, A. R. Sugg, et al.: Appl. Phys. Lett. **57** , 2844–2846 (1990).
7. J. M. Dallessase, N. Holonyak: Appl. Phys. Lett. **58**, 4 (1991).
8. M. J. Noble, J. H. Shin, K. D. Choquette: IEEE Phot. Tech. Lett. **10**, 475–477 (1998).
9. K. D. Choquette, K. L. Lear, R. P. Schneider, et al.: Appl. Phys. Lett. **66**, 25 (1995).
10. M. S. Unlu, S. Strite: J. Appl. Phys. **78**, 607 (1995).
11. N. N. Rao: Elements of Engineering Electromagnetics, 5th edn. New Jersey: Prentice Hall. pp. 615–618 (2000).

# Part XI

# X-ray Lasers

# Investigations on a Soft X-Ray Laser in the GRIP Geometry

K. A. Janulewicz,[1] J. Tümmler,[1] H. Stiel,[1] H. Legall,[1]
A. A. Bjeoumikhov,[2] N. Langhoff,[2] and P. V. Nickles[1]

[1] Max-Born-Institut fr Nichtlineare Optik und Kurzzeitspektroskopie, Max-Born-Strasse 2A, D-12489 Berlin, Germany
**nickles1@mbi-berlin.de**

[2] IfG—Institute for Scientific Instruments GmbH, Rudower Chaussee 29/31, D-12489 Berlin, Germany
**bjeoumikhov@ifg-adlershof.de**

**Abstract:** Soft X-ray laser have recently shown a dramatic reduction in the pump energy needs. Grazing incidence pumping geometry constitutes the last achieved milestone on the way to real repetitive soft X-ray lasers. We present the results of the experimental investigation on such a laser scheme using Mo as a slab target. The preplasma conditions are discussed in detail and compared with those of other experiments. Estimates of some of the output parameters and the development prospects for the new scheme are given as well. Additionally, the results of the measurements on wavelength nonselective capillary XUV optics are presented.

## 1 Introduction

Soft X-ray lasers have demonstrated enormous progress regarding reduction in the pump parameters during the last decade after demonstration of the transient inversion scheme [1–6]. Firstly, saturation at the wavelengths below 20 nm has been achieved with the energies as low as 2–3 J in a slab target geometry using as a lasant an abundance of Ni-like ions. This enabled some matured applications, especially in the short-wavelength interferometry [7, 8]. However, this arrangement suffered from nonfavorable energy deposition in the medium. Most of the delivered energy was stored in the high-density area close to the critical surface, and thus it was lost for the generation process. The latter was caused by the hindering effect of strong refraction. All this turned much attention to the media pumped by laser beams incident longitudinally or obliquely to the target surface/axis [9–11]. However, these arrangements suffered from limited length of the active medium caused by strong absorption of the pump radiation. Finally, a new pump geometry termed GRazing Incidence Pumping (GRIP) has been recently proposed and demonstrated [12, 13]. This scheme offers control over the position of the excitation volume by the choice of the grazing incidence angle. Generally, this volume is shifted from the target surface in the direction of lower (ne$\sim 10^{20}$ cm$^{-3}$) densities. Finally, this scheme, even if definitely not more efficient, has enabled lasing at the pump energy well below 1 J [4–6]. Moreover, the very low pump energy required has opened the possibility to use repetitive commercially available laser drivers working at 10 Hz. The scheme is in the beginning of its optimization, and the full applicable range of the pump parameters is under investigation. The most recent results reported in rocca05 show saturation down to a wavelength of 13.2 nm the amplification very close to saturation at 11.9 nm. The limited output of these lasers is easily compensated by opening of new possibilities in the scheme development by arrangement of a master-oscillator power-amplifier system in soft X-ray region (XMOPA). The XMOPA using an ultrashort pulse of high harmonics as a seed improves enormously the extraction efficiency and delivers well-defined and variable output parameters. The sketched development direction of the XRLs increases the chance to exceed a value of $10^{12}$ W $\cdot$ cm$^{-2}$ treated as a threshold for nonlinear optics in the soft X-ray region, one of the most interesting application for these devices. Achievement of the intensities in excess of this threshold value turns the attention to the beam focusability and searching for new devices facilitating it.

## 2 Experiment

The experimental arrangement of a GRIP scheme under investigation consisted of a Ti:sapphire laser delivering the pump pulses with a length between 3 and 9 ps and an energy of 400–600 mJ. These pulses were used for heating of a preplasma created by a 400–500-ps pulse with an energy of 200 mJ outcoupled from the laser system before compression. This radiation irradiated the target in the normal direction (Fig. 1). The width of the line focus was equal to 50 μm with a length of 7 mm. Delay between both pump pulses was optimized for each pump energy.

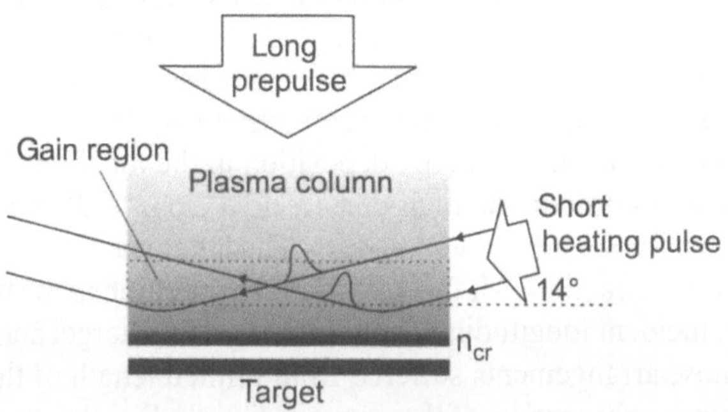


FIGURE 1. Irradiation geometry and the excitation principle in a GRIP experiment [6].

Molybdenum was used as a target material. The spectrum of the ouptut signal was obtained with a flat-field spectrometer with a Haradas grating of variable grooves spacing. As lasing is not the ultimate goal of the experiment and real applications have to confirm feasibility of such a source, we have characterized in the second part of the experiment some specific capillary optics. Such optics elements have the advantage to be wavelength nonselective, and then changing of the emission wavelength requires only replacing of the target. An incoherent source emitting about 13.4 nm was used in the tests.

# 3 Results

We have demonstrated strong and irrefutable lasing at 18.9 nm with a repetition rate of 10 Hz. Figure 2a shows the lasing line dominating the output spectrum. However, the laser worked in unsaturated regime contributing to the output signal fluctuations (Fig. 2b). The sequence of output pulses presented in Fig. 2b was registered at a repetition rate of 1 Hz due to limited acquisition rate of the CCD camera used. We have found that the pump pulse intensity, especially in the case of the prepulse, is of fundamental importance for the plasma creation and excitation. This was a result of analysis supported by simple numerical modeling of all experiments performed in the GRIP geometry. The basic preplasma parameters as the electron temperature, average ionization stage, and density were analyzed (see Fig. 3). It is seen that a low-intensity level of the preforming causes the average inonization stage and the corresponding density to be too low to support strong lasing. On the other hand, XMOPA using seeding by high harmonics (Fig. 4) offers much higher energy extraction efficiency as the whole length of the storage medium is traversed by an amplified signal at the saturation level. A simple estimate, using the formula $E_{STORED} = g_0 E_S V$ for the energy available in the medium, shows that under typical conditions we can expect up to 100 μJ

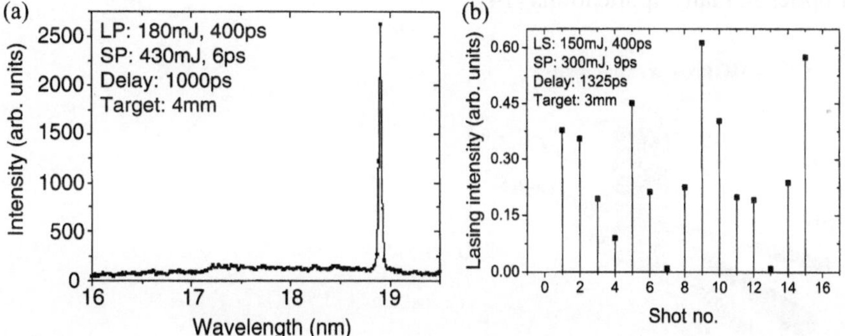


FIGURE 2. (a) Lasing line from Ni-like Mo at 18.9 nm; (b) sequence of the XRL shots at the repetition rate of 1 Hz [6].

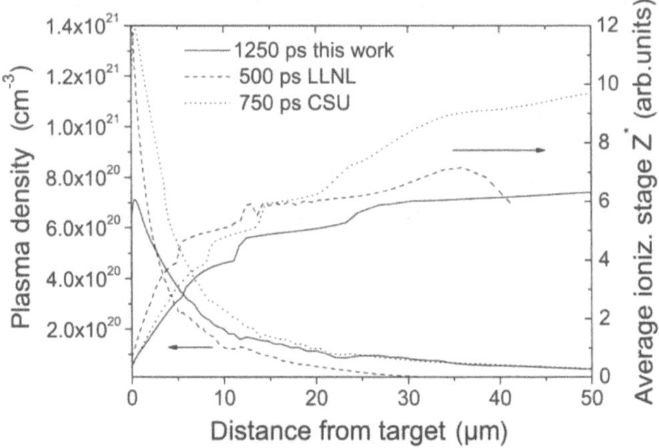


FIGURE 3. Plasma density and average ionizazion stage distributions in a preplasma of three realized experiments on GRIP XRL. LLNL, Lawrence Livermore National Laboratory; CSU, Colorado State University; MBI, Max Born Institute [6].

stored in the amplifying medium. The active volume dimensions are deciding for the amount of the extracted energy. In the experiment on glass capillary optics, a 10-Hz incoherent source with well-defined parameters was used. Two elements of the capillary optics were tested. Firstly, a glass plates stack as a very efficient beam splitter has shown a total efficiency of 80%, in comparison to the multilayer membranes with a total efficiency of less than 30%. Since many applications need focusing elements, we have tested a focusing capillary. Figure 3 shows the focal spot with about 27 μm FWHM. It should to be stressed that such focusing elements enable achievement of much tighter focal spots with a diameter of a few micrometers. Note that robustness, simplicity, and spectral nonselectivity in a broad range of the wavelengths constitute important advantages of these kind of optics in many applications [14].

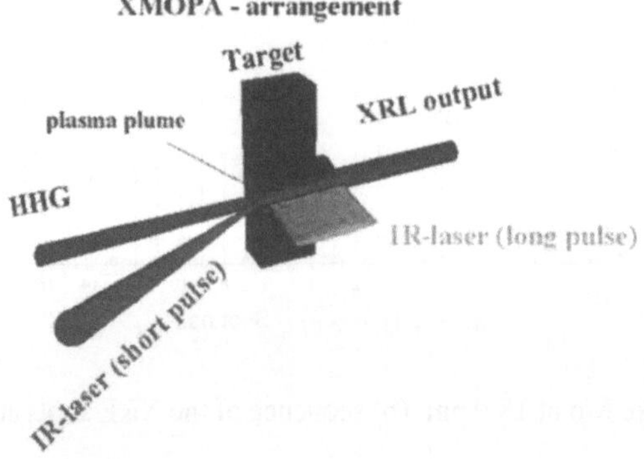


FIGURE 4. Sketch of the XMOPA scheme.

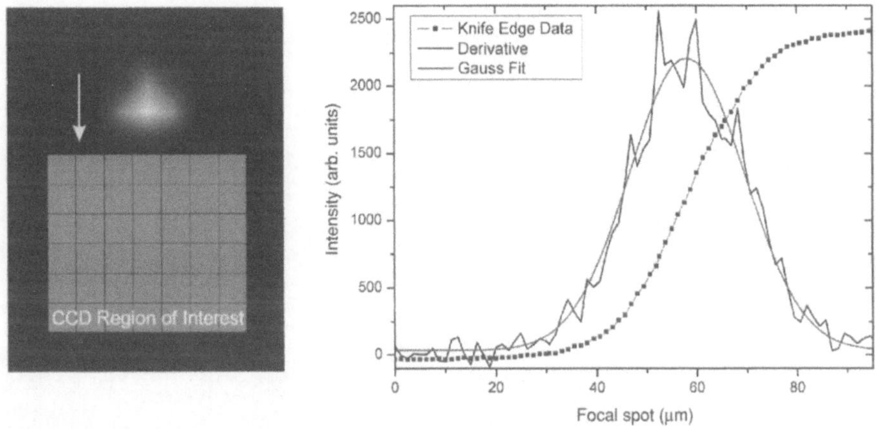


FIGURE 5. Focusing of a short-wavelength beam by a capillary optics lens.

## References

1. P. V. Nickles, V. N. Shlyaptsev, M. Kalachnikov, M. Schnürer, I. Will, and W. Sandner: Phys Rev. Lett. **78**, 2748 (1997).
2. J. Dunn, Y. Li, A. L. Osterheld, J. Nilsen, J. R. Honter, and V. N. Shlyaptsev: Phys Rev. Lett. **84**, 4834 (2000).
3. K. A. Janulewicz, A. Lucianetti, G. Priebe, W. Sandner, and P. V. Nickles: Phys. Rev. A, **68**, 051802(R) (2003).
4. B. M. Luther, Y. Wang, M. A. Larotondo, et al.: Opt. Lett. **30**, 165 (2005).
5. R. Keenan, J. Dunn, P. K. Patel, D. F. Prince, R. F. Smith, and V. N. Shlyaptsev : Phys. Rev. Lett. **94**, 103901 (2005).
6. J. Tümmler, K. A. Janulewicz, G. Priebe, and P. V. Nickles: Phys. Rev. E, **72**, 037401 (2005).
7. J. J Rocca, C. H. Moreno, M. C. Marconi, K. Kanizay: Opt. Lett. **24**, 420 (1999).
8. R. F. Smith, J. Dunn, J. Nilsen, et al.: Phys. Rev. Lett. **89**, 065004 (2002).
9. T. Ozaki, R. A. Ganeev, A. Ishizawa, T. Kanai, and H. Kuroda: Phys. Rev. Lett. **89**, 253902 (2002).
10. T. Ozaki, H. Nakano, and H. Kuroda: J. Phys. Soc. Jpn. **71**, 2963 (2002).
11. R. Tommasini, J. Nilsen, and E. E. Fill: Proc. SPIE. **4505**, 85 (2001).
12. V. N. Shlyaptsev, J. Dunn, S. Moon, et al.: Proc. SPIE. **5197**, 221 (2003).
13. R. Keenan, J. Dunn, V. N. Shlyaptsev, R. Smith, P. K. Patel, and D. F. Prince: Proc. SPIE. **5197**, 213 (2003).
14. V. A. Arkadiev, A. I. Kolomiitsev, M. A. Kumakhov, et al.: Sov. Phys. Usp. **32**, 271 (1989).

# Part XII

# High-intensity Lasers and Applications

# Characterization of a $10^{20}$ W · cm$^{-2}$ Peak Intensity by Focusing Wavefront-corrected 100-TW, 10-Hz Laser Pulses

Yutaka Akahane,[1] Jinglong Ma,[1] Yuji Fukuda,[1] Makoto Aoyama,[1] Hiromitsu Kiriyama,[1] Koichi Tsuji,[1] Yoshiki Nakai,[1] Yoichi Yamamoto,[1] Koichi Yamakawa,[1] Julia V. Sheldakova,[2] and Alexis V. Kudryashov[2]

[1] Advanced Photon Research Center, KANSAI Research Establishment, Japan Atomic Energy Research Institute, Kizu, Soraku-gun, Kyoto 619-0215, Japan
**akahane.yutaka@jaea.go.jp**

[2] Moscow State Open University, Sudostroitelnaya Str.18, Bld.5, Moscow, 115407 Russia

**Abstract:** An improvement of laser-focused peak intensity has been achieved in a JAERI 100 TW Ti:sapphire chirped-pulse amplifier chain with a feedback controlled adaptive optics system at a 10 Hz repetition rate. Independent measurements of optical parameters of the laser pulse and an experimental tunneling ionization ratio of He atom with laser energy scaling have practically confirmed an ultrarelativistic intensity of over $10^{20}$ W · cm$^{-2}$.

## 1 Introduction

The recent remarkable advances in the chirped-pulse laser amplification (CPA) technology enable us to construct 100-TW class Ti:sapphire laser systems with repetitive operation in modest-size laboratories [1, 2]. Their maximum laser focused peak intensity has reached fully relativistic intensity over $10^{19}$ W · cm$^{-2}$, and these systems are applied to various high-field physics experiments, such as monoenergetic electron generation with energies as high as 170 MeV from rare gas targets [3], neutron generation [4], and ion acceleration over MeV energies from solid targets [5] and intense keV X-ray line emission from atomic cluster targets [6].

From the point of view of high-intensity laser applications, focusability of the laser beam is one of the most important features of the high-peak power CPA system. On the general assumption that temporal and spatial shape of the laser focus are sech$^2$ and Gaussian, respectively, the peak intensity is determined as follows:

$$I_{\text{peak}} = \frac{2E_L}{\pi w^2 \tau}, \tag{1}$$

where $E_L$,$\tau$ and $\omega$ are energy of the laser pulse, FWHM pulse duration and 1/e$^2$ radius of the focus, respectively. When the 100-TW class laser pulse is focused by an f/3 parabola ideally, the maximum focused peak intensity, in the absence of any distortions, would be expected to reach the $10^{20}$ W · cm$^{-2}$ level. This is difficult to realize in typical high-peak power CPA laser systems for the following reason. Thermal lensing due to the high average-power operation and relatively low optical quality of the large diameter Ti:sapphire crystals used in the amplifiers, and low optical fidelity of the compressor gratings that can distort the wavefront of compressed laser pulses. Wavefront distortion results in focal plane aberrations when the laser pulse is focused, which appears in both the enlargement of the focal spot size and the lowering of the energy content in the spot. The focused peak intensity achieved relative to the ideal case is the Strehl ratio [7] and can be decreased to values below 0.2. Consequently, such intensities are limited to the $10^{19}$ W · cm$^{-2}$ level in typical repetition-rated CPA laser systems.

Recently, the adaptive optics have been applied to several high-peak power CPA laser systems for wavefront correction [8, 9]. With these devices in their multipass amplifiers and after compressors, near diffraction-limited focusing have been obtained with Strehl ratios as high as 0.8. More recently, a generation of $10^{20}$ W · cm$^{-2}$ intensity was reported and used for laser nuclear reaction within tightly focused 5 μm$^2$ area operating at 10 Hz [10], and a highest peak intensity of $0.7 \times 10^{22}$ W · cm$^{-2}$ has also been generated by focusing 0.1 Hz intermittent Ti:sapphire laser pulses with an f/0.6 parabola [11]. Though, ~f/1 tight-focusing geometry, e.g., a focal length of ~3 cm, is practically limited for laser application. Furthermore, the focusing and evaluation conditions of the laser beam in the former laser system was not reported in detail, and the latter system would benefit from a higher repetition rate that is beneficial for many applications. Based upon these considerations, we have implemented the adaptive optics system and multipurpose f/3 off-axis parabola into our JAERI 100-TW laser system in order to generate an ultrarelativistic intensity of over $10^{20}$ W · cm$^{-2}$ at a 10-Hz repetition rate.

In the meantime, the laser-focused peak intensity is conventionally calculated by independently measuring optical parameters, such as a pulse energy, duration, spot size, and energy content in the spot. In the practical high-field experiments, however, the one that is calculated from the results of independent measurements of optical properties (called an intensity calculated by 'Moptical measurements" in following sentences) was known to be uncertain at same place and same time. In order to obtain the same laser-focused peak intensity as "optical measurements" at the focal point, all the optical parameters should be satisfied simultaneously at the focal point in the experimental chamber. It is, however, difficult to confirm this satisfaction in nature only from optical measurements. On the other hand, the focused peak intensity is evaluated by the atom–field interaction. Optical-field ionization (OFI) of atoms is one of the most fundamental ones, and its ionization ratio is utilized as a reference of the laser peak intensity [12, 13]. Especially the tunnel-ionization yield of $He^+$ ion was precisely measured and compared with results calculated by the ADK (Ammosov-Delone-Krainov) model [14], based on a quasiclassical tunneling theory within ±15% accuracy [15]. Consequently, there are several researches in which they have used the OFI yield of $He^+$ as a reference for their laser intensity calibration. Larochelle and coworkers have used this technique in their photoionization experiment of xenon at the laser intensities from $10^{13}$ to $10^{16}$ W · cm$^{-2}$ [16]. More recently, Chowdhury and coworkers have observed the highly charged Ar ions in the fully relativistic intensity of over $10^{19}$ W · cm$^{-2}$ using multiterawatt 25 fs laser pulses with an intensity calibration by OFI of He [17]. They have used the intensity calibration with the energy scaling of the laser pulse. Although the laser-focused peak intensity was estimated to be $6\times10^{19}$ W · cm$^{-2}$ in their "optical measurements," the peak intensity calibrated by OFI of He was reported to be smaller by a factor of three. They assumed that the cumulative optical errors in their convoluted "optical measurements" were responsible for the degradation of the focused peak intensity. However, the roots of these optical errors are not clear quantitatively. McKenna and coworkers have also calibrated the focused intensity of $3.0 \pm 1.8 \times 10^{19}$ W · cm$^{-2}$ by measuring OFI yields of Ar ions instead of the $He^+$ ion with a factor of 1.5 uncertainty [18]. Therefore more careful examination and improvements of the experimental apparatus and procedure may be necessary to evaluate the laser intensities in the ultrarelativistic regime with high accuracy.

In this paper, we report on the generation and evaluation of the focused peak intensity of $1.2 \times 10^{20}$ W · cm$^{-2}$ operating at 10 Hz, which was confirmed not only by optical measurements but also by measuring OFI yield of He with improved experimental apparatus and procedure. By so doing, peak intensities calculated from independent measurements of optical properties and those determined from optical field ionization of He agree within 16%. This agreement is assumed in scaling to higher, ultrarelativistic intensities.

## 2 JAERI 100-TW Ti:sapphire Laser System

A four-stage Ti:sapphire CPA system [1] was used for all the experiments described in this paper. This laser system is capable of generating 100 TW, 20 fs laser pulses at a 10-Hz repetition rate. Seed pulses with 10 fs duration and at the center wavelength of 800 nm were generated from a mirror-dispersion controlled, self-mode locked Ti:sapphire oscillator. The seed pulses were then temporally expanded to 1 ns (FWHM) by the stretcher before amplification and then amplified in regenerative- and two multipass amplifiers up to ~2 J in this study. After amplification, the laser pulse was introduced into a beam attenuation line for energy adjustment. The beam attenuation line consisted of wedges, half waveplates, and polarizers with high optical qualities of $\lambda/10$, where $\lambda$ is the center wavelength of the laser pulse of 800 nm, the combination of which exhibits a net variable extinction from 1 to $10^{-6}$. These optics were placed after the last amplifier and before the pulse compressor in order to prevent accumulation of B-integral and additional high-order dispersions after compression. The laser pulse from the beam attenuation line was then upcollimated by a Galilean telescope and compressed in a grating-pair pulse compressor located in an evacuated chamber ($<10^{-4}$ Torr). The FWHM duration of the compressed laser pulse was measured to be $23 \pm 1$ fs. For the wavefront correction, the high-peak power laser pulse after pulse compression was reflected by the deformable mirror, and subsequently reflected from a beamsplitter (BS) with 98% reflectance and high optical quality of $\lambda/10$. The main part of laser pulse was introduced into an ultrahigh-vacuum (UHV, $6\times10^{-9}$ Torr) chamber passing through a 5-μm-thickness pellicle that separated UHV from vacuum in the compressor chamber. In the UHV chamber, the laser pulse was focused by the gold-coated f/3.2 off-axis parabola (OAP). The maximum laser pulse energy at the focal point was measured to be $600 \pm 30$ mJ by using an energy meter.

## 3 Wavefront Correction

Our adaptive optics system consists of a deformable mirror (DM), a wavefront sensor (WFS), and a personal computer (PC) for close-loop operation [19], which is shown in Fig. 1. The DM is based on a Bimorph structure, whose substrate consists of two-piezo disks, 31 electrodes, where the mirror is glued on the top of the substrate. The dimension of electrodes area is $40 \times 40$ mm$^2$, and arrangement of the electrodes is shown in Fig. 1b. A surface shape of each part of the mirror is deformed by applying a high voltage to each corresponding electrode independently. The clear aperture and stroke of the deformation of the mirror surface was 55 mm and $\pm4$ μm, respectively. The WFS is the Shack-Hartmann type, which consists of a $40 \times 40$ microlenslet array coupled with a charge-coupled device (CCD) camera. A distortion of the wavefront on each small part of the laser beam is converted into a positional variation in the focal plane. The accuracy and maximum dynamic phase range of the WFS are 0.1 and 10 μm, respectively. The PC calculates the wavefront from the sensor CCD image, determines the shape of the DM surface, and applies a high voltage to each electrode of the DM, which establishes feedback control in the adaptive optics system.

In the advance wavefront measurements of our laser pulse before and after pulse compression, the critical wavefront aberration in our laser system was found to be astigmatism on the pulse compressor gratings (see Fig. 2c), which had limited the maximum peak intensity of the laser pulses focused by the f/3.2 OAP to $2.6 \times 10^{19}$ W $\cdot$ cm$^{-2}$ corresponding to a low Strehl ratio of 0.20. To compensate for this aberration, the DM was placed in the evacuated beam line after pulse compressor.

The setup of wavefront correction system is shown in Fig. 1. The ultrashort laser pulse was reflected by the DM first and propagated to the 98% reflectance

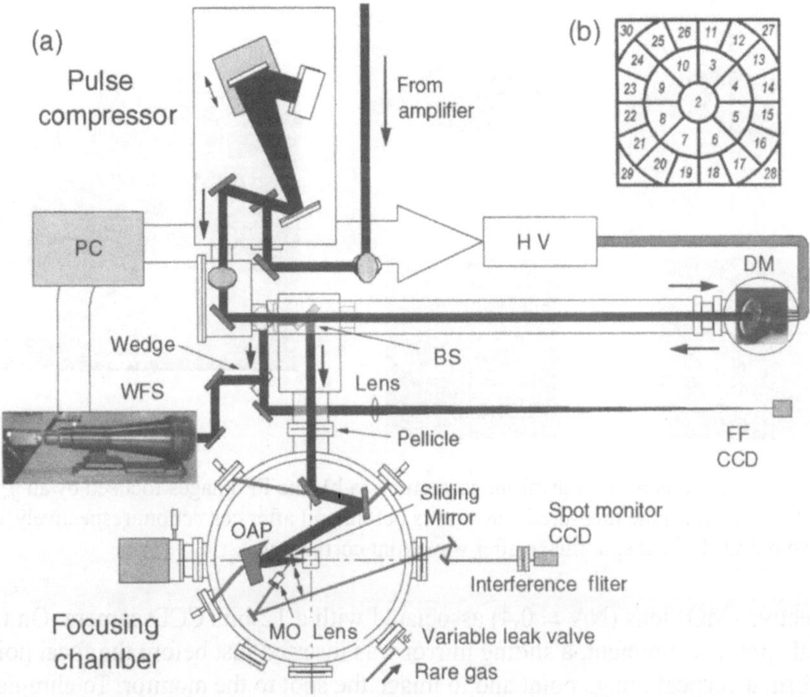


FIGURE 1. (a) Setup of the wavefront correction experiment. (b) Electrodes arrangement of the Bimorph deformable mirror.

BS. The leakage of the BS was split into two beams by a wedge that was placed out of vacuum and delivered to the WFS for the wavefront measurement and far-field (FF) monitor with long (f = +3.2 m) focal lens associated with a CCD camera, respectively. Our closed-loop adaptive optics system was capable to operate with higher repetition rates than that of the laser system 10 Hz, and the wavefront was corrected within a few seconds after several tens of feedbacks once the system was started to operate.

Figure 2 shows the measured wavefronts and FF images of the laser pulse before and after wavefront correction. The Strehl ratio of the laser pulse was significantly improved from 0.20 to 0.80 by correction, which appears clearly in the reduction of the large pedestal around the focal spot between the FF images (Figs. 2a and b). The RMS of wavefront distortion was reduced from 0.485 $\lambda$ to 0.102 $\lambda$ that can be seen in Figs. 2c and d, and astigmatism measured from wavefront data was significantly reduced to one-tenth by correction. From these results, it is concluded that the adaptive optics successfully corrected the wavefront of the laser pulse, and four times improvement of the focusability was expected at focus in the UHV chamber.

Then we measured the focal spot with the OAP in the chamber. As seen in Fig. 1, the spot image of the attenuated laser pulse was analyzed by a microscope

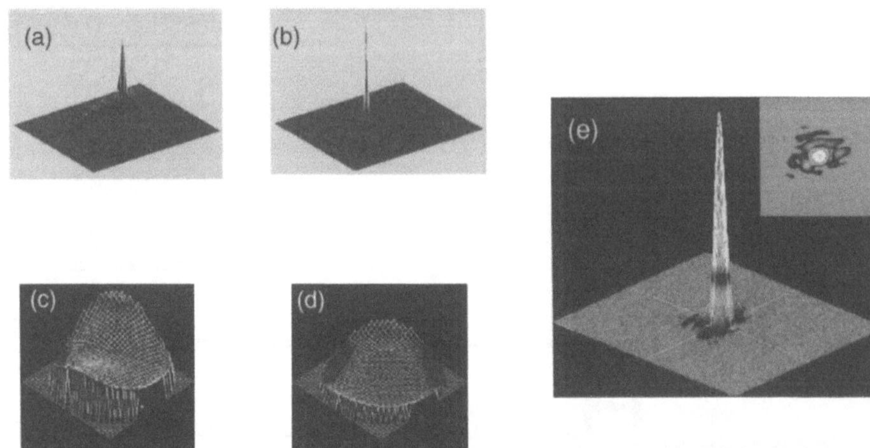


FIGURE 2. Results of the wavefront correction. (a,b) The FF images focused by an $f =$ +3.2 m lens, (c,d) the measured wavefronts before and after correction, respectively. (e) Measured OAP focal spot image after wavefront correction.

objective (MO) lens (NA = 0.4) associated with a 12-bits CCD camera. On the focal spot measurement, a sliding mirror was inserted just before the focal point to form a vertical image point and to image the spot to the monitor. To eliminate the chromatic aberration on the spot images, an interference filter ($\lambda_{center} = 800$ nm with 1.5 nm bandwidth) was inserted into the imaging path just before the CCD camera. To prevent air breakdown and optical damage of the lens and CCD camera by the high-intensity focused beam, the beam attenuation line was used. Since the laser pulse energy was attenuated after amplifier stages under full power operation, beam qualities, i.e., beam profile and wavefront distortion, were conserved.

The measured focal spot image after wavefront correction was shown in Fig. 2e. The $1/e^2$ size of the focal spot was $6.33 \pm 0.11$ μm, which is very close to the diffraction limited. The Strehl ratio was 0.72 in this case, which was slightly smaller than the previously expected one. This degradation is considered to originate from the quality of the optics placed after the 98% BS, such as the pellicle, guiding mirrors, and OAP. Based on the separate measurements of the laser energy and pulse duration, the laser-focused peak intensity was calculated to be $1.19 \pm 0.19 \times 10^{20}$ W $\cdot$ cm$^{-2}$ with the laser pulse energy of $600 \pm 30$ mJ.

## 4 "*In-situ*" Measurement of a Laser-peak Intensity

As mentioned in the previous section, we have shown from optical measurements that the laser intensity of over $10^{20}$ W $\cdot$ cm$^{-2}$ is accessible by wavefront correction. In the practical high-field experiments, however, the peak intensity calculated by optical measurements was known to be uncertain. To realize the calculated peak intensity, the measured optical parameters should be fulfilled simultaneously at the focal point in the experimental chamber. In principle, unfortunately, there is no way to confirm it by optical measurements *in situ*. Therefore the OFI yield of He as a function of laser intensity was measured in order to calibrate and confirm the focused peak intensity *in situ*, as discussed in Sect. 1.

A time-of-flight mass spectrometry (TOF-MS) was used to measure the ion yield. A pair of ion-extracting electrodes was set across the focal point, as shown in Fig. 3a. In these experiments, rarefied gas pressures of noble gasses in the UHV chamber were controlled to $\sim 10^{-7}$ Torr by a variable leak valve. The ions generated at the laser–atom interaction region were extracted by a static electric field and accelerated up to 2.2 keV. The length of the field-free region of the TOF-MS was set to be 982 mm. In these experiments, a mobile slit with a width of 300 μm was placed on the central axis of the TOF-MS tube 7 mm down from the last electrode plate with a ground potential. This slit made it possible to detect ions only from the tight-focused, high-intensity region of the laser focus. To detect the ions, a two-stage microchannel plate (MCP) was used, which has a rise time of 400 ps. TOF mass spectra were recorded by a digital oscilloscope with 500 laser shots accumulation for sufficiently good signal-to-noise ratio. The laser pulse energy was monitored by measuring either the 2% leakage through the 98% reflectance BS or the energy transmitted through the focal plane. It should be noted that the attenuated laser pulse was focused with the identical condition as described in Fig. 2e.

Regarding the OFI experiments, two critical issues were addressed to achieve a result close to that of optical measurements. One is the alleviation of the effects of surrounding long focal volume in the laser focus. In contrast to a solid target, the interaction volume in OFI studies of gaseous targets is an "extended" volume as determined by the peak laser intensity [13]. On the laser axis, there is a relatively long focal region before and after a best focal point that may have sufficient field strength to ionize atoms. Ions generated in this extended focal region can contribute significantly to the total ion yield [20]. And in such long interacting region, many contaminations in the UHV chamber such as $H_2O$, $N_2$, and hydrocarbon molecules are ionized and detected by MCP as embarrassing noise signals, which make the high-dynamic-range ion yield measurement difficult in TOF-MS. To solve this problem, we have inserted the mobile slit in the TOF tube for limitation of the detection volume. The other one is the temporal broadening of laser pulse at the focal point. The duration of the compressed pulse is usually measured by the autocorrelator located in an air environment. Although the autocorrelator is used to minimize its pulse duration, the process includes balancing dispersions of air, vacuum window, and optics in the autocorrelator, such as BS and a second-harmonic generation (SHG) nonlinear crystal that are not included in the main beam line. It means that the pulse duration is actually broadened from its minimum at the focal point and the focused peak intensity is decreased accordingly. To overcome this, we have minimized the pulse duration by adjusting the separation of the grating pair in the compressor to maximize OFI yields of various ions.

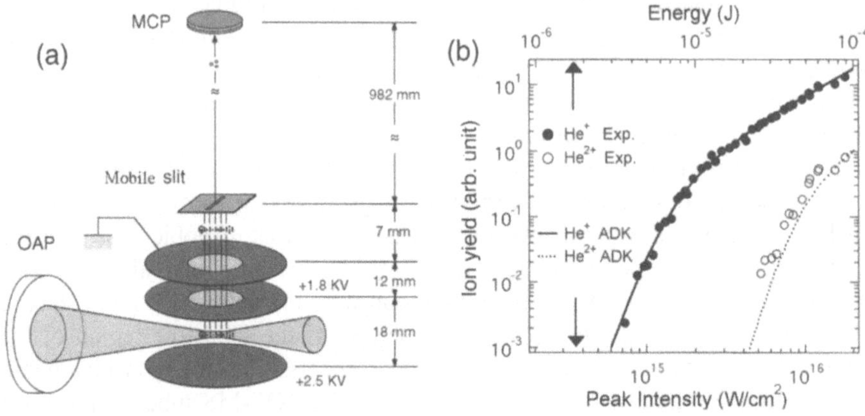


FIGURE 3. (a) Setup of the time-of-flight mass spectrometry for OFI experiments. (b) OFI yield of He as a function of laser intensity. The filled and open circles are the experimental ion yields of $He^+$ and $He^{2+}$, respectively. The solid and dotted line are the ion yields calculated from the ADK theory.

Figure 3b shows the final $He^+$ yield as functions of both laser energy and least-squares-fitted peak intensity. The factor of intensity calibration obtained from Fig. 3b was $2.02 \pm 0.30 \times 10^{20}$ W $\cdot$ cm$^{-2}$ per joule of energy. From the energy scaling of the laser pulse, the maximum peak intensity was calibrated to be $1.21 \pm 0.18 \times 10^{20}$ W $\cdot$ cm$^{-2}$ at the laser pulse energy of 600 mJ, which well agrees with the calculated intensity from optical parameters with an accuracy of 16%. This is the first result, to our knowledge, that a calculated intensity based on independent optical measurements has been confirmed by a direct OFI intensity calibration of He at the $10^{16}$ W $\cdot$ cm$^{-2}$ level for ultrashort pulse durations ($\sim$20 fs).

On the basis of the peak-intensity calibration confirmed above, an OFI experiment in the fully relativistic intensity of $\sim 10^{20}$ W $\cdot$ cm$^{-2}$ was performed. In this study, Kr gas was selected, which has many ionic charge states around the intensity regime. The setup of the experiment is identical as the He OFI. Figure 4 shows the typical TOF mass spectra of Kr ions at the gas pressure of $\sim 10^{-7}$ Torr and at the laser peak intensity of $0.8 \times 10^{20}$ W $\cdot$ cm$^{-2}$. These spectra were the sum averages of 2000 shots datum. In the mass spectrum shown in Fig. 4b, $Kr^{21+}$ peaks were not visible due to the peak overlap of the ionizing contaminations of $O^{4+}(C^{3+})$. The maximum observed Kr ion charge state was +24 (Fig. 4b), which is significantly higher than +19 observed in the previous OFI study with the laser intensity of $2.6 \times 10^{19}$ W $\cdot$ cm$^{-2}$ [21]. This result is a solid evidence of the increase of laser peak intensity by wavefront correction.

In the future study, we plan to develop an "*in-situ*" evaluation method of an ultrarelativistic laser field directly by measuring highly charged ion yields of heavy atoms as a function of laser intensity in the $10^{20}$ W $\cdot$ cm$^{-2}$ region. For laser intensities above the relativistic threshold near $10^{18}$ W $\cdot$ cm$^{-2}$, relativistic effects may alter intensity-dependent ion yields and the accuracy of simple ADK

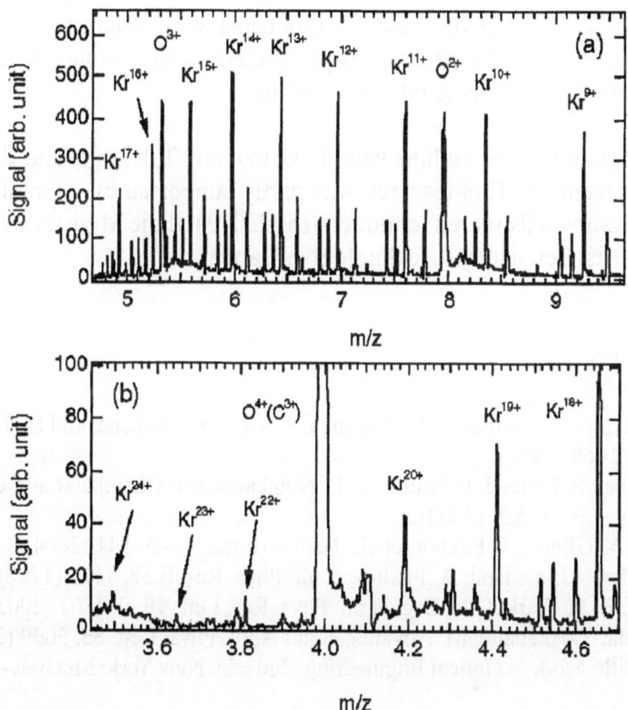


FIGURE 4. TOF spectra of multiple-charged ions of Kr at the laser peak intensity of $0.8 \times 10^{20}$ W · cm$^{-2}$.

model. We continue to investigate this critical subject matter, and we believe that "direct intensity metering" via laser-induced OFI can be extended to ultrastrong laser fields of ultrashort duration using currently measured ion yields of various rare gas atoms as a base [21, 22].

## 5 Summary

We have corrected the wavefront of the high-peak power laser pulses using the closed-loop adaptive optics system in the JAERI 100-TW, 10 Hz laser chain for the improvement of the focused peak intensity on targets. The Strehl ratio of the laser pulse was significantly improved from 0.20 to 0.80 by means of the Bimorph DM with the Shack-Hartmann wavefront sensor. The peak intensity was then enhanced to $1.2 \times 10^{20}$ W · cm$^{-2}$ by focusing with a f/3.2 OAP, which was confirmed not only by optical measurements of the laser pulse but also optical field ionization of He with improvements of experimental apparatus and procedure to limit the intensity-dependent volume expansion and overcome temporal broadening of the compressed pulse. A development of an "in-situ" evaluation method of the ultrarelativistic laser field directly is currently underway by measuring highly charged ion yields of various rare gas atoms as a function of laser intensity. The techniques described in this paper will benefit for a comprehensive characterization of experimental parameters in the studies of laser–matter interactions in the relativistic intensity regime.

**Acknowledgments.** The authors would like to thank T. Kimura and T. Tajima for their encouragement. This research was partly supported by a Grant-in Aid for Specially Promoted Research (contract #15002013) of the Ministry of Education, Culture, Sports, Science and Technology of Japan.

# References

1. K. Yamakawa, M. Aoyama, S. Matsuoka, T. Kase, Y. Akahane, and H. Takuma: Opt. Lett. **23**, 1468 (1998).
2. M. Pittman, S. Ferré, J. P. Rousseau, L. Notebaert, J. P. Chambaret and G. Chériaux: Appl. Phys. B **74**, 529 (2002).
3. J. Faure, Y. Glinec, A. Pukhov, et al.: Nature (London) **43**, 541 (2004).
4. G. Pretzler, A. Saemann, A. Pukhov, et al.: Phys. Rev. E **58**, 1165 (1998).
5. M. Zepf, E. L. Clark, F. N. Beg, et al.: Phys. Rev. Lett. **90**, 064801 (2003).
6. Y. Fukuda, Y. Akahane, M. Aoyama, et al.: Appl. Phys. Lett. **85**, 5099 (2004).
7. W. J. Smith: Modern Optical Engineering, 2nd edn. New York: McGraw-Hill, pp. 337 (1990).
8. F. Druon, G. Chériaux, J. Faure, et al.: Opt. Lett. **23**, 1043 (1998).
9. H. Baumhacker, G. Pretzler, K. J. Witte, et al.: Opt. Lett. **27**, 1570 (2002).
10. J. Magill, H. Schwoerer, F. Ewald, J. Galy, R. Schenkel, and R. Sauerbrey: Appl. Phys. B **77**, 387 (2003).
11. S.-W. Bahk, P. Rousseau, T. A. Planchon, et al.: Opt. Lett. **29**, 2837 (2004).
12. D. N. Fittinghoff, P. R. Bolton, B. Chang, and K. C. Kulander: Phys. Rev. Lett. **69**, 2642 (1992).
13. B. Chang, P. R. Bolton, D. N. Fittinghoff: Phys. Rev. A **47**, 4193 (1993).
14. M. V. Ammosov, N. B. Delone, and V. P. Krainov: Sov. Phys.-JETP **64**, 1191 (1986).
15. B. Walker, B. Sheehy, L. F. DiMauro, P. Agostini, K. J. Schafer, and K. C. Kulander: Phys. Rev. Lett. **73**, 1227 (1994).
16. S. F. J. Larochelle, A. Talebpour, and S. L. Chin: J. Phys. B **31**, 1201 (1998).
17. E. A. Chowdhury, C. P. J. Barty, and B. C. Walker: Phys. Rev. A **63**, 042712 (2001).
18. P. McKenna, K. W. D. Ledingham, I. Spencer, et al.: Rev. Sci. Instrum. **73**, 4176 (2002).
19. J. C. Dainty, A. V. Koryabin, and A. V. Kudryashov: Appl. Opt. **37**, 4663 (1998).
20. P. Hansch, M. A. Walker, and L. D. Van Woerkom: Phys. Rev. A **54**, R2559 (1996).
21. K. Yamakawa, Y. Akahane, Y. Fukuda, M. Aoyama, N. Inoue, and H. Ueda: Phys. Rev. A **68**, 065403 (2003).
22. K. Yamakawa, Y. Akahane, Y. Fukuda, et al.: Phys. Rev. Lett. **92**, 123001 (2004).

# Shaping Ultrafast Laser Pulses with Transient Optical Media

Randy A. Bartels and Klaus Hartinger

Department of Electrical and Computer Engineering, Colorado State University, Fort Collins, CO 80523-1320, USA
**Randy.Bartels@colorstate.edu**

## 1 Introduction

Standard ultrafast pulse shapers rely on linear spectral filtering to generate nearly arbitrarily shaped laser pulses by manipulating the spectral phase and/or amplitude of the input laser pulse. These pulse shapers have become indispensable tools in most ultrafast optical laboratories. The temporal shape of the laser pulse that can be crafted with a standard pulse shaper is limited by the bandwidth of the input laser pulse. Although additional bandwidth can be added through self-phase modulation (SPM), the process is nonlinear and requires intense laser pulses and tends to introduce additional noise and instability as a result of the nonlinear propagation.

We demonstrate a novel approach to pulse shaping that eliminates the spectral bandwidth limitations of traditional Fourier-domain pulse shaping. In this new approach, we propagate the pulse to be shaped in an optical medium that exhibits ultrafast variations in the index of refraction, i.e, a non-time-stationary (NTS) index of refraction. Linear propagation of a pulse through an NTS optical medium will modify the spectrum of the laser pulse. The precise spectral amplitude distribution and phase of the pulse when exciting the NTS optical medium depends on *both* the temporal shape of the index transient *and* the temporal phase and envelope of the ultrafast pulse incident on the NTS medium [1, 2]. In this paper, we experimentally demonstrate compression of the spectrum of a linearly chirped laser pulse. This molecular compression provides a method of producing transform-limited laser pulses with a tunable pulse duration without any energy loss. Experiments are in progress to demonstrate more complex pulse shaping with the NTS pulse shaper.

SPM has been used to temporally phase modulate ultrafast laser pulses for decades and is a form of NTS pulse shaping. The difficulty with SPM is that the temporal phase accumulated by the laser pulse to be shaped is directly proportional to the intensity profile of the laser pulse. The proportionality of the temporal phase to the intensity profile of the pulse severely limits the complexity of pulse shapes that can be produced with SPM pulse shaping. Furthermore, if one wishes to temporally compress laser pulses, the SPM process always broadens that bandwidth with a positive chirp, and thus temporal compression requires a lossy pulse compressor.

NTS pulse shaping completely decouples the temporal phase modulation and the laser pulse to be shaped. As a result, much more complex pulse shaping is possible. An additional benefit is that NTS pulse shaping with Impulsive Molecular Modulation (IMM) can be more efficient. The temporal phase imparted on a laser pulse through SPM is always positive; thus, SPM acting on a negatively chirped laser pulse tends to narrow the pulse spectrum. This property has been harnessed for temporal stretching of ultrafast laser pulses through SPM pulse propagation [3]. However, the output pulse shape from such an experiment cannot be controlled independently of the process because the temporal phase modulation imposed through SPM is proportional to the shape of the pulse. By contrast, IMM is a transient response of a molecular gas that imposes a strong temporal phase modulation on ultrafast pulses. We show that IMM can produce transform-limited pulses with tunable duration, and an independently controllable pulse shape. Alternatively, temporal pulse compression with IMM NTS pulse shaping can be accomplished with a transparent window, eliminating optical loss in the pulse compression [4].

## 2 Experiment

In this experiment, the pulse to be shaped is temporally chirped and made to propagate in an molecular gas with an ultrafast transient index of refraction induced by rotational IMM. The ultrafast temporal phase modulation of rotational IMM is due to a transient alignment of linear molecules in a gas induced by a strong alignment pump pulse. Normally, the random alignment of gas molecules in thermal equilibrium causes the macroscopic optical response of the gas to be isotropic, even if the molecules are intrinsically anisotropic in their polarizability. However, a strong, linearly polarized ultrashort pulse impulsively excites rotational energy states of the anisotropic molecules, creating a coherent rotational wave packet. The anharmonic spacing of the rotational eigenstates of the molecules causes this wave packet to dephase and rephase periodically. Classically, this corresponds to the molecules going in and out of alignment relative to the polarization of the pulse. Because the molecules are anisotropic, the transient alignment leads to ultrafast transients in the macroscopic polarizability and in turn the refractive index [4]. A laser pulse propagating in this transient refractive index $n(t)$ accumulates a temporal phase that is proportional to $n(t)$. As a result, the spectrum of the pulse will be modulated independently of its shape.

The temporal phase modulation induced by rotational IMM is independent of the pulse shape propagating through the molecular gas. This decouples the spectral modulation from the shape of the laser pulse. The combination of the temporal

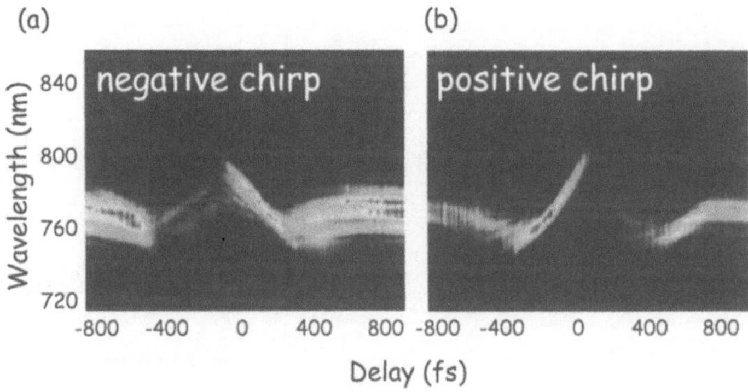


FIGURE 1. Spectral modulation of a probe pulse due to $n(t)$ as a function of the delay of the probe pulse with respect to the pump pulse.

chirp of the laser pulse that is sent into the molecular gas and the temporal shape of the IMM determines the spectral modulations experienced by the pulse. If the temporal duration of an ultrafast laser pulse propagating through the rotational IMM is comparable to the duration of the transient, the pulse spectrum will be either broadened (when the curvature of the IMM and the temporal phase are coincident) or narrowed (when the curvature of the IMM and the temporal phase are of opposite sign).

In the experiment, a 0.8-mJ laser pulse from a multipass Ti:sapphire amplifier is split into orthogonally linearly polarized pump and probe pulses. The pump is focused into an 1-m long, 250-μm diameter hollow core fiber, followed by a time-delayed probe pulse. The fiber is filled with $CO_2$ gas with pressures up to 800 Torr.

At the output of the fiber, we isolate the probe pulse with two reflective polarizers and couple the pulse into either a spectrometer or an SHG-FROG to measure the effect of the temporal modulation caused by the IMM. Figure 1a shows a measurement of the spectrum of a positively chirped probe pulse as the pump/probe delay is scanned through a rotational revival for 462 Torr of $CO_2$ gas. The probe pulse duration has increased from a duration of 40 to 170 fs by adjusting the output compressor of the amplifier. Because the duration of the probe pulse is comparable to that of the transient refractive index, the spectral width is smoothly modulated according to the curvature of the refractive index. At early times (negative relative delay in the figure), the sign of curvature of the input pulse temporal phase and the first half of the rotational revival are opposite, resulting in a compression of the spectral width to about half of its initial width. During the later half of the rotational revival, the commensurate signs of the index transient and the input pulse chirp lead to spectral broadening. We verify that the spectral broadening and compression are due to the IMM by changing the sign of the input pulse chip. As shown in Fig. 1b, switching the sign of the chirp

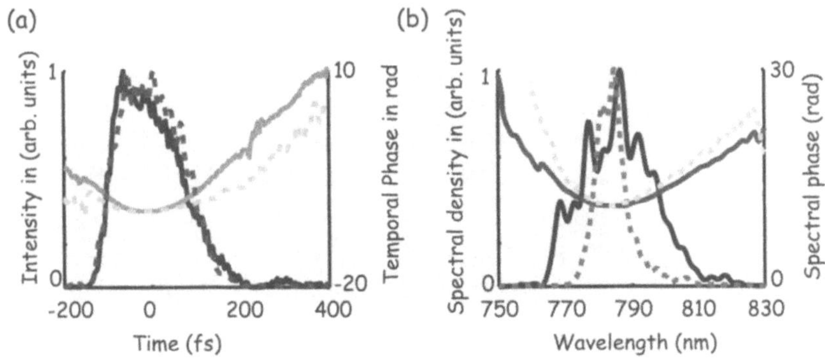


FIGURE 2. (a) Reconstructed spectra for unmodulated (blue) and modulated (red) temporal phase, (b) Reconstructed temporal and spectral pulse profiles.

of the input pulse inverts the times when spectral compression and broadening occur.

## 3 Results

Figure 2 compares the reconstructed temporal and spectral profiles of probe pulses with and without temporal phase modulation. The pulses were measured using SHG-FROG. The flatter temporal phase in Fig. 2a and the spectral compression in Fig. 2b as a result of the chirp compensation are clearly observable, indicating a nearly transform-limited pulse, while the temporal pulse shape is practically left unchanged.

## 4 Summary

In summary, we have demonstrated the technique of NTS pulse shaping with IMM from transiently aligned molecules. In this experiment, we demonstrate the generation of transform-limited pulses with a tunable pulse duration without energy loss to the pulse. Because the amount of temporal phase accumulation depends on the gas pressure and the amount of curvature of the refractive index over the duration of the pulse, the duration of the transform-limited pulse can be tuned over a broad range, starting from the transform limit of the ultrafast source to the duration of the refractive index with sufficient curvature. This duration is roughly 200 fs for $CO_2$ and can be varied by using other gases. We are currently experimentally exploring more complex pulse shapes with the NTS pulse shaper. Further development of this approach to ultrafast pulse shaping may open the possibility of complex pulse shaping over an extremely broad spectral range with a relatively narrow-bandwidth input laser pulse.

**Acknowledgments.** The authors gratefully acknowledge support for this work from the National Science Foundation CAREER Award ECS-0348068, the Office of Naval Research Young Investigator Award, the Beckman Young Investigator Award. One author (RAB) gratefully acknowledges generous support from a Sloan Research Fellowship.

## References

1. R. A. Bartels, N. L. Wagner, M. D. Baertschy, J. Wyss, M. M. Murnane, H. C. Kapteyn: Opt. Lett. **28**, 346 (2003).
2. K. Hartinger and R. A. Bartels: Pulse polarization splitting in a transient wave plate. Opt. Lett. **31**, 3526–3528 (2006).
3. B. R. Washburn, J. A. Buck, and S. E. Ralph: Opt. Lett. **25**, 445 (2000).
4. R. A. Bartels, T. C. Weinacht, N. Wagner, et al.: Phys. Rev. Lett. **88**, 013903 (2002).

# Part XIII

# Atoms and Molecules in High Field

# Quantum Interference in Aligned Molecules

C. Vozzi,[1] E. Benedetti,[1] F. Calegari,[1] J-P. Caumes,[1] G. Sansone,[1] S. Stagira,[1] M. Nisoli,[1] R. Torres,[2] E. Heesel,[2] N. Kajumba,[2] J. P. Marangos,[2] C. Altucci,[3] and R. Velotta[3]

[1] National Laboratory for Ultrafast and Ultraintense Optical Science, CNR– INFM, Dipartimento di Fisica, Politecnico, Milano, Italy
**mauro.nisoli@fisi.polimi.it**

[2] The Blackett Laboratory, Imperial College London, Prince Consort Road, London SW7 2BW, UK

[3] Coherentia–CNR–INFM and Dipartimento di Scienze Fisiche, Università di Napoli "Federico II," Napoli, Italy

## 1 Introduction

High-order harmonic generation (HHG) in aligned molecules is an important area in strong field ultrafast physics [1–3]. It has been shown that the dependence of HHG on the molecular structure can give new insights into measuring molecular structure on a subfemtosecond timescale [3]. A unique aspect of HHG in a simple molecular system is the possibility to observe quantum interference phenomena related to the shape of the molecular wavefunction [4, 5]. Recently this effect has been observed in HHG from aligned $CO_2$ molecules by Kanai et al. [6] and by Vozzi et al. [7].

In this work, we investigate the role of two-center interference in impulsively aligned $CO_2$ molecules. The harmonic response was measured from the 21st to the 49th order revealing the full shape of the destructive interference modulation: the experimental results can be well fitted by a simple model for the interference process. Moreover we demonstrate that such interference effects can be effectively controlled by changing the ellipticity of the driving laser field.

## 2 Quantum Interference in HHG Spectra

Pump (aligning pulse) and probe (driving pulse for HHG) pulses, generated by a Ti:sapphire laser system, are collinearly focused on a pulsed jet. The time delay between the pump and probe pulses is controlled by a stepper motor. The aligning pulse has an energy of ∼200 μJ and a duration of ∼60 fs; the energy of the 30-fs

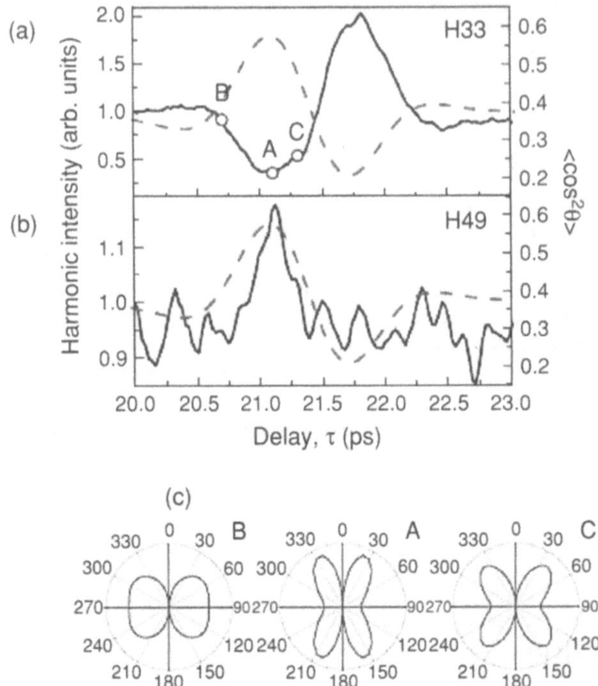


FIGURE 1. Harmonic yield in aligned $CO_2$ molecules at the half-revival of rotational wavepacket (a) for 33rd and (b) 49th order. Pump and probe pulses with parallel polarization. Dashed curves in (a) and (b) shows the calculated temporal evolution of the corresponding $< \cos^2 \theta >$. (c) Polar plot of the $\theta$ distribution for different pump-probe delays: $\tau = 20.7$, 21.1, and 21.3 ps.

probe pulse is ~250 μJ, and its polarization can be changed from linear to elliptical with a quarter-wave plate. The harmonic radiation is sent to the spectrometer by a grazing-incidence toroidal mirror. The spectrometer is composed of one toroidal mirror mounted at grazing incidence for stigmatic imaging with almost unitary magnification, and one variable line spaced grating, mounted after the mirror [8]. The resulting spectrum is stigmatic and almost flat even in a wide spectral region. The detector is a multichannel-plate intensifier with phosphor screen optically coupled to a low-noise fast-readout CCD camera.

We have first measured the evolution of the harmonic photon yield as a function of the delay, $\tau$, between pump and probe pulses, in aligned $CO_2$ molecules. Figures 1a and b show the evolution of the photon yield of 33rd and 49th harmonics, respectively, versus pump–probe delay, with parallel polarization of the two pulses (photon yield has been normalized to that measured in the gas without alignment beam). The photon yield is significantly modulated with a period given by the rotational period, $T_r$, of $CO_2$ ($T_r = 42.8$ ps). Figures 1a and b show the first half-revival. The calculated evolution of $< \cos^2 \theta >$ is also displayed, where $\theta$ is the angle between laser polarization and the internuclear axis. While the photon yield

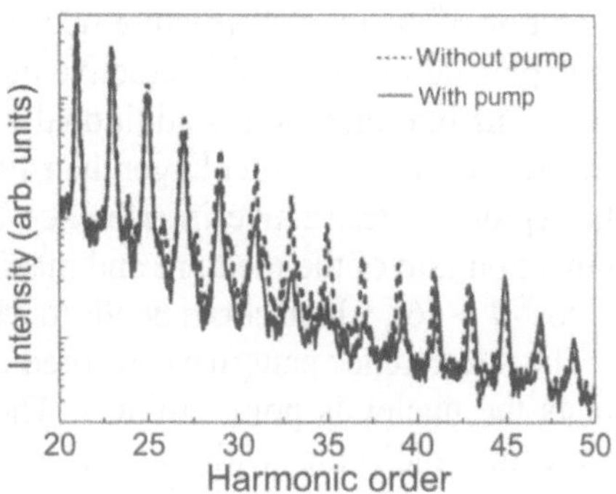


FIGURE 2. Measured harmonic spectra generated by 30-fs pulses with (solid curve) and without (dashed curve) pump beam. Pump–probe delay $\tau = 21.1$ ps.

of 33rd harmonic presents a temporal evolution reversed with respect to that of $< \cos^2 \theta >$, the photon yield of 49th harmonic presents a modulation following $< \cos^2 \theta >$.

We have then fixed the pump-probe delay in correspondence to the maximum of $< \cos^2 \theta >$, around the position of the first half-revival ($\tau = 21.1$ ps; point A in Fig. 1a), and we have measured the harmonic spectra with and without the pump beam. The results are shown in Fig. 2: a pronounced depression of the harmonic generation is visible around the 33rd order in the aligned case.

In order to evidence the effect, we have calculated the integrated intensity, $\Phi$, for each harmonic order. The circles in Fig. 3 show the ratio, $\Phi/\Phi_0$, between the harmonic intensities measured with ($\Phi$) and without ($\Phi_0$) the aligning pulse for $\tau = 21.1$ ps. A clear and broad minimum around the position of 33rd harmonic is evident. As observed by Kanai *et al.* [6], the measured inversion of the temporal

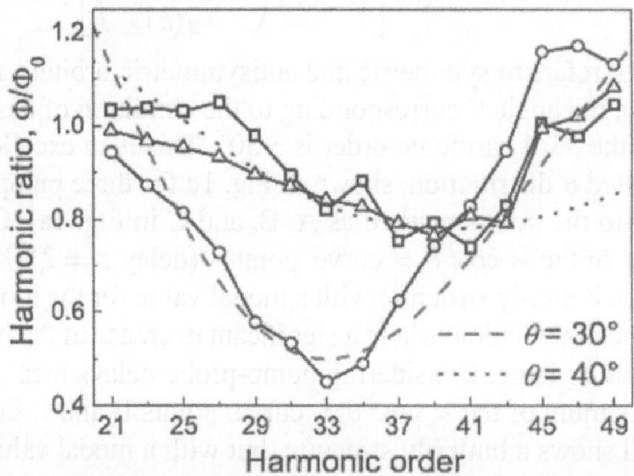


FIGURE 3. Ratio, $\Phi/\Phi_0$, between the harmonic intensities measured with ($\Phi$) and without ($\Phi_0$) the pump pulse, measured at three pump-probe delays: $\tau = 20.7$ ps (squares), 21.1 ps (circles), and 21.3 ps (triangles). The lines show the predicted modulation in harmonic yield for $\theta = 30°$ (dashed curve) and $\theta = 40°$ (dotted curve).

evolution of the revival structure with respect to that of $< \cos^2\theta >$ (Fig. 1a) can be explained in terms of destructive quantum interference in the recombination step. This interpretation is additionally supported by our observation of revival reversal at harmonic order larger than 49th, as displayed in Fig. 1b. In fact, outside the region of destructive interference the harmonic yield is dominated by the ionization rate of the medium and maximizes at the same pump-probe delays as $< \cos^2\theta >$ [6], which corresponds to when the molecules are aligned at $\theta = 30°$.

The interference pattern can be predicted by a simple physical argument which treats the nuclei as point emitters. The $CO_2$ molecule can be considered as a two-center molecule, with the corresponding point emitters placed on the two oxygen atoms. The conditions for interference in antisymmetric molecular orbital correspond to the simple conditions for destructive and constructive interference, respectively:

$$R\cos\theta = n\lambda_B, \tag{1}$$

$$R\cos\theta = (n + 1/2)\lambda_B, \tag{2}$$

where $n = 1, 2, \ldots, \theta$ is the angle between laser polarisation and the internuclear axis; $R$ is the internuclear separation (distance between the two O atoms in the $CO_2$ molecule: $R = 0.232$ nm); and $\lambda_B$ is the de Broglie wavelength of an electron, given by the following expression:

$$\lambda_B = h/\sqrt{2m_e(nh\nu_0 - I_P)}, \tag{3}$$

where $m_e$ is the electron mass; $h\nu$ is the emitted photon energy; and $I_P$ is the ionization potential ($I_P = 13.77$ eV). In this simple approach, the harmonic yield is modulated as

$$I(n,\theta) = I_0(\theta)\left[1 \pm \cos\left(\frac{2\pi R\cos\theta}{\lambda_B(n)}\right)\right], \tag{4}$$

where the sign $\pm$ refers to symmetric and antisymmetric orbitals, respectively.

Using Eq. (4), the angle $\theta$ corresponding to the condition of destructive interference around the 33rd harmonic order is $\sim 30°$. This is in excellent agreement with the calculated $\theta$ distribution, shown in Fig. 1c for three pump-probe delays (corresponding to the points marked as A, B, and C in Fig. 1a). Corresponding to the maximum of the $< \cos^2\theta >$ curve, point A (delay $\tau = 21.1$ ps), the $\theta$ distribution shows a butterfly structure with a modal value for the molecular axis of $\theta \simeq 30°$. The spectral window where a significant decrease of the harmonic yield is observed is rather large. Considering pump-probe delays located on opposite sides of the maximum of the $< \cos^2\theta >$ curve, points B and C in Fig. 1a, the $\theta$ distribution still shows a butterfly structure, but with a modal value increased to $\theta \simeq 40°$. In these cases, the harmonic order corresponding to the condition for destructive interference predicted by the model should increase.

This effect is clearly shown by the experimental curves in Fig. 3 (squares and triangles), which display the harmonic ratio $\Phi/\Phi_0$ measured at the pump-probe delays corresponding to the points B and C of Fig. 1a: the minimum is reached around the 39th harmonic order. The whole behavior of the ratio $\Phi/\Phi_0$ as a function of the harmonic order is well reproduced by the theoretical model, as shown by the dashed and dotted lines in Fig. 3, calculated assuming $\theta = 30°$ and $\theta = 40°$, respectively. In particular, the position of the minima predicted by this model are in close agreement with the measured data. This confirms that in the case of the $CO_2$ molecule, the effective electron wavelength, $\lambda_B$, in the recollision with the molecule is expressed by Eq. (3), i.e., the value expected for a continuum electron in the absence of the Coulomb field.

## 3 Control of Quantum Interference by the Ellipticity of the Driving Pulses

In order to further explore the role of intramolecular interference, we have measured the dependence of HHG upon ellipticity of the probe beam. Figure 4 shows the ellipticity, $\epsilon$, dependence of the photon yield of three harmonics in the case of aligned (solid curves) and unaligned (dashed curves) molecules, measured at a pump-probe delay $\tau = 21.1$ ps. Around the 33rd harmonic, i.e., for the harmonic orders corresponding to the destructive interference minimum, dramatic differences in the ellipticity dependence are measured in aligned and unaligned molecules: the HHG signal is much less influenced by ellipticity in the aligned sample compared to the unaligned sample. The harmonic yield of 33rd order barely drops out to values of $\epsilon > 0.25$. For the harmonic orders not strongly

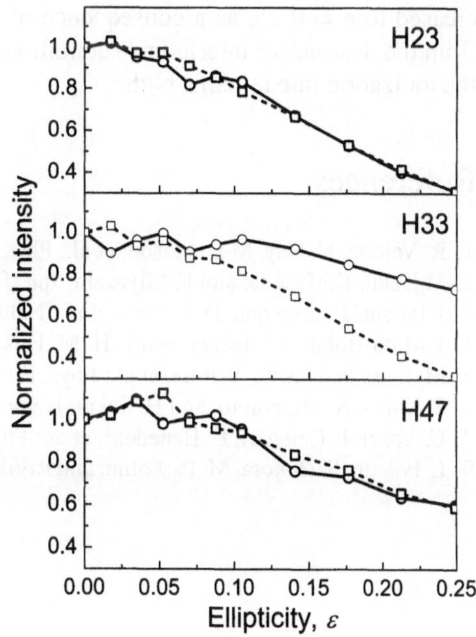


FIGURE 4. Ellipticity dependence of the harmonic yield for three different harmonic orders measured in aligned (solid curve) and not aligned (dashed curve) molecules normalized to the zero ellipticity case. Pump-probe delay $\tau = 21.1$ ps.

affected by destructive interference (below 25th and above 47th), there are only slight differences between the two cases.

A simple interpretation of the experimental results can be given in the framework of the intramolecular two-center interference. Upon increasing the probe–pulse ellipticity, the effective electron wavevector changes. A calculation based on the classical electron trajectories contributing to harmonic emission shows that the tilt of the wavefront of the recolliding electrons in the ellipticity range $\epsilon > 0.1$ can be sufficient to move the recombination step out of the destructive interference regime. Therefore, for low $\epsilon$ values the harmonic emission from aligned molecules is significantly reduced by the intramolecular interference, while at larger ellipticity values destructive interference plays a minor role and the ellipticity dependence will be more influenced by the increased ionization of the aligned molecules. These results show how the two-center interference can be controlled through the control of laser ellipticity. This may have application to controlling HHG yield, and this also points to a broader class of applications in the steering of strong-field molecular processes through ellipticity and alignment angle control.

## 4 Conclusions

In conclusion, we have measured the modulation of harmonic signal due to the rotational revival structure in impulsively aligned $CO_2$ molecules. The role of two-center quantum interference in the electron recombination step has been analyzed. Very anomalous ellipticity dependence is measured under these conditions, with the harmonic signal remaining almost constant even when the ellipticity is increased to $\epsilon = 0.25$, as a consequence of the recombination being tuned away from the destructive interference condition by the changed field ellipticity, while the ionization rate remains high.

## References


1. R. Velotta, N. Hay, M.B. Mason, et al.: Phys. Rev. Lett. **87**, 183901 (2001).
2. M.Kaku, K.Masuda, and K.Miyazaki: Jpn. J. Appl. Phys. **43**, L591 (2004).
3. J. Itatani, J. Levesque, D. Zeidler, et al.: Nature **432**, 867 (2004).
4. J. Muth-Böhm, A. Becker, and F. H. M. Faisal: Phys. Rev. Lett. **85**, 2280 (2000).
5. M. Lein, N. Hay, R. Velotta, et al.: Phys. Rev. Lett. **88**, 183903 (2002).
6. T. Kanai, S. Minemoto, and H. Sakai: Nature **435**, 470 (2005).
7. C. Vozzi, F. Calegari, E. Benedetti, et al.: Phys. Rev. Lett. **95**, 153902 (2005).
8. L. Poletto, S. Bonora, M. Pascolini, and P. Villoresi: Rev. Sci. Instrum. **75**, 4413 (2004).

# Ellipticity Dependence of High-order Harmonics Generated in Aligned Molecules

Tsuneto Kanai, Shinichirou Minemoto, and Hirofumi Sakai

Department of Physics, Graduate School of Science, The University of Tokyo, 7-3-1 Hongo, Bunkyo-ku, Tokyo 113-0033, Japan
**kanai@light.phys.s.u-tokyo.ac.jp**

Ellipticity dependence of harmonic generation from aligned molecules is found to be sensitive to the molecular alignment. The observed results give us rich information about the effects of electron trajectories on high-order harmonic generation.

## 1 Introduction

High-order harmonic generation (HHG) from atoms and molecules has been a subject of intense studies over decades for its potential applications as a coherent ultrashort radiation source in the extreme ultraviolet and soft X-ray regions [1–3]. A sample of aligned molecules [4–8] is an ideal quantum system to investigate the quantum phenomena associated with molecular symmetries. In the present study, this is illustrated by ellipticity dependence of HHG from aligned molecules. Recently, Shan et al. investigated the role of molecular symmetry on the ellipticity dependence of HHG [9]. However, they observed the ellipticity dependence of HHG not from aligned molecules but from randomly aligned ones and the random alignment of the molecules prevented them from studying detailed effects of the symmetry of the valence orbital. Here, we report the first experimental results demonstrating that the ellipticity dependence of HHG is sensitive to the molecular alignment and the valence orbital's symmetry and its geometrical shape. In addition, we observe that the destructive interference, which was recently demonstrated by our studies [10], also appears in the ellipticity dependence.

## 2 Experimental

We prepare femtosecond pump (for alignment) and probe (for HHG) pulses with a Michelson-type interferometer. An output from a Ti:sapphire-based chirped pulse amplification system with a pulse width of ~50 fs and a center wavelength of ~800 nm is split into two pulses. The first lineally polarized pulse is used as a pump to create rotational wavepackets and to induce nonadiabatic molecular alignment. The second pulse with elliptical polarization is delayed by a computer-controlled translation stage and is used as a probe to generate high-order harmonics. The intensity of the pump pulse is $\sim 6 \times 10^{13}$ W $\cdot$ cm$^{-2}$ and that of the probe is $\sim 2 \times 10^{14}$ W $\cdot$ cm$^{-2}$, which is well below the saturation intensity of molecular ionization. The generated harmonics are spectrally resolved by a grazing incidence monochromator and detected by an electron multiplier. The harmonic signals are accumulated by a digital oscilloscope and transferred to and processed in a personal computer.

## 3 Results and Discussions

Figure 1 shows the ellipticity dependence of the 23rd harmonics from aligned (squares) and antialigned (circles) $N_2$ molecules whose valence orbital has $\sigma_g$ symmetry. The pump–probe delay is fixed at 4.1 ps, which corresponds to the delay of half-revival when $N_2$ molecules are aligned [5, 10, 11]. As can be seen, the ellipticity dependence for antialigned $N_2$ molecules is different from that for aligned $N_2$ molecules. This is explained by considering the shape of the valence orbital, which ranges along the molecular axis, and the recombination probability of the recolliding electron. In order to consider the ellipticity dependence quantitatively, we introduce a parameter $\Delta\epsilon$ which is defined as a range of ellipticity of the probe pulse that can generate harmonics at an efficiency of more than 20% of those from the pulses with linear polarization. Using this parameter, Fig. 1 shows that $\Delta\epsilon_{\text{para}} \sim 0.20$ and $\Delta\epsilon_{\text{perp}} \sim 0.22$. That is, the $\Delta\epsilon$ from antialigned $N_2$ molecules is 1.1 times as large as that from aligned $N_2$ molecules. Qualitatively similar but quantitatively different ellipticity dependence was also observed for aligned $O_2$ and $CO_2$ molecules whose valence orbitals have $\pi_g$ symmetry. In our recent paper, we demonstrated that the destructive interference takes place according to the symmetry and the shape of valence orbitals [10]. In the case of $N_2$ molecules, we expect that the destructive interference takes place at around the 31st harmonics at the orientation angle of $\theta = 0°$. Blue squares in Fig. 2 shows the measured ellipticity dependence of the 31st harmonics from aligned

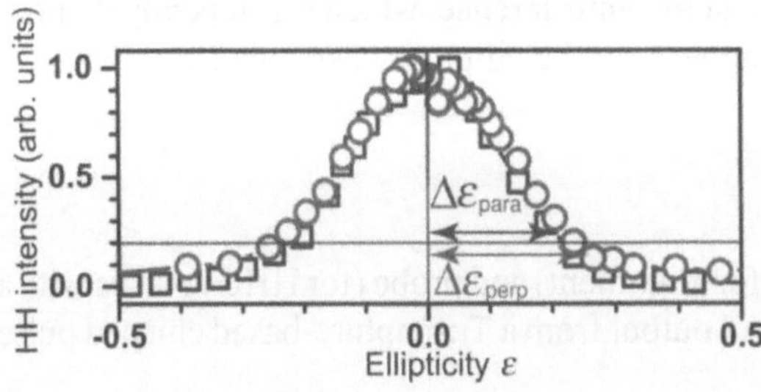


FIGURE 1. The ellipticity dependence of the 23rd harmonics from aligned (squares) and antialigned (circles) $N_2$ molecules.

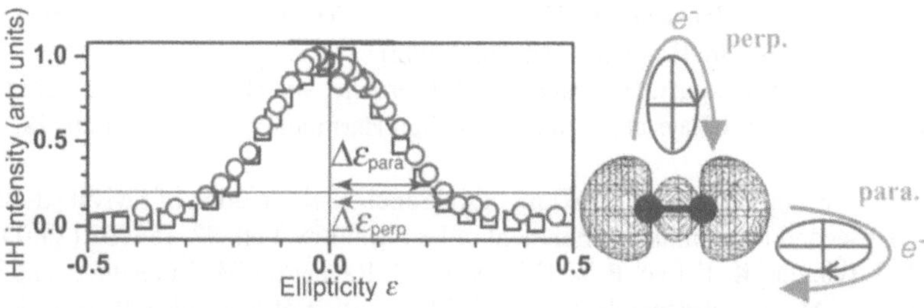


FIGURE 2. The ellipticity dependence of the 31st harmonics from $N_2$ (squares) and Ar (circles).

$N_2$. The ellipticity dependence of the 31st harmonics from Ar is also plotted in Fig. 2 for the purpose of comparison. In the case of aligned $N_2$, we observe a dip at ellipticity $\epsilon \sim 0$ and maximum at $\epsilon \sim 0.1$. However, in the case of Ar, no dip is observed at $\epsilon \sim 0$. The results observed for $N_2$ can be understood by developing the three-step model [12–14] as follows. First, the generation of the 31st harmonics by pulses with linear polarization is suppressed because of the destructive interference. However, as the ellipticity of the probe laser field becomes larger, the transverse field component gives the transverse component of the returning electron momentum, making the destructive interference less effective in the recombination process. When the polarization deviates too much from linear, the retuning electron can miss the parent ion and consequently the probability of harmonic generation rapidly drops to nearly zero. This observation should be compared to the destructive interference in double ionization of $C_6H_6$ [15]. Their observation comes from the nodal plane of the valence orbital and has nothing to do with the electron momentum. On the other hand, our observation comes from the destructive interference whose basic mechanism can be explained by the model of two-point emitters [16] and the conditions for the destructive and constructive interference depend on the harmonic order [10].

# 4 Summary

We observed that the ellipticity dependence of HHG is sensitive to the alignment of molecules. Since ellipticity dependence of HHG is a basis of the polarization gate method [17], using aligned molecules as a nonlinear medium can serve as a new route to control the pulse width of harmonics with the potential for the generation of attosecond pulses.

# References

1. M. Hentschel, R. Kienberger, C. Spielmann, et al.: Attosecond metrology. Nature (London) **414**, 509 (2001).

2. M. Drescher, M. Hentschel, R. Kienberger, et al.: Time-resolved atomic inner-shell spectroscopy. Nature (London) **419**, 803 (2002).
3. Z. Chang, A. Rundquist, H. Wang, M. M. Murnane, and H. C. Kapteyn: Generation of Coherent Soft X Rays at 2.7 nm Using High Harmonics. Phys. Rev. Lett. **79**, 2967 (1997).
4. F. Rosca-Pruna and M. J. J. Vrakking: Experimental observation of revival structures in picosecond laser-induced alignment of $I_2$. Phys. Rev. Lett. **87**, 153902 (2001).
5. I. V. Litvinyuk, K. F. Lee, P. W. Dooley, D. M. Rayner, D. M. Villeneuve, and P. B. Corkum: Alignment-dependent strong field ionization of molecules. Phys. Rev. Lett. **90**, 233003 (2003).
6. H. Sakai, C. P. Safvan, J. J. Larsen, K. M. Hilligs$\phi$e, K. Hald, and H. Stapelfeldt: Controlling the alignment of neutral molecules by a strong laser field. J. Chem. Phys. **110**, 10235 (1999).
7. R. Velotta, N. Hay, M. B. Mason, M. Castillejo, and J. P. Marangos: High-order harmonic generation in aligned molecules. Phys. Rev. Lett. **87**, 183901 (2001).
8. H. Sakai, S. Minemoto, H. Nanjo, H. Tanji, and T. Suzuki: Controlling the orientation of polar molecules with combined electrostatic and pulsed, nonresonant laser fields. Phys. Rev. Lett. **90**, 083001 (2003).
9. B. Shan, S. Ghimire, and Z. Chang: Effect of orbital symmetry on high-order harmonic generation from molecules. Phys. Rev. A **69**, 021404(R) (2004).
10. T. Kanai, S. Minemoto, and H. Sakai: Quantum interference during high-order harmonic generation from aligned molecules. Nature (London) **435**, 470 (2005).
11. J. Itatani, J. Levesque, D. Zeidler, et al.: Tomographic imaging of molecular orbitals. Nature (London) **432**, 867 (2004).
12. P. B. Corkum: Plasma perspective on strong field multiphoton ionization. Phys. Rev. Lett. **71**, 1994 (1993).
13. J. L. Krause, K. J. Schafer, and K. C. Kulander: High-order harmonic generation from atoms and ions in the high intensity regime. Phys. Rev. Lett. **68**, 3535 (1992).
14. M. Lewenstein, Ph. Balcou, M. Yu. Ivanov, A. L"Huillier, and P. B. Corkum: Theory of high-harmonic generation by low-frequency laser fields. Phys. Rev. A **49**, 2117 (1994).
15. V. R. Bhardwaj, D. M. Rayner, D. M. Villeneuve, and P. B. Corkum: Quantum interference in double ionization and fragmentation of $C_6H_6$ in intense laser fields. Phys. Rev. Lett. **87**, 253003 (2001).
16. M. Lein, N. Hay, R. Velotta, J. P. Marangos, and P. L. Knight: Role of the intramolecular phase in high-harmonic generation. Phys. Rev. Lett. **88**, 183903 (2002).
17. O. Tcherbakoff, E. M‘̀evel, D. Descamps, J. Plumridge, and E. Constant: Time-gated high-order harmonic generation. Phys. Rev. A **68**, 043804 (2003).

# Dissociative Ionization of Ethanol Using 400- and 800-nm Femtosecond Laser Pulses

H. Yazawa,[1] T. Shioyama,[1] F. Kannari,[1] R. Itakura,[2] and K. Yamanouchi[2]

[1] Department of Electronics and Electrical Engineering, Keio University, Japan
[2] Department of Chemistry, School of Science,The University of Tokyo, Japan
**kannari@elec.keio.ac.jp**

We have investigated the behavior of dissociative ionization of ethanol molecules under irradiation of 400- and 800-nm femtosecond laser pulses. When 800-nm pulses were irradiated, parent ions and fragment ions generated at C–C bond breaking were dominant. On the other hand, when 400-nm pulses were irradiated, fragment ions generated at C–O bond breaking showed significant increase. When irradiating both 400- and 800-nm pulses and changing the time delay between these pulses, different photofragmentation occurs depending on which pulse is irradiated first.

## 1 Introduction

Our previous experiments of dissociative ionization of ethanol with intense 800-nm femtosecond laser pulse at $4\text{PW} \cdot \text{cm}^{-2}$ [1] revealed that dissociation at C–O bond of singly charged ethanol molecules can be accelerated only when longer excitation laser pulses are applied. The nuclear wavepacket moves on the light-dressed potential energy surface (LD-EPS) defined by laser fields toward the critical energy-level crossing point, where the nonadiabatic transition to the repulsive energy level occurs. If the LD-EPS was substantially modified by pumping pulses, the motion of wavepackets could be guided to specific dissociation channels. For ethanol molecules, any changes of pulse amplitude shape or phase of the 800-nm femtosecond laser pulses induce less effect on the LD-EPS structure, except for their pulse width [1].

In this study, we focus our attention to the wavelength dependence of the dynamics of ethanol in ultrashort intense laser fields and investigate whether the

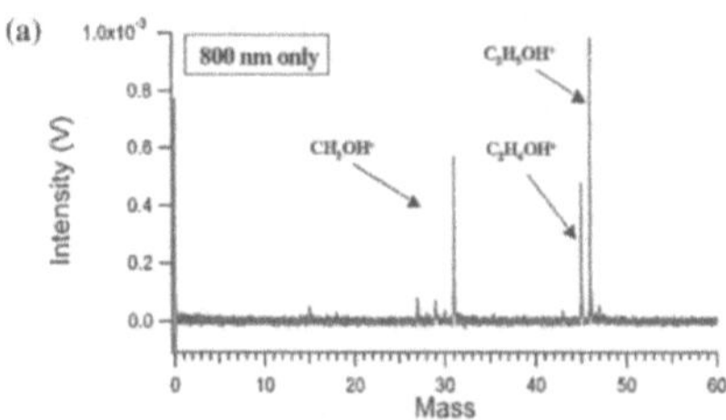


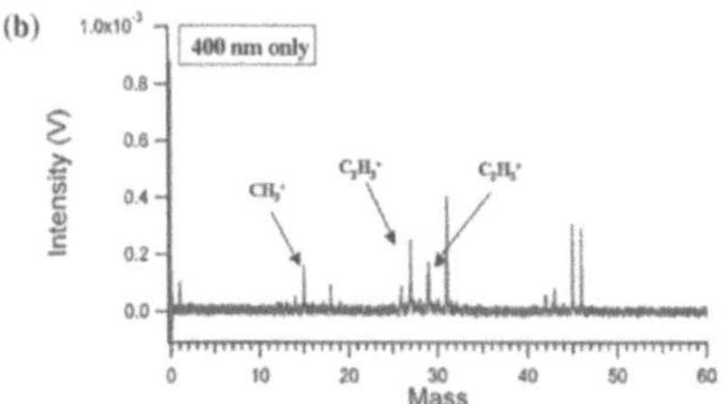


FIGURE 1. The TOF mass spectra obtained for ethanol molecules with intense laser pulse at 800- and 400-nm, respectively.

reaction behaviors with the different laser wavelength reflect the energy level manifolds of singly charged ethanol molecule.

## 2 Experiment

Femtosecond laser pulses with the center wavelength of 800 nm and the spectral width of 25 nm are shaped with a pulse shaper consisting of a computer-controlled liquid crystal spatial light modulation. The shaped pulses are amplified to 0.45 mJ with a chirped pulse amplifier (CPA). Second harmonic generation (SHG) pulses of 400-nm are generated by 1-mm thickness type-I BBO nonlinear crystal. The 800- and 400 nm pulses were once separated by a dichroic beam splitter. With a proper time delay between the 400 and 800 nm pulses, they are combined and focused to Time of Flight Mass Spectrometer (MS-TOF) by quartz lenses ($f =$ 300 mm). The focus power of 400- and 800-nm transform-limited pulses reaches to sub-0.1 PW $\cdot$ cm$^{-2}$ and 0.1 PW $\cdot$ cm$^{-2}$, respectively. The direction of the laser polarization is set to parallel to that of the detection axis of the mass spectrometer.

## 3 Results and Discussion

Figure 1 shows TOF mass spectra obtained for ethanol molecules when either an 800- or a 400-nm transform-limited pulse is irradiated. Parent ions $C_2H_5OH^+$ and $CH_2OH^+$ generated at the C–C bond breaking are dominant when an 800-nm pulse is irradiated. On the other hand, at a 400-nm pulse irradiation, fragment ions such as $CH_3^+$ and $C_2H_5^+$ are generated at the C–O bond breaking. Their ion yields are 0.5 times as large as $CH_2OH^+$. Therefore, significant multiphoton excitation takes place at the 400-nm pulse irradiation in the energy level manifolds of the singly ionized ethanol, whereas dissociative ionization at the C–O bond is induced

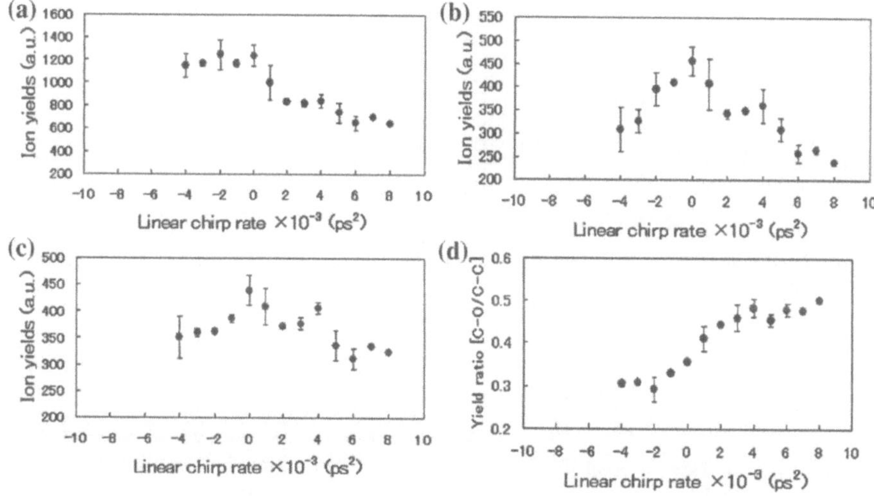


FIGURE 2. The linear chirp dependence of dissociative ionization of ethanol molecule in 400-nm fields. (a) $CH_2OH^+$, (b) $C_2H_5^+$, (c) $C_2H_5OH^+$, and (d) $[C_2H_5^+]/[CH_2OH^+]$.

only by the movement of nuclear packets to the critical nuclear distance along the grand-state energy potential of the singly ionized ethanol [1].

Figure 2 shows linear frequency chirp dependence of ion yields on dissociative ionization of ethanol molecule. Because the SHG conversion efficiency at a BBO decreases for longer laser pulses generated with linear frequency chirp rates, we kept the energy of the 400-nm pulse at 0.02 mJ for all the chirp rates. The temporal widths of 400-nm pulse are sub-90 fs at zero dispersion and sub-270 fs at the max chirp rate of $4 \times 10^{-}3ps^2$, respectively. In contrast to the experiment using 800-nm laser [1], dissociative ionization reactions at both the C–O and C–C bond breaking are enhanced with negative chirping. When plotting the relative ion yield ratio of $[C_2H_5^+]/[CH_2OH^+]$ (representing dissociation ratio [C–O/C–C]), it is shown that more C–O bond breakings take place relative to the C–C bond breakings at positive chirping. An enhancement factor of 1.7 is obtained only by the change of the sign of the chirping at $\pm 3 \times 10$.

Next, both the 800- and 400-nm pulses were irradiated with various relative time delays between two pulses. Figure 3 shows dependence of ion yields on the time delay. The positive delay indicates that the 800-nm pulse precedes the 400-nm pulse. Within ±100 fs time difference, temporal superposition of the two pulses generates parent ions more efficiently. Compared with a simple sum of the ion amounts obtained by the 800- and the 400-nm pulses, the ion amounts generated by the superposition of the two pulses is larger by a factor of two. The similar enhancements by a factor of two or more were obtained for $CH_2OH^+$ and $C_2H_5^+$. These are contribution from excess passes in the multiphoton processes involving both the 800- and 400-nm pulses. Regarding the ion yield ratios, especially $[C_2H_5^+]/[CH_2OH^+]$, the ion ratio drastically changes with the time difference

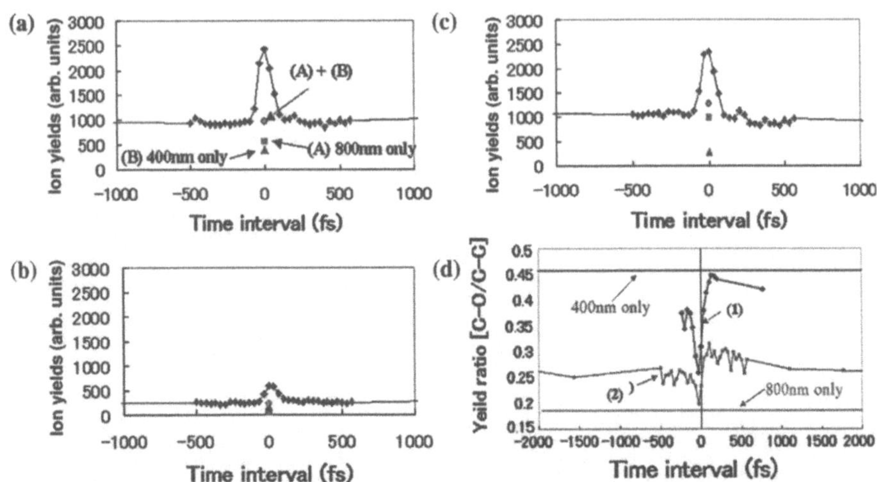


FIGURE 3. The relationship between the each ion yield and the pulse interval of 400- and 800-nm pulse. (a) $CH_2OH^+$, (b) $C_2H_5^+$, and (c) $C_2H_5OH^+$. (d) The relationship between the ion yield $[C_2H_5^+]/[CH_2OH^+]$ and the pulse interval of 400- and 800-nm pulse. Parent ion strength ratio between 400 nm and 800 nm are 0.71 (upper curve) and 0.30 (lower curve).

between the two pulses (Fig. 3d). With the 800- or 400-nm pulses alone, the ratios are 0.13 and 0.45, respectively. When the intensities of these pulses were decreased, both the ion ratios decreased slightly.

These results shows the possibility that a molecular control by the amplitude and phase by expanding the frequency resource of the excitation laser.

## References

1. R. Itakura, K. Yamanouchi, T. Tanabe, T. Okamoto, and F. Kannari: J. Chem. Phys. **119**, 4179 (2003).

# Coincidence Momentum Imaging of Two- and Three-body Coulomb Explosion of Ethanol in Intense Laser Fields

Ryuji Itakura, Takahiro Teramoto, Akiyoshi Hishikawa, and Kaoru Yamanouchi

Department of Chemistry, School of Science, The University of Tokyo, 7-3-1 Hongo Bunkyo-ku, Tokyo 113-0033
ita@chem.s.u-tokyo.ac.jp

**Summary.** Dissociative ionization of ethanol in intense laser fields is investigated by the coincidence momentum imaging method. Two-body Coulomb explosion channels with the hydrogen migration are identified in addition to a three-body Coulomb explosion channel. On the basis of the momentum vector distributions of the fragment ions, the dissociative ionization dynamics is discussed.

It was shown in our recent study on the dissociative ionization of ethanol that the branching ratio between the C–C bond breaking and the C–O bond breaking varies sensitively depending on the chirp rate of intense laser pulses [1]. If the fragment ions are detected by the conventional ion detection method, the ambiguity exists in the assignment of the counterpart fragment species. In the coincidence momentum imaging (CMI) method [2], all the fragment ions generated from a single parent ion are detected in coincidence, and consequently a definitive assignment of fragment pathway can be achieved. In the present study, the dissociative ionization of ethanol induced by an intense laser field ($\lambda = 800$ nm, $I \sim 10^{14}$ W $\cdot$ cm$^{-2}$, $\Delta t \sim 60$ fs) is investigated by the CMI method.

In the observed CMI maps, six different two-body Coulomb explosion channels from $C_2H_5OH^{2+}$

$$C_2H_5OH^{2+} \rightarrow CH_{3-m}^{+} + CH_{2+m}OH^{+} \qquad (m = -1, 0, 1) \quad (1a)$$

$$C_2H_5OH^{2+} \rightarrow C_2H_{5-n}^{+} + OH_{1+n}^{+} \qquad (n = 0, 1, 2) \quad (1b)$$

and one three-body channel from $C_2H_5OH^{3+}$

$$C_2H_5OH^{3+} \rightarrow CH_3^{+} + CH_2^{+} + OH^{+} \quad (2)$$

are identified.

TABLE 1. Anisotropy parameters and branching ratios for two-body Coulomb explosion pathways of $C_2H_5OH^{2+}$

| Hydrogen migration channel | | $\beta$ | Event number |
|---|---|---|---|
| | $m = -1$ | 0.212(2) | 3.04 |
| C–C breaking | $m = 0$ | 1.642(3) | 100 |
| | $m = 1$ | 1.64(3) | 0.14 |
| | $n = 0$ | 1.06(2) | 0.37 |
| C–O breaking | $n = 1$ | 0.505(3) | 0.59 |
| | $n = 2$ | 0.472(5) | 0.47 |

The observed angular distributions for all the two-body explosion channels are fitted by the formula, $A\{1 + \beta P_2(\cos\theta)\}$, where A and $\beta$ is a constant factor and an anisotropy parameter, respectively. The $\beta$ values for the pathways (1aa) and (1bb) are listed in Table 1. As shown in this table, the dominant pathway is the direct C–C bond breaking with no hydrogen migration. The pathway exhibiting the second largest yield is that accompanied by the hydrogen atom migration to form $CH_4^+$. The low $\beta$ value of this pathway indicates that the C–C bond is broken after a long-lived metastable precursor state with significant structural deformation is formed.

It can also be seen in this table that the $\beta$ values for the three C–O breaking channels are rather low. Even for the direct pathway ($n = 0$) with no hydrogen atom migration, the $\beta$ value is only $\beta = 1.06$. This indicates that longer time is required for the C–O bond breaking, i.e., the precursor state for the C–O bond breaking, which is expected to have the longer C–O distance, is long lived.

In the CMI map of the three-body explosion channel from a triply charged ethanol cation into $CH_3^+$, $CH_2^+$, and $OH^+$, two types of sequential pathways, Pathways I and II, are identified. In both pathways, $CH_3^+$ and $CH_2OH^{2+}$ are produced as the first step, and then $CH_2OH^+$ undergoes the second step dissociation to form $CH_2^+$ and $OH^+$.

In Pathway I, the second step dissociation of intermediate $CH_2OH^+$ exhibits little correlation with the momentum vector of $CH_3^+$. From the analysis of the angular distributions, the dissociation lifetime of the intermediate $CH_2OH^+$ is estimated to be comparable with the rotational period of $CH_2OH^+$. On the other hand, in Pathway II, the dissociation lifetime of the intermediate $CH_2OH^+$ is estimated to be much shorter than its rotational period of $CH_2OH^{2+}$.

## References


1. R. Itakura, K. Yamanouchi, T. Tanabe, T. Okamoto, and F. Kannari: Dissociative ionization of ethanol in chirped intense laser fields. J. Chem. Phys. **119**, 4179–4186 (2003).
2. H. Hasegawa, A. Hishikawa, and K. Yamanouchi: Coincidence imaging of Coulomb explosion of $CS_2$ in intense laser fields. Chem. Phys. Lett. **349**, 57–63 (2001).

# Temporal Coherent Control of Two-Photon Ionization by a Sequence of Ultrashort Laser Pulses

Kenichi L. Ishikawa[1] and Kiyoshi Ueda[2]

[1] Department of Quantum Engineering and Systems Science, Graduate School of Engineering, University of Tokyo, Hongo 7-3-1, Bunkyo-ku, Tokyo 113-8656 Japan
ishiken@q.t.u-tokyo.ac.jp

[2] Institute of Multidisciplinary Research for Advanced Materials, Tohoku University, Sendai 980-8577, Japan
ueda@tagen.tohoku.ac.jp

**Summary.** We study two-photon ionization of a hydrogen atom by a sequence of two 10-fs laser pulses whose photon energy is slightly lower than the ionization threshold, based on direct solution of the time-dependent Schödinger equation. Although the Rydberg wave packet formed by the first pulse quickly disintegrates before the Kepler time, it can be further ionized by the second pulse when the latter coincides with the return of a fragment of the wave packet. The ionization yield and the peak position can be coherently controlled by the time delay between the two pulses.

## 1 Introduction

Recently much effort has been devoted to laser control of quantum states with the goal to control chemical and molecular processes. The coherent nature of the laser light as well as quantum systems are used to favor a specific outcome when several output channels can occur simultaneously in the process.

Schemes with a sequence of two coherent laser pulses with a variable time delay $\tau$ are often used in the study of coherent control [1–5]. For example, Blanchet et al. [1] have considered the excited-state wave packet production by a sequence of two identical laser pulses and have shown that optical and quantum interferences can be distinguished from each other. Weinacht et al. [2, 3] have succeeded in retrieval of the amplitude and phase of a Rydberg wave packet formed by a tailored laser pulse through interference with the reference wave packet produced by a delayed pulse, which may be applied to information storage and retrieval [4, 5].

Apart from application to coherent control, different type of pump–probe schemes to detect the wave-packet motion have long been studied both theoretically and experimentally [6–8]. When a short pulse (typically picosecond pulse) is used to excite an atom to high-lying Rydberg levels, many levels are simultaneously and coherently excited to form a Rydberg wave packet, which is well localized in space and can come back to the nucleus in the classical Kepler orbit time. The returning wave packet could be detected, for example, by inducing transition to another (bound) level [7]. It has also shown that the influence of the relative phase of the two pulses on the total population of the wave packet can be seen as Ramsey fringes [9, 10].

While the pulse width considered in these studies is longer than 100 fs and often in the picosecond range, recent remarkable progress in laser technology has enabled the production of pulses shorter than 10 fs in the visible and infrared ranges [11, 12], and even attosecond pulses [13–16] have been reported in the extreme ultraviolet and soft X-ray ranges based on high-order harmonic generation technique. In this work, we theoretically study two-photon ionization of a hydrogen atom by a sequence of two identical ultrashort laser pulses, with a photon energy $\hbar\omega$ slightly below the ionization threshold $I_p = 13.605$ eV and a pulse width of 10 fs or even shorter, based on direct numerical solution of the time-dependent Schrödinger equation (TDSE). Especially we focus ourselves on the dynamics of a Rydberg wave packet formed by the first pulse and the ionization of the wave packet by the second pulse. The spectral bandwidth is so broad due to ultrashort pulse width that the wave packet disintegrates [6] before the Kepler orbit time. Nevertheless, the wave packet evolution itself is coherent and thus, we investigate the dependence of the ionization yield on the time delay $\tau$, or equivalently, the relative phase. This has not been studied for such a short pulse in previous work to our knowledge. Our results show that the disintegrated wave packet returns to the nucleus by fragments and when the second pulse coincides with it, the ionization yield oscillates with the relative phase between the two pulses. Hence, one may coherently control the two-photon ionization through a fine tuning of the delay between the two pulses.

## 2 Model

To study the interaction of a hydrogen atom initially in the ground state with a laser pulse, we numerically solve the one-electron TDSE in the length gauge

$$i\frac{\partial\Phi(\mathbf{r},t)}{\partial t} = \left[-\frac{1}{2}\nabla^2 - \frac{1}{r} + zE(t)\right]\Phi(r,t), \tag{1}$$

where $E(t)$ is the electric field of the pulse, linearly polarized in the $z$-direction. Its temporal profile is given by,

$$E(t) = E_s(t) + E_s(t-\tau), \tag{2}$$

where $\tau$ is the time delay between the pulses, and each single pulse has a Gaussian temporal profile $E_s(t)$:

$$E_s(t) = E_0 \exp\left[\frac{(2\ln 2)t^2}{T^2}\right]\cos\omega t, \tag{3}$$

where $E_0$ denotes the peak field amplitude, $T$ the full width at half maximum (FWHM) of the pulse duration, and $\omega$ the angular frequency. Equation (1) is numerically integrated using the alternating direction implicit (Peaceman-Rachford) method [17, 18]. In typical calculations, we use a grid with a maximum radius of 5000 arb. units and maximum number of partial waves $l_{\max} = 3$. The grid spacing is 0.25 arb. units, and the time step is 0.108 arb. units. The photoelectron spectra are determined with the help of a spectral analysis of the atomic wave function, as obtained immediately after the pulse. We used the technique developed in Ref. [19], which is based on the use of the window operator:

$$W(E_k, \gamma) = \frac{\gamma^4}{(H_{\text{at}} - E_k)^4 + \gamma^4}, \tag{4}$$

where $\gamma = 0.1$ eV (in our case) indicates the energy resolution of our analysis. We then defined the photoelectron spectra as $P(E_k) = \langle\Phi|W(E_k, \gamma)|\Phi\rangle$.

## 3 Dynamics of the Wave Packet

Let us first investigate the dynamics of a Rydberg wave packet generated by an ultrashort laser pulse. It is theoretically predicted [7] for the case of picosecond laser pulses that in a time-delayed two-photon experiment, where the first laser pulse generates a wave packet, which is then probed by a second short pulse, the two-photon transition probability shows peaks when the time delay is a multiple of the classical Kepler orbit time,

$$\tau_n = 2\pi n^3. \tag{5}$$

Subsequently, ten Wolde et al. [20] have observed returning wave packets of rubidium atoms around $n = 41$ in a direct pump–probe experiment with 6-ps laser pulses.

Figure 1 displays the temporal evolution of the $p$ wave population at $r <$ 50 arb.units when an hydrogen atom is subject to a 10-fs pulse with $\hbar\omega =$ 13.55 eV and a peak intensity of $10^{13}$ W $\cdot$ cm$^2$. Since the 10-fs pulse width translates to a spectral width of 0.18 eV, the excited wave packet has a broad spectrum centered at $n = 16$ but spanning approximately from $n = 8$ even to the continuum. As a consequence, the wave packet is not so well localized radially as in the case of picosecond laser pulses [7], but disintegrates [6] before the Kepler orbit time. Figure 2 illustrates the radial distribution of the wave packet at various delay times after its generation by the first pulse. The initially well-localized wave packet moves outward and spreads rapidly. Already at $\tau = 64$ fs, the inner part of

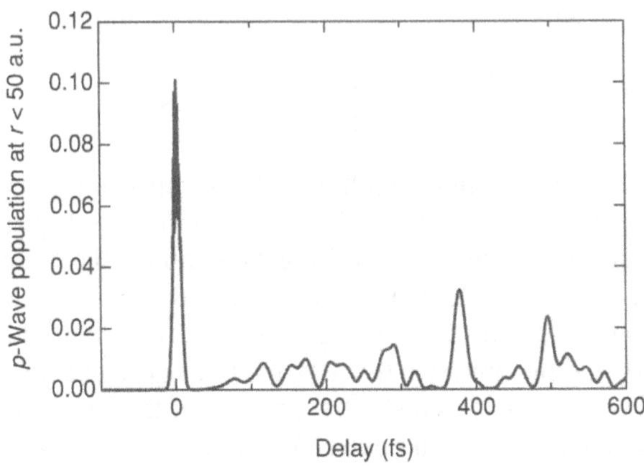


FIGURE 1. Temporal evolution of the population at $r < 50$ arb.units of the $p$-wave packet generated by the first pulse with $T = 10$ fs and $\hbar\omega = 13.55$ eV (the second pulse in Eq. (2) is suppressed).

the packet begins to be dispersed by the strong Coulomb potential [6] (Fig. 2a). At later time in Fig. 2b, the wave packet gets separated more clearly into two parts. At the outer region at $r > 160$ arb.units, a broad but well-formed wave packet moves outward and spread further. This part contains population in high-lying Rydberg and continuum states. On the other hand, we can see that the inner part bound at

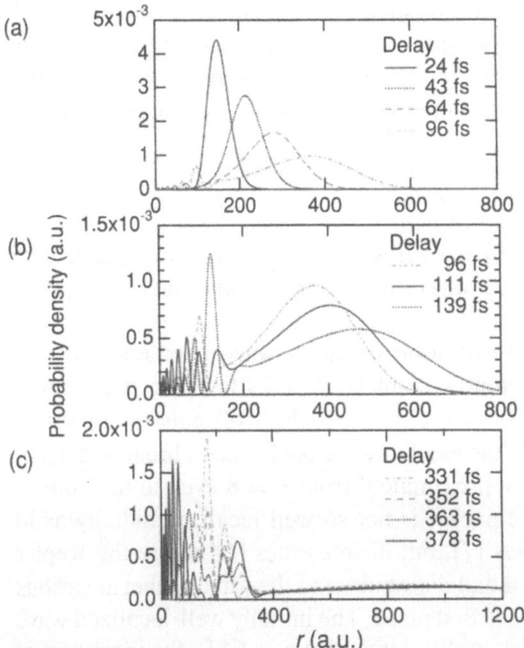


FIGURE 2. Radial probability density distribution of the $p$-wave packet generated by the first pulse with $T = 10$ fs and $\hbar\omega = 13.55$ eV (the second pulse in Eq. (2) is suppressed) at different time delays $\tau$, indicated in each panel.

TABLE 1. Energy eigenvalue and the Kepler orbit time $\tau_n$ calculated from Eq. (5) for several values of principal quantum number $n$

| $n$ | Energy (eV) | $\tau_n$ (fs) |
|---|---|---|
| 8 | −0.213 | 78 |
| 9 | −0.168 | 111 |
| 10 | −0.136 | 152 |
| 11 | −0.112 | 202 |
| 12 | −0.094 | 263 |
| 13 | −0.081 | 334 |
| 14 | −0.069 | 417 |

$r < 160$ arb.units oscillates as time passes from $\tau = 96$ to 139 fs. At even later time in Fig. 2c, a knee-like structure around $r = 310$ arb · units at $\tau = 331$ fs separates from the outgoing part, returns toward the nucleus, and forms a well-shaped Rydberg part for a while as can be seen at 160 arb. units $< r <$ 340 arb. units for $\tau = 378$ fs. Thus, the wave packet disintegrates into several parts and returns to the nucleus in fragments with different time delays, as can be seen in Fig. 1.

The values of the Kepler orbit time $\tau_n$ naively calculated with Eq. (5) for relatively small values of $n$ are listed in Table 1. These corresponds to first several peaks in Fig. 1. For example, it seems that small peaks situated at 77, 116, 154, and 205 fs correspond to the return of wave packets centered at $n = 8, 9, 10$, and 11, respectively. Figure 3 shows the energy spectra of electrons ejected by the second pulse with $\tau = 77.8$ fs (255 optical cycles), 111 fs (364 optical cycles), and 152 fs (498 optical cycles). We have performed simulations for a peak intensity of the second pulse of $10^{10}$ W · cm$^2$, which is 1000 times lower than the first pulse, in order to suppress direct two-photon ionization from the ground state. The peak position is 13.33 eV for $\tau = 77.8$ fs, 13.36 eV for $\tau = 111$ fs, and 13.40 eV for $\tau = 152$ fs, which are close to the theoretical values, 13.34 eV for $n = 8$, 13.38 eV for $n = 9$, and 13.41 eV for $n = 10$. This indicates that even in the case of ultrashort laser pulses for which the wave packet disintegrates rapidly, each fragment returns to the nucleus in the Kepler orbit time corresponding to its own central value of $n$, at least, at early stages.

## 4 Ionization by Double Pulse

In Fig. 1, a relatively large peak is present at the delay $\tau = 378$ fs, which corresponds to the Kepler orbit time for $n = 13.5$. As can be seen from the bottom panel of Fig. 2, at this time the wave packet is separated roughly into three parts: the inner dispersed part at $r < 160$ arb. units; the well-formed Rydberg part at 160 arb. units $< r <$ 340 arb.units; and, in addition, the outer continuum part at $r >$ 340 arb. units. Although the wave packet disintegrates in this manner, its evolution itself is completely coherent. Thus we would expect an interference

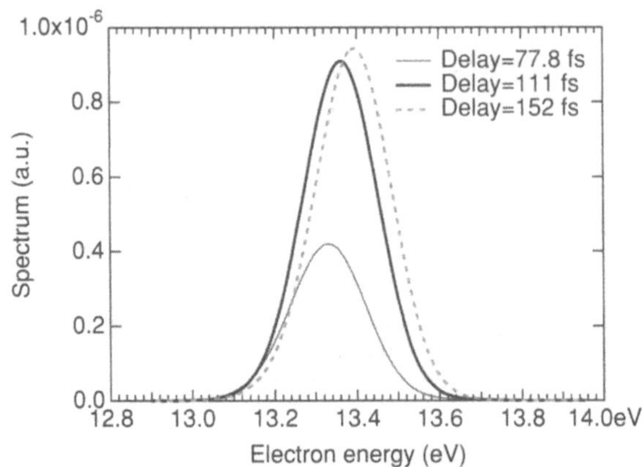


FIGURE 3. Energy spectra of electrons ejected by the second pulse at three different values of time delay $\tau$ indicated in the figure, corresponding to 255, 364, and 498 optical cycles, respectively. The peak intensity of the second pulse is set 1000 times weaker than that of the first pulse to suppress direct two photon ionization from the ground state and to extract only the stepwise contribution from the wave packet formed by the first pulse.

pattern in the two-photon ionization yield by the second pulse as a function of time delay $\tau$ or, equivalently, relative phase $\phi_{\text{rel}} = \omega\tau$ (modulo $2\pi$) between the first and second pulses.

We have performed simulations for a sequence of two equivalent pulses whose field is of the form Eq. (2), by varying $\tau$ between 1237 and 1238 optical cycles in such a way that $\phi_{\text{rel}}$ changes from 0 to $2\pi$. In Fig. 4a, we show the electron energy spectrum by the first and second pulse separately. We can clearly see that the peak height oscillates with $\phi_{\text{rel}}$. The result of similar calculations for $\tau \approx 111$ fs is shown in Fig. 4b. Again, the ionization yield oscillates with $\phi_{\text{rel}}$. Thus, the ionization rate is *coherently controlled.*

Note that in Fig. 4, the peak position also depends on $\phi_{\text{rel}}$. This can be understood as follows, though the broad spectrum of the wave packet extending to the continuum complicates thorough discussion. The wave packet generated by the first pulse is widely spread, as shown in Fig. 2, and only a part near the nucleus can interact with the second pulse. The inner part consists of the lower energy part of the spectrum. As a consequence, it is mainly the low-energy part of the wave packet that is affected by the interference effect, and thus, the peak position in Fig. 4 is higher upon destructive interference.

Strictly speaking, the interference is more directly related to $\omega_e\tau$ rather than to $\omega\tau$, where $\omega_e$ is the representative excitation frequency of the part interacting with the second pulse. As seen in Fig. 3, the spectral distribution of this part depends on $\tau$, and the value of $\phi_{\text{rel}}$ for constructive interference differs between Figs. 4a and b.

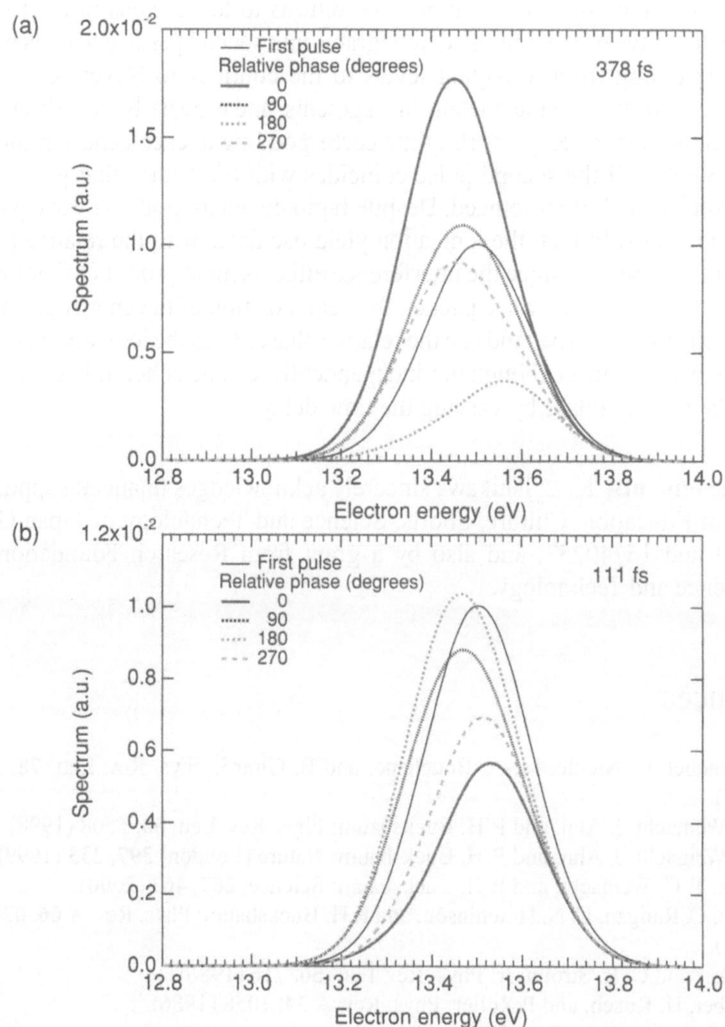


FIGURE 4. Two-photon ionization electron energy spectra by the second pulse with $T = 10$ fs and $\hbar\omega = 13.55$ eV with various relative phases $\phi_{\text{rel}}$ at (a) $\tau = 378$ fs and (b) $\tau = 111$ fs. The relative phase is controlled by varying the delay between 1237 and 1238 optical cycles for (a) and between 364 and 365 optical cycles for (b). Thin solid line denotes the electron energy spectra by the first pulse.

# 5 Conclusions

We have theoretically investigated the dynamics of the wave packet generated by an ultrashort laser pulse whose pulse width is 10 fs and photon energy is slightly below the ionization threshold, and also further ionization of the wave packet by the second pulse with the same pulse width and wavelength. Unlike the Rydberg wave packet formed by a longer pulse that returns to the nucleus in the Kepler orbit time, the wave packet rapidly disintegrates into several parts, due to its broad spectrum spanning from low-lying levels to the continuum. Nevertheless, the bound parts come back to the nucleus in fragments, and especially, at early stages, each part returns in the Kepler orbit time corresponding to each central principal quantum number. If the second pulse coincides with the return, that part of the wave packet can be further ionized. Despite rapid disintegration, the wave packet evolves coherently. In fact, the ionization yield oscillates with the relative phase between the two pulses. Since the interference effect is more prominent for lower-energy components of the wave packet, the peak position of the energy spectrum of ejected electrons also depends on the relative phase. Thus the ionization yield as well as the peak position, though not independently, can be coherently controlled through the relative phase by varying the time delay.

**Acknowledgments.** K. L. Ishikawa sincerely acknowledges financial support by Ministry of Education, Culture, Sports, Science and Technology of Japan Grant 15035203 and 15740252, and also by a grant from Research Foundation for Opto-Science and Technology.

## References

1. V. Blanchet, C. Nicole, M.-A. Bouchene, and B. Girard: Phys. Rev. Lett, **78**, 2716 (1997).
2. T. C. Weinacht, J. Ahn, and P. H. Bucksbaum: Phys. Rev. Lett. **80**, 5508 (1998).
3. T. C. Weinacht, J. Ahn, and P. H. Bucksbaum: Nature (London) **397**, 233 (1999).
4. J. Ahn, T. C. Weinacht, and P. H. Bucksbaum: Science, **287**, 463 (2000).
5. J. Ahn, C. Rangan, D. N. Hutchinson, and P. H. Bucksbaum: Phys. Rev. A **66**, 022312 (2002).
6. J. Parker and C. R. Stroud, Jr: Phys. Rev. Lett. **56**, 716 (1986).
7. G. Alber, H. Ritsch, and P. Zoller: Phys. Rev. A **34**, 1058 (1986).
8. R. W. Robinett: Phys. Rep. **392**, 1 (2004), and references therein.
9. L. D. Noordam, D. I. Duncan, and T. F. Gallagher: Phys. Rev. A **45**, 4734 (1992).
10. R. R. Jones, C. S. Raman, D. W. Schumacher, and P. H. Bucksbaum: Phys. Rev. Lett. **71**, 2575 (1993).
11. T. Brabec and F. Krausz: Rev. Mod. Phys. **72**, 545 (2000), and references therein.
12. A. Suda, M. Hatayama, K. Nagasaka, and K. Midorikawa: Appl. Phys. Lett. **86**, 111116 (2005).
13. M. Drescher, M. Hentschel, R. Kienberger, et al.: Nature (London) **419**, 803 (2002).
14. A. Baltuska, Th. Udem, M. Uiberacker, et al.: Nature (London) **421**, 611 (2003).
15. R. Kienberger, E. Goulielmakis, M. Uiberacker, et al.: Nature (London) **427**, 817 (2004).
16. T. Sekikawa, A. Kosuge, T. Kanai, and S. Watanabe: Nature (London) **432**, 605 (2004).

17. K. C. Kulander, K. J. Schafer, and J. L. Krause: Time-dependent studies of multiphoton processes. In: Atoms in Intense Laser Fields. M. Gavrila (ed.). New York: Academic Press, pp. 247–300 (1992).
18. J. L. Krause, K. J. Schafer, and K. C. Kulander: Phys. Rev. A **45**, 4998 (1992).
19. K. J. Schafer: Comp. Phys. Commun. **63**, 427 (1991).
20. A. ten Wolde, L. D. Noordam, A. Lagendijk, and H. B. van Linden van den Heuvell, Phys. Rev. Lett. **61**, 2099 (1988).

# Part XIV

# Postdeadline Paper

# 100-Attosecond Synchronization of Two-color Mode-locked Lasers by use of Optical Phase Locking

Dai Yoshitomi, Yohei Kobayashi, Masayuki Kakehata, Hideyuki Takada, and Kenji Torizuka

National Institute of Advanced Industrial Science and Technology (AIST)
d.yoshitomi@aist.go.jp

## 1 Introduction

Growing interest in ultrafast electronic processes in atoms, molecules, or condensed matters has driven shortening of the duration of ultrashort pulse lasers. Recent demonstration of attosecond pulse generation by high-order harmonics has opened a new frontier to so-called attosecond science [1,2]. Fourier synthesis of optically phase-locked multicolor pulses [3–5] is also an attractive method of attosecond pulse generation because of its scalability of repetition rate or pulse energy. It would be possible to synthesize arbitrary electric-field waveform including attosecond pulse shape by superimposing several phase-locked ultrashort pulses with separate spectral components which range far beyond an octave. In recent years, the optical phase locking among multicolor pulses has been realized by a femtosecond optical parametric oscillator [6,7] and two-color synchronized mode-locked lasers [8–10]. Obviously, stable phase locking and reproducible waveform synthesis require tight synchronization of pulses with a very low timing jitter. Until now, several groups have reported synchronization of two mode-locked lasers by active [11–14] and passive [15–18] schemes. In active scheme, the laser cavity is actively controlled with the electronic feedback circuits to minimize the relative timing jitter. Schibli et al. demonstrated active synchronization of Ti:sapphire and Cr:forsterite mode-locked lasers with a timing jitter as low as 300 as [14].

In contrast, passive synchronization is all-optical, in which the cross-phase modulation is utilized for synchronization [19,20]. The laser cavities are designed as the two beams cross on one of the laser crystals, in which the wavelengths of two lasers are shifted and the round-trip group delays vary with intracavity group delay dispersion. As the dispersion is negative in typical mode-locking conditions, the two pulses are self-synchronized [20]. The passive scheme is advantageous in that the feedback bandwidth is not limited by that of electronic circuits. We demonstrated passive synchronization of Ti:sapphire and Cr:forsterite two-color mode-locked lasers with a subfemtosecond timing jitter [8]. There are still two remaining causes of the jitter on the observation point. First, the path lengths to the observation point fluctuate by the environmental disturbance. Second, the spectral shifts induced by the mechanism of passive synchronization is a cause of the jitter along with the extracavity dispersion. In our previous work, we eliminated the remaining jitter by applying active control to the passively synchronized Ti:sapphire and Cr:forsterite mode-locked lasers and achieved a timing jitter of 126 as [21]. Recently, we also demonstrated long-term phase locking between the two-color lasers with a phase noise of 0.43 rad [22]. Since the relative optical phase slip of two lasers is significantly sensitive to a change in the cavity length difference, phase locking plays a role in stabilizing the cavity length fluctuation. Therefore, one can expect that the optical phase locking has an effect on suppression of the abovementioned spectral shifts and then on reduction of the timing jitter. However, the influence of the optical phase locking on the timing jitter has never been addressed to our knowledge.

In this paper, we demonstrate a timing jitter reduction of the passively synchronized two-color lasers by locking the relative optical phase slip to an rf reference. As a result of reduction by a factor of 1.7, we achieved an rms timing jitter of 123 as in a frequency range from 10 mHz to 1 MHz.

## 2 Passively Synchronized Two-color Lasers

Figure 1(a) shows the layout of passively synchronized Ti:sapphire and Cr:forsterite lasers [21]. Ti:sapphire and Cr:forsterite laser cavities are designed with their beams overlapping inside the Ti:sapphire crystal. The two laser pulses are self-synchronized by the cross-phase modulation. The size of the lasers is as small as $60 \times 25$ cm$^2$ for the purpose of stability. Chirped mirrors are used for dispersion compensation. The lasers and the whole experimental setup are shielded with steel boxes to prevent air flow and acoustic noise from coming inside. The base plate is made of low-thermal-expansion iron and is decoupled to the external vibration. The Cr:forsterite crystal is cooled to $-5$°C. Ti:sapphire laser is pumped with a 5-W, 532-nm light from a cw frequency-doubled Nd:$YVO_4$ laser (Spectra Physics Millennia Xs) and Cr:forsterite laser is pumped with a 9-W, 1064-nm light from a cw Nd:$YVO_4$ laser (Spectra Physics T40Z-106C) Typical mode-locked output powers of Ti:sapphire and Cr:forsterite are 250 mW and 50 mW, respectively. The repetition frequency is about 100 MHz.

## 3 Optical Phase Locking

The frequencies of Ti:sapphire ($f_{\mathrm{TiS}}$) and Cr:forsterite ($f_{\mathrm{CrF}}$) mode-locked pulse trains are expressed as

$$\begin{aligned} f_{\mathrm{TiS}} &= \delta_{\mathrm{TiS}} + m f_{\mathrm{rep}}, \\ f_{\mathrm{CrF}} &= \delta_{\mathrm{CrF}} + m' f_{\mathrm{rep}}, \end{aligned} \tag{1}$$

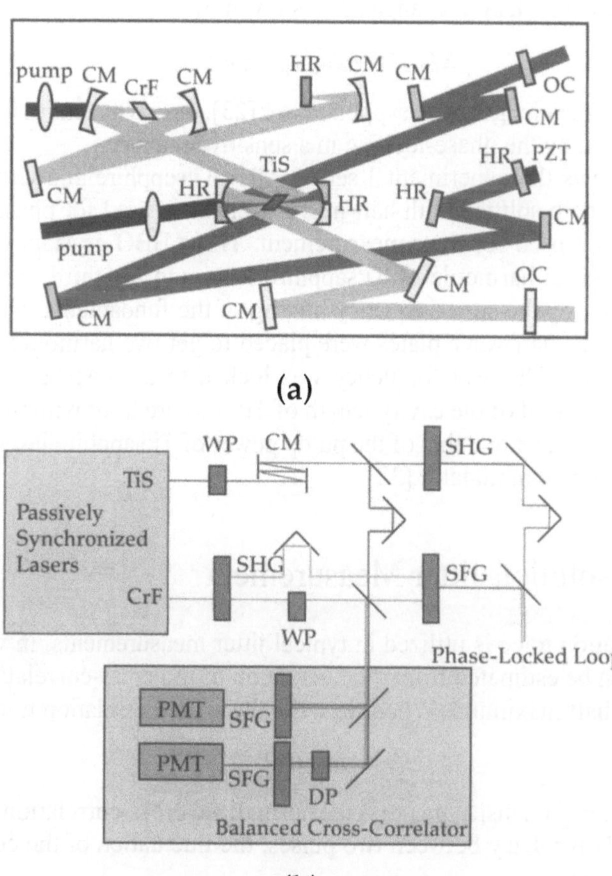


FIGURE 1. Experimental setup. (a) Passively synchronized two-color mode-locked lasers. TiS, Ti:sapphire crystal; CrF, Cr:forsterite crystal; CMs, chirped mirrors; HRs, high reflectors; OCs, output couplers; PZT, piezoelectric transducer. (b) Setup of jitter measurement and phase-locking. WPs, half-wave plates; SHGs, second harmonic generators; SFGs, sum-frequency generators; DP, delay plate.

respectively, where $f_{rep}$ is the repetition frequency, $\delta_{TiS}$ and $\delta_{CrF}$ are the carrier-envelope offset frequencies of Ti:sapphire and Cr:forsterite lasers, and $m$ and $m'$ are integers. Since the second harmonic of Ti:sapphire and the third harmonic of Cr:forsterite are in the same spectral range (~410 nm), the two harmonic frequency combs generate heterodyne beats. The beat frequencies are given by

$$f_{beat} = |2f_{TiS} - 3f_{CrF}| = |\Delta\delta + nf_{rep}|, \tag{2}$$

where $\Delta\delta = 2\delta_{TiS} - 3\delta_{CrF}$ and $n$ is an integer. The cavity length variation leads to a change in pulse-to-pulse slip of relative carrier-envelope phase and results in a change in beat frequency. Since the relation between the beat frequency shift

($\Delta f_{\mathrm{beat}}$) and the cavity length change ($\Delta l$) is given by

$$\Delta f_{\mathrm{beat}} = 4\Delta l f_{\mathrm{rep}}/\lambda_{\mathrm{TiS}}, \tag{3}$$

where $\lambda_{\mathrm{TiS}}$ is the wavelength of Ti:sapphire laser [23], the cavity length fluctuation can be suppressed by the phase-locking in a sensitive manner.

Figure 1b shows the experimental setup. After Ti:sapphire and Cr:forsterite laser pulses are both splitted with half mirrors, ones are used for phase locking and the others are used for jitter measurement. Three BBO crystals were used to obtain the second harmonic of Ti:sapphire laser and the third harmonic of Cr:forsterite laser by the sum-frequency mixing of the fundamental and its second harmonic. The half-wave plates were placed to get two harmonics with the same polarizations. The beat frequency was locked to a 40-MHz rf reference signal by active control of the cavity length of Ti:sapphire laser with the use of a piezoelectric transducer and that of the pump power of Ti:sapphire laser with the use of an electrooptic modulator [22].

## 4 High-Resolution Jitter Measurement

A cross-correlation trace is utilized in typical jitter measurements, in which the timing jitter can be estimated from the fluctuation of the cross-correlation signal at the slope of half maximum. When we write the cross-correlation trace as

$$I_{\mathrm{cc}} = I_0 g_{\mathrm{cc}}(\tau), \tag{4}$$

where $I_0$ is the peak intensity, $g_{\mathrm{cc}}(\tau)$ is the normalized cross-correlation function, and $\tau$ is the relative delay between two pulses, the fluctuation of the correlation signal is written as

$$\Delta I_{\mathrm{cc}} = I_0 \frac{dg_{\mathrm{cc}}(\tau)}{d\tau}\Delta\tau + g_{\mathrm{cc}}(\tau)\Delta I_0. \tag{5}$$

The first term represents the fluctuation caused by the jitter ($\Delta\tau$), while the second term represents the amplitude noise ($\Delta I_0$). If we choose a delay as $g_{\mathrm{cc}}(\tau) = 1/2$ and assume the typical parameters in our experiment (full width of correlation $\sim$ 70 fs, ampliutude noise $\sim$ 0.1%), the second term is comparable to the first term in a sub-100-as regime, which limits the resolution of this method.

To overcome this problem, we used the balanced cross-correlator, which was proposed and demonstrated by Schibli et al [14]. The balanced cross-correlator consists of two cross-correlators with different relative delays between two pulses, as shown in Fig. 1b, in which opposite slopes of correlation traces are used to monitor the fluctuation. A 3-mm-thick fused-silica delay plate was inserted on one branch to shift its delay to the opposite slope. The balanced cross-correlation is defined as the difference of two correlations, namely,

$$\begin{aligned} I_{\mathrm{bcc}}(\tau) &= I_{\mathrm{cc}}(\tau + \tau_0/2) - I_{\mathrm{cc}}(\tau - \tau_0/2) \\ &= I_0\{g_{\mathrm{cc}}(\tau + \tau_0/2) - g_{\mathrm{cc}}(\tau - \tau_0/2)\}, \end{aligned} \tag{6}$$

where $\tau_0$ is the relative delay given by the delay plate. The fluctuation of the signal is

$$\Delta I_{\mathrm{bcc}}(\tau) = I_0 \left\{ \left.\frac{dg_{\mathrm{cc}}(\tau)}{d\tau}\right|_{\tau+\tau_0/2} - \left.\frac{dg_{\mathrm{cc}}(\tau)}{d\tau}\right|_{\tau-\tau_0/2} \right\} \Delta\tau + \{g_{\mathrm{cc}}(\tau+\tau_0/2) - g_{\mathrm{cc}}(\tau-\tau_0/2)\}\Delta I_0. \quad (7)$$

Similarly, the first term represents contribution of the jitter ($\Delta\tau$) and the second term represents the amplitude noise ($\Delta I_0$). However, we can easily exclude the amplitude noise by choosing a delay to satisfy $g_{\mathrm{cc}}(\tau + \tau_0/2) = g_{\mathrm{cc}}(\tau - \tau_0/2)$, which is $\tau = 0$ for symmetric correlation functions. In other words, since the amplitude noise affects the two correlations in the same way when the condition is satisfied, the effects are cancelled by taking the difference of two correlations. The two correlation signals were detected simultaneously with two photomultiplier tubes and measured with a multichannel digital oscilloscope. The balanced signal was obtained by calculated difference of two traces.

## 5 Results

Figure 2 shows the timing jitter results measured with observation bandwidths of 1 kHz, 400 kHz, and 1 MHz. As can be seen in Fig. 2a, we observed a clear evidence of suppression of the drift and fluctuation by the phase locking at 1-kHz bandwidth. As a result, an rms jitter of 194 as was reduced to 87 as. Likewise, we observed decrease of an rms jitter from 108 to 96 as at 400-kHz bandwidth (Fig. 2b) and that from 101 to 88 as at 1-MHz bandwidth (Fig. 2c), although we found no significant evidence of the jitter suppression at the two fastest bandwidths.

Figure 3 shows the beat frequency variation measured at the same time as the 1-kHz jitter measurement shown in Fig. 2a. We observed a drift in beat frequency, which is similar to the drift in timing shown in Fig. 2a in the case without phase locking. The fact suggests that a cavity length change is a dominant cause of the timing jitter. The measured beat frequency shift of 5 MHz corresponds to a cavity length change of about 10 nm according to Eq. (3). Since the optical delay produced by the path length change is only 30 as, the path length change is not the cause of the measured timing drift of about 500 as. The intracavity path length change is immediately compensated by the passive synchronization mechanism. We rather attribute the timing drift to the spectral shift induced by the cavity length change along with the passive synchronization. The spectral shift causes the relative delay outside the cavity because of the extracavity group delay dispersion, as mentioned in Section 1.

Figure 4 shows the power spectral density profiles directly calculated from the time-domain data in Fig. 2 and the integrated rms values. The timing jitter is effectively suppressed in the low-frequency region, especially lower than 1 Hz, whereas no remarkable suppression is seen in the high-frequency region.

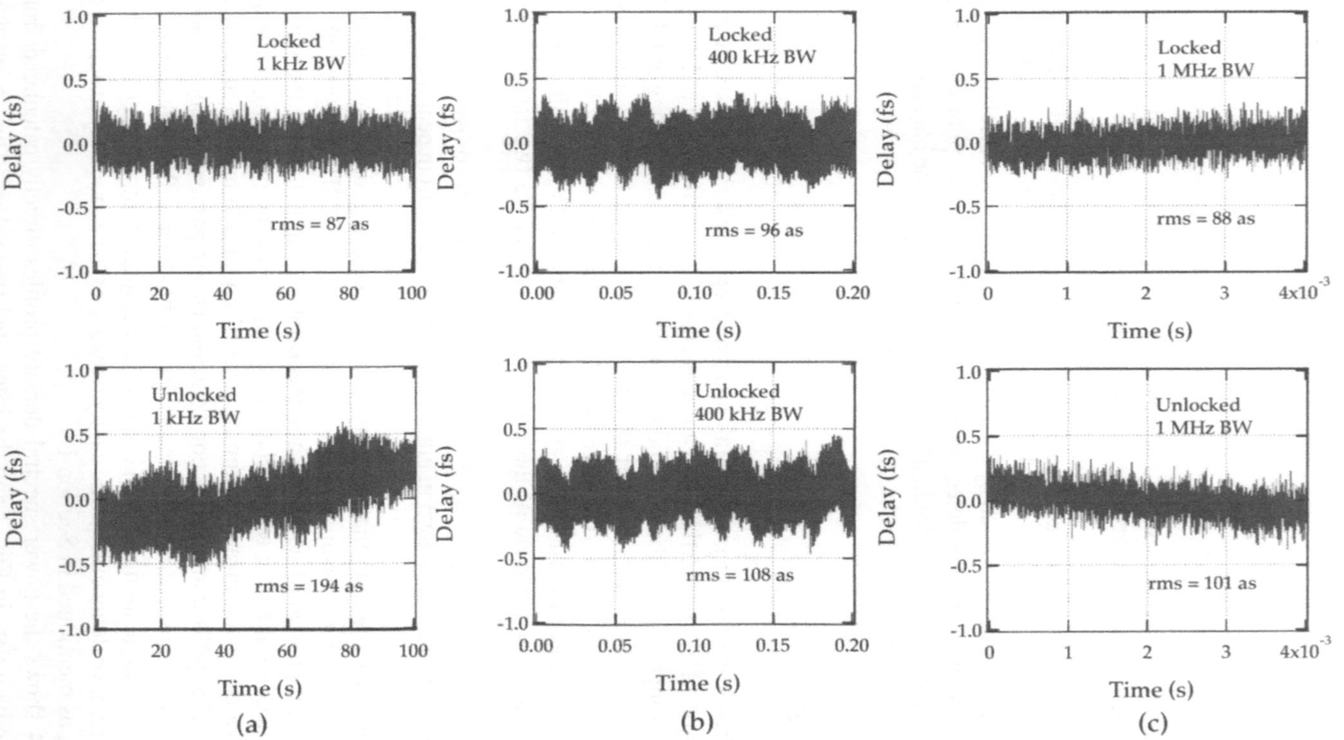


FIGURE 2. Measured timing jitter at bandwidths of (a) 1 kHz, (b) 400 kHz, and (c) 1 MHz. The top and bottom figures are the results with and without phase locking, respectively.

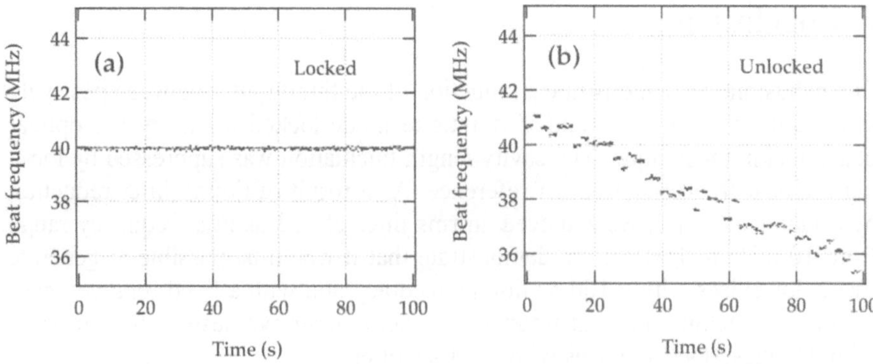


FIGURE 3. Beat frequency variation at a bandwidth of 1 kHz (a) with and (b) without phase locking.

The results suggest that the piezoelectric transducer control contributes to the jitter reduction, because the piezoelectric transducer has lower bandwidth than the electrooptic modulator. The fact is well convincing, because the jitter is more related to the cavity length fluctuation than to the carrier-envelope phase slip.

Finally, the rms jitter integrated over a frequency range from 10 mHz to 1 MHz was 213 as without phase locking, which was reduced to 123 as by phase locking. The reduction factor was approximately 1.7. The obtained rms jitter is comparable to our previous result with active jitter control in Ref. [21]. The scheme presented in this paper requires a simpler setup than the previous one, because it allows simultaneous locking of timing and phase with one active control.

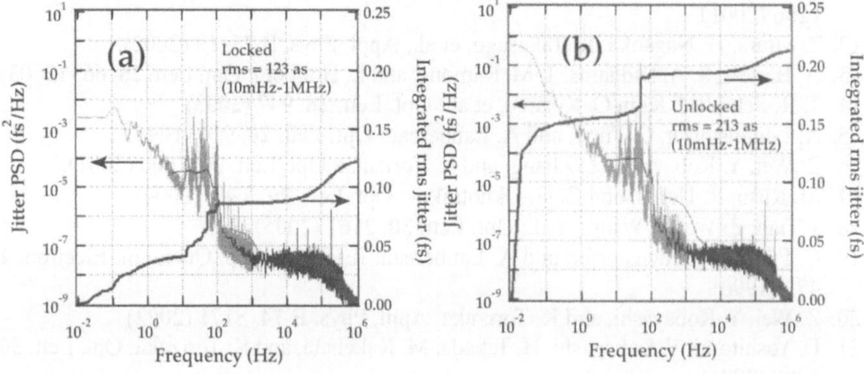


FIGURE 4. Power spectral density and integrated rms jitter (a) with and (b) without phase locking. The gray scales correspond to respective time-domain data with different bandwidths.

## 6 Conclusion

In conclusion, we demonstrate a reduction of the timing jitter between passively synchronized Ti:sapphire and Cr:forsterite mode-locked lasers by the optical phase-locking technique. The cavity length fluctuation was suppressed by locking the beat frequency to an rf reference. As a result of timing jitter reduction by a factor of $\sim 1.7$, we achieved an rms jitter of 123 as in a frequency range from 10 mHz to 1 MHz. We demonstrate that it would be possible to generate two-color pulses with a 100-as timing accuracy and with a fixed relative phase relation by locking the beat frequency to dc. Fourier synthesis is now feasible with the phase-locked pulses with a 100-as jitter.

## References

1. M. Hentschel, R. Kienberger, Ch. Spielmann, et al.: Nature **414**, 509 (2001).
2. T. Sekikawa, A. Kosuge, T. Kanai, and S. Watanabe: Nature **432**, 605 (2004).
3. T. W. Hänsch: Opt. Comm. **80**, 71 (1990).
4. K. Shimoda: Jpn. J. Appl. Phys. **34**, 3566 (1995).
5. M. Y. Shverdin, D. R. Walker, D. D. Yavuz, G. Y. Yin, and S. E. Harris: Phys. Rev. Lett. **94**, 033904 (2005).
6. Y. Kobayashi, H. Takada, M. Kakehata, and K. Torizuka: Appl. Phys. Lett. **83**, 839 (2003).
7. Y. Kobayashi, H. Takada, M. Kakehata, and K. Torizuka: Opt. Lett. **28**, 1377 (2003).
8. Y. Kobayashi, K. Torizuka, and Z. Wei: Opt. Lett. **28**, 746 (2003).
9. A. Bartels, N. R. Newbury, I. Thomann, L. Hollberg, and S. Diddams: Opt. Lett. **29**, 403 (2004).
10. J. Kim, T. R. Schibli, L. Matos, H. Byunn, and F. X. Kärtner: In Joint Conference on Ultrafast Optics V and Applications of High Field and Short Wavelength Sources XI, paper M3-5 (2005).
11. R. K. Shelton, L. Ma, H. C. Kapteyn, M. Murnane, J. L. Hall, and J. Ye: Science **293**, 1286 (2001).
12. T. Miura, H. Nagaoka, K. Takasago, et al.: Appl. Phys. B **75**, 19 (2002).
13. A. Bartels, S. A. Diddams, T. M. Ramond, and L. Hollberg: Opt. Lett. **28**, 663 (2003).
14. T. R. Schibli, J. Kim, O. Kuzucu, et al.: Opt. Lett. **28**, 947 (2003).
15. A. Leitenstorfer, C. Fürst, and A. Laubereau: Opt. Lett. **20**, 916 (1995).
16. Z. Wei, Y. Kobayashi, Z. Zhang, and K. Torizuka: Opt. Lett. **26**, 1806 (2001).
17. M. Rusu, R. Herda, and O. G. Okhotnikov: Opt. Lett. **29**, 2246 (2004).
18. J. Tian, Z. Wei, P. Wang, et al.: Opt. Lett. **30**, 2161 (2005).
19. C. Fürst, A. Leitenstorfer, and A. Laubereau: IEEE Sel. Top. Quantum. Electron. **2**, 473 (1996).
20. Z. Wei, Y. Kobayashi, and K. Torizuka: Appl. Phys. B **74**, S171 (2002).
21. D. Yoshitomi, Y. Kobayashi, H. Takada, M. Kakehata, and K. Torizuka: Opt. Lett. **30**, 1408 (2005).
22. Y. Kobayashi, D. Yoshitomi, M. Kakehata, H. Takada, and K. Torizuka: Opt. Lett. **30**, 2496 (2005).
23. Z. Wei, Y. Kobayashi, and K. Torizuka: Opt. Lett. **27**, 2121 (2002).

# Complete Automatic Phase Compensation for the Generation of A-few-cycle Pulses

Keisuke Narita, Muneyuki Adachi, Ryuji Morita, and Mikio Yamashita

Department of Applied Physics, Hokkaido University, Kita-13, Nishi-8, Sapporo, 060-8628 Japan
**mikio@eng.hokudai.ac.jp**

**Abstract:** An automatic feedback spectral-phase compensation system using the wavelet transformation method for the M-SPIDER signal analysis has been developed. The system is completely free from the manual operation and generates 9.6-fs pulses after chirp compensation.

Recently, we have developed a feedback (FB) spectral-phase compensation system for the generation of quasimonocycle optical pulses [1–3]. The system consists of a 4-f phase compensator with a spatial light modulator (SLM), a highly sensitive M-SPIDER, and personal computers for the phase analysis and the SLM driving. However, it does not compensate automatically for the spectral phase, but needs the manual FB operation in some parts. In this paper, we demonstrate for the first time that the application of the wavelet transformation (WT) method for the analysis of the M-SPIDER signal [4] instead of the conventional Fourier transformation (FT) method enables the automatic FB compensation of the spectral phase for ultrabroadband chirped pulses without any manual judgment.

Figure 1a shows the difference of the analytical procedure of the M-SPIDER signal $D(\omega)$ between FT and WT methods. In the former method, the following six steps are required: (1) the inverse FT of $D(\omega)$ ($D(\tau) = \mathrm{IFT}[D(\omega)]$), (2) the inverse FT of replica spectra $R_1(\omega)$ and $R_2(\omega)$, i.e., ($R_1(\tau) = \mathrm{IFT}[R_1(\omega)]$ and $R_2(\omega) = \mathrm{IFT}[R_2(\omega)]$), (3) the determination of minimum and maximum time values ($\tau_1$ and $\tau_2$) of the squared filter function to pick up the so-called +ac component $D_{\mathrm{ac}}(\tau)$ from $D(\tau)$, (4) filtering of $D_{\mathrm{ac}}(\tau)$, (5) the FT of $D_{\mathrm{ac}}(\tau)$ ($D_{\mathrm{ac}}(\omega)$), and (6) the calculation of the argument $\theta(\omega) = \phi(\omega) - \phi(\omega - \Omega) + \tau\omega$ of $|D_{\mathrm{ac}}(\omega)| \exp\{i\theta(\omega)\}$. Especially, when pulses have the complicated intensity spectrum $I(\omega)$ with fine structures (see Fig. 1b: an example of $D(\tau)$ of the tapered fiber output [5,6]), the difficult work for the step (3) is demanded for determining

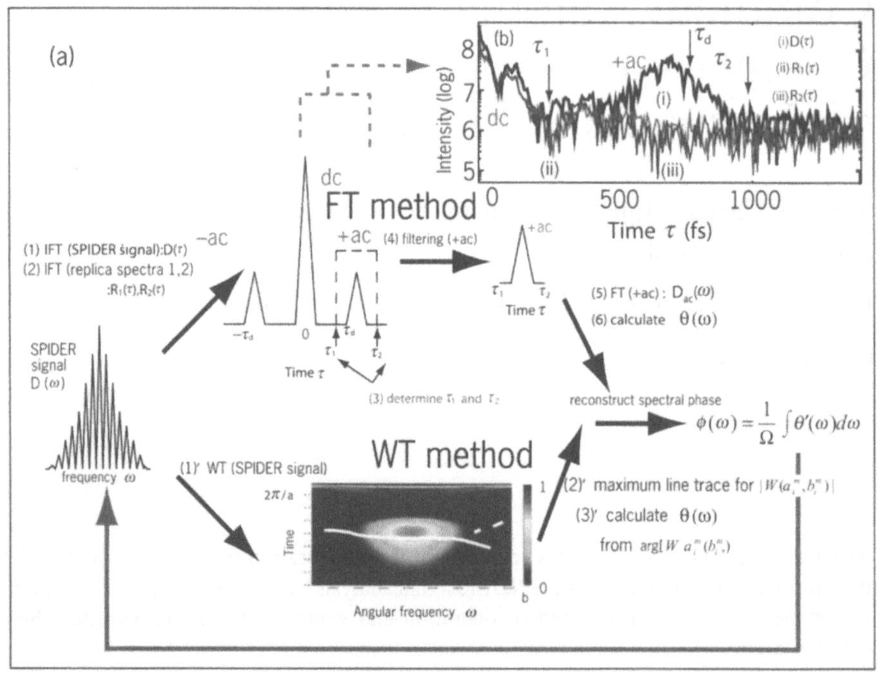


FIGURE 1. (a) Procedures of spectral phase reconstruction using FT and WT methods. (b) Practically measured example of $D(\tau)$ including dc and +ac components, $R_1(\tau)$ and $R_2(\tau)$, for tapered fiber output [5,6].

correct $\tau_1$ and $\tau_2$ after watching and comparing curves $D(\tau)$, $R_1(\tau)$, and $R_2(\tau)$ on the TV monitor, and then $\tau_1$ and $\tau_2$ are determined manually. Incorrect values of $\tau_1$ and $\tau_2$ lead to the incorrect spectral phase $\phi(\omega)$. On the other hand, in the latter method, following three steps with no manual determination are enough: (1) WT of $D(\omega)$ ($W(a, b) = \mathrm{WT}[D(\omega)]$), (2) the calculation of points $(a_i^m, b_i^m)$ where $|W(a_i^m, b_i^m)|$ becomes maximum for different $a_i$ s ($\partial|W(a, b)|/\partial a = 0$), and (3) the calculation of the argument $\theta(\omega)$ of $|W(a_i^m, b_i^m)| \exp\{i\theta(\omega)\}$. In addition, the WT method enables us to avoid uncertainty in phase retrieval and to lead to the correct $\phi(\omega)$ even in the case of the complicated and noisy $I(\omega)$.

We carried out the complete automatic experiment of the FB spectral-phase compensation using the WT method for the M-SPIDER signal analysis (Fig. 2). After being split, pulses with a 440-mW average power from a 12-fs Ti:sapphire laser at a 75-MHz repetition rate (a spectral broadening from 681 to 901 nm at the center wavelength of 786 nm: the curve (i) in Fig. 3c) propagated through a BK7 glass with a 7.5-mm length. Other split pulses with a 180-mW average power were employed as reference chirped pulses for the M-SPIDER. After passing through the BK7 glass, chirped pulses were automatically compensated for by the compensation system with the WT analysis. Figures 3a,b, the curve (ii) in Fig. 3c, and the curve (i) in Fig. 3d show the M-SPIDER signal before phase compensation,

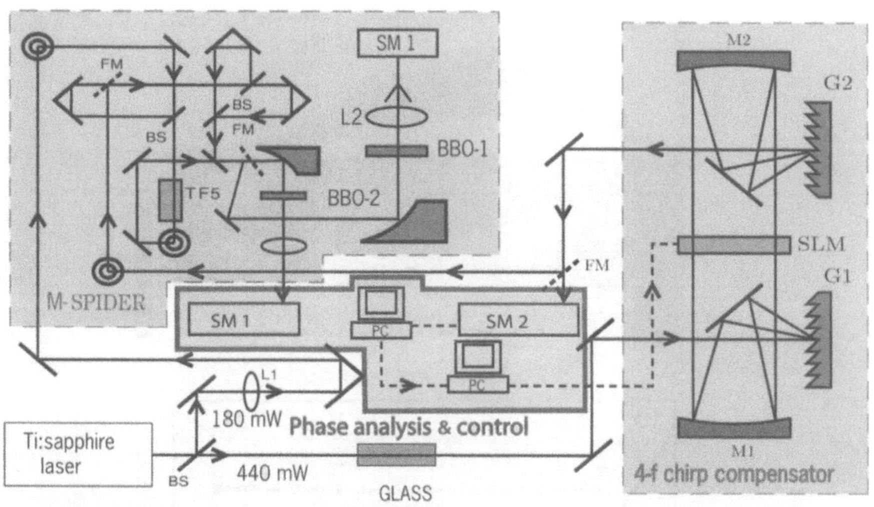


**Ti:sapphire laser**
pulse duration 12 fs
repetition rate 75 MHz
center wavelength 786 nm
average power 620 mW

BS : beam splitter
M : mirror
L : lens
G : grating
FM : flipper mirror

BBO-1 : Type I
BBO-2 : Type II
GLASS : BK7 lengh 7.5 mm
SM : spectrometer
TF5 : dispersive glass

FIGURE 2. Experimental setup.

the corresponding absolute function $|W_{BF}(a,b)|$ of the two-dimensional wavelet image of $W_{BF}(a,b)$, its spectral phase $\phi_{BF}$, and its temporal intensity profile $I_{\mathrm{BF}}(t)$, respectively (the delay time $\tau_d = 868$ fs and the spectral shear $\Omega = 53.3$ rad·THz). Here, $|W_{\mathrm{BF}}(a,b)|(i = \mathrm{BF}$, 1stFB, 2ndFB, and 3rdFB) is expressed by

$$W_i(a,b) = \frac{1}{\sqrt{a}} \int_{-\infty}^{\infty} D_i(\omega) \cdot \psi\left(\frac{\omega - b}{a}\right) d\omega$$

$$\psi(x) = \frac{1}{2\sqrt{\pi}\sigma} \exp\left(-\frac{x^2}{\sigma^2}\right) \exp(-ix)$$

with the dilation factor $a$, the translation factor $b$, and the window width $\sigma$ in the Gabor wavelet [4]. Curves (iii), (iv), and (iv) in Fig. 3c show spectral phases after first, second, and third FB compensations, respectively. Curve (v)′ in Fig. 3c is the enlarged version of curve (v). For one FB loop, it takes about 15 s for the WT analysis and then about 15 s for the SLM control at present, after one M-SPIDER signal was measured within ~40 s (three automatic rotations of the spectrometer grating are needed because of ultrabroadband pulses).

The spectral phase before compensation varies over 100 rads, but the spectral phase after third FB compensation is almost completely flattened. The group delay dispersion (GDD) $\ddot{\phi}\,(\omega_0) = +820\ \mathrm{fs}^2$ at the center wavelength, which was

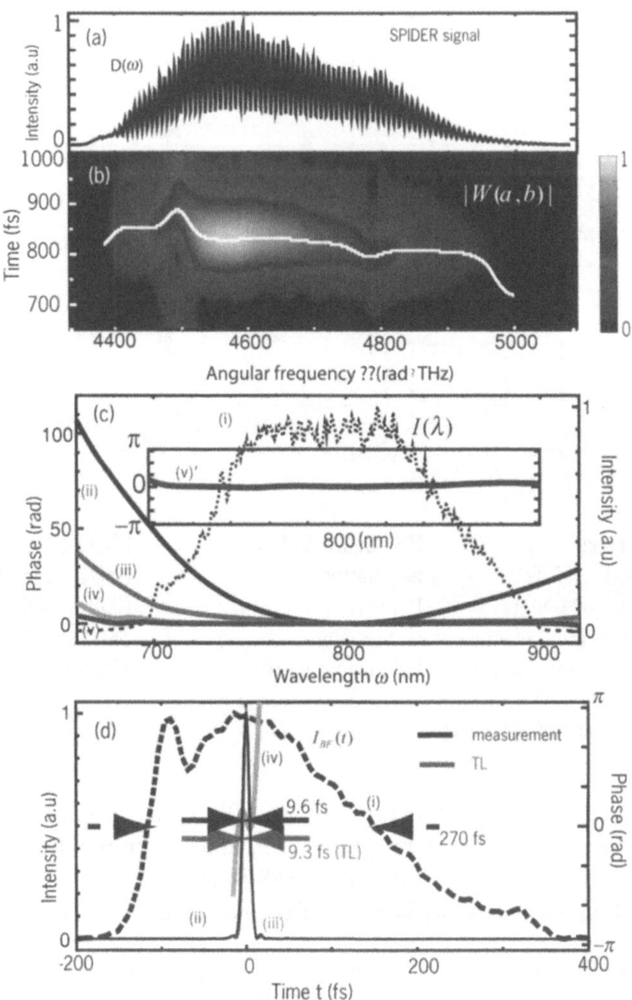


FIGURE 3. (a) M-SPIDER signal before compensation. (b) Its 2-dimensional map of $|W(a, b)|$. (c) Intensity spectrum; dotted curve (i): spectral phase before compensation; brown curve (ii): spectral phase after first compensation; red curve (iii): spectral phase after second compensation; green curve (iv): spectral phase after third compensation; blue curve (v): its enlarged one; curve (v)′. (d) Reconstructed temporal intensity profile before compensation; dashed curve (i): temporal intensity profile after third compensation; black curve (ii): temporal intensity profile of TL pulses; red curve (iii): time-dependent phase after third compensation; green curve (iv).

obtained from the spectral phase before compensation (curve (ii) in Fig. 3c), was reasonable compared with the GDD of the BK7 glass and other optical components. Curves (i), (ii), (iii), and (iv) in Fig. 3d show temporal intensity profiles before FB, after third FB, and of transform limited (TL) pulses and the time-dependent phase $\varphi(t)$ after third FB, respectively. Results indicate that strongly chirped pulses with a 270-fs duration and an asymmetric structure were compressed to 9.6 fs, which is very close to the TL pulse duration of 9.3 fs.

We believe that the developed self-recognition type of the spectral phase FB technique with complete automatic controllability which permits us to apply the technique to compensation for the unknown, complicated spectral phase exceeding a one-active bandwidth is significantly useful for scientists in other fields.

# References


1. K. Yamane, Z. Zhang, K. Oka, R. Morota, M. Ymashita, and A. Suguro: Optical pulse compression to 3.4 fs in the monocycle region by feedback phase compensation, Opt. Lett. **28**, 2258–2260 (2003).
2. K. Yamane, T. Kito, R. Morita, and M. Yamashita: 2.8-fs transform-limited optical-pulse generation in the monocycle region. In: Conference on Laser and Electro-Optics (CLEO) /International Quantum Electronics Conference (IQEC) 2004, Postdeadline Paper Book CPDP2 (Opt. Soc. Am.) (2004).
3. K. Yamane, T. Naoi, R. Morita, K. Sugiyama, M. Kitano, and M. Yamashita: Carrier-envelop-phase stabilized 3.3-fs optical pulses in the monocycle region. In: Conference on Laser and Electro-Optics (CLEO) 2005, JThE18 (Opt. Soc. Am.) (2005).
4. Y. Deng, Z. Wu, L. Chai, et al.: Wavelet-transform analysis of spectral shearing interferometry for phase reconstruction of femtosecond optical pulses, Opt. Exp. **13**, 2120–2126 (2005).
5. M. Adachi, M. Hirasawa, A. Suguro, et al.: Spectral-phase characterization and adapted compensation of strongly chirped pulses from a tapered fiber, Jpn. J. Appl. Phys. **42**, L24–L26, (2003).
6. M. Adachi, K. Yamane, R. Morita, and M. Yamashita: Photonic-crystal-fiber pulse compression by direct feedback of spectral phase without the Taylor expansion, IEEE Photon. Technol. Lett. **16**, 1951–1953 (2004).

# Carbon Nanotube Based Saturable Absorber Mirrors and Their Application to Ultrashort Pulse Generation

T. R. Schibli, K. Minoshima, H. Kataura, E. Itoga, N. Minami, S. Kazaoui, K. Miyashita, M. Tokumoto, and Y. Sakakibara

National Institute of Advanced Industrial Science and Technology, AIST, 1-1-1 Umezono, Tsukuba, Ibaraki 305-8563, Japan
t.schibli@aist.go.jp

**Abstract:** We demonstrate passive mode locking of solid-state lasers by saturable absorbers based on carbon nanotubes (CNT). These novel absorbers are fabricated by spin coating a polymer doped with CNTs onto commercial dielectric laser mirrors. We obtain broadband artificial saturable absorber mirrors with ultrafast recovery times without the use of epitaxial growth techniques, and the well-established spin-coating process allows the fabrication of devices based on a large variety of substrate materials. First, results on passive mode locking of Nd:glass and Er/Yb:glass lasers are discussed. In the case of Er/Yb:glass, we report the—to our knowledge—shortest pulse generated in a self-starting configuration based on Er/Yb:bulk-glass: 68 fs (45 fs Fourier-limit) at 1570-nm wavelength at a pulse-repetition rate of 85 MHz.

## 1 Introduction

Just a few years after the first demonstration of lasing in a rod of ruby it was realized that lasers could generate pulses much shorter than the round-trip time of the laser cavity when a specific dye was brought into the laser resonator. It was known that these dyes reveal an optical absorption that decreases with increasing optical energy flux. The insertion of such saturable absorbing dyes into the laser resonator forced the laser to operate in a pulsed—("mode-locked") rather than in a continuous—mode. Up to now a large variety of saturable absorbers for the ultrashort pulse generation were developed with properties much more suitable for the today's needs compared to these early dye-based absorbers.

For the ultrafast pulse generation, saturable absorbers with ultrafast recovery times are needed [1]. Up to date, artificial saturable absorbers for femtosecond pulse generation are mainly based on semiconductors (SESAM: semiconductor saturable absorber mirror) with ultrafast recovery times [2–4]. However, such absorbers have to be fabricated by very complex and costly processes such as metal-organic vapor-phase epitaxy (MOVPE) or metal-organic chemical vapor deposition (MOCVD), and further treatment is often required for reducing the recovery time of the absorbing layer [5]. Substrate removal is another technically challenging process that is required if the SESAM ought to be combined with a broadband dielectric or metallic mirror.

In recent years, several groups [6–9] have reported saturable absorption in carbon nanotubes (CNT) with ultrafast recovery times ($\sim$1 ps). With this new type of saturable absorbing material, first-mode locking of Er fiber lasers was soon demonstrated [10]. For solid-state lasers, Q-switched mode locking using CNT dissolved in heavy water has been reported [11] but continuous mode locking with CNT-based absorbers has not yet been demonstrated.

For solid state lasers, the requirements for the saturable absorbers (such as device uniformity, insertion loss, controllability of the saturable absorption, and thermal durability) are usually much more stringent than those for fiber lasers, and those requirements seem to be very difficult to be met with conventional CNT materials made by the deposition of bundled and entangled CNTs. Recently, Sakakibara et al. [12] demonstrated that carbon nanotube-polymer composite materials are well suited for reproducible construction of mode-locked Er fiber lasers, and the generation of sub-200-fs from a fiber oscillator was demonstrated [13]. The merit of such CNT–polymer composite materials is that they can be fabricated as large area thin films with excellent optical uniformity and fine controllability of the saturable absorption. In addition, the CNTs in these materials can be suspended in an isolated state. This isolation seems advantageous for self-starting the mode locking process [1], because isolated CNTs have a long-lived (several picoseconds or more) saturation recovery component in addition to the ultrafast ($\sim$1 ps) saturation recovery component [14]. These properties of the CNT–polymer composite materials also make them well suited for mode lockers for solid-state lasers. In this paper we describe the fabrication of an artificial saturable absorber mirror structure based on such composite materials that can be used in similar ways as SESAMs, and we discuss our first applications of these novel CNT-based saturable absorber mirrors in various solid-state lasers based on Nd- and Er/Yb:glass.

## 2 CNT-Based Saturable Absorber Mirrors

To be able to build a bulk laser in a similar configuration to that of a SESAM-based cavity we developed reflective-type saturable absorbers based on the above-mentioned CNT–polymer composite materials. Our current absorbers consist of commercially available dielectric mirrors based on $TiO_2/SiO_2$ on a fused silica substrate with a broad reflectivity bandwidth. These mirrors were coated with a thin layer (a few microns thick) of the CNT–polymer composite material by spin coating. For the polymers we used polyimide and carboxymethyl cellulose (CMC).

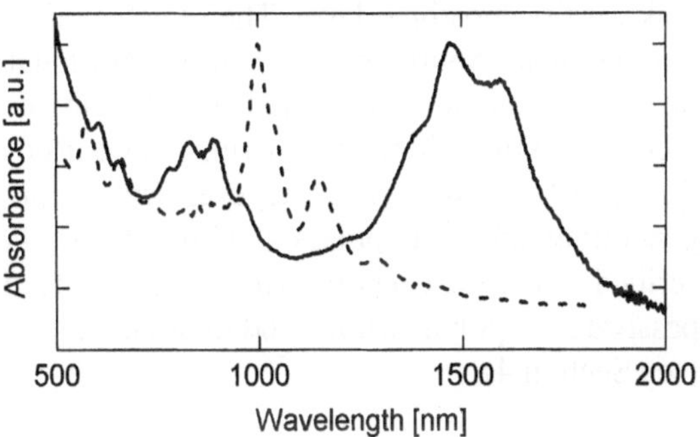


FIGURE 1. Absorption spectra of the CNT–polymer composite films. (Solid line) CNTs made by the laser ablation method dispersed in polyimide for use in Er/Yb:glass lasers. (Dashed line) CoMoCAT CNTs dispersed in CMC for use in Nd:glass lasers.

To achieve efficient saturation of the CNTs, the wavelength of the CNT absorption should coincide with the emission wavelength of the laser. Because the wavelength of the CNT absorption is a function of tube diameter [15], we adopted two types of CNT materials with different diameter distributions. With a mixture of slightly different nanotubes, broad absorption bands can be realized. For 1.5 μm use, CNTs were prepared by the laser ablation method [16]. For 1.0 μm use, commercial CoMoCAT CNTs (SouthWest NanoTechnologies Inc.) were purchased. Both types of CNTs were sonicated in solution and then mixed with polymer solutions. Figure 1 shows the absorption spectra of cast films made from these solutions. The amount of saturable absorption of each sample was controlled by the concentration of CNTs in the polymers and the rotation speed of the spin coater. Thin single layers of CNT–polymers were coated onto the $TiO_2/SiO_2$ dielectric mirrors designed for use in Nd- and Er/Yb:glass based solid lasers.

Despite this technically simple fabrication process we obtain broadband, reflective type saturable absorber mirrors with ultrafast recovery times. Thanks to the well-established spin-coating process, high-quality large area coatings onto a large variety of substrates become possible and samples in transmission, reflection, or partially reflecting samples (output coupling absorbers) can be obtained on a huge variety of optical surfaces. This technique might even allow for mode locking of monolithic cavities such as microchip lasers or nonplanar ring oscillators (NPRO) by directly applying the coating to the laser crystal.

Owing to the excellent surface quality and durability of these spin-coated composite materials (particularly being true for polyimide), postprocessing, such as applying antireflection (AR) coatings or partial reflection (PR) or high reflection (HR) coatings on the spin-coated polymer layer, is possible. In cases where very thick absorber layers are required, AR coatings might be desirable to avoid interference effects between the polymer–air and polymer–mirror interfaces that could lead to strong variations of the dispersion of these saturable absorber mirrors comparable to a Gires-Tournois interferometer. AR, PR, or HR coatings might also be used to tailor the saturation properties of the absorber and such coatings might be of interest for applications, where the absorber layer is directly brought onto the surface of a laser crystal. In such cases, the laser crystal could serve as the heat sink for the absorbing layer. The direct coating of the laser crystal would allow very compact cavity designs, and no additional focusing on the absorber would be required as the laser mode is usually already tightly confined inside the gain medium. In the following experiments we used absorbers without any additional AR or HR coatings. However, even without any such coatings we succeeded in generating sub-70 fs pulses at 1570 nm in a laser based on Er/Yb:glass as described in the next section. However, a protective coating might be required to passivate the polymer films and to protect the CNTs against atmospheric oxygen (see Section 4).

## 3 Ultrashort Pulse Generation in Solid-State Lasers

To test these absorbers in bulk lasers, we have built several cavities based on Nd:glass (QX/Nd, Kigre Inc.) and Er/Yb:glass (QX/Er, Kigre Inc.). For both types of laser materials we chose astigmatic compensated X-fold cavities with an additional focus on one of the end mirrors, which was replaced by one of the CNT-based saturable absorber mirrors (see Fig. 2). In the case of the Er:glass laser, the focal spot on the saturable absorber was $\sim$120 μm in diameter. This resulted in peak intensities of about 35 to 700 $\mathrm{MW \cdot cm^{-2}}$ depending on the pulse duration (68 fs – 2.3 ps) and the average intracavity power (0.5 – 3 W). To control the intra-cavity dispersion we could replace the mirror M3 by a GTI mirror (Gires-Tournois interferometer) if required. However, in the case of the Er:glass laser we obtained the shortest pulses without any such dispersion compensation since the laser glass provided sufficient anomalous dispersion at the emission wavelength to obtain soliton-like pulse shaping. The Er:glass laser was directly diode pumped by a pigtailed single transversal and single longitudinal-mode 980-nm laser diode (2900-series LD, JDS Uniphase Corp.). This diode delivered up to 500 mW of output power through a polarization maintaining fiber. However,

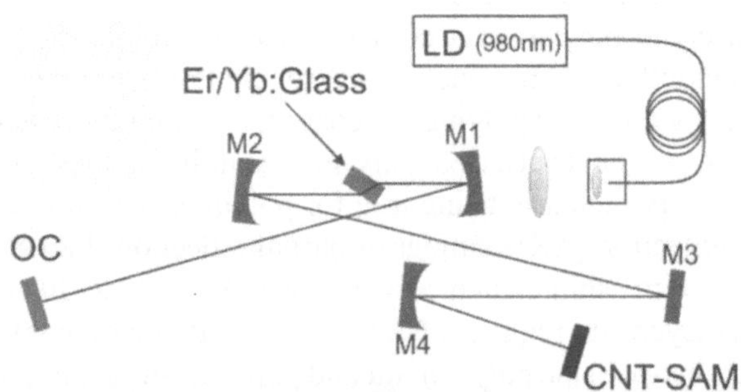


FIGURE 2. Experimental setup of the Er/Yb:glass laser. OC: output coupler; M1–M4: standard Bragg mirrors; CNT-SAM: saturable absorber mirror based on carbon nanotubes; LD: pigtailed laser diode for pumping the Er/Yb:glass (QX/Er, Kigre Inc., 4.8-mm path-length).

because of the limited transmission of the coupling-mirror M1 and the collimating optics, only about 400 mW of pumping power was available at the position of the laser glass. The Nd laser was pumped by a tapered laser diode (Eagleyard photonics EYP-TPL-0808) that delivered up to about 1.5 W of pumping power at 805-nm wavelength in a near diffraction limited beam ($M^2 \approx 1.5$).

For both types of lasers (Nd:glass and Er/Yb:glass) we obtained very stable self-starting mode locking with pulse durations in the femtoseconds and picaseconds regimes. However, we achieved higher performance in the Er-based (68 fs) laser compared to the Nd-based laser (~200 fs). In this report we shall therefore focus on the description of the Er-based laser.

The first sample we produced had a saturable absorption of about 0.4% and about 1–2% insertion loss. With this sample we already achieved reliable self-starting mode locking of the Er:glass laser shown in Fig. 2 with pulse durations around 100 fs at 1560 nm, with an average output power of ~10 mW from cavities with 90–200-MHz fundamental pulse-repetition rates. In a 500-MHz cavity, we observed thermal damage of the absorber due to the higher heat-load caused by the higher average power. For such high repetition rates, we might therefore need to use substrates and mirror coatings with higher thermal conductivities compared to the present fused-silica mirrors and/or thermal conductivity-enhancing codopants in the polymer layer (see below).

In order to achieve even shorter pulse durations, we opted for a sample with higher saturable absorption. We achieved this by increasing the CNT concentration in the polyimide layer without changing its thickness. Because of the higher absorption of this sample the repetition rate could no longer be increased to 200 MHz. However, we obtained reliable and stable self-starting mode locking at 85 MHz pulse-repetition rates with intracavity powers around 1W. With this second sample we readily obtained shorter pulses. Without any extracavity chirp compensation, we currently obtain pulse durations as short as 68 fs (45 fs Fourier limit) and spectra covering more than 200 nm at −30 dBc (see Fig. 3) from the 85-MHz cavity shown in Fig. 2. This is the—to our knowledge—shortest pulse [17]

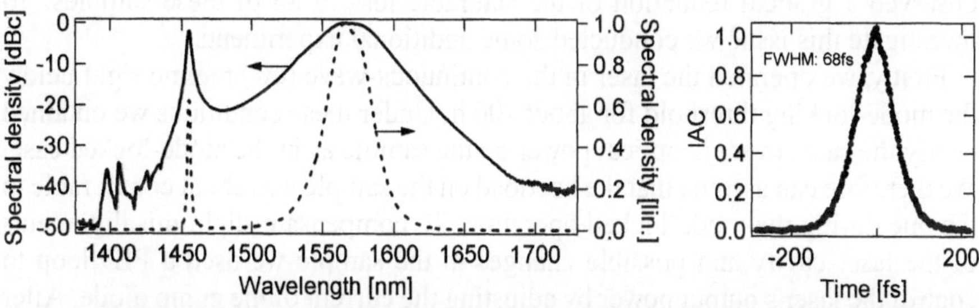


FIGURE 3. (Left side) Optical spectrum emitted by the Er:Yb:glass laser mode locked by the CNT-based saturable absorber mirror on a linear (dotted line) and logarithmic (solid line) scale. The Fourier-limit of the spectrum corresponds to 45 fs. (Right side) Background-free auto-correlation. The solid line is a sech2 fit with a corresponding FWHM pulse duration of 68 fs.

generated in an Er/Yb:bulk-glass laser even though we did not employ any intra- or extracavity dispersion compensation. It is interesting to note that these short pulses were obtained without the use of Kerr-Lens mode locking (KLM). This was made possible by the ultrashort recovery time of the CNTs. Currently the pulse duration is limited by the reflectivity bandwidth (~1450–1700 nm) and the resulting dispersion of the cavity mirrors. The sharp spike on the short wavelength side is due to the dispersion roll-off of those mirrors. With mirrors with a broader bandwidth even shorter pulses might be obtained with the very same saturable absorber layer.

As mentioned above, one might want to increase the thermal conductivity of the polymer film by codoping the polymer with thermally enhancing dopants or to increase the thermal conductivity of the substrate material in order to scale up the output power or the repetition rate of these lasers. For the third absorber we therefore codoped the CNT-doped polymer with 50 wt% of alumina. With this absorber we succeeded in producing up to 80 mW of average output power from the same 85-MHz cavity in the single-pulse regime. This value was solemnly limited by the available pumping power of 400 mW corresponding to an optical-to-optical quantum efficiency of more than 30% at an output-coupling ratio of 3%. We think that the combination of the alumina codoping and the use of substrate materials with higher thermal conductivities will greatly increase the range of applications of these absorbers. However, even the current samples with alumina codoped polyimide on a fused-silica substrate can already withstand several watts of power with spot sizes as small as $10^{-4}$ cm$^2$ corresponding to heating densities exceeding 0.5 MW $\cdot$ cm$^{-3}$.

## 4 Long-Term Stability of CNT Saturable Absorbers

Because of the excellent adhesion and mechanical properties of the polyimide films we never observed delamination or cracking of the absorber layer nor could we find any limitations of the self-life. However, during the first experiments we observed a gradual reduction of the saturable loss in all of these samples. To investigate this issue we conducted some additional experiments.

Firstly, we operated the laser in the continuous-wave (cw) regime right below the mode-locking threshold for about 100 h. Under these conditions we obtained nearly the same average optical power on the sample as in the mode-locked case. We therefore can assume that the heatload on the sample was about comparable to the one during the mode-locked operation. To compensate slight misalignments of the laser cavity and possible changes in the sample we used a PID loop to control the laser's output power by adjusting the current of the pump diode. After 100 h of continuous operation we slightly increased the pump power to achieve mode-locked operation. However, we did not find any significant difference between this preilluminated sample position and a non-preilluminated position on the same sample. We therefore concluded that the heating due to the average power or the average intensity was not the cause for the slow degradation of the saturable absorption.

Next, we operated the laser with different pulse durations from 2.3 ps down to about 100 fs. In order to keep the average power on the sample constant we could either vary the intracavity dispersion or the position of the laser glass. If we, for instance, slightly moved the laser glass out of the focus in the X-fold cavity we obtained less nonlinear phase-shifts, which lead to longer pulses at the same pulse energies. During these experiments, we found that the rate of the reduction of the saturable loss of the sample nonlinearly depends on the peak intensity on the absorber. Up to date we cannot yet fully identify the exact process of this degradation. However, we have reasons to believe that reactive oxygen $O_2(^1D_g)$ (also known as "singlet oxygen") might destroy the CNTs. Singlet oxygen can be generated from atmospheric oxygen, which is usually in the much less reactive triplet state $O_2(^3S_g^-)$, by electronic energy transfer from photoexcited sensitizers. The required energy for this transition is approximately 94 kJ/mol, corresponding to roughly 1 eV/molecule. The energy of the photons of the laser at 1.55 μm is at about 0.8 eV per photon. Therefore, a multiphoton excitation of either the CNTs (for instance, through excited-state absorption) or of the polymer or one of its additives (through multiphoton absorption) would be sufficient to convert a rather stable triplet oxygen molecule from the atmosphere into a highly reactive singlet oxygen molecule that might react with the CNTs.

To verify this assumption we blew a weak jet of nitrogen gas toward the sample. As soon as we used this nitrogen rich atmosphere we could not observe any degradation of the saturable absorption even under strong illumination intensities (75-fs pulses; corresponding to $0.5 \sim 1\ \mathrm{GW \cdot cm^{-2}}$ peak intensities at 85-MHz pulse-repetition rate) over extended periods of time (>30 h). This is a strong indication that the degradation was caused by the combination of multiphoton excitation and the presence of atmospheric oxygen. We found that if at least one of these two factors is removed the samples do not show any degradation or ageing.

It is well known that gas molecules can penetrate and pass polymer films. Therefore, inorganic layers are often used to cover the polymer films in order to form gas barriers. Fortunately, an additional thin-film coating of dielectric materials on the surface of the polyimide absorber layer is possible due to the robustness and the high thermal durability of polyimide. Such a protective coating could simultaneously serve as an antireflection coating for the absorber. This finally paves the way toward robust and high-performance artificial saturable absorbers based on polymer-embedded carbon nanotubes.

## 5 Conclusions

In conclusion, we have introduced a novel type of artificial saturable absorber mirrors based on polymer-embedded CNTs. These mirrors can be produced in a much less costly way than the current sate-of-the-art semiconductor-based saturable absorber mirrors (SESAM). Further, they enable the fabrication of new types of broadband saturable absorber mirrors with ultrashort recovery times without the requirement of the very challenging substrate-removal and ion-implantation processes. In this paper we demonstrated a few applications for these saturable absorber mirrors and we succeeded in producing 68-fs pulses at 1570 nm directly from a diode-pumped all-Bragg-mirror Er/Yb:glass laser. In the end we discussed the initial stability issues of these CNT–polymer based saturable absorber mirrors and we presented a way how to obtain long-term stable saturable absorber mirrors based on polymer-embedded carbon nanotubes.

**Acknowledgments.** This research was gratefuly supported in parts by the "Grants-in-Aid for Scientific Research" from the Japan Society for the Promotion of Science (JSPS) and by the Industrial Technology Research Grant Program in '03 from New Energy and Industrial Technology Development Organization (NEDO) of Japan. T. R. Schibli was supported by the JSPS postdoctoral fellowship program for foreign researchers.

## References

1. E. P. Ippen: Principle of passive mode locking. Appl. Phys B **58**, 159–170 (1994).
2. L. R. Brovelli, U. Keller, and T. H. Chiu: Design and operation of antiresonant Fabry-Perot saturable semiconductor absorbers for mode-locked solid state lasers. J. Opt. Soc. Am. B **12**, 311–322 (1995).
3. U. Siegner, R. Fluck, G. Zhang, and U. Keller, 'Ultrafast high-intensity nonlinear absorption dynamics in low-temperature grown gallium arsenide,' Appl. Phys. Lett. **69**, 2566–2568 (1996).
4. B. Collins, K. Bergmann, S. T. Cundiff, et al.: 'Short cavity Erbium/Ytterbium fiber lasers mode-locked with a saturable bragg reflector,' IEEE J. Sel. Top. Quantum Electron. **3**, 1065 (1997)
5. H. H. Tan, C. Jagadish, M. J. Lederer, et al.: 'Role of implantation induced defects on the response time of semiconductor saturable absorbers,' Appl. Phys. Lett. **75**, 1437–1439 (1999)
6. Y.-C. Chen, N. R. Raravikar, L. S. Schadler, et al.: 'Ultrafast optical switching properties of single-wall carbon nanotube polymer composites at 1.55 μm,' Appl. Phys. Lett. **81**, 975–977 (2002).
7. J-S. Lauret, C. Voisin, G. Cassabois, et al.: 'Ultrafast carrier dynamics in single-wall carbon nanotubes,' Phys. Rev. Lett. **90**, 057404 (2003).
8. S. Tatsuura, M. Furuki, Y. Sato, I. Iwasa, M. Tian, and H. Mitsu, 'Semiconductor carbon nanotubes as ultrafast switching materials for optical telecommunications,' Adv. Mater. **15**, 534 (2003).
9. Y. Sakakibara, S. Tatsuura, H. Kataura, M. Tokumoto, and Y. Achiba, 'Near-infrared saturable absorption of single-wall carbon nanotubes prepared by laser ablation method,' Jpn. J. Appl. Phys. **42**, 494–496 (2003).
10. S. Y. Set, H. Yaguchi, Y. Tanaka, et al.: Mode-locked fiber lasers based on a saturable absorber incorporating carbon nanotubes, OSA Trends in Optics and Photonics

(TOPS). In: Optical Fiber Communication Conference, Technical Digest, Postconference Edition (Optical Society of America), Washington, DC, Vol. 86, p. 44 (2003).
11. N. N. Il'ichev, E. D. Obraztsova, S. V. Garnov, and S. E. Mosaleva, 'Nonlinear transmission of single-wall carbon nanotubes in heavy water at wavelength of 1.54 μm and self-mode locking in a Er3+:glass laser obtained using a passive nanotube switch,' Quant. Electron. **34**, 572–574 (2004).
12. Y. Sakakibara, A. G. Rozhin, H. Kataura, Y. Achiba, and M. Tokumoto, Jpn. J. Appl. Phys. **44**, 1621 (2005).
13. A. G. Rozhin, Y. Sakakibara, S. Namiki, M. Tokumoto, and H. Kataura: Sub-200 fs pulsed erbium-doped fiber laser using a carbon nanotube- polyvinylalcohol modelocker. In submitted for publication.
14. G. N. Ostojic, S. Zaric, J. Kono, et al.: 'Interband recombination dynamics in resonantly excited single-walled carbon nanotubes,' Phys. Rev. Lett. **92**, 117402 (2004).
15. Carbon Nanotubes: Synthesis, Structure, Properties, and Application, M. S. Dresselhaus, G. Dresselhaus, P. Avouris, eds. Berlin: Springer-Verlag (Topics in Appl. Phys.) Vol. 80 (2001).
16. H. Kataura, Y. Kumazawa, Y. Maniwa, et al.: Carbon **38**, 1691 (2000).
17. F. J. Grawert, J. T. Gopinath, F. Ö. Ilday, et al.: '220-fs erbium-ytterbium:glass laser mode locked by a broadband low-loss silicon/germanium saturable absorber,' Opt. Lett. **30**, 329–331 (2005).

# Generation of Extreme Ultraviolet Continuum Radiation Driven by Sub-10-fs Two-color Field

M. Kaku,[1] Y. Oishi,[1,2] A. Suda,[1] F. Kannari,[2] and K. Midorikawa[1]

[1] RIKEN, 2-1 Hirosawa, Wako-shi, Saitama 351-0198, Japan
[2] Keio University, 3-14-1 Hiyoshi, Kohoku-ku, Yokohama 223-0061, Japan
mkaku@riken.jp

**Abstract:** We observed extreme ultraviolet (XUV) continuum radiation generated by using a sub-10-fs two-color pulse. When a laser field consisting of fundamental and its second harmonic pulses with parallel polarization was applied to high-order harmonic generation in argon, a continuum spectrum centered at 30 nm was obtained with an energy as high as 10 nJ. The broadband radiation indicates the possibility of generating powerful single attosecond pulses in the XUV region.

## 1 Introduction

The high-order harmonic generation (HHG) provides a high-brightness coherent light source in the extreme ultraviolet (XUV) region. By using HHG, the generation of both a train of attosecond pulses and a single attosecond pulse has been demonstrated [1–3]. The properties of HHG can be understood in terms of a semiclassical model [4]. Under an intense laser field, an atom undergoes tunneling ionization close to the maximum of driving field, whereupon the ionized electron is accelerated away from the ion. Then, the electron is driven back to the parent ion and recombined with it to generate an attosecond burst of XUV light. Since this process occurs every half-cycle of the driving field, an attosecond pulse train is generated by using a conventional multicycle driving pulse. To generate a single attosecond pulse, the driving pulse duration must be as short as possible. A polarization gating technique was proposed for a half-cycle HHG [5] and broadband XUV radiation was observed, which corresponds to a Fourier-transform-limited pulse duration of 200 as [6]. In this method, however, only leading and trailing edges of the fundamental pulses can contribute to HHG.

As an alternative approach, we apply a two-color method to HHG for an efficient single attosecond pulse generation. The two-color HHG using relatively long conventional pulses has so far been studied both theoretically [7] and experimentally [8, 9], where the most pronounced feature is the appearance of even-order harmonic components in addition to the conventional odd-orders because of the collapse of inversion symmetry in the light-matter system. One can expect that when the driving pulse duration decreases to the sub-10-fs regime, the bandwidth of each harmonic component broadens and overlaps with neighbors to form a less-structured continuum. In the time domain, the characteristics are understood as reshaping of the pulse envelope, resulting in a shortening of effective pulse duration contributing to HHG. Here we consider a synthesized electric field $E(t)$ consisting of a fundamental and its second harmonic (SH) fields, expressed as

$$\begin{aligned} \boldsymbol{E(t)} &= \boldsymbol{E}_1 \exp\left(-2\ln 2\left(\frac{t}{\tau_1}\right)^2\right)\cos(\omega t+\phi) \\ &\quad + \boldsymbol{E}_2 \exp\left(-2\ln 2\left(\frac{t-\Delta t}{\tau_2}\right)^2\right)\cos\left(2\omega t+\phi+\Delta\phi\right) \end{aligned} \tag{1}$$

where the subscript $i$ (1 or 2) denotes the fundamental or SH field, respectively. $E_i$ is the electric field amplitude, $\tau_i$ is the full width at half maximum (FWHM) pulse duration assuming a Gaussian profile, the carrier frequency for the fundamental, $\Delta\tau_i$ is the time delay between the two pulses, $\phi$ is the absolute phase, and $\Delta\phi$ is the relative phase between them. Figure 1 shows one of the examples of the square of synthesized electric field $|\ E(t)\ |^2$, where $E_1$ and $E_2$ are set at 0.96 and 0.04, respectively. $\omega$ is $2.36 \times 10^{15}$ rad/s corresponding to a wavelength of 800 nm, both $\tau_1$ and $\tau_2$ are 10 fs and all others ($\Delta t$, $\phi$, and $\Delta\phi$) are set at zero. As can be seen around the peak, the amplitude of the nearest neighbors

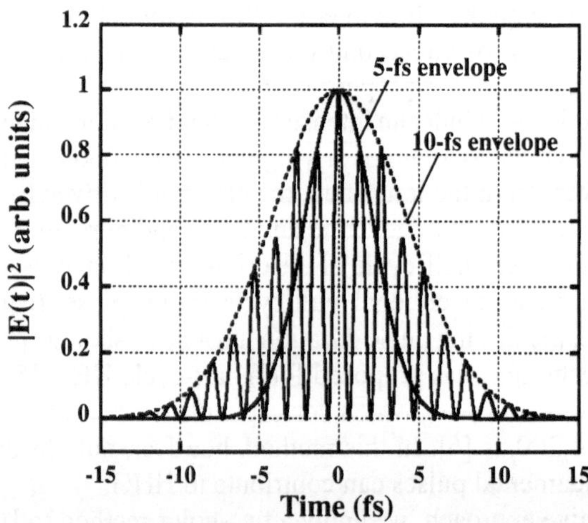


FIGURE 1. Temporal profile of the square of a synthesized electric field consisting of a fundamental and its second harmonic pulses. The envelopes of 10-fs and 5-fs pulses are also shown.

on both sides is well suppressed by adding small amount of the SH field to the fundamental field, resulting in a reshaped envelope which is equivalent to a 5-fs pulse in this restricted time (-2 to 2 fs). As a result, the number of optical cycles contributed to HHG is limited and the harmonic radiation can be confined to a time interval much shorter than the fundamental pulse. Since the high-order harmonics are generated near the peak of the driving field in the two-color method, one can expect an efficient conversion to the XUV radiation. From the standpoint mentioned above, the intense two-color pulses generated based on sub-10-fs multi-mJ Ti:sapphire laser system with a pressure gradient hollow fiber technique [10] can be applied to the broadband XUV emission toward the generation of single attosecond pulses.

In this paper, we report the generation of continuum high-order harmonic radiation in the XUV region by using sub-10-fs two-color laser pulses. When a laser field consisting of sub-10-fs fundamental and SH pulses with parallel polarization is used, the harmonic spectrum shows a continuum near the cut-off. We also observe harmonic energy in the XUV region as high as 10 nJ. The broadband radiation in the XUV region indicates the possibility of generating single attosecond pulses.

## 2 Experiment

Figure 2 shows the experimental setup for the HHG using two-color pulses. The experiment was performed using a Ti:sapphire laser chirped-pulse amplification (CPA) system with a pressure-gradient hollow fiber [10]. The CPA system at 800 nm produced the output energy of 5 mJ with a pulsewidth of 20 fs at a repetition rate of 1 kHz. The output pulses from the CPA system were further shortened by means of pressure-gradient hollow fiber pulse compression technique, resulting in a pulse duration of 9 fs with an energy of 2 mJ.

The fundamental pulse at the center wavelength of 800 nm was focused by a 150-cm focal-length lens with a broadband (600–1000 nm) antireflection coating into a HHG gas cell installed in a vacuum chamber. The gas cell, which was 2-cm length, had 0.5-mm diameter pinholes at both sides to isolate the vacuum and the gas-filled regions by differential pumping. As an interaction medium, argon was introduced from a reserve chamber into the gas cell. The generated high harmonics were separated from the fundamental by using a silicon beam splitter with a reflectivity of 50% at 30 nm. The harmonic spectrum was observed using a microchannel plate (MCP) mounted on a flat-field grazing-incidence spectrometer. A CCD camera detected a fluorescence image from a phosphor screen placed behind the MCP.

The optical elements for generating two-color pulses were closely aligned straight in front of the entrance of the HHG chamber, as illustrated in Fig. 2. A small amount of the fundamental pulse was frequency-doubled by a beta-barium borate ($\alpha$-BBO) type-I SHG crystal (300-μm thick) and subsequently the remaining fundamental and orthogonally-polarized SH pulses were passed through

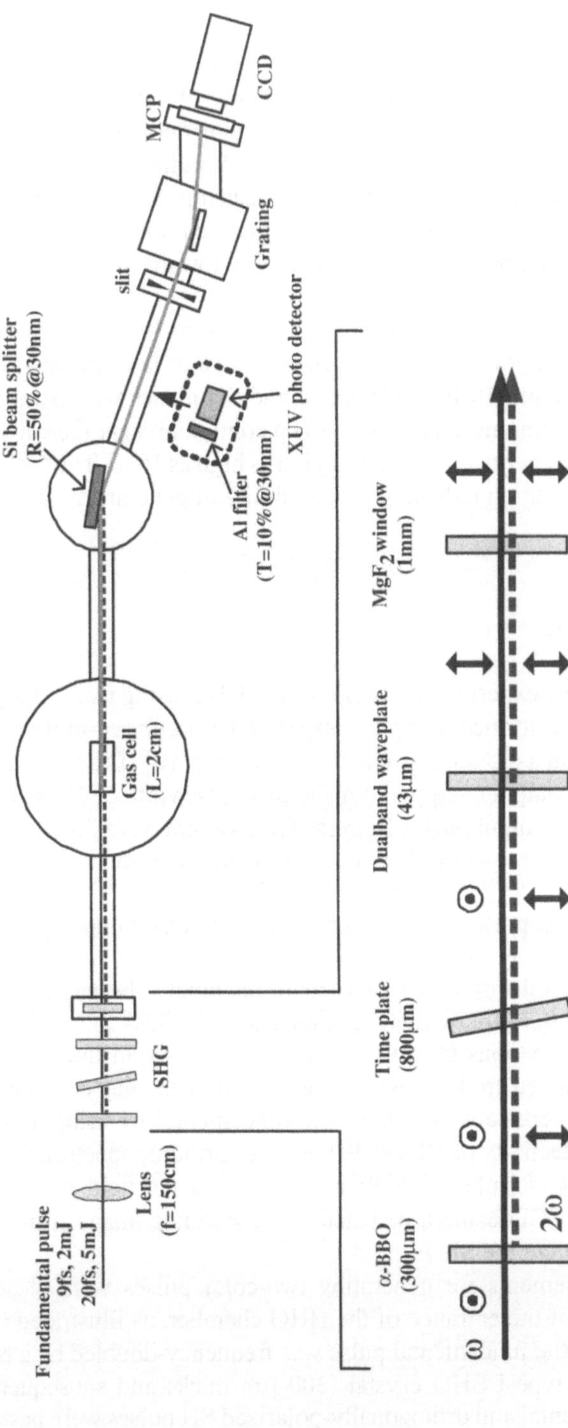


FIGURE 2. Experimental setup for two-color HHG.

a $\beta$-BBO time plate (800-μm thick) which adjusts a temporal overlap between the fundamental and SH pulses at the gas cell. The relative phase between them was also adjusted by slightly rotating the time plate. Then, the delay-controlled two-color pulses passed through a quartz dual-band wave plate (43-μm thick) which rotates the polarization direction of the fundamental by 90° while keeping that of the SH unchanged. Thus, the polarizations are parallel to each other. Finally, the two-color pulses passed through an $MgF_2$ window (1-mm thick) in the HHG vacuum chamber. We confirmed that dispersion for the fundamental pulse was well compensated and the duration was 9 fs with and without SHG. On the other hand, the SH pulse duration was 35 fs estimated from the dispersion induced by optical materials at 400 nm. The spatial overlap of the two pulses at the gas cell can be achieved in this configuration, and we actually confirmed that the waist positions for both pulses agreed to within 3 mm. The spot diameter of the SH pulse at the focus was approximately 100 μm, which was a bit smaller than the fundamental (130 μm). To measure the absolute harmonic energy, a 200-nm-thick aluminum filter with a transmittance of 10% at 30 nm and a calibrated XUV photodetector were inserted after the silicom beam splitter.

## 3 Results and Discussion

In a preliminary experiment on HHG, we used 20-fs pulses directly from the Ti:sapphire CPA system. To maximize the harmonic intensity, the pressure of argon in the reserve chamber was optimized to be 28 torr. Figure 3 shows harmonic spectra in the XUV region generated using (a) fundamental pulse alone and (b) two-color pulse with adding an SH pulse of 0.3-mJ energy. The spectrum obtained using one-color driving pulse shows a typical absorption-limited harmonic spectrum maximized at around the 27th–29th harmonics. By adding the SH pulse to the fundamental followed by fine-tuning of the relative phase between them, the harmonic spectrum equally contained the odd- and even-order

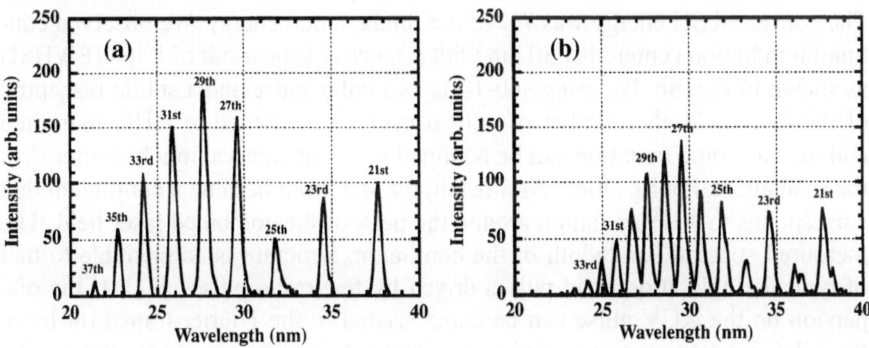


FIGURE 3. Harmonic spectra generated by using (a) one-color 20-fs pulse and (b) two-color pulse.

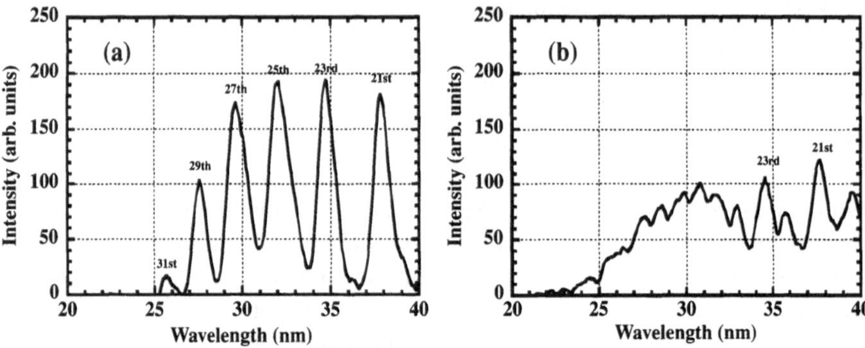


FIGURE 4. Harmonic spectra generated by using (a) one-color sub-10-fs pulse and (b) two-color pulse.

components as shown in Fig. 3b. We confirmed that the even-order components disappeared when the polarizations were set at perpendicular to each other by rotating the dual-band wave plate. This is because the intensity of the SH pulse is too low to affect the HHG in the perpendicular polarizations. Of course, the SH pulse alone cannot generate high-order harmonics in this spectral region. In both the one-color and two-color HHG, the harmonic energy measured after passing through the Al filter was 2.5 nJ at the XUV detector and the corresponding energy generated at the gas cell was evaluated to be 50 nJ.

Then, we used sub-10-fs fundamental pulses from the hollow-fiber pulse compressor. Figure 4a shows a harmonic spectrum generated using only sub-10-fs fundamental pulse at 800 nm. The optimum pressure of argon was 10 torr in this case. The harmonic spectrum has a discrete structure corresponding to odd multiples of the fundamental frequency. The spectral bandwidth of the individual harmonic component was broad due to the driving pulse duration as short as 9 fs.

Figure 4 b shows a harmonic spectrum generated using a two-color laser field consisting of a sub-10-fs fundamental and SH pulses with parallel polarization. The converted SH energy was 7% of the fundamental energy. We observed continuum radiation centered at 30 nm with a spectral bandwidth of 8 nm (FWHM) as shown in Fig. 4b. By using sub-10-fs two-color pulse and a subtle reshaping of the laser field, the number of optical cycles contributed to HHG is limited and the harmonic radiation can be confined to a time interval much shorter than the fundamental pulse alone. As a result, the spectrum became a continuum that corresponds to XUV radiation around the peak of the combined laser field. The measured spectral bandwidth of the continuum structure is comparable to that of typical single attosecond pulses driven by few-cycle pulses [11]. If the dispersion on the XUV pulse can be compensated to the Fourier-transform-limit, the pulse duration corresponding to the obtained spectrum would be 200 as. The harmonic energy within the spectrum limited by the Al filter was measured to be 0.5 nJ, from which the generated energy in the gas cell was evaluated as

10 nJ, which is comparable to the energy generated by using only fundamental pulse.

## 4 Summary

In conclusion, we have demonstrated a continuum high-order harmonic radiation in the XUV region driven by using sub-10-fs two-color laser pulses. When the laser field consisting of fundamental and its SH pulses with parallel polarization was used, we observed continuum radiation centered at 30 nm with a spectral bandwidth of 8 nm (FWHM). The generated energy is as high as 10 nJ which is close to an absorption limited value in the phase-matched HHG. In this way, the HHG driven by a two-color field has a great potential for shorting the XUV pulse along with the enhancement of pulse energy. This technique also allows extension of cutoff wavelength toward the water-window region, since the depletion of the neutral atoms by ionization in the leading part of the pulse can be eliminated by the use of two-color technique.

**Acknowledgments.** M. Kaku is grateful to the Special Postdoctoral Researchersí Program of RIKEN. Y. Oishi was supported by the Junior Research Associate Program of RIKEN.

## References

1. P. M. Paul, E. S. Toma, P. Breger, et al.: Science **292**, 1689 (2001).
2. M. Hentschel, R. Kienberger, Ch. Spielmann, et al.: Nature **414**, 509 (2001).
3. R. Kienberger, E. Goulielmakis, M. Uiberacker, et al.: Nature **427**, 817 (2004).
4. P. B. Corkum: Phys. Rev. Lett. **71**, 1994 (1993).
5. P. B. Corkum, N. H. Burnett, and M. Y. Ivanov: Opt. Lett. **19**, 1870 (1994).
6. B. Shan, S. Ghimire, and Z. Chang: J. Mod. Optic. **52**, 277 (2005).
7. H. Eichmann, A. Egbert, S. Nolte, et al.: Phys. Rev. A **51**, R3414 (1995).
8. S. Watanabe, K. Kondo, Y. Nabekawa, A. Sagisaka, and Y. Kobayashi: Phys. Rev. Lett. **74**, 2692 (1994).
9. I. J. Kim, C. M. Kim, H. T. Kim, et al.: Phys. Rev. Lett. **94**, 243901 (2005).
10. Y. Oishi, A. Suda, F. Kannari, and K. Midorikawa: Rev. Sci. Instrum. **76**, 093114 (2005).
11. T. Brabec and F. Krausz: Rev. Mod. Phys. **72**, 545 (2000).

# Ultrawideband Regenerative Amplifiers via Intracavity Acousto-Optic Programmable Gain Control

Thomas Oksenhendler,[1] Daniel Kaplan,[1] Pierre Tournois,[1] Gregory M. Greetham,[2] and Frédéric Estable[2]

[1] FASTLITE, Bât 403, Ecole Polytechnique, 91128 Palaiseau, France
**thoksen@fastlite.com**
[2] AMPLITUDE TECHNOLOGIES, 2, rue du Bois Chaland, 91029 EVRY, France

## 1 Introduction

Chirped-pulse amplification (CPA) currently produces multiterawatt pulses of approximately 30-fs duration which are used for a variety of high-intensity ($>10^{18}$ W $\cdot$ cm$^{-2}$) and high-field applications. Higher intensities can be reached either by higher energies or shorter pulse durations. The high cost of raising the energy favors the shorter pulse duration approach.

Shorter pulse durations (sub-20 fs) in CPA lasers have been demonstrated by a reduction of the gain narrowing in two ways: First, the precompensation of the gain narrowing by the use of an acousto-optic programmable dispersive filter (AOPDF) before the amplifier [1], second, the introduction of spectral filters, such as Fabry–Perot etalons [2], birefringent filters in the laser cavity. The AOPDF approach is limited because large enhancements of the bandwidth require strong reduction of the energy of the seed pulses. The drawbacks of the intracavity spectral filters are the lack of programmability and the generation of artifacts.

In this paper we report the development of a new intracavity acousto-optic programmable gain control filter (AOPGCF) for ultrawideband regenerative amplification. We have developed a new Brewster cut acousto-optic collinear tunable filter that accurately controls the spectral losses of a linear regenerative cavity. The spectral losses are introduced through partial diffraction of the unwanted spectral components. The spectral transmission of the filter is thus modified by the diffraction : Total diffraction of a wavelength corresponds to zero transmission and no diffraction of a wavelength corresponds to a total transmission of this wavelength. To optimize diffraction efficiency and spectral resolution, the AOPGCF is based on the same collinear beams interaction geometry as the AOPDF [3]. The incident, diffracted, and acoustic beams are nearly collinear in energy directions leading to a long interaction length, i.e., a high-diffraction efficiency and a high resolution. As the physical constants of the crystal used ($TeO_2$) are well known, the calibration is carried out either by an acoustic frequency tuning or a crystal tilt in the main interaction plane. The Brewster cut of optical faces avoids insertion losses and spurious pulse replica.

The main drawback of this device is the own dispersion of the 25-mm-long $TeO_2$ crystal (12406 fs$^2$, 8114 fs$^3$, 4700 fs$^4$ at 800 nm for 25 mm) which is amplified by 24 passes through the crystal in the regenerative cavity. A simple tuning of the compressor allows to compensate for the second order but not for the huge third order ($\simeq$ 200000 fs$^3$) and higher order terms. To optimally compress the pulse we have used a high-resolution AOPDF system (DAZZLER HR-800 from Fastlite).

## 2 Experimental Setup

The layout of the complete laser is shown in Fig. 1. The oscillator is an ultra-broadband Femtolasers oscillator (>100-nm FWHM) for a good injection on the complete bandwidth. The stretcher is a conventional Offner type stretcher with a 1200-mm$^{-1}$ grating. The optical components limit the bandwidth to approximately 110 nm. At the output of the stretcher, we introduce a high-resolution AOPDF system (DAZZLER HR-800) to compensate for the phase over a temporal excursion of 8 ps. After the Faraday isolator, the AOPDF diffracted pulses are

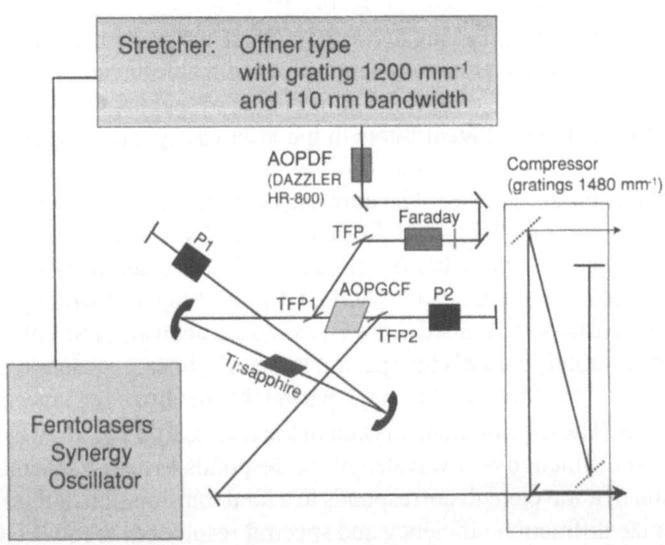


FIGURE 1. Schematic of the complete laser setup. TFP, thin film polarizer; P, Pockels cell; AOPGCF, acousto-optic programmable gain control filter.

injected into the cavity through the thin film polarizer TFP1. Pulsing the Pockels cell (P1) to its quarter wave voltage traps a single pulse inside the cavity. At each round trip in the cavity, the pulse is passing twice through the Ti:sapphire rod and the AOPGCF. The diffracted beam is lost and only the nondiffracted beam is amplified. After a number of passes, determined by the build-up time, the pulse is ejected from the cavity through the thin-film polarizer TFP2 by application of an electrical pulse to the second Pockels cell (P2). The output pulse is then recompressed by a 1480 lines/mm gratings compressor with a limited bandwidth of about 105 nm.

## 3 Characterization of AOPGCF

Many applications of sub-20-fs high-power laser systems require also a good stability and a high contrast ratio. We have characterized for each AOPGCF setup, the spectrum, the stability, the contrast ratio, and the temporal intensity. The contrast ratio has been measured by a high dynamic range third-order cross-correlator (SEQUOIA from Amplitude Technologies), the temporal intensity by a SPIDER (SPIDER 20–80 fs from APE).

In every experiment, the measured spectral phase is fedback to the AOPDF to flatten the phase. The contrast measurement by high dynamic range third-order cross-correlator is of prime interest because of known spurious pulses apparitions in the former techniques using Fabry–Perot etalons or birefringent filters and the absolute necessity of a good amplified spontaneous emission (ASE) contrast ratio for such high-power laser. The stability is measured by a photodiode and an oscilloscope.

The free-running spectrum of the laser, without AOPGCF inside the cavity and with AOPDF only phase modulation,, is measured to be 33-nm FWHM, the pulse duration is 29-fs FWHM, the contrast ratio is about $10^7$, and stability is 2.6% rms (Fig. 2). An optimization of the seed pulse by the AOPDF amplitude and phase shaping enlarges the spectrum up to 65-nm FWHM, corresponding to 26.2-fs FWHM pulse, with a contrast ratio about $10^6$ and a stability of 7.5 %rms. The deterioration of the contrast ratio and the stability are due to the important decrease in the seed pulse energy introduced by the AOPDF amplitude modulation.

The optimizations of the laser using the AOPGCF for spectral amplitude and the AOPDF for spectral phase optimization necessitate only simple, direct, few steps algorithms. No blind algorithm optimizations are needed for both spectral amplitude and phase optimizations.

The amplitude optimization is obtained by the direct feedback of the spectrum to the AOPGCF. The output-pulse-measured spectral amplitude is directly used as the spectral amplitude to be diffracted by the AOPGCF. Thus, the peaks are more attenuated and the spectrum converges onto a square shape. In fact, our goal is to determine the spectral losses that completely flatten the global gain of the amplifier. For 100-nm bandwidth, this process needs only a few steps (<20). The final spectral losses thus determined can be used as long as the cavity does

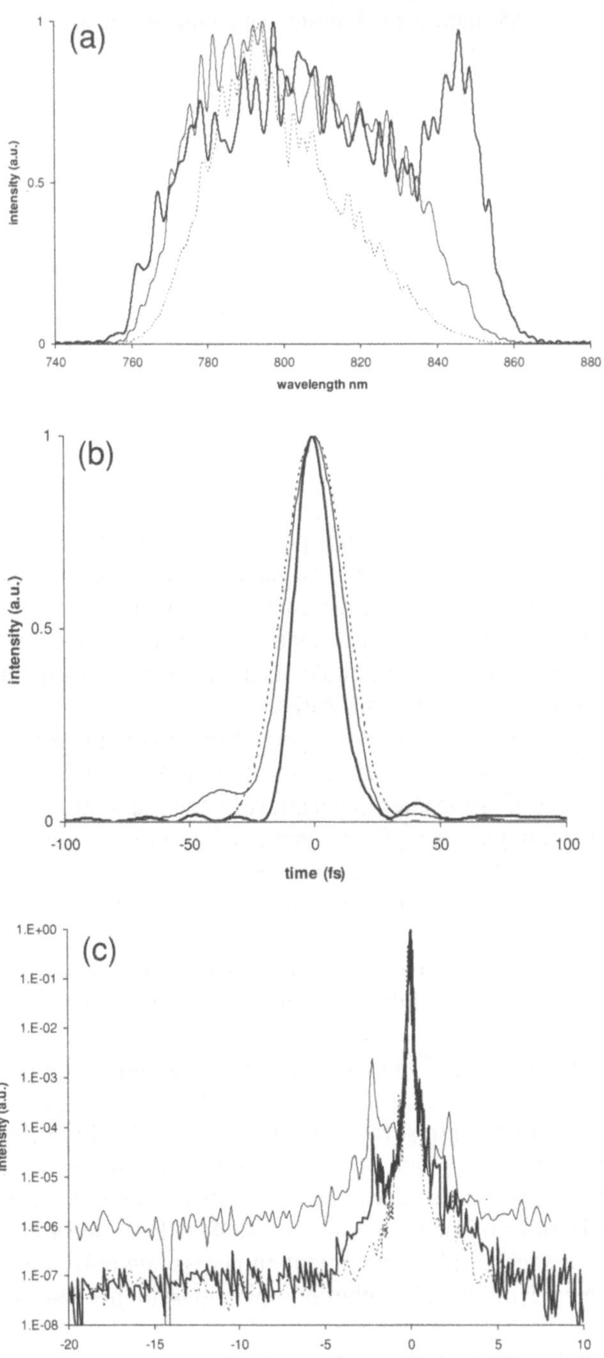


FIGURE 2. (a) Spectra, (b) temporal intensities, and (c) high dynamic range third-order cross-correlator measurements for free-running system without any amplitude modulation (dotted curve: 33-nm FWHM, 28-fs FWHM, $10^7$ contrast ratio), with AOPDF amplitude modulation (thin solid curve: 65-nm FWHM, 25-fs FWHM, $10^6$ contrast ratio), and with AOPGCF optimization (thick solid curve: 80-nm FWHM, 18.6-fs FWHM, $10^7$ contrast ratio).

not change. We have checked that day to day fluctuations are small enough to not require the complete reoptimization. With the spectral losses determined, the amplifier acts as a flat gain amplifier except for its saturation. Arbitrary spectral amplitude shape is then obtained through 4 or 5 steps spectral amplitude shaping with the AOPDF. All these optimizations were semiautomatically done in this experiment, necessitating approval of the end-user. Complete automation of the optimization should lead to less than a minute optimization process on a 10-Hz laser.

The flat phase optimization was simply obtained by introducing in the AOPDF the opposite of the phase measured by the SPIDER. For very large phase variations, more than one loop was necessary.

The key point of these two optimizations is the calibration of the control systems (AOPDF and AOPGCF) and the measurement devices (spectrometer and SPIDER [4]).

The bandwidth is limited by the compressor grating size to 105 nm. Thus, the best result we obtain is really close to this bandwidth (80-nm FWHM). The pulse is sub-20 fs (18-fs FWHM), clean on its leading edge, contrast ratio as good as without regenerative pulse shaping ($10^7$), and the stability better (1.2 %rms).

Theoretically, we expect these results because the ASE contrast ratio depends principally upon the injection level [5] and the stability of the laser could be largely improved by fine tuning of the global cavity losses [6]. The stability should be improved even further but we have not optimized the global losses for this purpose.

The output pulse energy was limited to about 500 μJ to avoid any damage in the AOPGCF crystal. Higher energy level necessitates a bigger spot size in the crystal decreasing the simplicity and stability of the cavity. Joule level should be possible by post amplification through multipass amplifier. Gain narrowing is then precompensated by amplitude shaping of the 500-μJ pulse.

## 4 Conclusion and Prospects

In conclusion, we have developed a new intracavity AOPGCF. We have demonstrated broader amplified spectrum of 500 μJ, sub-20-fs optical pulses without contrast degradation and with stability enhancement. Even shorter pulses in the range of 15 fs should be obtained by changing the limiting bandwidth optical components (Faraday isolator, gratings, mirrors, etc.). The device is fully software controlled and optimization of the laser bandwidth is done completely by the software. Complete automation of the optimization is possible and requires only few steps (<30). Its flexibility makes possible any kind of amplitude shaping as two-color femtosecond pulses, for example.

## References

1. F. Verluise, V. Laude, Z. Cheng, C. Spielmann, and P. Tournois: Amplitude and phase control of ultrashort pulses by use of an acousto-optic programmable dispersive filter: pulse compression and shaping. Opt. Lett. **25**, 575–577 (2000).

2. C. P. J. Barty, T. Guo, C. Le Blanc, et al.: Generation of 18-fs, multiterawatt pulses by regenerative pulse shaping and chirped-pulse amplification. Opt. Lett. **21**, 668–670 (1996).
3. D. Kaplan and P. Tournois: Theory and performance of the acousto-optic programmable dispersive filter used for femtosecond laser pulse shaping. J. Phys. IV France **12**, 69–75 (2002).
4. T. Oksenhendler, P. Rousseau, R. Herzog, O. Gobert, M. Perdrix, and P. Meynadier: 20Hz femtosecond laser amplifier optimization using an AOPDF pulse shaper and a SPIDER. CLEO, San Francisco, USA (2000).
5. V. V. Ivanov, A. Maksimchuk, and G. Mourou: Amplified spontaneous emission in a Ti:sapphire regenerative amplifier. Appl. Opt. **42**, 7231–7234 (2003).
6. F. P. Strohkendl, D. J. Files, and L. R. Dalton: Highly stable amplification of femtosecond pulses. JOSA B **11**, 742–749 (1994).

# Demonstration of Pulse Switching with $>10^{11}$ Prepulse Contrast by Cascaded Optical Parametric Amplification

Constantin Haefner,[1] Igor Jovanovic,[2] Benoit Wattelier,[3] and C. P. J. Barty[2]

[1] Nevada Terawatt Facility, University of Nevada, Reno, Nevada 89506, USA
[2] Lawrence Livermore National Laboratory, 7000 East Ave, Livermore, California 94550, USA
[3] Phasics, Campus de l'Ecole Polytechnique, Bat.404, 91128 Palaiseau Cedex, France
**haefner@unr.edu**

## 1 Introduction

The use of chirped-pulse amplification (CPA) in high-energy systems allows the generation of high-peak power and ultrahigh intensity, and has led to intensities greater than $10^{20}$ W · cm$^{-2}$ [1–3]. These systems are of especially high interest for exploring high-energy dense plasma physics and for fast ignition in inertial confinement fusion [4]. In these systems the control of the prepulses is crucial as their intensity might perturb the target before the main pulse arrives. Plasma generation thresholds of solid targets are generally of the order of $10^{10}$ W · cm$^{-2}$ and hence require a contrast ratio exceeding $10^{10}$ for a $10^{20}$ W · cm$^{-2}$ system. Optical parametric chirped-pulse amplification (OPCPA), a CPA technique which relies on the three-wave mixing process in a nonlinear crystal, has received much attention due to its superior characteristics for broadband amplification in high-power laser system front-ends. In fact, OPCPA operated near degeneracy and thus providing high gain while maintaining broad spectral bandwidth is increasingly replacing Ti:sapphire front-ends in Nd:glass short-pulse lasers [5–8]. With respect to the prepulse contrast ratio, an OPCPA system offers enormous advantages. High-gain regenerative amplifiers usually produce a train of prepulses caused by the cavity leakage due to the limited extinction ratio of the intracavity polarizer, and they thus require additional postamplification temporal pulse cleaning techniques (e.g. [9, 10]). OPCPA amplifies only a single pulse out of the oscillator pulse train resulting in a prepulse contrast enhancement equal to the total saturated gain, provided that the pump pulse duration is sufficiently short so that there is no temporal overlap of the pump pulse with preceding or subsequent pulses. Notwithstanding the significant prepulse contrast enhancement, OPCPA does not remove prepulses; it merely reduces their relative peak intensity, typically on the order of $10^8$ in a large laser system. Oscillator prepulses amplified by subsequent laser amplification stages could still preionize the target. In this article we report for the first time the measurement of the extreme contrast enhancement and optical switching using cascaded optical parametric amplification (COPA). The technique is implemented by reconfiguration of a two-stage OPCPA into a phase-conjugated COPA system. Its feasibility has been described and demonstrated before [11, 12]. The method removes any pre- and postpulses in the neighborhood of the main, amplified pulse. Here we present for the first time instrument-limited experimental measurement of the prepulse contrast $1.4 \times 10^{11} : 1$ at a 30-mJ level using cascaded electrooptical modulators to protect the photodiode from damage by the main pulse. We also show for the first time the excellent recompressibility of the COPA pulse to almost the system optimum.

## 2 The Principle of Cascaded Optical Parametric Amplification (COPA)

The central idea of the COPA technique is that of the permutation of signal and idler in a multistage optical parametric amplifier, meaning that instead of the signal we amplify the idler in subsequent stages of amplification. The idler is generated in the parametric mixing process of the pump and the signal beam and therefore carries no prepulses. The central wavelength of the idler is shifted with respect to the signal wavelength if the process is not ideally degenerate. Also, the chirp gets reversed and the temporal phase conjugated: The sign of even orders of dispersion gets inverted while the sign of the odd orders is maintained. Thus, a second OPA process using again the idler pulse is used to produce a pulse which has the original temporal and spectral properties of the seed pulse. This is assuming ideal conditions such as single frequency pump and flat spectral gain in the OPAs. If the pulse duration of the pump is less than half of the temporal spacing of two seed pulses from the oscillator, all the pre- and postpulses are removed. However, since the idler carries the accumulated phase errors of the pump in the first and second OPA process, the recompressibility may be limited for nonideal pump pulses.

## 3 Measurements and Results

Our experimental setup is based on a simple reconfiguration of a previously described double-stage OPCPA system [13]. In the first OPA process, a signal and an idler pulse are generated by three-wave mixing with central wavelengths of 1053

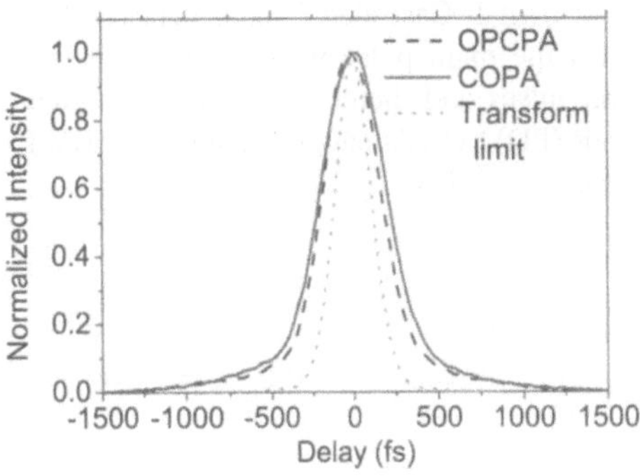


FIGURE 1. Autocorrelation of OPCPA and COPA. The dotted line shows the calculated Fourier transform limit pulse of the OPCPA spectrum.

and 1075 nm, respectively. The idler is separated from the signal pulse and amplified in the second OPA to a 50-mJ level. The idler of the second stage is the output of the COPA system and is consequently recompressed in a grating compressor. The spectrum is identical to the output spectrum of the OPCPA system under the same pump conditions. We recompressed both the pulse of the OPCPA and the COPA system and characterized it using a second-order, dispersion-balanced intensity autocorrelator. The autocorrelation shows pulse duration of 420 fs in OPCPA mode and 460 fs in COPA mode, shown in Fig. 1.

The slight pulse duration increase of $\sim$10% is attributed to the transferred phase aberrations from pump to idler. The calculated transform-limited pulse duration corresponding to the OPCPA output spectrum is 190 fs. The factor of approximately 1.7 difference between theoretical and measured pulse duration has been noted before in this system and is caused by chromatic and spherical aberrations in the pulse stretcher. The excellent recompressibility of the COPA pulse close to the optimum of the system justifies the COPA technique of fast optical switching and prepulse removal.

Demonstration of a large prepulse contrast requires a high dynamic range prepulse measurement technique. We have chosen a relatively simple setup using a sensitive photodiode and a cascade of fast Pockels cells to protect the photodiode from damage by the main pulse (Fig. 2). Pockels cells were operated so that the main pulse was rejected, while the prepulses were transmitted to the photodiode. The attenuation of the main pulse was $1.1 \times 10^5$ and was sufficient to protect the diode from damage by the main pulse. We first determined the contrast ratio for the OPCPA system. With the Pockels cells turned on we measured the amplitude of the prepulse. Calibrated neutral density filters were then inserted and the Pockels cells were turned off. Filters were chosen such that the amplitude of the signals was nearly identical throughout the measurement to avoid errors due to small photodiode response nonlinearities. We have also averaged each signal over 100 pulses to determine signal fluctuations introduced by the laser itself, Pockels cell drivers, beam pointing, or other sources. The prepulse contrast was $1.4 \times 10^8$, which is approximately equivalent to the gain of the system. We reconfigured the OPCPA setup to the COPA configuration as described before. We first recorded

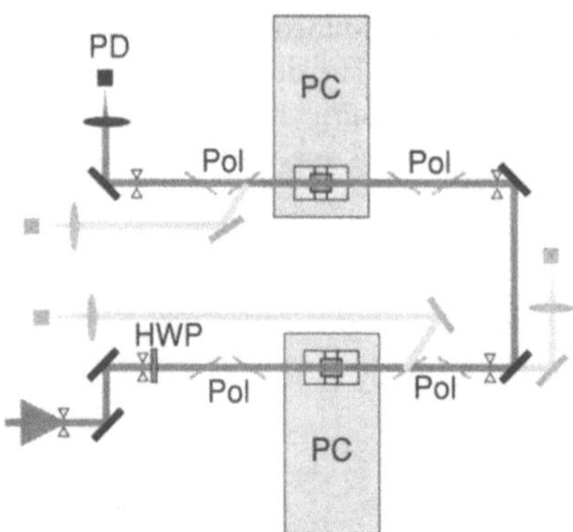


FIGURE 2. Cascaded Pockels cells (PC) reject the main pulse while transmitting the prepulses and thereby protect the photodiode (PD) from damage (Pol, polarizer; HWP, half-wave plate).

the maximum amplitude of the pulse with the Pockels cell turned off, while using neutral density filters to provide sufficient attenuation of the intensity on the photodiode. In the next step we turned the Pockels cell on and removed the filters. As expected, we could not detect any prepulses. The baseline was measured with and without COPA turned on to correct for the noise level. The noise level was $6.2 \pm 0.05 \times 10^{-3}$ on the normalized scale of Fig. 3.

The main sources of noise are the digitizer and crosstalk effects in the oscilloscope. We determined the contrast ratio by comparing the pulse amplitude with the leading baseline ahead after subtracting the noise level and obtained an instrument-limited contrast ratio of $1.4 \times 10^{11}$. Fig. 2 shows the comparison between OPCPA and COPA contrast measurement. The measurement dynamics were limited by the maximum available energy determining the signal to noise ratio.

## 4 Conclusion

We have experimentally shown for the first time that the COPA technique [11, 12] indeed allows extreme prepulse contrast enhancement. Our measured

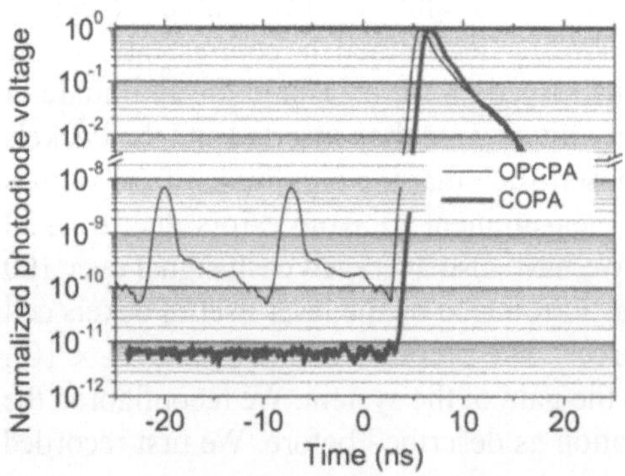


FIGURE 3. (Thin line) OPCPA amplifies an isolated oscillator pulse, but does not remove prepulses; (thick line) COPA removes all oscillator prepulses. The COPA pulse is shifted by 1 ns for better distinction.

instrument-limited prepulse contrast was $1.4 \times 10^{11}$ : 1 at a 30-mJ level. In principle, infinite contrast enhancement is possible by the COPA technique and is only limited by the pump pulse time duration. Within this window the COPA contrast is limited by the usual parametric fluorescence. Simple modification of a multistage OPCPA system results in its operation in a COPA mode, particularly for systems operating near degeneracy. In the case of nondegenerate OPCPA, angular dispersion of the idler beam requires careful relay imaging through each pair of phase-conjugation stages, but results in a collimated beam if the ratio of the noncollinear angle to the idler angular dispersion is preserved in two OPAs. Since the COPA technique does not rely on the nonlinear effects on the pulse itself, it is extraordinarily versatile. COPA can be operated at any energy level, with either stretched or compressed pulses. It can be used as a preamplification stage or as a high-contrast, large-bandwidth pulse selector. A particularly attractive implementation of the scheme is to operate in conjunction with quasi-phase-matched OPA [14], where a simple low-energy contrast enhancement can be accomplished utilizing a single OPA crystal.

This work was performed under the auspices of the US Department of Energy by the University of California, Lawrence Livermore National Laboratory under Contract No. W-7405-Eng-48 and supported by DOE/NNSA under UNR grant DE-FC52-01NV14050.

## References

1. M. D. Perry, D. Pennington, B. C. Stuart, et al.: Petawatt laser pulses. Opt. Lett. **24**, 160–162 (1999).
2. M. Pittman, S. Ferré, J. P. Rousseau, L. Notebaert, J. P. Chambaret, and G. Chériaux: Design and characterization of a near-diffraction-limited femtosecond 100-TW 10-Hz high-intensity laser system. Appl. Phys. B **B74**, 529–535 (2002).
3. M. Aoyama, K. Yamakawa, Y. Akahane, et al.: 0.85-PW, 33-fs Ti:sapphire laser. Opt. Lett. **28**, 1594–1596 (2003).
4. M. Tabak, J. Hammer, M. E. Glinsky, et al.: Ignition and high gain with ultrapowerful lasers. Phys. Plasmas **1**, 1626–1634 (1994).
5. Y. Izawa, Y. Kitagawa, H. Fujita, et al.: Petawatt laser system for fast ignitor studies at ILE, Osaka University. In: 19th International Atomic Energy Agency (IAEA) Fusion Energy Conference (FEC), Osaka, Japan (2002).
6. V. Bagnoud, et al.: Optical parametric chirped-pulse amplifier as the front end for the Omega EP laser chain. In: Third International Conference on Inertial Fusion Sciences and Applications (IFSA 2003), Monterey, CA, USA (American Nuclear Soc.) (2004).
7. I. Jovanovic, B. J. Comaskey, C. A. Ebbers, R. A. Bonner, D. M. Pennington, and E. C. Morse: Optical parametric chirped-pulse amplifier as an alternative to Ti:sapphire regenerative amplifiers. Appl. Opt. **41**, 2923–2929 (2002).
8. I. N. Ross, J. L. Collier, P. Matousek, et al.: Generation of terawatt pulses by use of optical parametric chirped pulse amplification. Appl. Opt. **39**, 2422–2427 (2000).
9. D. Homoelle, Alexander L. Gaeta, V. Yanovsky, and G. Mourou: Pulse contrast enhancement of high-energy pulses by use of a gas-filled hollow waveguide. Opt. Lett. **27**, 1646–1648 (2002).

10. A. Jullien, O. Albert, F. Burgy, et al.: $10^{1}0$ temporal contrast for femtosecond ultraintense lasers by cross-polarized wave generation. Opt. Lett. **30**, 920–922 (2005).
11. I. Jovanovic, C. P. J. Barty, C. Haefner, and B. Wattellier: Opt. Lett. **31**, 787–789 (2006).
12. B. Wattellier, I. Jovanovic, and C. P. J. Barty: Cascaded-optical parametric amplification for extreme contrast enhancement. In: Conference on Lasers and Electro-Optics (CLEO), Technical Digest (Optical Society of America) Washington, DC, p. 74 (2002).
13. I. Jovanovic, C. Brown, B. Wattellier, et al.: Precision short-pulse damage test station utilizing optical parametric chirped-pulse amplification. Rev. Sci. Instrum. **75**, 5193–5202 (2004).
14. I. Jovanovic, C. G. Brown, C. A. Ebbers, C. P. J. Barty, N. Forget, and C. Le Blanc: Generation of high-contrast millijoule pulses by optical parametric chirped-pulse amplification in periodically poled $KTiOPO_4$. Opt. Lett. **30**, 1036–1038 (2005).

# Part XV

# High-order Harmonic Generation

# Generation of Exceptionally Strong Harmonics from He in an Orthogonally Polarized Two-color Laser Field

Chang Hee Nam, I Jong Kim, Chul Min Kim, Hyung Taek Kim, and Gae Hwang Lee

Department of Physics and Coherent X-ray Research Center, KAIST, 373-1, Kuseong-dong, Yuseong-gu, Daejeon 305-701, Korea
**chnam55@kaist.ac.kr**

Highly efficient high-harmonic generation was achieved using a two-color laser field that consisted of the fundamental and its second harmonic. The harmonics generated in an orthogonally polarized two-color field were stronger than those obtained in the fundamental field by more than two orders of magnitude. Conversion efficiency as high as $5 \times 10^{-5}$ was obtained for the 38th harmonic (21.6 nm) from helium driven by the two-color laser field with a proper relative phase. This enhancement is explained by analyzing the electron behavior in the two-color field. In addition, we confirmed the effects of mixed field by controlling the time delay between the two-color laser fields.

## 1 Introduction

High-order harmonic generation (HHG) is a coherent interaction process between atoms and driving laser field and has a promising potential as a soft X-ray/ extreme-ultraviolet light source with ultrashort pulse duration and superb coherence [1,2]. Recently a number of studies have focused on increasing the conversion efficiency and photon flux of harmonics by maximizing the number of atoms using a long interaction medium; harmonic generation can now be achieved in the absorption-limited level [3–6]. However, these optimized harmonics are limited in the regions around 53, 30, and 13 nm and harmonics in other wavelength regions are not as strong as in these cases, either due to strong absorption of harmonics by neutral atoms or due to the limit of highest harmonic order by applied laser intensity. The application of a two-color laser field for HHG has been a fascinating topic of research since various kinds of synthesized electric field, prepared by varying polarization, relative phase, and intensity ratio between the fundamental and the second-harmonic (SH) fields, could be applied to atoms. These parameters have significant influence on HHG. Theoretically, it was predicted that, when the polarizations of the fundamental and the SH fields were parallel, harmonic signal could be enhanced by more than two orders of magnitude, compared to that generated just with the fundamental laser field [7]. Experimentally, even-order harmonics were previously observed due to breaking of inversion symmetry, signifying mixing effect of two-color field and signal increase by an order of magnitude was demonstrated [8]; however, dramatic enhancement as predicited by theory was not achieved.

## 2 Experimental Results

We report the generation of exceptionally strong harmonics in the 20-nm region (38th order) from helium (He) obtained by applying a two-color laser field. Two-color HHG experiments were performed using the experimental setup shown in Fig. 1. A high-power femtosecond laser produced 27-fs pulses at 10 Hz with a spectral bandwidth of 42 nm centered at 820 nm [9], which were then focused into a He gas jet (0.5-mm diameter) using a spherical mirror (f = 60 cm). Emitted harmonics were detected using a flat-field extreme ultraviolet spectrometer equipped with a back-illuminated X-ray charge coupled device. For SH generation, a beta-barium borate crystal (BBO; 100-μm thick) was placed between the focusing mirror and gas jet, so after the BBO crystal, the laser field consisted of

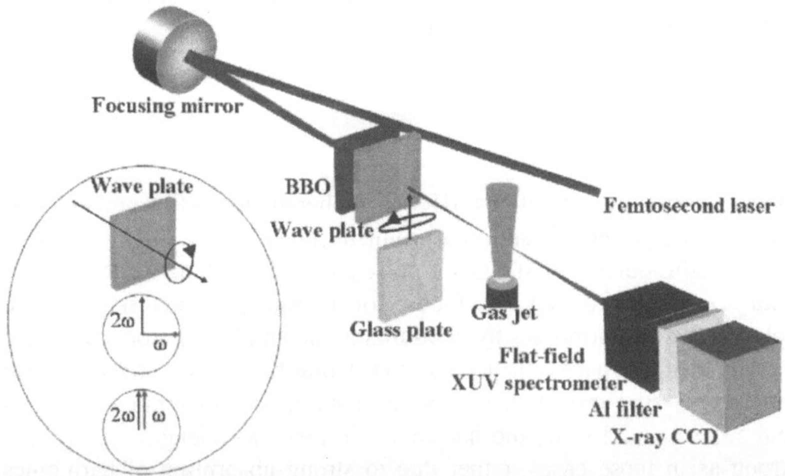


FIGURE 1. Experimental setup for high-harmonic generation in a two-color laser field. The polarization of the fundamental ($\omega$) field was controlled using a wave plate. For control of the relative phase between the fundamental and second harmonic fields, a 150-μm-thick glass plate was inserted.

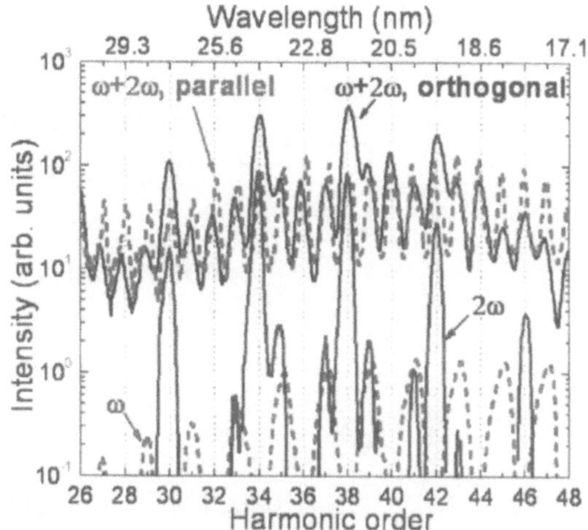


FIGURE 2. Harmonic spectra from He atoms generated in the fundamental ($\omega$), second harmonic ($2\omega$), and two-color ($\omega + 2\omega$) laser fields. In the two-color field, the cases of parallel and orthogonal polarizations are given.

both the SH and the residual fundamental laser fields. The polarization of the SH field generated in this case was orthogonal to that of the fundamental laser field. With a special quartz wave plate that acts as a half-wave plate for the fundamental and a full-wave plate for SH, we were able to continuously rotate the polarization of the fundamental laser field, while maintaining the polarization of the SH. This simple setup allows two-color harmonic generation in parallel or orthogonal polarization between the fundamental and SH fields.

Using a two-color laser field consisting of the fundamental and its SH from the femtosecond laser, we obtained highly efficient high-order harmonics from He. When the polarization between the fundamental and SH was orthogonal, generated harmonics were more intense than those with the parallel polarized two-color case. Compared to the one-color case of the fundamental field only, the signal enhancement was more than 100-fold, as shown in Fig. 2. High-harmonic generation is greatly affected by the shape of the synthesized electric field that sensitively depends on the relative phase ($\phi$) between the fundamental and SH fields. By rotating a thin dispersive glass plate, we could control the relative phase. The harmonics generated in the two-color field (for both parallel and orthogonal polarizations) are periodic with respect to the relative phase shift of it. Our results confirmed this periodicity for both polarization conditions, which, in turn, proves that the harmonics were indeed generated in the two-color field. By controlling their relative phase, we obtained further enhancement in HHG, reaching a conversion efficiency of $5 \times 10^{-5}$ and photon energy of 150 nJ/shot at the 38th harmonic (21.6 nm).

## 3 Analysis of Electron Behavior

To gain physical understanding of the harmonic generation in the orthogonally polarized two-color laser field, we analyzed the behavior of electron in the

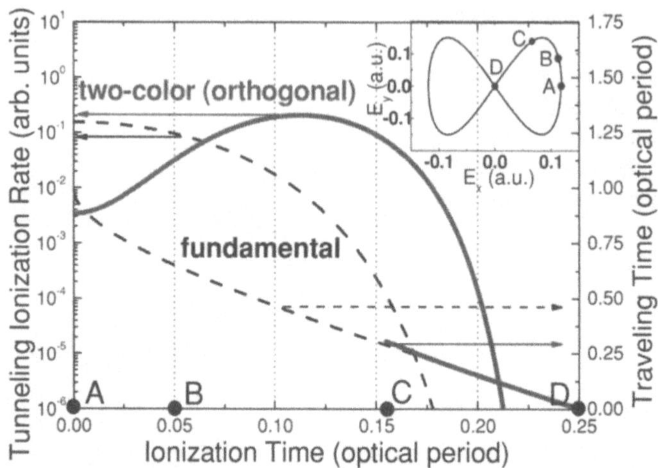


FIGURE 3. Tunneling ionization rate and traveling time of the electron in the fundamental field and orthogonally polarized two-color field. The inset shows the Lissajous diagram of the electric field ($I_\omega = 5 \times 10^{14}\mathrm{W} \cdot \mathrm{cm}^{-2}$, $I_{2\omega} = 8 \times 10^{14}\mathrm{W} \cdot \mathrm{cm}^{-2}$) A, B, C, and D correspond to 0.00, 0.05, 0.16, and 0.25 optical cylces, respectively.

synthesized electric field. Values were chosen to match the experimental parameters, setting the same duration for the fundamental and SH pulses. With the relative phase $\phi = \pi/2$, the Lissajous diagram of the synthesized electric field exhibits a bow-tie shape, as shown in the inset of Fig. 3a. As a result, the ionized electron can recollide with the nucleus, similar to the usual one-color case. The electron recollision can occur when the electron is ionized only during the straight section of the electric field (between the points C and D) and the traveling time is then short ($< 0.3$ optical period). Thus the harmonics are generated mainly from the short quantum path [11]. The electrons ionized in this time period are also the main contributors to phase-matched harmonic generation even in the fundamental field, because the wave packet spreading is small and the resulting harmonic phase is insensitive to laser intensity variation. In this case, the tunneling ionization rate, calculated using the Ammosov-Delone-Krainov formula [10], for the two-color field is much larger than that for the fundamental field as shown in Fig. 3, i.e. the electron wave packet at the time of ionization is significantly denser. Consequently, the orthogonally polarized two-color field can generate harmonics much more strongly than the fundamental field can, agreeing with the results shown in Fig. 2. According to the single-atom calculation, these behaviors are feasible only for the relative phase around $\phi = \pi/2$ [11, 12].

The strong enhancement of all-order harmonics is dominated by the mixing effect in a two-color laser field. The two-color laser field with a suitable relative phase can select the short quantum path component which has denser electron wave packet. As a result highly efficient high-order harmonics are obtained. However, there is a time delay between the two laser fields due to dispersion in optical elements such as the BBO crystal, wave plate, and the glass plate. While passing through these materials, due to group velocity mismatch (GVM), the temporal overlap between the two laser fields is reduced. By considering the laser parameter ($\lambda = 820$ nm), phase matching angle ($\theta = 28.2°$), and crystal length (L = 100 μm), the time delay between the peaks of the two fields becomes about 17.8 fs after BBO crystal. The wave plate used for controlling polarization and the glass plate for controlling relative phase introduce a further time delay of about 20 fs. When the two-color fields are partially overlapped, the entire pulse envelop could be separated into three regions: One region where the two fields overlap in time, the other regions are where the two fields exist independently. In the region where only the fundamental field exists, odd-order harmonics of fundamental are generated. In the overlap region all order harmonics are enhanced by more than two orders of magnitude. In the region where only SH field is present, the 2(2n + 1)th orders are additionally enhanced.

Inserting a time plate (TP) [13], we could adjust the temporal overlapping between the two-color laser fields. TP is a handy optical element that can substitute for a Michelson-type interferometer setup used for the combination of two-laser fields. The change of harmonic structure was measured by controlling the time delay using TP. In Fig. 4b we selected the maximum harmonic intensity for a given time delay obtained by scanning the relative phase. Figure 4a shows the modulation of 2(2n + 1)th and (2·2n)th harmonics with respect to the relative phase between two-color laser fields in the condition of normal incidence to the TP. The modulation period is $\pi$, consistent with the shape of the combined two-color field with relative phase change [12]. As the time delay was adjusted from 0 to 25 fs, the temporal overlapping of the two laser fields became better. After the time delay of 25 fs, the SH field came before the fundamental field. The most prominent feature in Fig. 4b is the increase of purely mixing orders such as 36th, 40th, and 44th with good temporal overlapping. Strong harmonics were initially observed at 2(2n + 1)th orders (Fig. 4a), but with better temporal overlap, purely mixing 2(2n)th orders also became prominent, i.e., all even harmonics were strong. Beyond this temporal overlap, SH field is advanced and the intensity of purely mixing order is reduced. Thus we could confirm the effects of the mixed field by controlling the relative phase and time delay between the two-color laser fields.

# 4 Conclusion

Using an orthogonally polarized two-color laser field, we have achieved highly efficient high-harmonic generation in He. With high SH intensity in the two-color field, a strong enhancement of harmonics from He was achieved in the 20-nm region. The modulation of harmonics obtained with the control of relative phase between the fundamental and SH fields clearly shows the harmonic generation in the two-color field. The physical origin for the strong enhancement of the harmonics was explained from the analysis of the electron behavior in the two-color

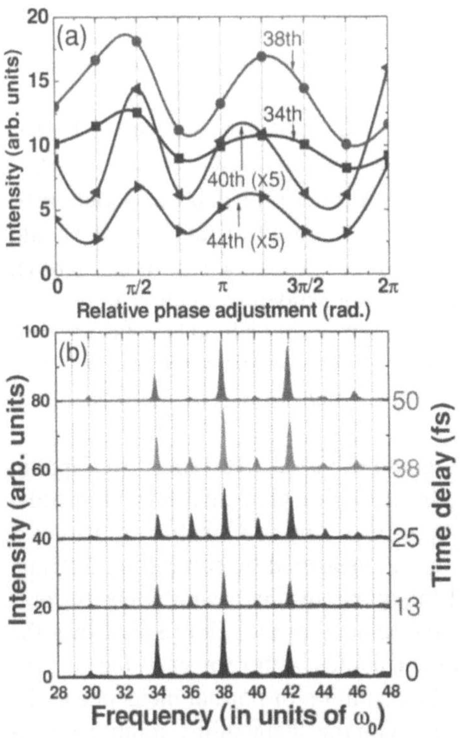


FIGURE 4. (a) Modulation of 2(2n + 1)th and 2(2n)th harmonics attained by adjusting the relative phase in the case of the zero time delay (normal incidence to the time plate). (b) Change of the spectral structure of high-order harmonics obtained by controlling the time delay between two-color laser fields.

field. We also confirmed the effects of mixed field by controlling the relative phase and the time delay between two-color laser fields. The relative simplicity and effectiveness of this two-color method for strong harmonic generation will be of great use for practical applications.

This research was supported by the Korea Science and Engineering Foundation through the Creative Research Initiative Program.

## References


1. D. G. Lee, J.-H. Kim, K.-H. Hong, and C. H. Nam: Coherent control of high-order harmonics with chirped femtosecond laser pulses. Phys. Rev. Lett. **87**, 243902 (2001).
2. D. G. Lee, J. J. Park, J. H. Sung, and C. H. Nam: Wave-front phase measurements of high-order harmonic beams by use of point-diffraction interferometry. Opt. Lett. **28**, 480 (2003).
3. E. Constant, D. Garzella, P. Bréger, et al.: Optimizing high harmonic generation in absorbing gases: model and experiment. Phys. Rev. Lett. **82**, 1668 (1999).
4. M. Schnurer, Z. Cheng, M. Hentschel, et al.: Absorption-limited generation of coherent ultrashort soft-x-ray pulses. Phys. Rev. Lett. **83**, 722 (1999).
5. E. Takahashi, Y. Nabekawa, and K. Midorikawa: Low divergence coherent soft x-ray source at 13 nm by high-order harmonics. Appl. Phys. Lett. **84**, 4 (2004).

6. H. T. Kim, I J. Kim, D. G. Lee, et al.: Optimization of high-order harmonic brightness in the space and time domains. Phys. Rev. A **69**, 031805(R) (2004).
7. D. B. Milosevic and W. Sandner: Extreme-ultraviolet harmonic generation near 13nm with a two-color elliptically polarized laser field. Opt. Lett. **25**, 1532–1534 (2000).
8. H. Eichmann, A. Egbert, S. Nolte, C. Momma, and B. Wellegehausen: Polarization-dependent high-order two-color mixing. Phys. Rev. A **51**, 3414(R) (1995).
9. Y. H. Cha, Y. I. Kang, and C. H. Nam: Generation of a broad amplified spectrum in a femtosecond terawatt Ti:sapphire laser by a long-wavelength injection method. J. Opt. Soc. Am. B **16**, 1220 (1999).
10. M. V. Ammosov, N. B. Delone, and V. P. Krainov: Tunnel ionization of complex atoms and of atomic ions in an alternating electromagnetic field. Sov. Phys. JETP **64**, 1191 (1986).
11. I. J. Kim, C. M. Kim, H. T. Kim, et al.: Highly efficient high-harmonic generation in an orthogonally polarized two-color laser field. Phys. Rev. Lett. **94**, 243901 (2005).
12. C. M. Kim, I. J. Kim, and C. H. Nam: Generation of a strong attosecond pulse train with an orthogonally polarized two-color laser field. Phys. Rev. A **72**, 033817 (2005).
13. C. Wan: United States Patents, Tunable Time Plate, ♯ 5,852,620 (1998).

# Single-shot Measurement of Fringe Visibility of 13-nm High-order Harmonic

Yutaka Nagata, Kentaro Furusawa, Yasuo Nabekawa, and Katsumi Midorikawa

RIKEN, Laser technology laboratory, 2-1 Hirosawa, Wako-shi, Saitama 351-0198 Japan
ynagata@riken.jp

**Abstract:** We demonstrate single-shot measurement of spatial coherence of 13-nm high-order harmonic beam in a Young's double-slit experiment. The visibility of 0.96 was obtained with optimal phase-matching condition for the 13-nm harmonics. To our knowledge, this is the first demonstration of the fringe visibility measurement of the 13-nm harmonic beam with a single shot. This result shows that the 13-nm harmonic beam is useful for applications in imaging and microscopy.

## 1 Introduction

Recently, high-order harmonics (HHs) are recognized as a highly coherent, intense light source, which are attractive for many applications [1], such as nonlinear optics [2, 3], imaging, and interferometry, in the extreme ultraviolet (EUV) region. Optical systems of EUV lithography required fabrication tolerances (wave-front error) of the order of 0.3 nm. To measure such tolerances, at-wavelength interferometric inspection of optical system is required. Because of the lack of another coherent light sources, undulator beamlines are used for inspections [4, 5]. However, photon number per pulse is too low for single-shot measurement, amount of mechanical vibration should be reduced to less than measurement accuracy during the measurement. Single-shot testing makes it possible to simplify a measurement apparatus, because an adverse effect could be avoided, such as pointing stability of a light source and vibration of a sample and measurement apparatus. High-energy 13-nm harmonics are candidate for the single-shot measurement. The measurement of the fringe visibility of harmonics was demonstrated in the EUV regions with millions of pulses in a Young's double pinhole experiment. We have already succeeded in increasing the output energy of HHs up to 40 nJ/pulse at the 57th and the 59th harmonic [6]. In this letter we show the first demonstration of a single-shot spatial coherence measurement of the 13-nm harmonic.

## 2 Phase-matched High-order Harmonic Generation

High-order harmonic generation was performed using a 26-fs Ti:sapphires laser at 790 nm with energies up to 200 mJ. We used a deformable mirror to compensate the phase distortion of the pump pulse. Because the energy concentration of the pump pulse at the focus point was improved, the energy conversion efficiency and the beam quality of the HHs were also improved. The laser was loosely focused into the differential-pumping interaction cell, which had two pinholes on each end, filled with Ne gas by an R = 5m concave mirror. In this experiment, the pressure of the gas and the interaction length were fixed to 7.8 torr and 50 mm, respectively. The harmonics radiation was sampled by a grazing incidence EUV spectrometer placed 5 m from the exit pinhole of the cell. To optimize the phase-matching condition for the 13-nm harmonics (57th and 59th), the diameter and the energy of the pump pulses were adjusted using a motorized aperture. As a result, the input energy of the pump pulse became about 40 mJ. A typical high-order harmonics spectrum and the beam profile of the 57th harmonic are shown in Fig. 1. The spectrum has a peak around 13 nm and it disappeared around 11 nm. The effective interaction intensity of the pump laser pulse was estimated about $4 \times 10^{14}$ W $\cdot$ cm$^{-2}$ at the focus point. The output energy and the beam divergence of the 57th harmonic were estimated as 40 nJ and 0.35 mrad (FWHM), respectively. The Gaussian-like profile and the low beam divergence of the 57th harmonic support the fact that the phase-matching condition was satisfied along the propagation axis of the pump pulse. From these results, highly spatial coherence of the harmonic beam was expected for the 13-nm harmonics.

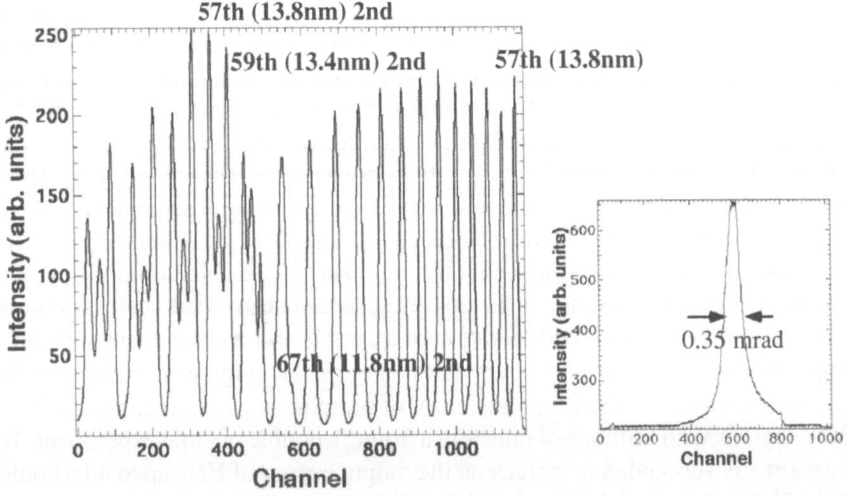


FIGURE 1. Typical spectrum of high-order harmonics in Ne and spatial profile of the 57th harmonic beam.

## 3 Young's Double-slit Experiment for 13-nm Harmonics

The spatial coherence of the 13-nm harmonics was measured through the double-slit interference technique. The interference image was captured with an EUV CCD camera (Andor, Inc., 13.5-μm pixel size). Because the quantum efficiency of the CCD camera for the pump pulse is roughly same as that for EUV light, the pump laser pulse should be attenuated more than seven orders of magnitude.

### 3.1 *Attenuation of Pump Pulse Using High-throughput, High Damage-threshold Beam Separator*

The intensity of the pump pulse copropagated with HHs was much higher than the damage threshold intensity of a Mo/Si multilayer mirror and EUV filters. We have already developed a high-throughput and high-damage-threshold beam separators (BSs) for HHs with wavelengths longer than 25 nm uses Si and (or) SiC plates set at Brewster's angle with respect to the pump wavelength [7]. We extended this procedure to shorter wavelength using a 10-nm-thick niobium nitrogen (NbN) film deposited on a Si substrate [8]. The attenuation rate for the pump pulse was about 0.01 at Brewster's angle of 78°. The attenuation rate was not sensitive to the incident angle between 76° and 80°. The calculation results can be explained by the complex refractive index of NbN for the pump pulse. The reflectivity for the HHs at the Brewster's angle was measured using a pair of BSs, which was placed 3 m after the exit pinhole and was aligned in parallel. Typical harmonic spectra with and without BSs and the reflectivities for the HHs are shown in Fig. 2. The BS has high reflectivities from 12 nm to 25 nm at the Brewster's angle for the pump pulse. No damage was found on the surface of BSs after these experiments. The damage

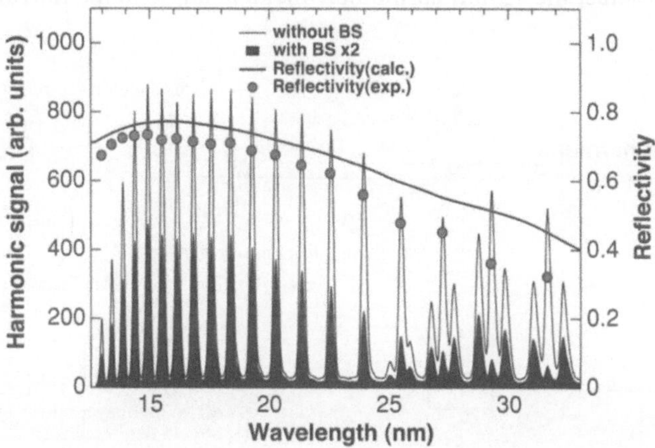


FIGURE 2. High-order harmonic spectra observed with and without NbN beam separators and reflectivity of the separator at pump pulse's Brewster angle of 12°. (Filled circles) measured reflectivity and (solid line) calculated reflectivity.

threshold intensity is guaranteed higher than 0.25 TW · cm$^{-2}$. The attenuation rate of the pump pulse is not enough for the measurement. But intensities of the attenuated pump pulse were lower than the damage threshold intensity of Mo/Si mirror and EUV filters. Reflectivities of the BS for the 13-nm harmonics were 0.7. The use of EUV optics became possible by using this BS without large energy loss.

## 3.2 Experimental Setup for Single-shot Measurement of Interference Image

For the evaluation of the spatial coherence of the 13-nm harmonics, the slit separation should be comparable to the diameter of the harmonics. When pixel numbers of the CCD chip in an interference fringe spacing are insufficient, a fringe visibility in appearance is degraded. We estimate the pixel numbers to reproduce high visibility. It is about 20 pixels, which corresponds to 260 μm for the CCD chip. The fringe spacing is almost the same as the space given by $\lambda L/d$, where $\lambda$ is wavelength of harmonic pulse, $L$ is the distance between the double slit and detector, and $d$ is the double-slit separation. Because the diameter of the harmonics is about 1.3 mm at the position of the BS, a distance required to measure the high visibility becomes 10 m or more. Because this distance is not realistic in our experimental space, a Mo/Si multilayer concave mirror was used to reduce the diameter of the 13-nm harmonics. The double slit is moved on the optical axis of the harmonics. Measurement system, which is shown in Fig. 3, was designed for a single-shot measurement of the fringe visibility. At first, we optimized the phase-matching conditions for the 13-nm harmonics generation without mirrors. The beam separator, which was used to attenuate the pump pulse, was placed at 3 m after the exit pinhole with an incident angle of 78.5°. Mo/Si multilayer mirror was used to select the 13-nm harmonics. Incident angle to the mirror was set to

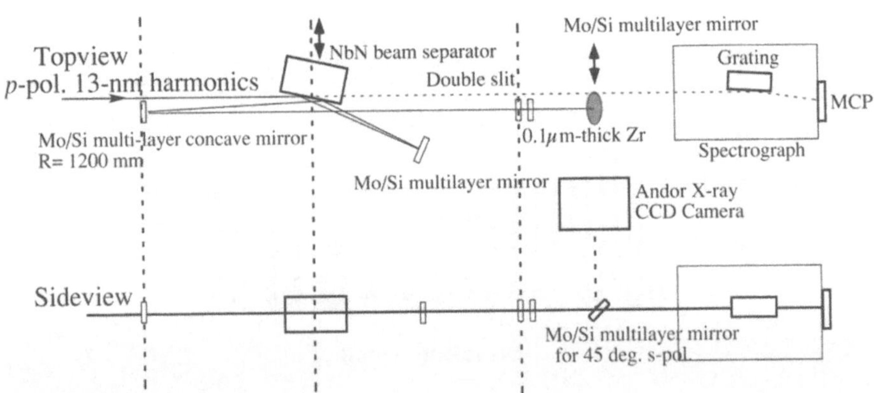


FIGURE 3. Experimental setup for single-shot measurement of spatial coherence of the 13-nm harmonics with Young's double-slit configuration.

0.5°. The reflected harmonics were led to the Mo/Si multilayer concave mirror with a focal length of 600 mm using the same BS. A 20-μm-thick Ni foil with two slits was placed after the focus points. The width of slits and slit separation designed for single-shot measurement were 6 μm and 30 μm, respectively. The Ni-foil with double slits was mounted on a motorized $x$–$y$–$z$ stage to adjust the relative position between the harmonic beam and the pump pulse and to change the beam diameter of the harmonics. Transmits of the slits for the 13-nm harmonics were a few percent. A 0.1-μm-thick Zr filter, which was used to remove the residual pump laser pulse, was located 10 mm after the slits. The final steering mirror placed 200 mm after the slits led the harmonic beam to the CCD camera. Surface figures of the mirrors and the BS were λ/20 PV at 633 and 0.5-nm rms. The interference image was recorded with the CCD camera placed 63 cm after the focus point.

## 4 Single-shot Measurement of Interference Image of 13-nm Harmonics

Figure 4 shows typical interference image and intensity profile of the interference fringe of the 13-nm harmonics captured with a single-shot for different diameter of the harmonic beam. The 13-nm harmonics was sampled at 50%, 60%, and 81% of the beam diameter (FWHM) which was varied from 37 to 60 μm by changing the distance between the focus point and the slits from 22.5 to 37.5 mm. The fitting curves of the interference fringes, which were also shown in Fig. 4, were calculated for the Gaussian beam without wavefront error. As a result of the low divergence of the 13-nm harmonic beams from each slit and wave-front tilt at the slits, the airy patterns overlap in only a limited region and the modulation depth of the fringe in the outside of fringe decreases. Therefore, only the central fringes can be used to evaluate the visibility of the interference fringes. The intensity profile was obtained by the integration of the intensity along the vertical axis. The fringe visibility is defined by $V = (I_{max} - I_{min})/(I_{max} + I_{min})$, where $I_{max}$ and $I_{min}$ are the maximum and minimum intensity of the intensity pattern, respectively. The fringe visibility at 50%, 60%, and 81% of the beam diameter was 0.96, 0.95, and 0.94, respectively. The modulation depth of the observed interference fringes was well reproduced by the theoretical results even in the outside of the fringe. These results show that high signal-to-noise ratio could be achieved with a single-shot measurement. Figure 5 illustrated the fringe visibility of the 13-nm harmonics as a function of slit separation($D_{slit}$)/beam diameter($d_{13nm}$). The visibility decreased with an increase of $D_{slit}/d_{13nm}$. These results are supported by the calculation results. Insufficient pixel numbers of the CCD chip in an interference fringe also decreases the fringe visibility in appearance. The modulation depth of the fringes depends on the correlation between the local phases of the wavefront of the beam in the two slits. Random variations in the phase between the two slits will degrade the fringe visibility. Surface figure of the mirror was 0.5 nm rms and 1.6 λ PV

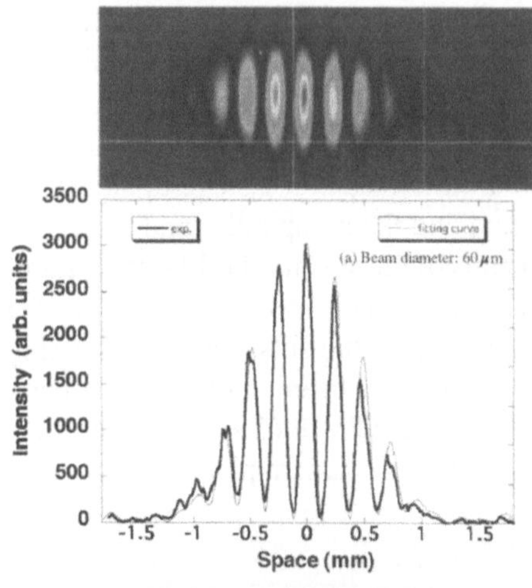


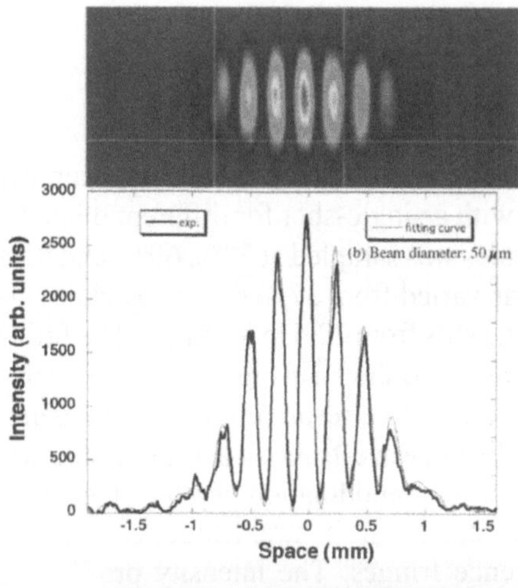


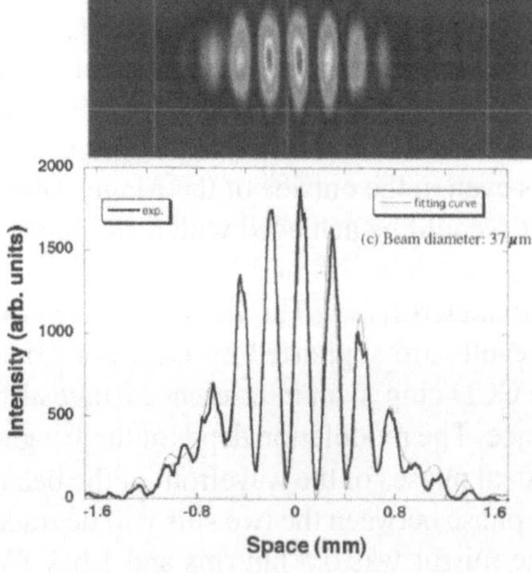


FIGURE 4. Typical CCD image (top) and intensity profile (bottom) of the interference patterns of the 13-nm harmonics which was sampled at (a) 50%, (b) 60%, and (c) 81% of the beam diameter (FWHM).

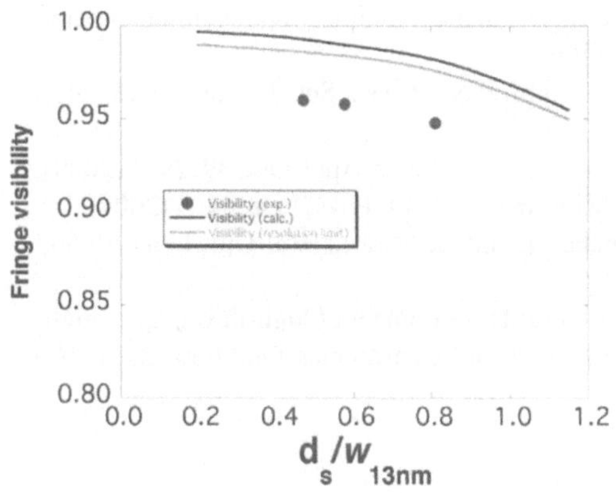


FIGURE 5. Fringe visibility of 13-nm harmonics as a function of slit separation/beam diameter. Theoretical results are also shown.

at 13 nm. In this experimental configuration, the harmonic beam was reflected 5 times with the EUV optics and transmitted the Zr filter with wrinkles and waves. This is the cause of the degradation of fringe visibility.

# 5 Conclusions

In conclusion, we have measured spatial coherence of the 13-nm harmonic beam generated under the phase-matching conditions with a single shot. The low divergence of 0.35 mrad for the 13-nm harmonics supports the fact that the phase-matching condition is satisfied along the propagation axis of the pump pulse. In our experimental conditions, fringe visibility in appearance was restricted less than 0.98 for fully spatial-coherence beam. The wavefront of the 13-nm harmonics was distorted by surface roughness of the EUV optics. Therefore, the fringe visibility of the 13-nm harmonics before the EUV optics was guaranteed 0.96 or more. To our knowledge, this is the first demonstration of spatial-coherence measurement of the 13-nm harmonic beam with a single shot. Such a highly spatial coherent, intense 13-nm harmonic source is useful for imaging, interferometry, and nonlinear optics.

**Acknowledgments.** This research has been supported by Grant-in-Aid for Scientific Research (A) no. 16206010 from Ministry of Education, Culture, Sports, Science, and Technology, Japan.

# References

1. A. L'Huilleir, D. Descanmps, A. Johansson, J. Norin, J. Mauritsson, and C. G. Wahlstrom: Eur. Phys. J. D **22**, 3 (2003).
2. Y. Nabekawa, H. Hasegawa, E. J. Takahashi, and K. Midorikawa: Phy. Rev. Lett. **94**, 043001 (2005).

3. T. Sekikawa, A. Kosuge, T. Kanai, and S. Watanabe: Nonlinear optics in the extreme ultraviolet. Nature (London) **432**, 605 (2004).
4. E. Tejnil, K. A. Goldberg, S. Lee, et al.: Appl. Phys. J. Vac. Sci. Technol. B **15**, 2455 (1999).
5. P. Naulleau, K. A. Goldberg, E. M. Gullikson, and J. Bokor: Appl. Opt. **39**, 2941 (2000).
6. E. J. Takahashi, Y. Nabekawa, and K. Midorikawa: Appl. Phys. Lett. **84**, 4 (2004).
7. E. J. Takahashi, H. Hasegawa, Y. Nabekawa, and K. Midorikawa: Opt. Lett. **29**, 507 (2004).
8. Y. Nagata, Y. Nabekawa, and K. Midorikawa: Development of high-throughput, high-damage-thresold beam separator for 13nm high-order harmonics. Opt. Lett., **31**, 1316–1318 (2006).

# High-order Harmonic Generation from the Laser Plasma Produced on the Surface of Solid Targets

R. A. Ganeev, M. Suzuki, M. Baba, and H. Kuroda

The Institute for Solid State Physics, The University of Tokyo, 5-1-5 Kashiwanoha, Kashiwa 277-8581, Japan
msuzuki@issp.u-tokyo.ac.jp

We demonstrate the generation of high harmonics (up to the 65th order, $\lambda = 12.24$ nm) of a Ti:sapphire laser radiation after the propagation of femtosecond laser pulses through the low-excited plasma produced by a picosecond prepulse radiation on the surface of different targets. High-order harmonics generating from the surface plasma of most targets showed a plateau pattern. The harmonic generation in these conditions is assumed to occur due to the interaction of the femtosecond pulses with the ions. The conversion efficiencies at the plateau region were varied between $1 \times 10^{-7}$ to $8 \times 10^{-6}$, depending on the target. The main contribution to the limitation of harmonic generation efficiency and cutoff energy was attributed to the self-defocusing of main pulse. A considerable restriction of the 27th harmonic generation was observed at different focusing conditions in the case of chromium plasma. Our observation of the resonance-induced enhancement of a single harmonic ($\lambda = 61.2$ nm) at a plateau region with the efficiency of $8 \times 10^{-5}$ in the case of In plasma can offer some expectation that analogous processes can be realized in other plasma samples in the shorter wavelength range where the highest harmonics were achieved.

## 1 Introduction

The laser plasma has long been used as a medium for harmonic generation. The high-order harmonic generation (HHG) in visible and extreme ultraviolet (XUV) ranges has been investigated using the laser plasma formed by an optical breakdown in gases and at the surface of solid targets [1–6]. The method for generating harmonics from solid target atoms apart from solid surface (see, for example, [7], which uses a completely different mechanism) is based on the preparation of a gas-like ablation plume via laser-surface interaction with further propagation of short laser pulse through this medium.

Kubodera et al. [2] suggested that the laser-produced plasma was very effective in terms of preparing nonlinear media for the HHG regardless of the physical and chemical characteristics of the targets. The application of various ions provided important information on their role for the HHG. However, the maximum observed order of harmonics reported in those studies was limited to the 27th one [3] due to some concurred effects in high-excited plasma. No plateau pattern for high harmonics was reported in those studies.

The peculiarity of the harmonic generation from the surface plasma in comparison to the conventional gas-jet technique includes its capability of generating the plasma with high density, long length, and variable conditions of plasma excitation. This technique also gives new degrees of freedom that can be used to optimize the HHG. The use of any elements in the periodic table that can be formed as solid targets may disclose many interesting peculiarities for the generation of coherent XUV radiation. The optimization of plasma conditions can lead to further growth of HHG efficiency. In comparison to the high-excited plasma, the low-excited plasma could possess some attractive properties (i.e., less reabsorption of generated harmonics, less influence of free electrons on phase-matching and self-defocusing, etc.) and then can increase both the conversion efficiency and cutoff energy of harmonics.

In this paper, we demonstrate the efficient generation of coherent XUV radiation after the propagation of femtosecond pulse through the low-excited plasma produced on the surface of different solid targets. The harmonics up to the 65th order and the conversion efficiencies at the plateau region up to $8 \times 10^{-6}$ were achieved in these studies. We show our results on the spectral modification of harmonic distribution for some plumes. A strong resonance enhancement of single harmonics will also be discussed.

## 2 Experimental Arrangements

The pump laser used in this research consisted of chirped-pulse amplification Ti:sapphire laser (Spectra-Physics, Tsunami + TSA10F) operated at a 10-Hz pulse repetition rate, whose output was further amplified using a three-pass amplifier. First the preplasma was produced by the 210-ps prepulse on the solid target surface. Then the 150-fs main pulse was focused in the preplasma. The beam waists of focused femtosecond radiation were measured to be 60 and 18 μm in the cases of 200 and 100 mm focal length lenses, respectively. The focal size of prepulse was 600 μm. The intensity of picosecond prepulse, $I_{pp}$, on the target surface was varied between $7 \times 10^9$ and $8 \times 10^{10}$ W $\cdot$ cm$^{-2}$. The maximum intensity of the femtosecond main pulse is $I_{fp} = 1 \times 10^{15}$ W $\cdot$ cm$^{-2}$. The contrast ratio of main pulse was measured to be $10^{-5}$. The time delay between the prepulse and main pulse was from 8 to 100 ns. The central wavelength of laser pulse was 796.5 nm. The low-order harmonics were analyzed using a vacuum monochromator (Acton Research Corporation, VM-502) and photomultiplier tube. The high-order harmonic spectra were measured using grazing incidence spectrometer with

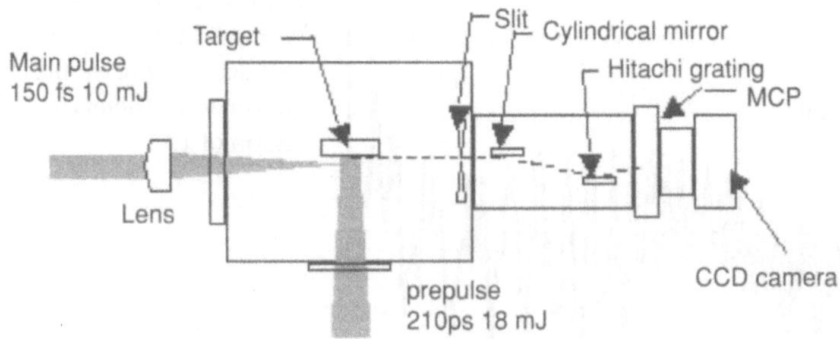


FIGURE 1. Experimental setup. First, the prepulse was focused on the target surface. The preplasma was produced by prepulse. Then the main pulse was focused in the preplasma. Delay between the prepulse and the main pulse was from 3 to 94 ns.

Hitachi-grating 1200 grooves/mm. The gold-coated grazing incidence cylindrical mirror was used for image translation from the laser plasma to the detector plate (Fig. 1). The spectra was detected by microchannel plate with phosphor screen and recorded by CCD camera. The details of the spectrometer calibration have been reported in [8].

## 3 Results and Discussion

Most of our studies were carried out using the orthogonal setup. We used different solid targets (Ag, B, In, C, W, Se, Mo, Nb, Si, Zr, Mg, Cd, Pd, Al, Cr, Ta) for the preparation of suitable laser plumes for our experiments. The choice of targets was stipulated by our aim to analyze the influence of atomic number, ionization potentials, and other parameters of samples on the HHG from different plumes. Our efforts were concentrated on the determination of the optimal conditions for achieving the maximum harmonic order (i.e., cutoff energy) as well as maximum conversion efficiency for high harmonics using the low-excited laser plasma.

The plateau pattern of high-order harmonic distribution appeared in these experiments. In particular, the high harmonics up to the 63rd order at a wavelength of 12.6 nm were observed in the experiments with boron plasma at loosely focused conditions ($b = 6$ mm, $L_p = 0.6$ mm, $L_p$ is the plasma sizes). The HHG from the B plume appeared to be similar to those observed in gas-jet experiments, with the characteristic shape of plateau for the harmonics exceeding 17th order (Fig. 2). The conversion efficiencies were varied in the range of $10^{-4}$ (for the 3rd harmonic) to $10^{-7}$ (for the harmonics at the plateau region). The plateau disappeared when the prepulse energy was increased above 10 mJ. In that case the highest harmonic observed was the 19th one.

The higher harmonics (up to the 65th order, $\lambda = 12.24$ nm) were observed at the conditions when the confocal parameter of focused radiation was changed ($b = 1.2$ mm) to be close to the plasma sizes. However, in that case we observed

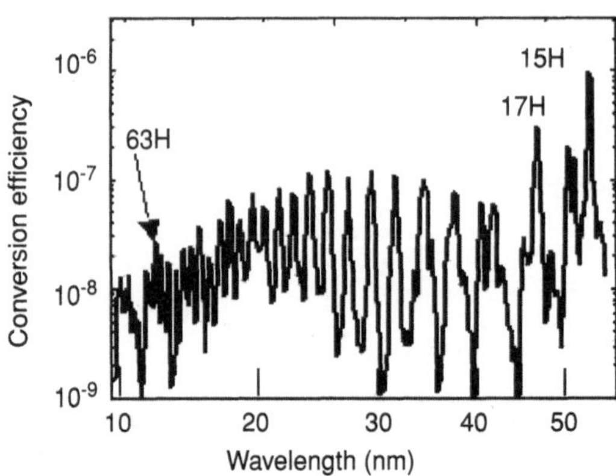


FIGURE 2. 63rd order (12.24 nm) harmonic was observed from the low-density boron plasma with conversion efficiency at plateau region of $10^{-8}$.

the instability of HHG in the terms of the plasma conditions. Most of our further presented studies were carried out at $b = 6$ mm.

The plateau pattern appeared in the case of the HHG from most of the plasma samples used in these experiments, whereas in some cases (W, C, Mg, Se) we observed a steady or even steep decrease of conversion efficiency for each next harmonic (Fig. 3). The origin of this phenomenon is not clear and further studies are required for the clarification of the difference between the harmonic distributions and cutoff energies observed from various targets.

We observed a direct relation between the cutoff energy of the harmonics generated from different targets and the second ionization potential of atoms. Figure 4 shows a linear dependence between these two parameters for different targets. The plateau pattern was not observed in the cases of C, Se, W, and Mg plumes. The conversion efficiencies in the plateau region were varied between $1 \times 10^{-7}$ and $8 \times 10^{-6}$, depending on the target. The application of the targets with high second ionization potential (Li, Na, and K) probably could further extend a cutoff energy at optimal plasma conditions. The investigation of alkali metals for the HHG was reported previously in [1–3] at the conditions of high-excited plasma. However, the maximum reported harmonic orders in those experiments

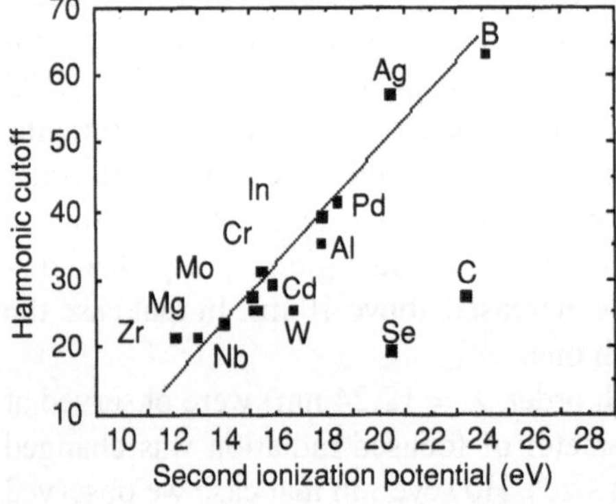


FIGURE 3. Dependence between the cutoff harmonic and ionization potential of ions (second ionization potential).

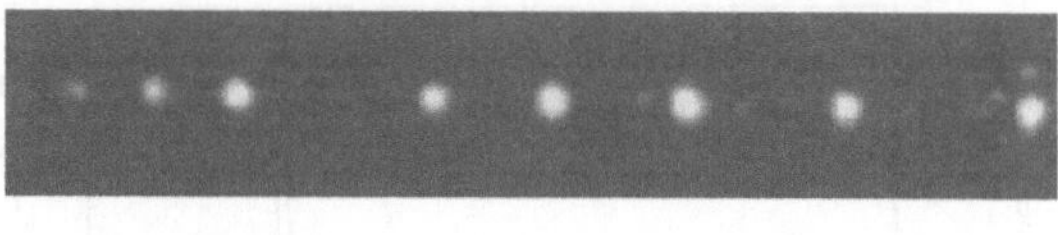

FIGURE 4. Harmonic spectrum from chromium laser plasma in XUV range. 27th order harmonic is disappeared using chromium laser plasma.

were considerably lower than the ones observed in present studies, probably due to some saturation processes in such plasma. For the confirmation that the observed harmonic generation was originated from the nonlinear process involving a single-atom or single-ion response rather than the collective movement of electron cloud we carried out the polarization measurements of HHG. The quarter-wave plate was installed in front of the focusing lens to vary the polarization of main beam from linear to elliptical, circular, and finally again linear. We analyzed the harmonic output at the different angles of rotation of this plate.

While most of targets showed a prolonged plateau-like distribution of high harmonics, some of them demonstrated very interesting features that were not reported previously in the gas-jet experiments. In particular, in the case of chromium plasma we registered a considerable suppression of 27th harmonic in comparison to the neighbor harmonics (Fig. 4). Such a feature was observed in the cases of different focusing geometries using both the 200- and 100-mm focal length lenses.

The investigation of the conditions of enhanced conversion efficiency for some group of harmonics in the plateau region was carried out in various studies (see, for example, [9] and references therein). The aim of those studies was to optimize the phase-matching conditions for the increase of the yield of coherent XUV photons during the gas-jet experiments. Different approaches were proposed for these purposes, in particular the use of the hollow-core fibers and thin tubes [9–11].

However, in our studies we observed another process, a considerable restriction of a single harmonic in the plateau region, whereas the neighbor harmonics yield was relatively equal. A possible mechanism explaining this effect could be related with an influence of free electrons and depletion of neutrals and singly charged ions that led to a decrease of coherence length for some specific harmonics. At the same time, a propagation effect could not completely suppress a single harmonic. Probably, the target properties play a crucial role on a considerable suppression of 27th harmonic intensity in the case of chromium plasma. There are no known transition lines of Cr ions in the vicinity of 27th harmonic ($\lambda = 29.5$ nm). More detailed analysis need to be carried out for the understanding of the processes leading to the exclusion of a certain harmonic from a whole spectrum in the plateau region.

Most intriguing picture appeared in the case of indium plasma. We observed the anomalously strong 13th harmonic radiation ($\lambda = 61.2$ nm), which by almost two orders of magnitude exceeded the neighbor ones. The experimental conditions were analogous to the ones used in the case of Cr plasma. These studies were repeated using the different focusing conditions and plasma geometries. The conversion efficiency of 13th harmonic was measured to be $8 \times 10^{-5}$, while the

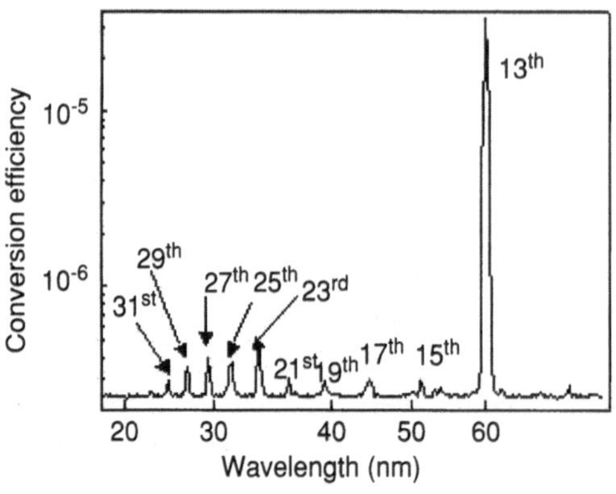


FIGURE 5. Spectrum of the high-order harmonic generation was obtained using the indium plasma. The conversion efficiency of 13th harmonic is $8 \times 10^{-5}$.

conversion efficiency in the region of plateau was varied between $7 \times 10^{-7}$ and $2 \times 10^{-6}$ (Fig. 5).

For the confirmation that this strong line in the range of 61 nm was originated from the nonlinear process rather than from a simple reexcitation of plasma line we investigated the influence of 13th harmonic output on the polarization characteristics of the main pulse. Again as in aforementioned experiments, the 25 degree rotation of quarter-wave plate led to the complete disappearance of 61.2-nm emission, as it should be assuming the origin of HHG, while the excited lines of plasma spectrum observing at the different polarizations of main beam remained unchanged. Another test of this process was carried out by the variation of the wavelength of main beam to analyze whether the excited transitions from indium plasma influence on a plateau pattern of harmonic distribution. We were able to change the central wavelength of the output radiation of our laser from 770 to 820 nm. The 13th harmonic output was decreased with the displacement of the fundamental wavelength from 796 nm to the short-wavelength region. At the same time we observed a strong enhancement of the 15th harmonic at the wavelength of main pulse of 775 nm, whereas other harmonics remained relatively constant (see Fig. 6).

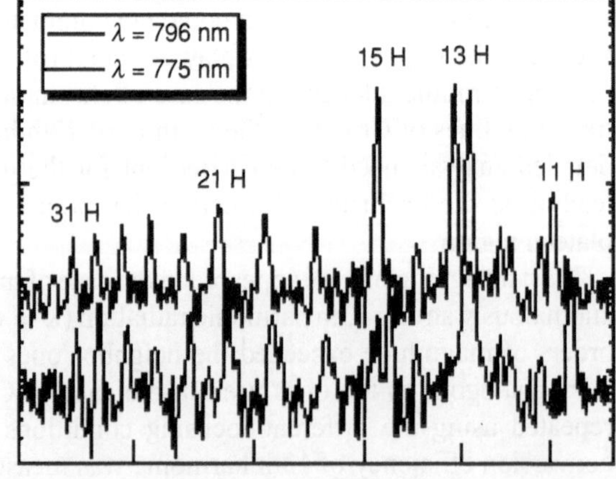


FIGURE 6. Harmonic distribution from the indium plasma in the cases of different central wavelength of the main laser pulse.

Again as in the case of Cr plasma, there are probably some transitions between the high-level states of In ions, still unobservable in previous studies, which could play an important role in the enhancement of the nonlinear susceptibility in the vicinity of 53 and 61 nm. Currently, we continue these studies and try to find other plumes where such an enhancement of harmonic output could be realized.

## 4 Conclusion

In conclusion, we demonstrated the generation of high harmonics (up to the 65th order, $\lambda = 12.24$ nm) after the propagation of femtosecond laser pulses through the low-excited surface plasma produced by a prepulse radiation. The high-order harmonics generated from most plumes showed a plateau pattern. The harmonic generation in these conditions occurred due to the interaction of femtosecond pulses with the ions. The conversion efficiency for the harmonics generated in the plateau region was varied in the range of $10^{-7}$ to $8\times10^{-6}$, depending on the target. The main contribution to the limitation of HHG efficiency was caused by the free-electron-induced self-defocusing of main beam. Our studies showed that the physical characteristics of targets play a decisive role in the terms of conversion efficiency, plateau appearance, and specific properties of separate harmonics in the plateau region.

A considerable restriction of the 27th harmonic generation was observed at different focusing conditions in the case of chromium plasma. We also analyzed the observed limitation of harmonic cutoff and attributed it to the self-defocusing of main pulse by free electrons. We showed the linear relation between the second ionization potential of atoms and the cutoff energy of harmonics generated at the plateau region. Our studies of various solid targets revealed that the choice of target has a decisive influence on the highest achievable harmonic order and conversion efficiency, as well as on the appearance of plateau.

Our observation of the resonance-induced enhancement of a single harmonic ($\lambda = 61.2$ nm) at a plateau region with the efficiency of $8 \times 10^{-5}$ in the case of In plume can offer some expectation that analogous processes can be realized in other plasma samples in the shorter wavelength range where the highest harmonics were achieved. It is difficult to realize such an approach in gas puffs due to the necessity of the preparation of specific conditions for the excitation of appropriate levels of nonlinear medium prior to the laser-matter interaction.

**Acknowledgments.** The authors thank T. Ozaki for the help in installation of the microchannel plate and I. A. Kulagin for the calculations of the influence of the delay and main pulse intensity on the low-order harmonic intensity.

## References

1. Y. Akiyama, K. Midorikawa, Y. Matsunawa, et al.: Generation of high-order harmonics using laser-produced rare-gas-like ions. Phys. Rev. Lett. **69**, 2176–2179 (1992).

2. S. Kubodera, Y. Nagata, Y. Akiyama, et al.: High-order harmonic generation in laser-produced ions. Phys. Rev. A. **48**, 4576–4582 (1993).
3. C. G. Wahlström, S. Bordström, J. Larson, and S. G. Pettersson: High-order harmonic generation in laser-produced ions using a near-infrared laser. Phys. Rev. A. **51**, 585–591 (1995).
4. S. Meyer, H. Eichmann, T. Menzel, et al.: Phys. Rev. Lett. **76**, 3336–3339 (1996).
5. R. A. Ganeev, V. I. Redkorechev, and T. Usmanov: Optical harmonics generation in low-temperature laser-produced plasmas. Opt. Cpmmun. **135**, 251–256 (1997).
6. S. Banerjee, A. R. Valenzuela, R. C. Shah, A. Masksimchuk, and D. Umstadter: High-harmonic generation in plasmas from relativistic laser-electron scattering J. Oot. Soc. Am. **B20**, 182–190 (2003).
7. U. Teubner, G. Pretzler, T. Schlegel, K. Eidmann, E. Forster, and K. Witte: Anomalies in high-order harmonic generation at relativistic intensities. Phys. Rev. A. **67**: 013816 (2003).
8. R. A. Ganeev, M. Baba, M. Suzuki, and H. Kuroda: High-order harmonic generation from silver plasma. Phys. Lett. A. **339**, 103–109 (2005).
9. A. Rundquist, C. G. Durfee, Z. H. Chang, et al.: Phase-matched generation of coherent soft X-rays. Science **280**, 1412–1415 (1998).
10. Y. Tamaki, Y. Nagata, M. Obara, and K. Midorikawa: Phase-matched high-order-harmonic generation in gas-filled hollow fiber. Phys. Rev. A. **59**, 4041–4044 (1999).
11. M. Schnürer, Z. Cheng, M. Hentschel, et al.: Absorption-limited generation of coherent ultrashort soft-X-ray pulses. Phys. Rev. Lett. **83**, 722–725 (1999).

# Scaling of keV HHG Photon Yield with Drive Wavelength

Ariel Gordon, Christian Jirauschek, and Franz X. Kärtner

Department of Electrical Engineering and Computer Science, and Research Laboratory of Electronics, Massachusetts Institute of Technology, 77 Massachusetts Avenue, Cambridge, MA 02139, USA.

## 1 Introduction

It is well known that the cutoff energy of high harmonic generation (HHG) increases linearly with the pump intensity and quadratically with pump wavelength [1]. It is also well known that quantum diffusion suppresses the HHG photon yield by a factor that decreases cubically with the pump wavelength [1]. This cubical scaling was one of the reasons why most of the HHG studies focused on 0.8 μm rather than longer wavelength drive pulses (LWDP), in spite of the increased cutoff energy associated with the latter.

Many HHG works are dedicated to exploiting the phenomenon for building a coherent X-ray source, and much effort has been devoted to increasing the photon energies. Only recently have researchers been able to reach the long-desired keV photon energy range [2]. However, as the photon energy is increased, the photon yield rapidly decreases.

In this chapter we present an explanation to the decrease in photon yield as the cutoff energy is increased [3]. We show that the HHG photon yield from a single helium atom with a 0.8 μm pump drops exponentially as the cutoff is pushed forward by increasing the pump intensity. The yield drops by about 5.5 orders of magnitude per keV, starting at about 500 eV. We therefore come to the conclusion that the potential of HHG with Ti:sapphire pumps in terms of photon energy is already nearly exhausted.

The exponential drop of the photon yield is, for LWDP, shifted to higher energies and is slower. In particular, above 1 keV the photon yield from 1.5–3 μm exceeds the one from a 0.8 μm pump by many orders of magnitude for a neutral helium target, and even more so with other neutral target gases. This property stems from the single atom response and is of absolutely fundamental origin. Before we explain and derive our yield formula, we present right at the beginning Fig. 1, where the photon yield for different energies is compared:

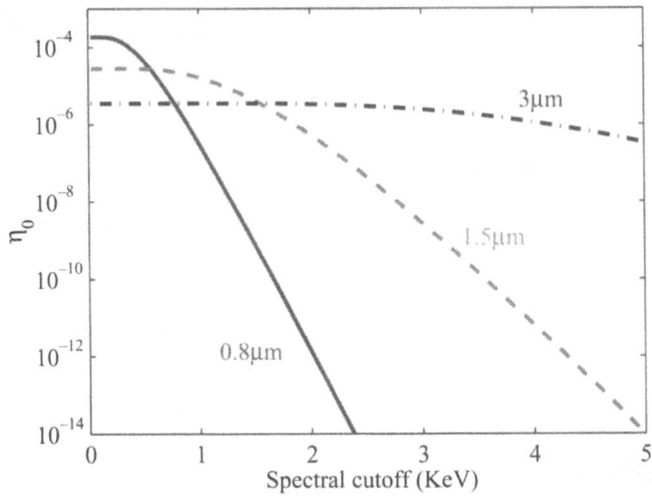


FIGURE 1. Normalized efficiency $\eta_0$ of a single helium atom for different pump wavelengths as function of the spectral cutoff energy (determined by the pump intensity). The normalized efficiency is defined in (Section 2) and is proportional to the single-atom photon yield.

## 2 Scaling of the Yield

We now explain the origin of our result and derive the formula plotted in Fig. 1. It has been long recognized that depletion of the ground state poses a major limitation on HHG. In particular, it was realized that short drive pulses are necessary for high photon energies, or else the ground state is depleted before the main peak of the pulse is reached. Indeed the highest photon energies are generated with few cycle pulses.

However, at the fields currently applied in HHG, significant depletion of the ground state occurs within one cycle. Even if the ground state is fully populated at the beginning of an ideal single cycle pulse, if the pulse entirely ionizes the target atom, no HHG will occur. This is because HHG is the interference phenomenon between the returning electron and its part remaining at the ground state. If the ground state is empty upon return of the electron, no interference occurs.

To quantify this effect one can define the normalized efficiency $\eta_0$:

$$\eta_0 = \omega^3 |a(t_1)a(t_2)|^2$$

where $a(t)$ is the amplitude of the ground state as function of time, $t_1$ ($t_2$) is the time where the most energetic electron trajectory initiates (returns to the parent atom), and $\omega$ is the drive pulse carrier frequency. We claim that the photon yield is proportional to $\eta_0$, since $\omega^3$ accounts for quantum diffusion, and since the intensity of harmonics generated by the vicinity of the most energetic trajectory is proportional to the population of the ground state at the initial and final times

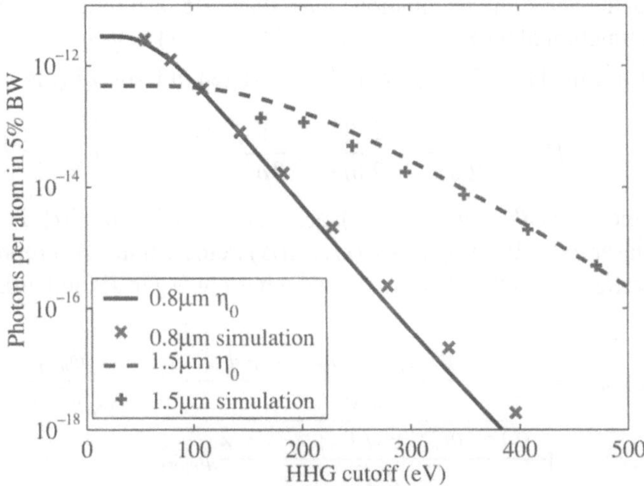


FIGURE 2. HHG photon yield as function of the spectral cutoff for hydrogen, a 0.8-μm and 1.5-μm source, a single cycle driving pulse, and no initial depletion of the ground state. The points were obtained from simulations of the three-dimensional Schrödinger equation. The solid curves are $\eta_0$ obtained from tabulated static ionization rates for hydrogen [5]. $\eta_0$ was multiplied by a coefficient (the same for both curves) such that it would match the simulated yields at high cutoff energies.

of this trajectory. $a(t)$ can be computed using the quasistatic approximation using tabulated static ionization rates [4].

In order to test the predictive power of $\eta_0$ we compared it to single atom yields obtained from a numerical solution of the Schrödinger equation of a hydrogen atom in a *single cycle* drive field. The results are displayed in Fig. 2, showing quantitative agreement between $\eta_0$ and the simulated rates.

## 3 Numerical Method

Simulating single-atom HHG becomes a more and more computationally demanding problem as the drive wavelength is increased. Most of the algorithms in the literature use a spherical harmonics representation of the wavefunction. However, the excursion of the electron away from the parent ion increases quadratically with the wavelength, the wavefunction becomes more and more prolonged in the direction of the field, and spherical harmonics become less and less natural a basis: more and more angular momentum channels ($\ell$-s) are required. The centrifugal potential in each channel increases quadratically with $\ell$, which introduces major numerical difficulties due to singularities.

In order to simulate HHG with LWDP we therefore had to apply a different algorithm. For LWDP the volume occupied by the ionized electron is cylinder-like, which makes the cylindrical coordinates a natural choice. The cylindrical symmetry of the Schrödinger equation then reduces it to a 2D one. Let $\psi(x, y, z, t)$ be the wavefunction of the electron. Putting $\psi(x, y, z, t) = \phi(\rho, z, t)/\sqrt{\rho}$, where $\rho \equiv \sqrt{x^2 + y^2}$, the HHG Hamiltonian for hydrogen in terms of $\phi$ is

$$H = -\frac{1}{2}\frac{\partial^2}{\partial z^2} - \frac{1}{2}\frac{\partial^2}{\partial \rho^2} - \frac{1}{8\rho^2} - \frac{1}{r} - zE(t),$$

where atomic units ($\hbar = m = e = 1$) are being used and $E(t)$ is the time-dependent laser field. With $\phi_{m,n} \equiv \phi(m\Delta, n\Delta)$ being a finite grid representation of $\phi$, a discretization of the Hamiltonian which is correct to second order in space is

$$H_\Delta \phi_{m,n} = \frac{\phi_{m+1,n} + \phi_{m-1,n} + \phi_{m,n+1} + \phi_{m,n-1}}{2\Delta^2} - \frac{\phi_{m,n}}{\Delta\sqrt{m^2 + n^2}} + \\ + \frac{\sqrt{1 + m^{-1}} + \sqrt{1 - m^{-1}} - 2}{2\Delta^2}\phi_{m,n}.$$

Note especially the last and rather nontrivial term, which is the correct way to discretize the $1/\rho^2$ potential term in $H$ [6]. The trivial discretization $1/(\Delta m)^2$ leads to entirely incorrect results. While many works use implicit algorithms for computing the time evolution of the wavefunction, we used the explicit scheme often referred to as the "leapfrog scheme" [7]. This scheme has second-order accuracy in space and time just like the commonly used implicit schemes; however, it is simpler and faster than the latter, since it does not require matrix inversion. The price to pay is that the stability of the scheme is not unconditional. Rather, the time step should be small enough to meet a certain stability criterion. However,

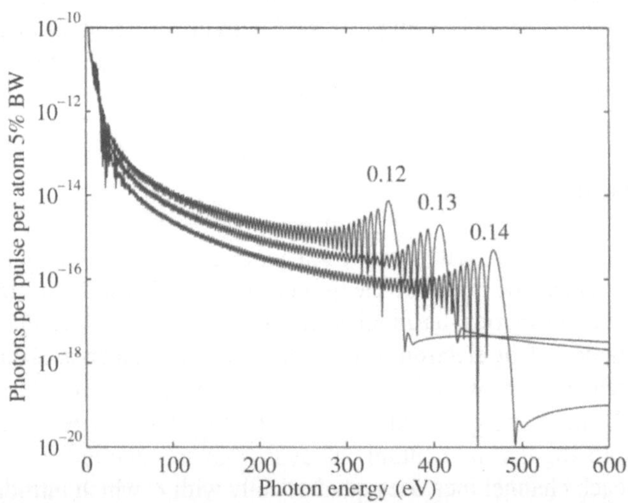


FIGURE 3. Simulated HHG spectra for a 1.5-m single-cycle drive pulse and a single hydrogen atom. The number near each curve denotes the amplitude of the electric field in atomic units.

precision requirements anyway dictate a time step which is even smaller than the stability criterion does.

Figure 3 shows some examples of simulated HHG spectra in hydrogen with a 1.5-μm single-cycle pulse. To our knowledge, no simulated spectra for such a long wavelength drive exist in the literature. The results were obtained with a spatial step size $\Delta = 0.06$ au, temporal step size of $1.8 \times 10^{-4}$, 10,000 points in the $z$ direction, and 2000 points in the $\rho$ direction. The calculation of each spectrum lasted about a week on a 3-GHz PC.

## 4 Summary

We have shown that the single atom HHG response to LWDP (1.5–3 μm) outperforms the traditional 0.8-μm pump by many orders of magnitude in terms of photon yield for photon energies above 1 keV. We conclude that LWDP are the most promising direction for significant HHG above the 1 keV barrier.

## References

1. M. Lewenstein, Ph. Balcou, M. Yu. Ivanov, et al.: Phys. Rev. A. **49**, 2117 (1994).
2. J. Seres, E. Seres, A. J. Verhoff, et al.: Nature **433**, 596 (2005).
3. A. Gordon and F. X. Kärtner: Opt. Exp. **13**, 2941 (2005).
4. A. Scrinzi, M. Geissler, and T. Brabec: **83**, 706 (1999).
5. M. V. Ivanov: J. Phys. B **34**, 2447 (2001).
6. K. C. Kulander, K. R. Sandhya Devi, and S. E. Koonin: Phys. Rev. A **25**, 2968 (1982).
7. C. Leforestier, R. H. Bisseling, C. Cerjan, et al.: J. Comput. Phys. **94**, 59 (1991).

# Self-compression of an Ultraviolet Optical Pulse Assisted by Raman Coherence Induced in the Transient Regime

Yuichiro Kida, Shin-ichi Zaitsu, and Totaro Imasaka

Department of Applied Chemistry, Graduate School of Engineering, Kyushu University, 6-10-1, Hakozaki, Higashi-ku, Fukuoka 812-8581, Japan
**y-kida@cstf.kyushu-u.ac.jp**

## 1 Introduction

An ultrashort laser pulse has recently been used in many types of applications, such as observations of a dephasing process of a molecule and a monitoring of an ultrafast chemical reaction. Several techniques are demonstrated to generate such an ultrashort pulse. One of the techniques utilizes high-order stimulated Raman scattering (HSRS). In this approach, spectrally discrete sidebands are generated, in addition to the original spectral component. The resultant spectrum can be spread from the ultraviolet to the near-infrared regions and is capable of generating a subfemtosecond optical pulse [1]. Impulsive stimulated Raman scattering (ISRS) [2] is a type of stimulated Raman scattering which is used by researchers in Max Born Institute to generate an ultrashort optical pulse [3,4]. A high-energy laser pulse, whose pulse width is shorter than a period of molecular motion (vibration or rotation), is used to excite coherent molecular motion. The induced coherent motion broadens the spectrum of a temporally delayed pulse (probe pulse). Furthermore, the phases of the newly created spectral components in this regime can easily be controlled, and therefore the pulse width of the temporal profile is compressed [3]. The temporal structure of the modulated probe pulse consists of a train of compressed pulses, when the probe pulse longer than the period of molecular motion is employed [3]. Whereas a single pulse is obtained by decreasing the probe pulse width below this time period [4].

In our laboratory, compression of a laser pulse based on spectral broadening by means of stimulated Raman scattering and subsequent four-wave mixing has been studied [1,5–7]. A hydrogen gas is utilized as a Raman medium because of its high Raman gain. In a recent study, we have investigated the generation of a 10-fs pulse in the ultraviolet region, based on frequency-doubling and subsequent compression using a 100-fs Ti:sapphire laser system widely used for various applications. This compression technique based on ISRS cannot successfully be used in the present experimental condition, since the pulse width of the laser is longer than the period of molecular rotation (57 fs for ortho-hydrogen, 94 fs for para-hydrogen) and vibration (8 fs) for molecular hydrogen. There is an alternative approach that makes it possible to excite coherent molecular motion. This is called transient stimulated Raman scattering (TSRS), a more well-known type of stimulated Raman scattering than ISRS. In TSRS, a laser pulse whose pulse width is longer than the period of coherent molecular motion and is shorter than the dephasing time is employed. In our approach, a high-energy pump laser pulse is utilized in order to excite coherent rotation of ortho-hydrogen for subsequent modulation of a relatively weak probe pulse propagating after a certain time of delay. A difference between our approach and the approach that uses ISRS [3,4] is existence of a threshold for the energy of a pump pulse to excite molecular coherent motion. The threshold does not exist in ISRS [2], while certain pulse energy is needed in TSRS as is widely known. This threshold, however, does not induce a serious problem, since the energy of the pump pulse used herein is strong enough to exceed the threshold.

In this report, a pulse compression technique, consisting of TSRS and pump-probe configuration, is outlined. The potential of this pulse compression technique is discussed for obtaining the maximum efficiency for the generation of Raman emission. The experimentally characterized temporal profile of the modulated pulse is shown, and the temporal structure is explained in detail. Finally, this report is summarized.

## 2 Experimental Setup

The experimental setup used in our technique is shown in Fig. 1. A laser beam obtained from a commercial Ti:sapphire laser regenerative amplifier system (Thales Concerto, 784 nm center wavelength, 100 fs pulse width, 1.6 mJ maximum energy,

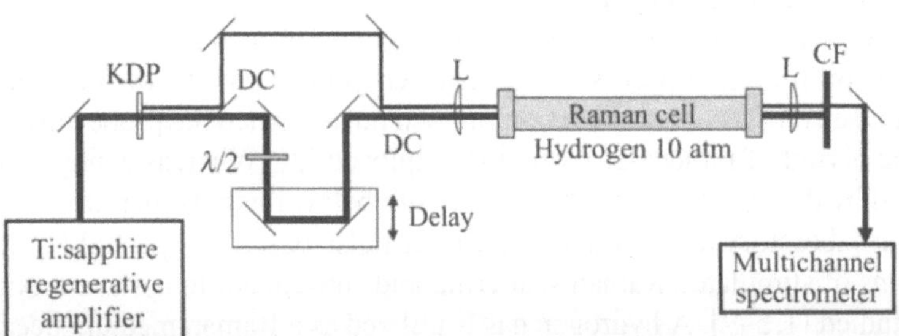


FIGURE 1. Experimental setup: KDP, type I frequency-doubling crystal; DC, dichroic mirror with high reflectivity for the second harmonic emission; λ/2, half-wave plate; L, focusing lens; CF, color glass filter with high absorptivity for the near-infrared beam.

1 kHz repetition rate) is used. The width of the pulse is stretched to about 200 fs by adding negative group delay dispersion using a pulse stretcher consisting of a pair of gratings. The beam is introduced into a KDP crystal to generate a second harmonic beam. The near-infrared laser beam and the harmonic beam transmitted from the crystal are used as a pump beam and a probe beam, respectively. These two beams are separated into two parts spatially using a dichroic mirror and then the time delay between two laser pulses is controlled using an optical delay line. The beams are recombined spatially using another dichroic mirror and focused into a Raman cell (80 cm long) filled with a hydrogen gas at 10 atm. The pump beam is removed from the transmitted beam using a color glass filter. The spectrum of the probe beam is then measured using a multichannel spectrometer.

## 3 Results and Discussion

### *3.1 Influence of Cross Phase Modulation*

When the pump pulse and the probe pulse is temporally overlapped with each other, cross phase modulation (XPM) occurs in the hydrogen gas. In this case, the probe pulse is modulated to provide a broad spectrum, and the efficiency for the generation of the rotational Raman emission can be reduced. However, this modulation can be controlled by changing the delay between the two pulses, because XPM occurs only when the two pulses temporally overlap each other. In order to estimate the delay required to reduce XPM to negligible levels, the variation in the spectrum of the output probe pulse, which is caused by changing the time delay of the input probe pulse with respect to the input pump pulse, was investigated. The result is shown in Fig. 2. Minus signs of the delays in Fig. 2 depict the case that the probe pulse was propagating prior to the pump pulse.

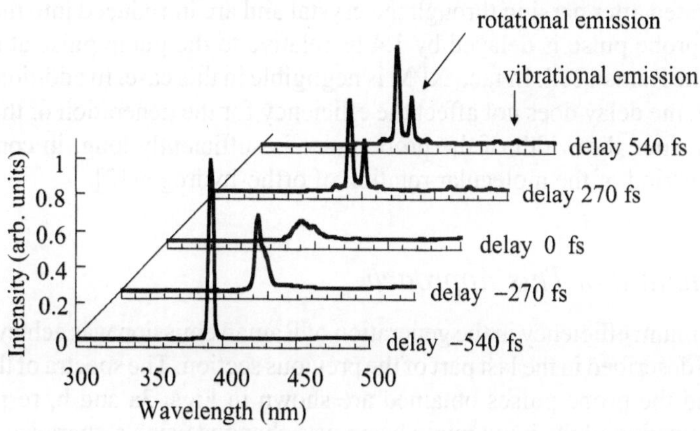


FIGURE 2. Spectrum of the probe pulse at different delays between the pump and probe pulses.

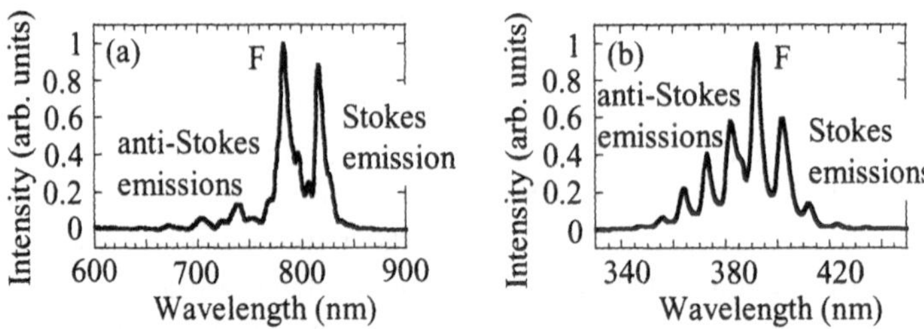


FIGURE 3. Spectra of the output pump pulse (**a**) and probe pulse (**b**). F indicates the fundamental line.

In the case of −540 fs, the pump pulse and the probe pulse are temporally separated from each other. No spectral change by XPM and no Raman emission are observed in the spectrum of the output probe pulse. When the delay is −270 fs, the probe pulse is partially overlapped with the pump pulse and the spectrum is slightly broadened by XPM only to longer wavelengths, because of the negative frequency chirp of the input pulse. The complete overlap is achieved in the case of 0 fs, and XPM becomes maximum. This broadens the spectrum of the probe pulse to both longer and shorter wavelengths, while no appreciable Raman emission in the probe pulse is observed. When the pump pulse propagates preceding the probe pulse (in the cases of 270 fs and 540 fs), no influence of XPM appears in the spectrum of the probe pulse. Alternatively, rotational Raman emission of ortho-hydrogen is observed at longer delays (270 fs, 540 fs). On the other hand, vibrational Raman emission is observed at around 450 nm above 270 fs. These facts suggest that XPM is negligible under these conditions. Furthermore, it should be noted that spectral components newly created by the interaction with hydrogen consist of only the Raman emission.

In the experiment, the pump and probe beams were spatially separated from each other after passing the beam through a KDP crystal. When these beams are not separated after passing through the crystal and are introduced into the Raman cell, the probe pulse is delayed by 1.4 ps relative to the pump pulse at the focal point in the Raman cell. In fact, XPM is negligible in this case. In addition, a small change in the delay does not affect the efficiency for the generation of the Raman emission since the width of the probe pulse is sufficiently long, in comparison with the period of the molecular rotation of ortho-hydrogen [2].

## 3.2 *Potential of This Approach*

The maximum efficiency in the generation of Raman emission was achieved using the setup described in the last part of the previous section. The spectra of the output pump and the probe pulses obtained are shown in Figs. 3a and b, respectively. The polarization of the input pump beam was changed using a quarter-wave plate and was adjusted to be elliptical. Strong Stokes and relatively weak anti-Stokes emissions are observed in the spectrum of the output pump pulse. The difference in the peak heights for the Stokes and anti-Stokes emissions arises from a coupling between these two types of emissions, which is one of the distinct features in stimulated Raman scattering. High conversion efficiency from the fundamental to Stokes emission suggests large coherent rotation of ortho-hydrogen induced by the pump pulse. Up to eight Raman lines are generated in the output probe pulse and the spectral width is expanded to 28 nm, which corresponds to a 7-fs pulse in the case of a transform-limited pulse.

A similar approach to our technique has been reported by Bartels et al., in which a slightly longer pump pulse than the period of molecular vibration is used [8]. This is called displacive ISRS. In their approach, the pump pulse modulated by self-steeping has a structure varying temporally faster than the vibrational period and can excite Raman coherence impulsively. The spectrum of the pump pulse is broadened smoothly and extended toward high frequencies. In contrast to the case of displacive ISRS, the spectrum of the output pump pulse obtained in our technique (Fig. 3a) does not have a smooth profile but have a structured one. Furthermore, the fundamental peak in the spectrum is mainly extended to lower frequencies (Stokes emission) and the pulse train formed by the fundamental and Stokes emissions induces coherent molecular motion.

## 3.3 *Temporal Profile of the Modulated Pulse*

Self-diffraction frequency-resolved optical gating (SD-FROG) [9] is a useful means for the measurement of an ultraviolet pulse. A typical temporal structure of the probe pulse characterized using an SD-FROG system is shown in Fig. 4b. The spectrum of the pulse measured by means of a multichannel spectrometer is shown in Fig. 4a. Three rotational Stokes lines and the same number of rotational anti-Stokes lines are generated, in addition to the fundamental line. The third-order Raman emissions are very weak, in comparison with the other emission lines; the peak heights are approx. 1% of the fundamental peak. Inverse Fourier transformation of this spectrum provides a train of 11-fs pulses with a period of 57 fs under the assumption that the pulse is transform-limited. However, the width experimentally obtained is 23 fs, as is shown in Fig. 4b. This value is twice as large as the calculated one (11 fs). This is associated to a negative frequency chirp,

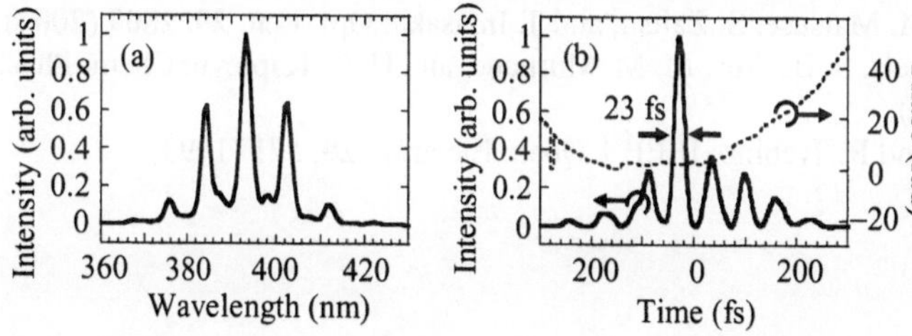


FIGURE 4. (**a**) Spectrum of the modulated probe pulse measured by a multichannel spectrometer; (**b**) Intensity (solid line) and phase (dotted line) versus time retrieved from the SD-FROG trace.

which is recognized by the quadratic phase curve shown as a dotted line in Fig. 4b. A glass plate, which causes a positive group delay dispersion, will compensate for this frequency chirp and should compress the pulse to 11 fs. The temporal structure of the compressed pulse in Fig. 4b is a train of pulses. Generation of a single pulse is possible by a combination of several types of molecular motions, e.g., rotations of ortho-hydrogen and para-hydrogen [6]. This approach reduces subpulses, the spacing of which is related to each molecular motion.

## 4 Summary

The combination of TSRS and pump–probe technique reported herein makes it possible to compress a relatively longer pulse than the molecular motion. In this approach, a long pulse is used to excite coherent molecular motion, which is followed by subsequent spectral broadening of a delayed probe pulse. In this study, a hydrogen gas is used as a Raman medium because of its nature of high Raman activity. Eight Raman sidebands are obtained under the optimum condition, being capability of generating a 7-fs optical pulse. The temporal profile of the pulse measured consists of a pulse train, and a pulse width of 23 fs is achieved without any extra phase control. For further compression, it is suggested to use a high-energy picosecond pulse as a pump pulse. This produces strong coherent motion in a Raman medium through a TSRS process. In the near future, our approach requiring a simple, less-expensive, easy-to-operate 100-fs Ti:sapphire laser (1 kHz) will provide a 10-fs pulse and can be used in numerous applications such as fundamental studies of ultrafast phenomena.

## References


1. H. Kawano, Y. Hirakawa, and T. Imasaka: Appl. Phys. B **65**, 1 (1997).
2. Y. X. Yan, E. B. Gamble, and K. A. Nelson: J. Chem. Phys. **83**, 5391 (1985).
3. M. Wittmann, A. Nazarkin, and G. Korn: Phys. Rev. Lett. **84**, 5508 (2000).
4. N. Zhavoronkov and G. Korn: Phys. Rev. Lett. **88**, 203901 (2002).
5. S. Yoshikawa and T. Imasaka: Opt. Comm. **96**, 94 (1993).
6. H. Otsuka, T. Uchimura, and T. Imasaka: Opt. Lett. **29**, 400 (2004).
7. Y. Kida, M. Matsuse, S. Zaitsu, and T. Imasaka: Opt. Lett. **29**, 2809 (2004).
8. R. A. Bartels, S. Backus, M. M. Murnane, and H. C. Kapteyn: Chem. Phys. Lett. **374**, 326 (2003).
9. D. Kane and R. Trebino: IEEE J. Quan. Electron. **29**, 571 (1993).

# Part XVI

# Ultrafast Amplifier

# Generation of High-energy Few-cycle Pulses by Filament Compression

C. P. Hauri, J. P. Rousseau, F. Burgy, G. Chériaux, and R. López-Martens

Laboratoire d'Optique Appliquée, ENSTA-Ecole Polytechnique, Chemin de la Hunière, F-91761 Palaiseau Cedex, France
**hauri@ensta.fr**

## 1 Introduction

Recent developments in intense few-cycle laser pulse generation have triggered many breakthrough experiments in high-field science and have paved the way toward time-resolved spectroscopy on the attosecond ($10^{-18}$ s) timescale [1–3]. In particular, it was shown that intense driving pulses with only two to three reproducible field oscillations are a prerequisite for controlling the generation of isolated attosecond pulses by high-order harmonic generation in rare gases [4]. Until recently, intense few-cycle pulses could only be obtained through spectral broadening of amplified femtosecond laser pulses in a gas-filled, hollow-core fiber, followed by chirped mirror compression [5, 6]. Although successful, the hollow-fiber compression technique limits the pulse energy available for experiments to a few hundred microjoules. However, exploring laser–matter interactions in the relativistic intensity ($>10^{18}$ W $\cdot$ cm$^{-2}$) regime requires significantly higher focused pulse energies [5]. When confined in space and time to a volume of a few $\lambda^3$, femtosecond laser pulses can reach relativistic intensities with pulse energies close to 1 mJ [6]. This leads to strong nonlinear laser–plasma interactions such as relativistic reflection, deflection, and compression of a few-cycle femtosecond pulse down to the attosecond regime with an efficiency up to 10% [7]. One major requirement for this generation technique is the availability of carrier-envelope phase (CEP) stabilized few-cycle pulses with microjoule energy. We therefore explored a novel scheme for high-energy, CEP-stable few-cycle pulse generation, which is easy to handle and which overcomes the energy barrier of the hollow fiber approach. It is based on self-guided beam propagation and provides spectral broadening to bandwidths up to 140 THz. The measured pulse spectra support Fourier transform-limited durations of less than 6 fs (few-cycle) with diffraction-limited beam quality.

## 2 Self-guided Propagation

Discovered more than 10 years ago by Braun et al. [8], self-channeling of high-peak power lasers in a gaseous medium is a highly nonlinear process. When the critical power, $P_{cr}$, is exceeded, the laser starts to self-focus and generate a plasma which then tends to defocus the beam. It has been shown that a broad manifold of physical effects such as Kerr-lens-induced self-focusing, plasma-induced defocusing, diffraction, and others can lead to a competitive dynamical equilibrium that clamps the laser intensity and the diameter of the propagating beam to about 100 μm. This leads to the formation of a filament that can persist tens of centimeters or even kilometers, depending on initial laser and background gas parameters as well as the focusing conditions. During filamentary propagation, both the pulse temporal profile and the pulse spectrum are significantly modified due to group velocity dispersion, self-steepening, and self-phase modulation. Once self-guided beam propagation is established, the filament core intensity is typically clamped to around $10^{13}$–$10^{14}$ $\mathrm{W \cdot cm^{-2}}$ (for argon) and is surrounded by an energy reservoir of several hundreds of micrometers in diameter. Because intensity clamping sets in during filamentary propagation, one can assume that the product $\tau \times \epsilon$ of the pulse duration $\tau$ and the pulse energy $\epsilon$ remains roughly constant: the shorter the pulse in the core, the less energy it contains. This turns out to be a major limitation for producing few-cycle pulses at the microjoule level by long (20–40 cm) filaments. The shortest pulses demonstrated so far by twofold filamentation in argon are 5.1 fs long with an energy of only 180 μJ [9, 10]: a temporal pulse compression of factor 7 can be achieved by seeding the double-plasma compressor with a 35-fs, 0.68-mJ pulse. This result is comparable both in pulse duration and energy to the one demonstrated with the hollow-fiber pulse compressor by Nisoli et al. [5]. In our experiments, however, we show that the limitation in energy due to intensity clamping in the filament can be overcome by reducing the length of self-guided propagation to just a few Rayleigh length. We show that comparable spectral broadening can be obtained by a single, 5- to 8-cm-long filament in argon. We achieved this by increasing the input laser energy and by carefully choosing the focusing parameters of the higher energy driving laser pulse.

## 3 Generation of Few-cycle Bandwidths with Microjoule Energy

The high-energy filament compressor setup, shown in Fig. 1, comprises a single filamentation step. We implemented it on two different dual-stage CPA laser systems, one delivering sub-40-fs pulses with energies up to 8 mJ at 1 kHz, and a second one delivering 25-fs pulses with energies up to 5 mJ at 10 Hz. The multi-microjoule pulses are loosely focused into an argon gas cell. The focusing

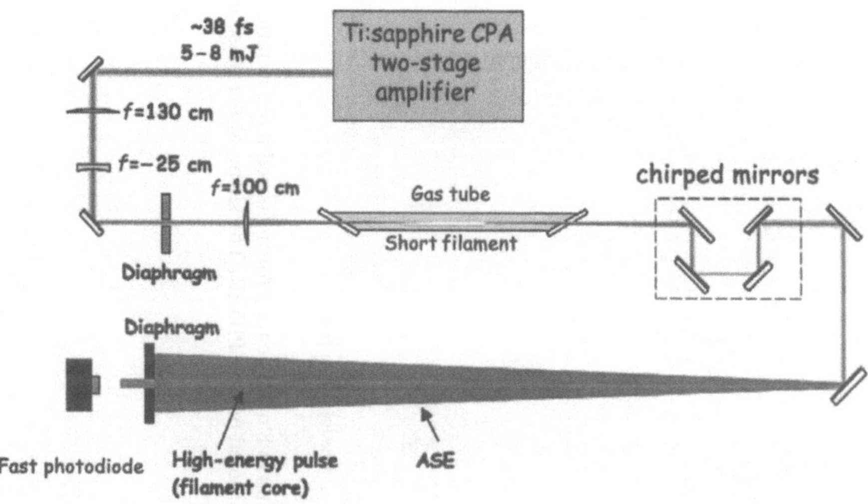


FIGURE 1. High-energy filament compressor setup.

conditions are chosen such that self-channeling takes place over distances less than 10 cm. A diaphragm is used to control the output pulse wavefront and energy. A few bounces on chirped mirrors provide the necessary compression of the pulses that are then temporally characterized using SPIDER. The diverging output pulses are spatially filtered in order to measure the real energy contained in the filament core. Figure 2 shows the input laser pulse characteristics from the kilohertz as reconstructed from SPIDER measurements. The spectral phase of the pulses was modified using an AOPDF (Dazzler) [11] to give the best pre-to-main pulse contrast. This has been shown to be instrumental in providing optimal conditions for filamentary propagation [9, 10, 12]. The optimal pulse duration was 38 fs, which is close to the transform limit of 35 fs. The spectral phase was slightly negatively chirped to compensate for telescope lenses and the entrance window of the gas cell.

The broadest spectrum and best pulse quality (energy stability and beam profile) were obtained by adjusting the input pulse energy with the diaphragm to just over 2 mJ. The energy contained in the broadband filament core was measured to be above 1 mJ. After spectral broadening in the filament, the intrinsic positive chirp of the output pulses was compensated by six bounces on broadband chirped mirrors (Layertec GmbH). A temporal reconstruction of the generated pulses obtained by SPIDER is shown in Fig. 3 . The spectrum covers a range from 670 to 890 nm and supports a transform-limited pulse duration of 9.8 fs. The reconstructed spectral phase shows uncompensated second and higher order phase introduced by the double-chirped mirrors, which leads to an actual pulse duration of 19.8 fs. Shorter pulses can be obtained by improving the phase compensation characteristics of the chirped mirrors. Broader spectra could be generated by

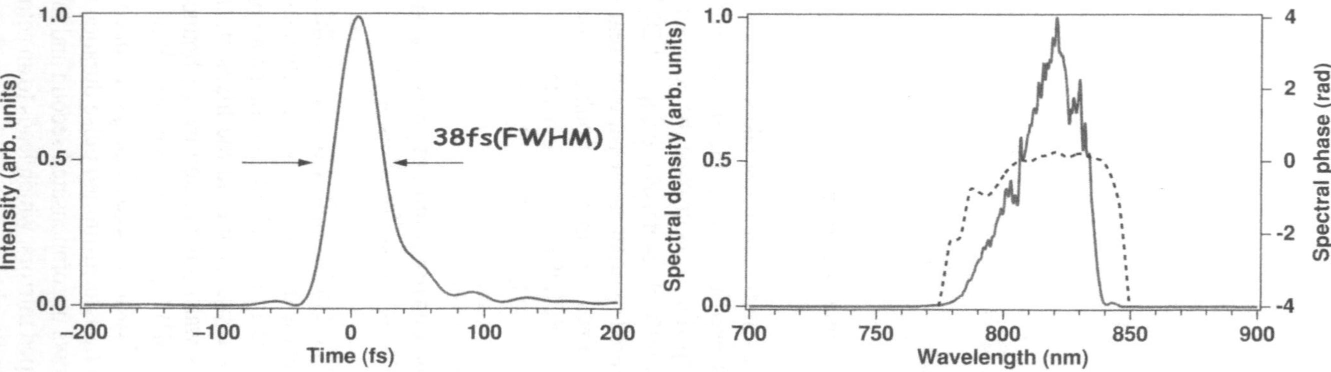


FIGURE 2. Left: 38-fs pulse delivered by a 1-kHz dual-stage amplifier system with an energy up to 8 mJ. The transform-limited pulse duration is 35 fs. Right: Corresponding spectrum (solid line) and spectral phase (dashed line) reconstructed using SPIDER. The shape of the pulse is tailored with the AOPDF (Dazzler) for optimum filament compressor output.

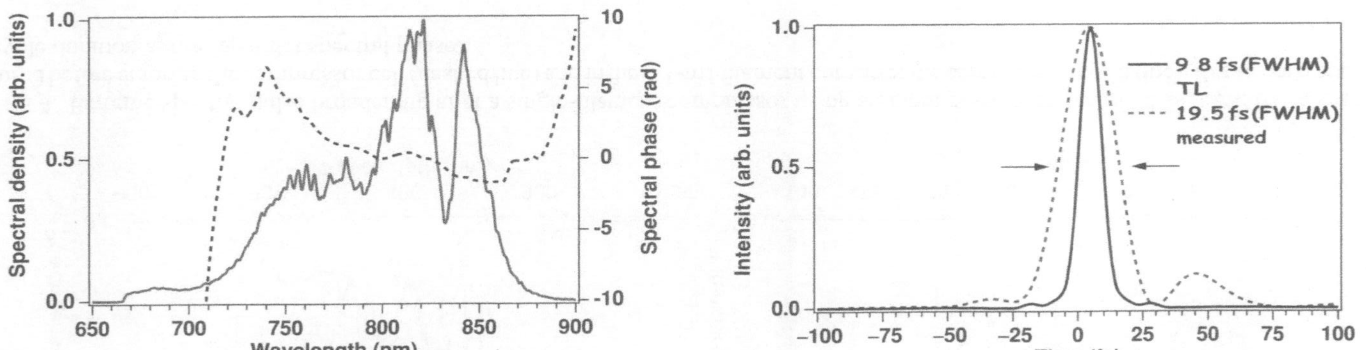


FIGURE 3. SPIDER measurement of the 1-mJ pulses generated in the filament compressor using 38-fs, 2-mJ, 1-kHz pulses. Left: spectrum (solid line) generated in the filament core. The spectral phase (dashed line) features higher order phase distortions introduced by the broadband chirped mirrors used in the experiments. Right: reconstructed temporal profile of the generated pulses indicating a 19.8 fs duration although the spectrum supports transform-limited pulse durations of 9.8 fs.

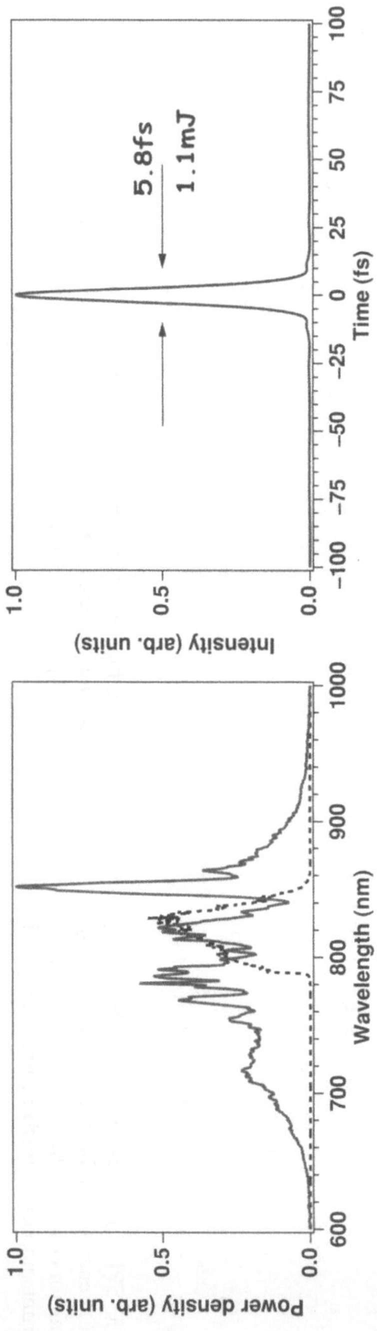


FIGURE 4. Extreme spectral pulse broadening after a single-filament compressor using an input pulse duration of 25 fs. Left: pulse spectrum measured before entering the compressor cell (dashed line) and in the 1.1-mJ filament core after the compressor (solid line). Right: corresponding few-cycle duration assuming a flat spectral phase.

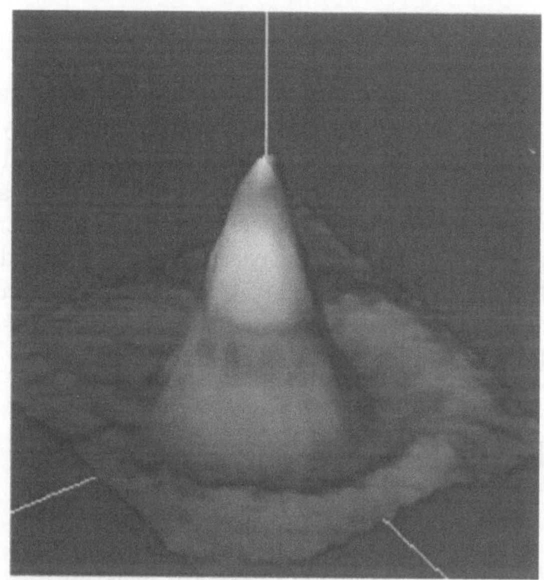

FIGURE 5. Measured spatial profile of the 1.1-mJ, 140-THz-wide filament core obtained with 3.1-mJ, 25-fs input pulses.

seeding the filament compressor with the 5-mJ, 25-fs pulses from our 10-Hz laser system. In this configuration, the best output pulse characteristics were achieved by increasing the input pulse energy to 3.1 mJ. As shown in Fig. 5, we could produce pulses with 140-THz-wide spectra corresponding to transform-limited pulse durations of 5.8 fs. However, the instability of the laser system precluded any reliable SPIDER measurement of the pulse temporal characteristics. Output filament energies in excess of 1 mJ were measured. In order to assess the energy content of the filament core, we blocked the surrounding reservoir with a diaphragm in the far field. Figure 5 shows a typical output spatial beam profile characterized using an 8-bit CCD camera. The filament core carried an energy of 1.1 mJ in a good spatial mode. We believe such pulses could be compressed down to the few-cycle limit while retaining most of the energy by using the appropriate chirped mirror sequence. In addition to pulse broadening, we investigated the possibility of increasing the peak-to-ASE (amplified spontaneous emission) ratio of the pulses at the ouput of the filament compressor. Effectively, the longer ASE pedestal at the input pulse leading edge is not intense enough to induce any self-focusing and will therefore be more divergent than the more intense peak of the pulse that experiences self-guiding. The diaphragm, shown in Fig. 1, acts therefore as a spatial filter. Preliminary results obtained with a fast photodiode placed at approximately 1 m after the filament show a fourfold enhancement of the peak-to-ASE ratio. This measurement was performed by physically clipping the laser beam at moderate propagation distance from the filament channel. A third-order autocorrelation measurement after significantly longer propagation distances is needed to assess the true increase in contrast ratio of the compressed pulses.

## 4 Conclusion

We demonstrate the extreme nonlinear spectral broadening of amplified femtosecond pulses using a technique based on filamentary propagation. In contrast with the hollow-core fiber compression technique that is limited to output energies of a few hundred microjoules, we show preliminary results indicating that filament compression is scalable in energy to the microjoule level. For 3.1-mJ input pulses, output pulses with energies up to 1.1 mJ and spectra supporting transform-limited durations down to 5.8 fs could be obtained, with excellent mode quality and improved peak-to-ASE contrast. Such laser sources are needed for intense laser–matter physics in the relativistic regime.

## References

1. Hentschel et al.: Nature **414**, 509 (2001).
2. Drescher et al.: Nature **419**, 803 (2003).
3. Kienberger et al.: Nature **427**, 817 (2004).
4. Baltuska et al.: Nature **421**, 611 (2003).
5. M. Nisoli, S. De Silvestri, O. Svelto, et al.: Opt. Lett. **22**, 522 (1997).
6. S. Sartania, Z. Cheng, M. Lenzner, et al.: Opt. Lett. **22**, 1562 (1997).
7. D. H. Reitze, S. Kazamias, F. Weihe, et al.: Opt. Lett. **29**, 86 (2000).
8. A. Braun, G. Korn, X. Liu, D. Du, J. Squier, G. Mourou: Opt. Lett. **20**, 73 (1995).
9. C. P. Hauri, W. Kornelis, F. W. Helbing, et al.: Appl. Phys. B **79**, 673 (2004).
10. C. Hauri, A. Guandalini, P. Eckle, W. Kornelis, J. Biegert, U. Keller: Opt. Express **13**, 7541 (2005).
11. F. Verluise, V. Laude, Z. Cheng, Ch. Spielmann, P. Tournois: Opt. Lett. **25**, 575 (2000).
12. Hauri et al.: *Intense Ultrashort Laser Pulse Generation for Nonlinear Laser–Matter Interactions*, PhD thesis, ETH Zurich, 2005, pp. 79–93.

# High Average Power, 7.5-fs Blue Source at 5 kHz

Xiangyu Zhou, Teruto Kanai, Taro Sekikawa, and Shuntaro Watanabe

Institute for Solid State Physics, University of Tokyo, 5-1-5 Kashiwanoha, Kashiwa 277-8581, Japan
zxy@issp.u-tokyo.ac.jp

High average power (1 W), 7.5-fs blue pulses have been generated at 5 kHz from broadband frequency doubling of a spectrally broadened Ti:sapphire laser by using adaptive phase control with genetic algorithm.

## 1 Introduction

Intense ultrashort pulses at a high repetition rate have opened the way to attosecond pulses in the XUV and soft X-ray region. Isolated attosecond pulses were demonstrated in the harmonic continuum of around 13 nm by using the carrier-envelope phase stabilized, few-cycle pulses [1]. We have recently generated isolated attosecond pulses from a single harmonic by another method where the harmonic was generated in the nonadiabatic regime by the rising part of driving pulses and shut off by full ionization [2]. In this scheme, an ultrashort pulse, short-wavelength laser played an essential role because the pulse rise should be steep and the spectral separation between the neighboring harmonics should be wide enough to prevent the spectral overlap.

We reported the generation of subterawatt sub-10-fs blue pulses by broadband frequency doubling (BFD) [3] and compensated the pulse front distortion due to a single lens BFD system [4]. However, the shortest pulse obtained was limited to 8.3 fs mainly because of the fundamental pulse width [2,3]. Obviously the spectral broadening along with adaptive phase compensation is necessary for further pulse shortening.

In this paper, we show how the spectral width of a 5-kHz Ti:sapphire laser system was broadened by the spectral control with two etalons and spatial masks in the regenerative amplifier consisting of broadband chirped mirrors. The dispersion over the wide spectral range was compensated by a deformable mirror (DM) along with a genetic algorithm (GA), resulting in a pulse width of 15 fs. The pulse width is the shortest in the chirped pulse amplification (CPA) systems with a regenerative amplifier [5], although the similar [6] or the shorter pulses [7] were obtained by multipass amplifiers alone. Next, the BFD system in addition to the Ti:sapphire laser was phase-compensated by using the self-diffraction (SD) intensity as the feedback signal into the GA. The pulse width was 7.5 fs. The average power of the second harmonic (SH) was 1 W with a fundamental input of 7 W. This average power is higher by almost two orders of magnitude than those obtained by using hollow fibers in the short-wavelength region [8, 9].

In this paper, we describe a 5-kHz CPA laser system in Section 2.1, spectral broadening and phase compensation in the fundamental beam in Section 2.2, and phase compensation in BFD in Section 3.

## 2 Adaptive Phase Control in the Fundamental Beam

### *2.1 5-kHz CPA Laser System*

A 5-kHz CPA Ti:sapphire system with an adaptive phase controller was used in this experiment. The 5-kHz amplifier system is shown schematically in Fig. 1 [10, 11], which consists of a mode-locked oscillator, an Offner-type stretcher, a regenerative amplifier, a four-pass amplifier, a single-pass amplifier, and a compressor. The oscillator generates 12-fs (FWHM) pulses at 75 MHz. The pulses were stretched to ~400 ps by an Offner-type stretcher with a single 1200 grooves/mm grating and sent to the regenerative amplifier. The regenerative amplifier consists of an *x*-folded cavity with two flat mirrors and two concave mirrors. The cavity of the regenerative amplifier contains a Pockels cell driven by a triode unit with a pre-bias module, a 12-mm-long Ti:sapphire crystal, a thin-film polarizers, and a LAH64 glass-prism pair. The prism pair was used to compensate the high-order

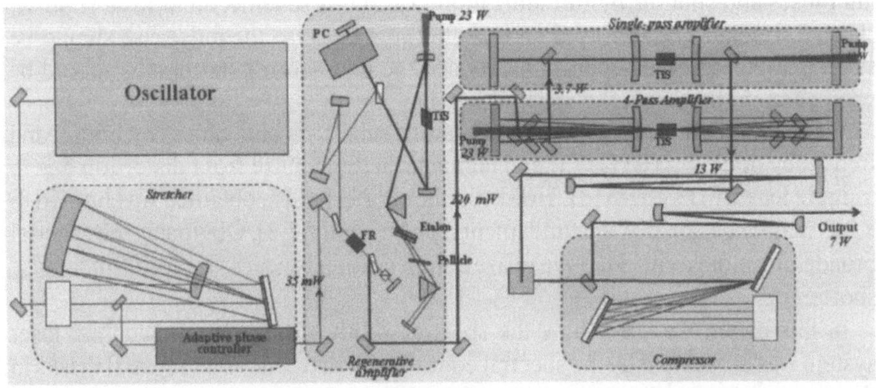


FIGURE 1. Schematic diagram of a 5-kHz amplifier system. PC: Pockels cell, FR: Faraday rotator, TiS: Ti:sapphire. The system consists of a mode-locked oscillator, an Offner-type stretcher, a regenerative amplifier, a four-pass amplifier, a single-pass amplifier, and a compressor.

dispersions. The Ti:sapphire crystal was pumped by the SH of a Q-switched Nd:YAG laser at 5 kHz. The single pulse was extracted after 15 round trips from the regenerative amplifier and sent to the multipass amplifier. The 30-mm-long Ti:sapphire crystals in the two amplifiers at a later stage were pumped by the SH of a Q-switched Nd:YAG laser. After four-pass and single-pass amplification, the pulse was finally sent to the compressor. The second- and third-order dispersions were balanced over the given spectrum by the compressor along with the prism pair in the regenerative amplifier. The pulse width was 22 fs [10, 11].

## 2.2 *Spectral Broadening and Adaptive Phase Control in the Fundamental Beam*

The broader spectral width and the compensation of higher order dispersion are necessary to realize the further pulse shortening. The spectrum of the pulse is from 720 nm to 870 nm after the stretcher, although the edge of the spectrum is lost because of the limited width of the folding mirrors. However, the spectral narrowing occurs mainly in the regenerative amplifier because of gain narrowing and the limited bandwidth of the total reflectors (from 750 nm to 857 nm above 99% reflectance). We developed broadband chirped mirrors with the spectral range from 700 nm to 880 nm above 99% reflectance (Fig. 2a). Then the four mirrors in the regenerative amplifier were replaced by the broadband chirped mirrors. The group-delay dispersion (GDD) of the chirped mirror is about $-60$ fs$^2$ over the above spectral range (Fig. 2b). The negative GDD is not important because the total GDD is much larger ($1.77 \times 10^6$ fs$^2$). The reason why we adopted the chirp mirror is that it gave a high reflectance and flat GDD over a spectral range wider than a dispersion-less mirror.

A 2-μm-thick pellicle beam splitter was added to the air-spaced etalon whose space is controlled by a piezoelectric actuator to improve the spectral broadening [5, 12]. The pellicle and etalons were inserted between the prisms because there was no other space to accommodate them. In addition, spatial masking was performed for spectral shaping between the prism pair and the total reflector,

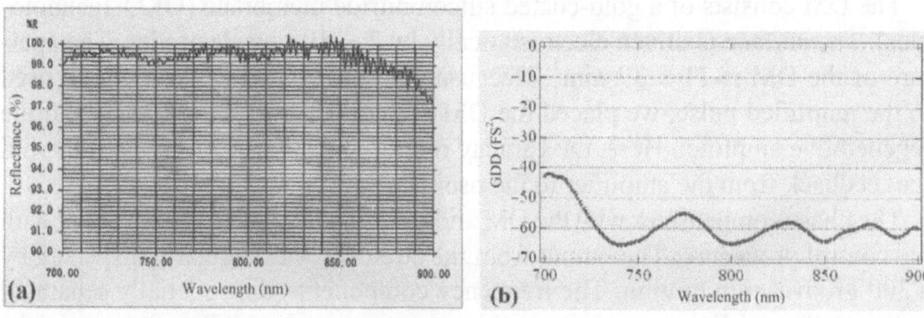


FIGURE 2. The reflectance and the GDD of a broadband chirped mirror: (a) The measured reflectance. The reflectance is above 99% from 700 nm to 880 nm. (b) The calculated GDD of the chirped mirror.

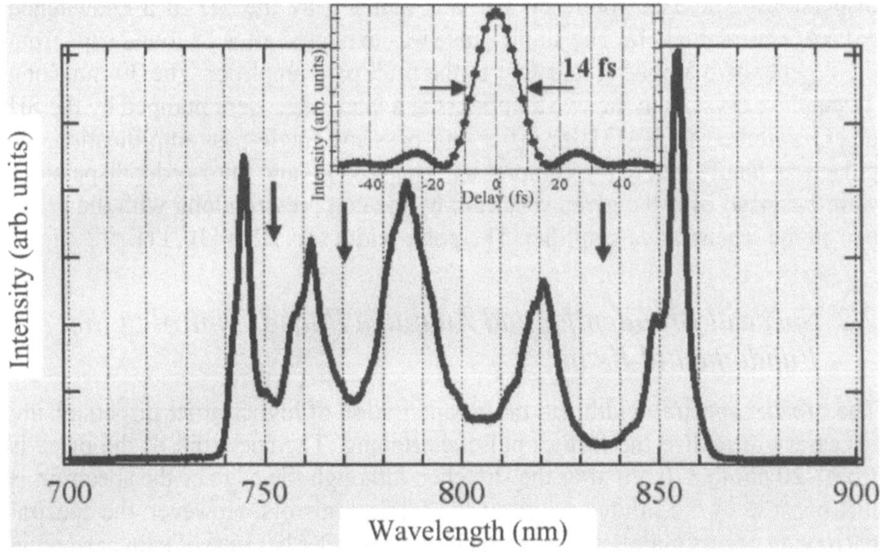


FIGURE 3. Spectrum of the amplified pulse. The FWHM of the amplified spectrum is about 115 nm. The arrows correspond to the positions of needles. The Fourier transform of the spectrum gives a pulse width (FWHM) of 14 fs (inset).

where the spectrum is spatially separated with a parallel beam, in the regenerative amplifier. Three needles were inserted in the positions corresponding to the arrows in Fig. 3. The spectrum of the regenerative amplifier is shown in Fig. 3 along with the transform-limited pulse shape. The amplified spectrum covers from 737 nm to 865 nm, resulting in a spectral width of 115 nm (FWHM). The Fourier transform of the spectrum gives a pulse width (FWHM) of 14 fs.

The phase distortion over such a wide spectrum cannot be compensated by the grating pair in the compressor and the prism pair in the generative amplifier. The dispersion including higher order over the wide spectrum was compensated by using a DM in this experiment [6, 13].

The DM consists of a gold-coated silicon nitride membrane (OKO Technologies). The surface is driven electrostatically by $2 \times 19$ row electrodes. The aperture of the DM is $11 \times 39$ mm$^2$. Because the gold-coated DM cannot be used to the amplified pulse, we placed the DM system between the stretcher and the regenerative amplifier. Here, the Faraday rotator and polarizer were used to void the feedback from the amplifier to the oscillator.

The phase compensator with the DM and a grating is shown in Fig. 4 along with the control procedure. The output from the stretcher was angularly dispersed by a 300 grooves/mm grating. The frequency components were spatially separated horizontally and Fourier-transformed by a concave mirror ($R = 0.6$ m) on the DM. After the reflection by the DM, the beam returned to the concave mirror and grating and was collimated to the initial spot. The phase of the reflected beam was modulated by the DM.

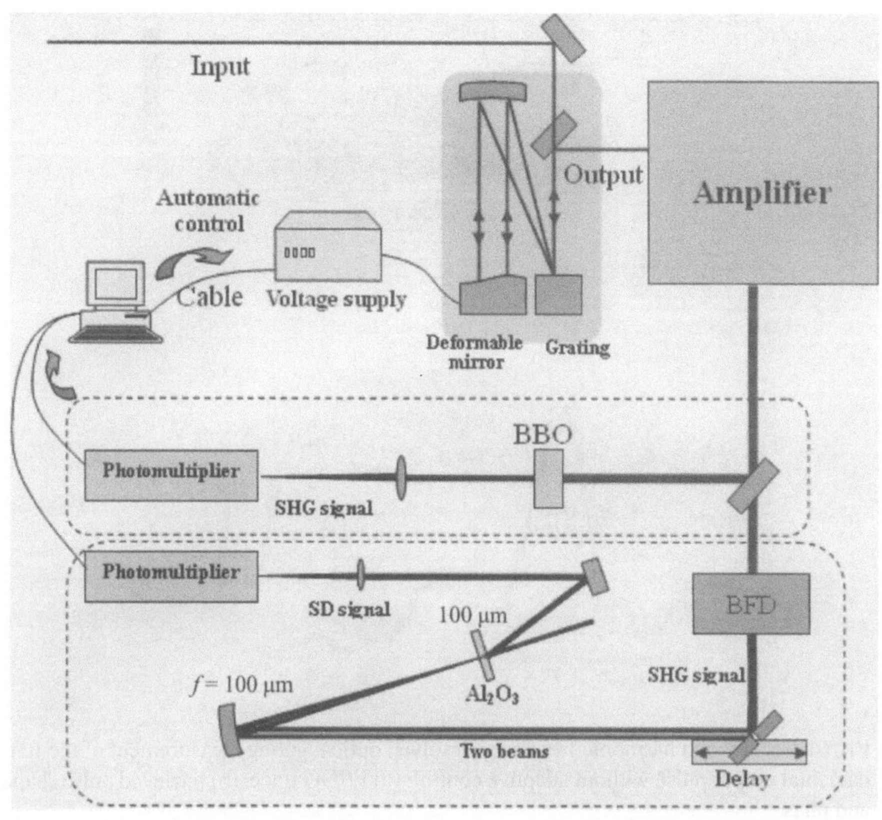


FIGURE 4. Experimental setup for adaptive phase control. The DM was used as a phase modulator to compensate the higher order dispersion along with an iterative GA. For the feedback to the GA, the algorithm sampled the parameter space to maximize the SH and SD intensities for the fundamental and SH pulses, respectively.

The voltages of the actuator electrodes can be controlled from 0 V to 285 V. Both ends of the mirror are fixed on the base. When all of the actuators are set to 0 V, the surface of the DM is almost flat. The advantage of the DM over a liquid-crystal modulator is smooth phase modulation. However, an individual actuator cannot be controlled independently from the neighboring actuators, and cannot be adjusted by the applied voltage to the calculated shape. Then we adopted a GA [6, 13–15] to search the best surface shape of the DM. Because the surface of the DM can only be pulled by the actuators, the mirror cannot be set to be a convex mirror. Thus, the initial position of the mirror surface was set up to a concave mirror with all the actuators set to 100 V.

For the feedback to the GA, the amplified pulse was focused by a lens into a 50-μm-thick BBO crystal, generating an SH signal. The signal detected by a photomultiplier was digitized and sent to a computer. The algorithm sampled the parameter space to maximize the SH output. The best solution (10 parents)

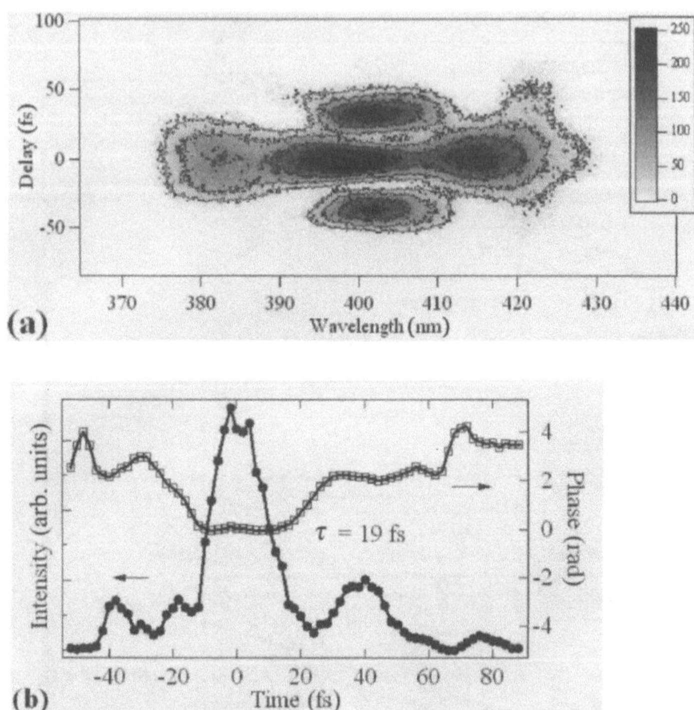


FIGURE 5. Second harmonic frequency-resolved optical gating measurement of the fundamental output pulse without adaptive control: (a) FROG trace, (b) retrieved pulse shape and phase.

generated the next set of possible solutions (100 children) in each iteration. The actuator voltages and the fitness value were the only parameters needed by the GA.

The convergence of the GA was not good because of the fluctuation in the laser intensity. Then we adopted two steps to operate the GA program. We increased the mutation quantity (variable from 0 to 255) and the mutation probability (variable from 0 to 100%) in the mutation step for coarse search. This also avoids the fall in the local maxima. Second, we reduced the mutation quantity and the mutation probability for fine search, and operated the program with the solution of the first step as the initial population.

After running the optimization algorithm for about 40 min, the SH signal increased and began to stagnate. Figures 5 and 6 show second harmonic generation (SHG) frequency-resolved optical gating (FROG) measurements [16] of the output pulse shapes without and with adaptive control, respectively. In both figures, subparts a and b show the SHG-FROG trace and the retrieved pulse shape with the associated phase, respectively. After adaptive phase control, the phase became flatter and the side bumps were reduced, resulting in the reduction of the pulse width from 19 fs to 15 fs. The measured pulse width of 15 fs is a little longer than

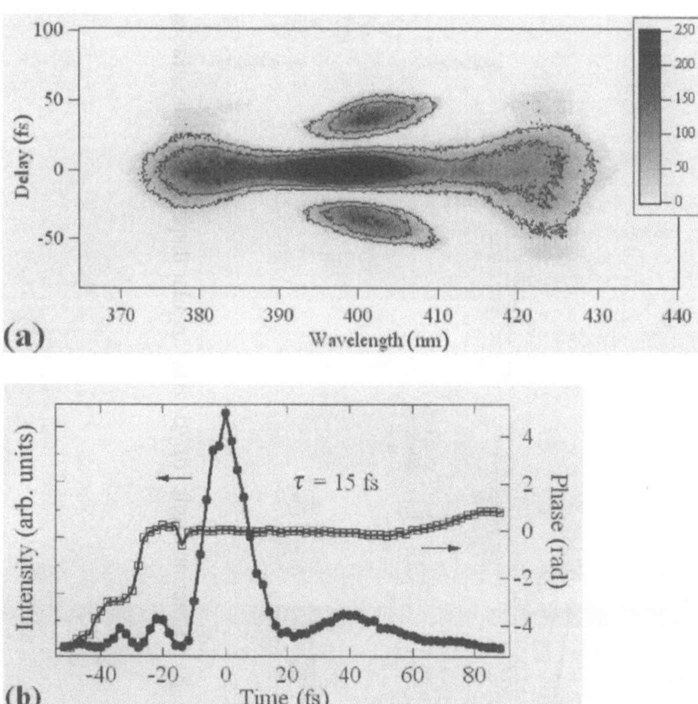


FIGURE 6. Second harmonic frequency-resolved optical gating measurement of the fundamental output pulse with adaptive control: (a) FROG trace, (b) retrieved pulse shape and phase.

the transform limit (14 fs). This may be due to the incomplete phase compensation or due to the broadband component of amplified spontaneous emission from the regenerative amplifier.

# 3 Adaptive Phase Control in the SH Beam

## 3.1 Broadband Frequency Doubling

If the whole spectral components of a square spectrum are frequency-doubled, the frequency width of the SH becomes twice that of the fundamental, making the SH pulse width half of the fundamental. This is also valid for the spectrum shown in Fig. 3 having sharp edges at both spectral ends. The BFD scheme was proposed and demonstrated [3, 17, 18]. The experimental setup of BFD is shown in Fig. 7. The dispersion of phase-matching angle ($d\theta/d\lambda$) is given by the angular dispersion ($d\beta/d\lambda$) of the grating (G1) with a magnification factor ($M$), i.e., $d\theta/d\lambda = (d\beta/d\lambda)/M$, where $M = 0.2$. A telescope with two concave mirrors was used for image relay instead of a single lens to eliminate the pulse-front distortion due to aberration [4]. The optical configuration after the BBO crystal

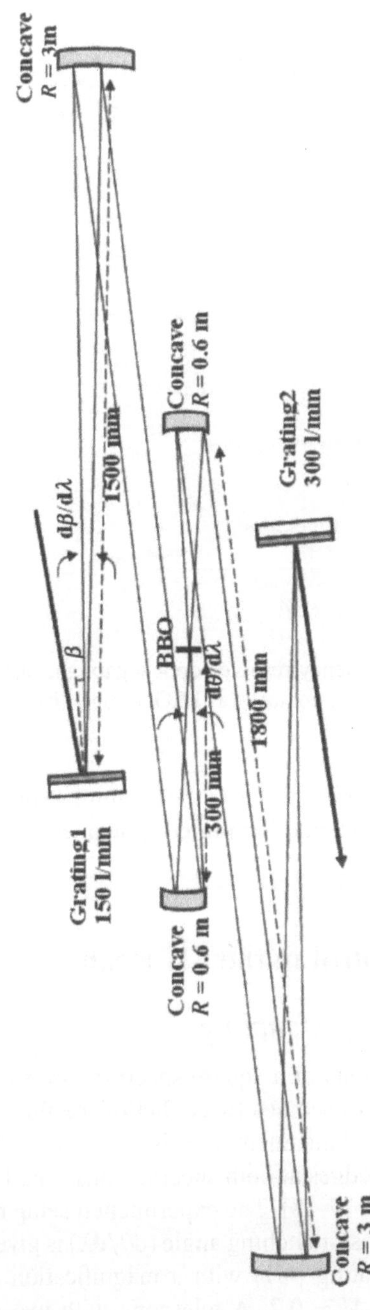


FIGURE 7. Schematic of modified broadband frequency-doubling system. $\beta$: diffraction angle; $d\beta/d\lambda$: the angular dispersion of the grating; $d\theta/d\lambda$: the wavelength dispersion of the phase-matching angle.

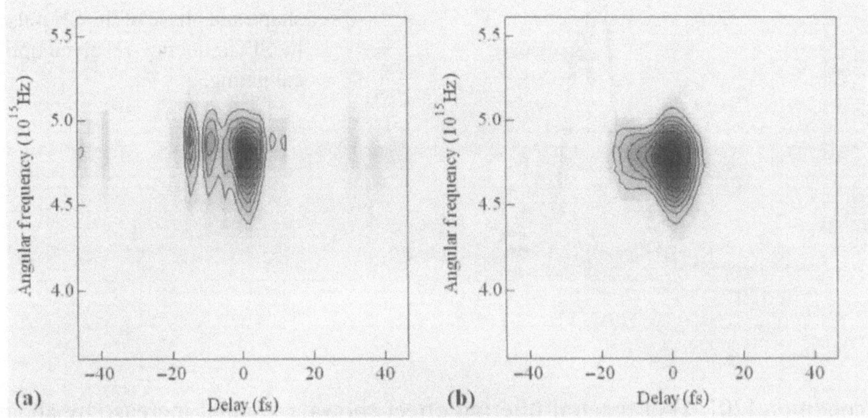


FIGURE 8. The measured (a) and retrieved (b) SD-FROG traces of the SH pulse.

is symmetry to that before the BBO crystal except for the twice greater groove density of G2, which compensates the angular dispersion. The thickness of the BBO crystal was 300 μm or 150 μm.

## *3.2 Adaptive Phase Control in the SH*

The full spectral width of a 15-fs pulse as shown in Fig. 6 is about 100 nm (from 750 nm to 850 nm), while the full acceptable spectral widths of BFD are 127 nm and 179 nm, respectively, for 300-μm and 150-μm-thick BBO crystals. Then the second-order dispersion of the phase-matching angle was not taken into account. In BFD, the extra dispersion from BBO crystal, mirrors, and gratings is added to the fundamental Ti:sapphire system. Then adaptive compensation should be done in the total system. As shown in Fig. 4, the total self-diffraction intensity in a 100-μm-thick sapphire plate was used as the feedback signal to the GA. The procedure of phase compensation with a GA is the same as in fundamental beam as described above. The SD spectrum can be fed to the GA and may give us a more robust search algorithm. However, this was not attempted in this paper. The relative delay between the two beams was always fixed so as to give the maximum signal and is not changed by the form of DM because it does not change the spatial shape of the beam. The coarse adjustment was done by changing the grating separation in the compressor of a Ti:sapphire system.

Frequency-resolved optical gating was performed by using the SD signal from a sapphire plate [8, 9, 19]. Figures 8a and b show the measured and the retrieved FROG traces, respectively. The RMS error was 2%. Figure 9 shows the retrieved pulse shape with the corresponding phase. The angle of two beams was 2.9° and the confocal parameter was 50 μm with the beam diameter of 1 cm. The wavelength dependence of the conversion efficiency from the input to SD signal was calculated by combining the $\omega^2$ dependence with phase matching in this

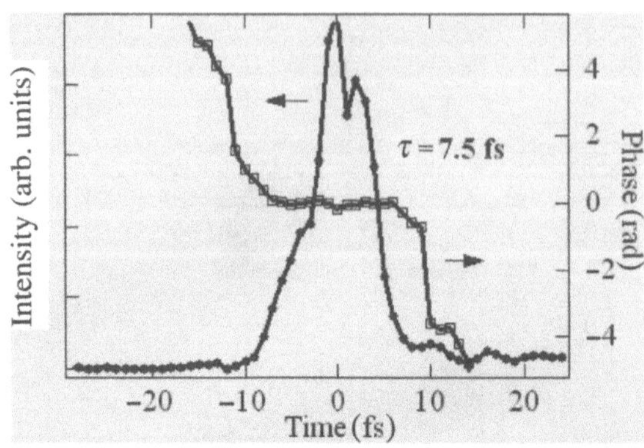


FIGURE 9. The retrieved pulse shape and phase of the SH pulse by SD frequency-resolved optical gating.

condition [20]. This spectral filtering effect shows a gradual increase by about 20% from 430 nm to 370 nm for the interaction lengths of 100 μm (the crystal thickness) and 50 μm (the confocal parameter). The correction of this effect does not change the retrieved pulse shape. However, this effect becomes more serious with decreasing pulse width. The pulse width of 7.5 fs is half of the fundamental pulse width (15 fs) from Fig. 9. The retrieved spectrum reflects quite well the typical measured spectrum.

The output power from BFD was 1 W by using a 300-μm-thick BBO crystal when the input power was 7 W with a beam diameter of 12 mm. The output powers and the pump powers at the intermediate stage are already shown in Fig. 1.

## 4 Conclusion

We have developed a broadband Ti:sapphire amplifier system with an adaptive phase controller. Broadband chirped mirrors were used in the regenerative amplifier. The output spectrum from the regenerative amplifier was shaped with a thin-film polarizer etalon and a pellicle beam splitter in addition to a spatial mask. A DM was used as a phase modulator to compensate the higher order dispersion along with an iterative GA. The pulse width was 15 fs. The SH was generated by broadband frequency doubling. The phase distortion of the total system was compensated by using an SD intensity as a feedback signal to the GA. Frequency-resolved optical gating measurements revealed that the pulse duration of the SH was 7.5 fs. This scheme will be extended to a 1-kHz [21] and a 10-Hz system [22] to obtain subterawatt and terawatt peak powers.

## References

1. R. Kienberger, E. Goulielmakis, M. Uiberacker, A. Baltuska, V. Yakovlev, F. Bammer, A. Scrinzi, T. Westerwalbesloh, U. Kleineberg, U. Heinzmann, M. Drescher, and F. Krausz: Nature **427**, 817 (2004).
2. T. Sekikawa, A. Kosuge, T. Kanai, and S. Watanabe: Nature **432**, 605 (2004).

3. T. Kanai, X. Zhou, T. Sekikawa, S. Watanabe, and T. Togashi: Opt. Lett. **28**, 1484 (2003).
4. T. Kanai, X. Zhou, T. Liu, A. Kosuge, T. Sekikawa, and S. Watanabe: Opt. Lett. **29**, 2929 (2004).
5. Y. Nabekawa, Y. Shimizu, and K. Midorikawa: Opt. Lett. **27**, 1265 (2002).
6. E. Zeek, R. Bartels, M. Murnane, H. C. Kapteyn, and S. Backus: Opt. Lett. **25**, 587 (2000).
7. H. Takada, M. Kakehata, and K. Torizuka: *Ultrafast Optics IV.* F. Krausz, G. Korn, P. Corkum, and I. A. Walmsley (eds). Berlin: Springer Verlag, p. 85 (2004).
8. O. Duhr, E. T. J. Nibbering, and G. Korn: Opt. Lett. **24**, 34 (1999).
9. C. G. Durfee III, S. Backus, H. C. Kapteyn, and M. M. Murnane: Opt. Lett. **24**, 697 (1999).
10. Y. Nabekawa, T. Togashi, T. Sekikawa, S. Watanabe, S. Konno, T. Kojima, S. Fujikawa, and K. Yasui: Appl. Phys. B **70**, S171 (2000).
11. Y. Nabekawa, T. Togashi, T. Sekikawa, S. Watanabe, S. Konno, T. Kojima, S. Fujikawa, and K. Yasui: Opt. Express **5**, 318 (1999).
12. K. Yamakawa and C. P. J. Barty: Opt. Lett. **28**, 2402 (2003).
13. E. Zeek, K. Maginnis, S. Backus, U. Russek, M. M. Murnane, G.Mourou, and H.Kapteyn: Opt. Lett. **24**, 493 (1999).
14. O. Albert, H. Wang, D. Liu, Z. Chang, and G. Mourou: Opt. Lett. **25**, 1125 (2000).
15. D. Meshulach, D. Yelin, and Y. Silberberg: J. Opt. Soc. Am. B **15**, 1615 (1998).
16. D. J. Kane and R. Trebino: IEEE J. Quantum Electron. **29**, 571 (1993).
17. G. Szebo and Z. Bor: Appl. Phys. B **50**, 51 (1990).
18. O. E. Martinez: IEEE J. Q. E. **25**, 2464 (1989).
19. T. S. Clement and A. J. Taylor: Opt. Lett. **20**, 70 (1995).
20. A. Baltuska, M. S. Pshenichnikov, and D. A. Wiersma: *Frequency-Resolved Optical Gating: The Measurement of Ultrashort Laser Pulses.* R. Trebino (ed). Boston: Kluwer Academic Publisher, p. 286 (2002).
21. Y. Nabekawa, Y. Kuramoto, T. Togashi, T. Sekikawa, and S. Watanabe: Opt. Lett. **23**, 1384 (1998).
22. J. Itatani, Y. Nabekawa, K. Kondo, and S. Watanabe: Opt. Comm. **134**, 134 (1997).

# Downchirped Regenerative Amplification of Femtosecond Laser Pulses at 100 kHz Repetition Rate

Kyung-Han Hong, Tae Jun Yu, Sergei Kostritsa, Jae Hee Sung, Il Woo Choi, Young-Chul Noh, Do-Kyeong Ko, and Jongmin Lee

Femto-Science Group, Advanced Photonics Research Institute, Gwangju Institute of Science and Technology, 1 Oryong-dong Buk-gu Gwangju 500-712, Republic of Korea, **khhong@gist.ac.kr**

## 1 Introduction

Ti:sapphire laser technology has provided practical solutions of solid-state femtosecond light source with high average power as well as high peak power. Especially, compact high-average-power femtosecond lasers with a high repetition rate (kilohertz range) are good tools for basic and applied science. Typical high-repetition-rate Ti:sapphire amplification lasers using electro-optic pulse selection method operate at 1–10 kHz [1–3] with an energy of ~mJ, whereas those using acousto-optic pulse selection method can operate at 100–250 kHz with an energy of ~ μJ. As for the 100–250 kHz Ti:sapphire amplification system, the regenerative amplifier setup suggested by Norris [4] has been a conventional configuration. Several features of the lasers based on Norris's amplifier configuration can be described as follows: First, they are pumped by a continuous-wave (CW) green laser and intracavity Q-switched for the suppression of prelasing. Second, the high-repetition-rate pulse injection and dumping is achieved by the acousto-optic cavity dumper composed of a Brewster-angle Bragg cell and an RF driver. Third, the dispersion control of the amplification system has been achieved either without [4, 5] or with [6] a chirped-pulse amplification (CPA) technique [7]. In the case of non-CPA systems, pulses are naturally broadened and compressed by use of prism pairs, whereas, in the case of CPA system, transmission gratings are used for both stretching and compression. Adaptive control has been also applied for the non-CPA system to compress the laser pulse down to 35 fs [8]. The compression efficiency for those systems is 60–80%.

Despite higher repetition rate, 100–250 kHz systems have no advantage with average power because of much lower energy per pulse or even lower peak power: so far the highest energy and peak power reported at 100 kHz is 7 μJ and 0.2 GW [8]. Thus, the enhancement of the energy per pulse will make this kind of lasers a more valuable tool not only for high-average-power applications such as micromachining and medical surgery but also for high-peak-power applications such as high-order harmonic generation [9] and above-threshold ionization study. For the generation of higher peak power at 100 kHz, we demonstrate an efficient acousto-optically switched regenerative amplification system, which is modified from the conventional scheme in terms of pumping source and dispersion compensation method. First, we employed a Q-switched pump laser, instead of a CW one, to obtain higher pumping fluence and simplify the cavity by removing an intracavity Q-switcher. Use of a pulsed source allows more efficient pumping than CW source because the portion of the pump energy participating in the population inversion process is higher than that of CW source. Also, accumulated dispersion and spectral narrowing due to an intracavity Q-switcher can be reduced. Second, we applied downchirped pulse amplification (DPA) technique [10] for higher compression efficiency.

In a DPA configuration, the laser pulse is stretched with negative dispersion elements (downchirp), such as prism pair and grating pair, and then compressed by a positive dispersion element, which is just an optical glass block. The concept of DPA is essentially the same as that of CPA, but it was recently demonstrated at a 10 kHz system [10]. The main advantage of this technique is its high compression efficiency (>95%) and simple compressor alignment. Since the pulse duration and phase are insensitive to the compressor alignment, the DPA technique is also a good approach for the amplification of carrier-envelope phase stabilized pulses [11, 12]. It should be noted, however, that a material compressor is more suitable to low-peak-power pulses than high-peak-power pulses due to third-order nonlinear effect at the compressor. Therefore, the DPA technique is an optimal method of dispersion control for the development of a 100-kHz fs laser whose peak power is relatively low compared to that of 1–10 kHz lasers. In this chapter, we describe the configuration and operational characteristics of a 100-kHz DPA Ti:sapphire laser system producing an energy of 28 μJ and a pulse duration of 45 fs.

## 2 Design of a DPA Regenerative Amplification Laser

### *2.1 Regenerative Amplifier*

A homemade mirror dispersion controlled sub-10-fs Ti:sapphire laser is used as a front end femtosecond oscillator. The laser pulses have the energy of 5 nJ with a repetition rate of 90 MHz. Before stretching these laser pulses, we have to carefully design a regenerative amplifier because the dispersion characteristic of the regenerative amplifier changes the stretcher and compressor configurations due to a relatively large number of roundtrips inside the amplifier.

Basic design parameters of an amplifier, such as output energy, the spot size of a pump source at a gain medium, and the number of roundtrips, are mainly determined by a pumping energy. We used a diode-pumped, frequency-doubled, Q-switched Nd:YAG laser (HSQG5000, Golden-light Co.)

operating at 10–100 kHz, in contrast to a CW pump laser used in previous 100–250 kHz regenerative amplifiers [4–6]. The energy of the pump laser is 200 μJ at 100 kHz and its pulse duration is ~400 ns. If the spot size at the Ti:sapphire crystal is 150 μm in diameter, the pumping fluence becomes 1.1 J/cm$^2$, which is slightly higher than the saturation fluence of Ti:sapphire (0.9 J/cm$^2$). Calculation based on Frantz and Nodvick [13] model shows that the amplification in saturation regime allows the output energy of 55 μJ and that the number of roundtrips required for the gain saturation is about 28, where the experimentally measured energy loss of 90% at the stretcher is considered. Actually, we estimate the output energy to be 33 μJ assuming that the dumping efficiency of a Bragg cell is 60%. The nonlinear phase (B-integral) accumulated inside the amplified is about 0.7 rad when the pulse is stretched to 20 ps. Thus, the stretched pulse duration of ~20 ps should be safe against optical damage as well as nonlinear effect.

To realize the calculated parameters, we have designed a regenerative amplifier similar to a X-folded cavity-dumped laser oscillator as shown in Fig. 1a. Both Ti:sapphire crystal and Bragg cell (BC of Fig. 1) are installed with a Brewster angle in a confocal geometry. The seed beam and amplified beam are diffracted at the Bragg cell in the vertical plane with a small angle (~50 mrad). They are either injected to or dumped from the cavity via a folding mirror (FM of Fig. 1). Double pass scheme at the Bragg cell is used for higher dumping efficiency. The crystal temperature is kept <10°C by a thermo-electric cooler to minimize the thermal lens effect. The crystal mount holds the top and the bottom of the Ti:sapphire crystal with a vertical dimension of 2 mm. It is expected that the thin crystal enhances the thermal dissipation from the crystal to the cooled mount. To maximize the absorption at the crystal, we have installed a re-focusing mirror (RM of Fig. 1) for the pump beam. The pump beam absorption is enhanced from 75% to 94% with this mirror. The chirped mirrors (CM1–4 of Fig. 1) are for the dispersion-compensated broadband amplification. Net group-delay dispersion (GDD) per roundtrip of the regenerative amplifier is only −5 fs$^2$, which makes the accumulated GDD insensitive to the number of roundtrips. The Faraday isolator and polarizing beam splitter (FI and PBS of Fig. 1) are for the discrimination between injected and dumped beams.

Injection and dumping signals for driving Bragg cell as well as the pump laser trigger signal are synchronized to the mode-locked pulse train generated by the laser oscillator. Figure 2 illustrates the synchronization scheme. The 100 kHz trigger signals are generated by frequency diving the 90 MHz RF signal from the oscillator pulse train with a synchronous countdown and a digital delay pulse generator. The RF signal of the Bragg cell is also generated using the harmonics of 90 MHz pulse train. As a result, injection and dumping triggers, pump laser trigger, and RF signal of Bragg cell are all synchronized to the mode-locked pulse train.

## *2.2 Dispersion Control of a DPA System*

Dispersion calculation is very important with a DPA laser design because the compressor, i.e., a glass block with a finite length, has no other degree of freedom

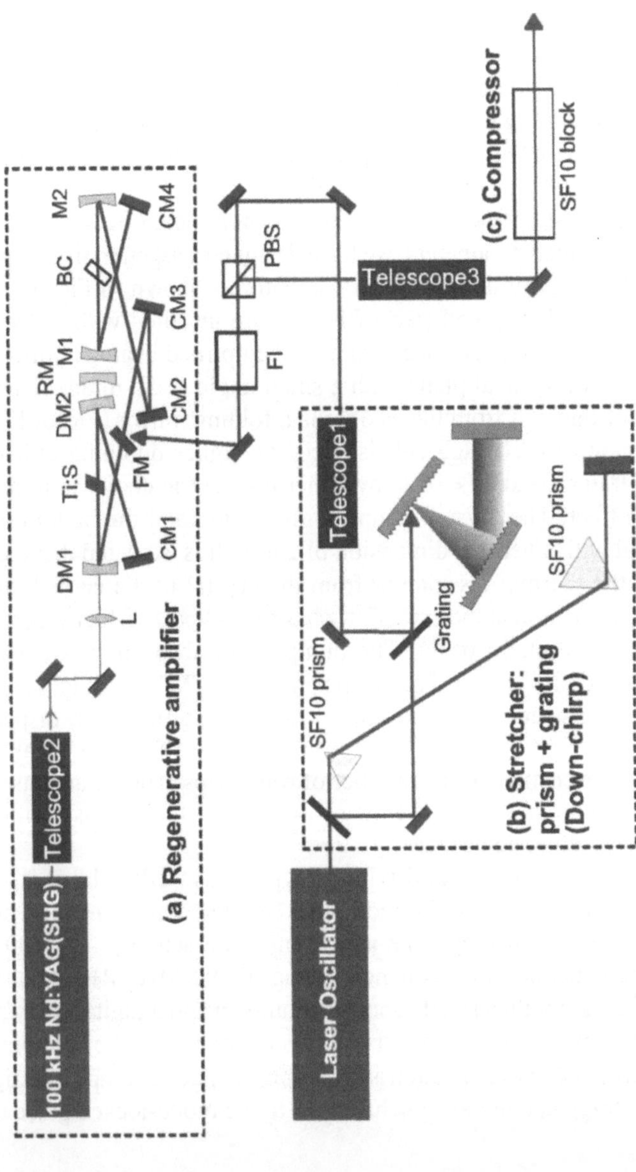


FIGURE 1. Schematic diagram of 100-kHz DPA Ti:sapphire amplification system. (a) Regenerative amplifier, (b) pulse stretcher, and (c) pulse compressor. L, lens; DMs; dichroic mirrors; Ms, dielectric mirrors; RM, re-focusing mirror; FM, folding mirror; CMs, chirped mirrors; FI, Faraday isolator; PBS, polarizing beam splitter; Ti:S, Ti:sapphire crystal.

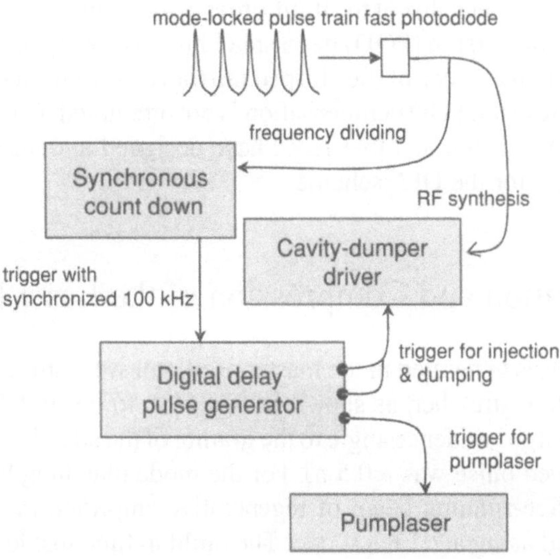


FIGURE 2. Synchronization scheme for the regenerative amplifier.

than the combination of several blocks with a different length to control the laser pulse duration. To minimize the inconvenience of changing the stretcher parameter or compressor length, we have to carefully calculate the dispersion of all the optical elements. It is already known that a stretcher consisting of a prism pair and a grating pair is good for the dispersion compensation up to the third-order term in a DPA configuration. We used a SF10 prism pair and a 600 grooves/mm holographic grating pair as a stretcher and a SF10 glass block as a compressor.

Table 1 shows the calculated dispersion of the laser system up to the fourth-order term. Additionally considered optical elements are Faraday isolator, polarizing beam splitter, Bragg cell, Ti:sapphire crystal, and chirped mirrors. The dispersion compensation is considered up to the third-order term (see Table 1). As for the

TABLE 1. Calculated dispersion values of each optical element for the design of DPA laser system

| Optical element | Function | GDD ($fs^2$) | TOD ($fs^3$) | FOD ($fs^4$) |
|---|---|---|---|---|
| SF10 prism pair (separation of 2 × 1380 mm) | stretcher | −38200 | −115300 | −196800 |
| Grating pair (600 gr/mm, incidence angle of 70°, separation of 58 mm) | stretcher | −47800 | 49700 | −75700 |
| AR-coated SF10 block (421-mm long) | compressor | 67000 | 42600 | 11400 |
| Other materials | amplifier | ~19000 | ~23000 | ~−2200 |
| Total | | ~ 0 | ~ 0 | −263300 |

chirped mirrors, the net value of the third-order dispersion (TOD) is zero, whereas the fourth-order dispersion (FOD) has a small but unknown positive value. However, we set the FOD values of the chirped mirrors as zero for the convenience of calculation because the FOD compensation is not attempted in this system. Based on the parameters shown in Table 1, we have designed and installed a stretcher and a compressor for the DPA scheme.

## 3 Amplification and Compression of the Laser Pulses

The 5-nJ · fs pulses generated in the master oscillator were strongly downchirped by a prism-grating stretcher, as shown in Fig. 1(b), to be stretched up to 20 ps. Because of the large incidence angle to the grating of the stretcher (70°), the actual energy of the seed pulse was <0.5 nJ. For the mode matching between the seed beam and the free-running beam of regenerative amplifier, the seed beam was down-collimated using a 2:1 telescope. The buildup time inside the regenerative amplifier was minimized by cavity alignment and seed beam alignment. The dumped output energy of amplified pulses is maximized by controlling the trigger timing and the phase of RF signal to the Bragg cell.

The number of roundtrips for the saturation of the regenerative amplifier was 30–33. The time sequence of the pump beam, the intracavity pulse train, and the amplified and dumped beams is shown in Fig. 3. With the pump energy of 200 μJ, we have obtained an amplified energy of 30 μJ at maximum. Both the number of roundtrips and the output energy were closely matched to the calculated results. The rms jitter of shot-to-shot energy of the amplified pulses is about 3%. The main limitation of the amplified energy is its relatively low dumping efficiency.

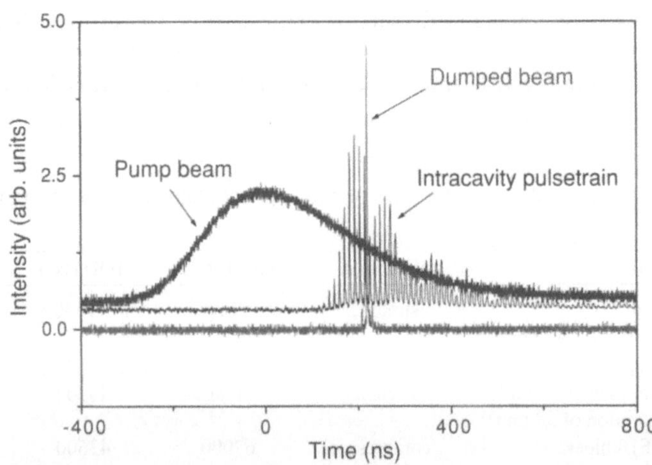


FIGURE 3. Typical time sequence among pump beam, intracavity pulse train, and amplified and dumped seed beams.

It may be enhanced by more tightly focusing the laser beam to the Bragg cell, but it can cause damage on the Bragg cell. Estimated focused size at the Bragg cell is comparable to that at the Ti:sapphire crystal (~150 μm).

Laser spectra from the master oscillator, after the stretcher, and after the compressor are shown in Fig. 4(a). The bandwidth of the stretcher (180 nm) is limited by the prism size, while the spectral loss of the long-wavelength part is due to the grating diffraction efficiency. Gain narrowing was severe because the gain was high enough ($10^5$) and so the spectral bandwidth in full-width at half maximum (FWHM) of the amplified spectrum is only 26 nm, centered at 780 nm. However, this spectrum still supports the transform-limited pulse duration as small as 30 fs (FWHM) due to the very long tales covering the spectral range from 740 nm to 850 nm. We believe that the broad spectral range results from the removal of the Q-switcher inside the regenerative amplifier cavity and the use of broadband chirped mirrors.

Amplified pulse is up-collimated using a telescope to avoid any nonlinear effect and is compressed by 2 anti-reflection-coated SF10 blocks with a compression efficiency of 95% (total 3 passes). The resultant output energy after compression was 28 μJ. The pulse duration was monitored at an intensity autocorrelator and minimized by changing both the path length through the SF10 medium and the grating separation of the stretcher. The shortest pulse duration measured was 45 fs in FWHM, as shown in Fig. 4(b), which was 1.5 times larger than the transform-limited one. As a result, the peak power of the pulses reached 0.6 GW at 100 kHz. To our knowledge, the energy and the peak power are the highest ever reported with a Ti:sapphire amplification system operating at 100-kHz repetition rate. Moreover, this system is capable of generating 1-GW pulses at 100 kHz, considering the bandwidth of the optical spectrum.

Further compression close to 30 fs is possible by optimizing the dispersion of the laser system. We found that 45-fs pulses were obtained when the grating separation was 60 mm and the SF10 medium length was 380 mm. These parameters did not exactly match to the designed ones, i.e., 58 mm and 421 mm, respectively, as shown in Table 1. Thus, either more accurate calculation of the system dispersion or better optical alignment according to the designed parameters will allow a better compression for the generation of higher peak power pulses. However, the pulse compression is still limited by FOD. In addition, spectral shaping and adaptive phase control technique will make it possible to generate even sub-30 fs pulses.

## 4 Conclusions

We have demonstrated the downchirped regenerative amplification of femtosecond Ti:sapphire laser pulses at 100 kHz repetition rate. For higher energy and peak power, the acousto-optic regenerative amplifier is pumped by a pulsed green laser instead of a CW laser, and dispersion compensated by broadband chirped mirrors. The DPA compressor allows a compression efficiency of 95% along with

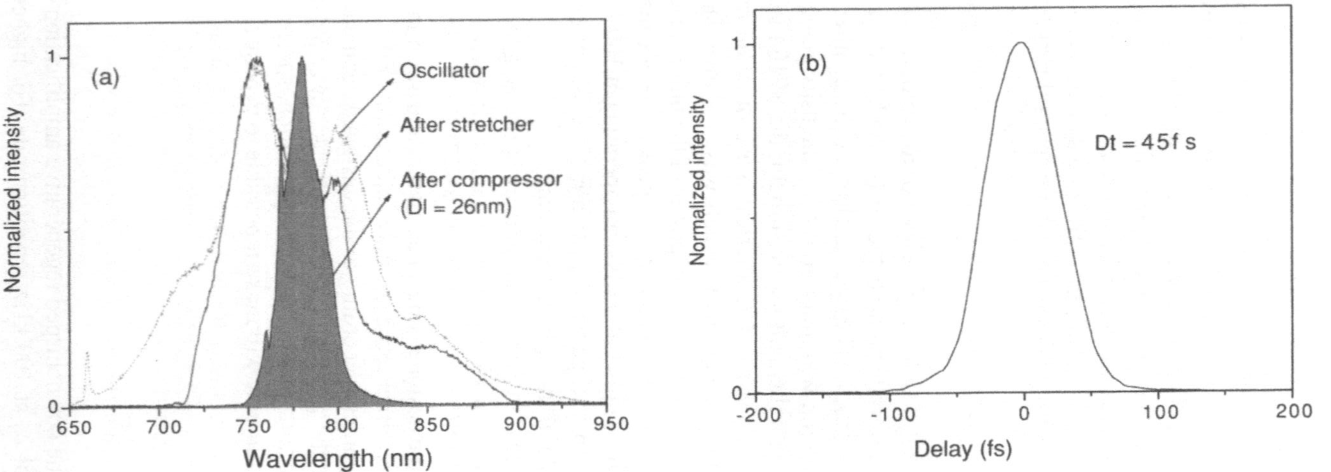


FIGURE 4. (a) Optical spectra of the pulses from master oscillator, after stretcher, and after compressor and (b) intensity autocorrelation measurement of compressed pulse.

simple alignment. As a result, 28-μJ, 45-fs pulses have been obtained from this system, and further compression close to 30 fs is possible by accurate dispersion compensation. The 100 kHz femtosecond high-power DPA laser system will be useful for a variety of research fields such as the generation of high-order harmonics, carrier-envelope phase locking, and micromachining.

## References


1. C. P. J. Barty, C. L. Gordon III, and B. E. Lemoff: Opt. Lett. **19**, 1442 (1994).
2. J. Zhou, C.-P. Huang, M. M. Murnane, and H. C. Kapteyn: Opt. Lett. **20**, 64 (1995).
3. N. Zhavoronkov and G. Korn: Opt. Lett. **29**, 198 (2004).
4. T. B. Norris: Opt. Lett. **17**, 1009 (1992).
5. See the data sheet of RegA series of Coherent Incorporation. http://www.coherentinc.com.
6. J.-K. Rhee, T. S. Sosnowski, T. B. Norris, J. A. Arns, and W. S. Colburn: Opt. Lett. **19**, 1550 (1994).
7. D. Strickland and G. Mourou: Opt. Commun. **56**, 219 (1985).
8. F. Lindner, G. G. Paulus, F. Grasbon, A. Dreischuh, and H. Walther: IEEE J. Quan. Elec. **38**, 1465 (2002).
9. F. Lindner, W. Stremme, M. G. Schatzel, F. Grasbon, G. G. Paulus, H. Walther, R. Hartmann, and L. Struder: Phys. Rev. A **68**, 013814 (2003).
10. D. M. Gaudiosi, A. L. Lytle, P. Kohl, M. M. Murnane, H. C. Kaptey, and S. Backus: Opt. Lett. **29**, 2665 (2004).
11. M. Kakehata, H. Takada, Y. Kobayashi, and K. Torizuka: Opt. Exp. **12**, 2070 (2004).
12. Y. S. Lee, J. H. Sung, C. H. Nam, T. J. Yu, and K.-H. Hong: Opt. Exp. **13**, 2969 (2005).
13. L. M. Frantz and J. S. Nodvick: J. Appl. Phys. **34**, 2346 (1963).

# Temporal Contrast Enhanced to $10^{-10}$ for Petawatt Class Femtosecond Lasers

A. Jullien,[1,2] O. Albert,[2] J. Etchepare,[2] F. Augé-Rochereau,[2] J.-P. Chambaret,[2] G. Chériaux,[2] S. Kourtev,[3] N. Minkovski,[3] and S. M. Saltiel[3]

[1] Thales Laser, RD 128, Domaine de Corbeville, 91400 Orsay, France
**Aurelie.Jullien@ensta.fr**

[2] Laboratoire d'Optique Appliquée, Ecole Nationale Supérieure de techniques avances, Ecole Polytechnique, Centre National de la Recherche Scientifique UMR 7639, 91761 Palaiseau Cedex, France
**Olivier.Albert@ensta.fr**

[3] Faculty of Physics, University of Sofia, 5 J. Bourchier Boulevard, BG-1164, Sofia, Bulgaria

**Abstract:** We present a new nonlinear solid-state cleaner for the contrast enhancement of femtosecond pulses. The contrast reaches $10^{-10}$ for mJ input pulse, making it a reliable way to solve the contrast issue for Petawatt class lasers in a double chirped-pulse amplification scheme. This filter is based on cross-polarized wave generation in nonlinear crystals. It takes advantage of nonlinear properties associated with third-order tensor elements in $BaF_2$ cubic crystal to improve the temporal contrast of femtosecond laser pulses. Furthermore, it does not affect the spectral shape or the phase of the cleaned pulse. It also acts as an efficient spatial filter.

## 1 Introduction

Petawatt class femtosecond lasers are currently under construction. With intensity of $10^{21}$ W $\cdot$ cm$^2$ or more on target, they are opening the way to relativistic physic experiments. Intensity is not the only requirement of those ultraintense lasers, the problem of contrast needs to be addressed. The contrast of a femtosecond pulse is the intensity ratio between the main femtosecond pulse and its picosecond and nanosecond pedestal. This pedestal is due to amplified spontaneous emission (ASE), pre- or postpulses generated in the amplifiers, and spectral phase distortions. In most high-power Ti : $Al_2O_3$ laser systems with chirped-pulse amplification (CPA), this pedestal is about 6 orders of magnitude below the main pulse in intensity. This is enough to blow or modify the target before the main femtosecond pulse could interact with it. A contrast of $10^{-9}$ is a minimum requirement for a Petawatt femtosecond laser.

Many techniques have been proposed for improving contrast of ultraintense femtosecond laser systems. One way is to benefit from the inherent high contrast of femtosecond oscillator by using a high-output energy oscillator reducing the needed amplification by the preamplifier [1–3]. As most of the ASE is produced in preamplifier, several methods have been investigated to reduce or suppress the preamplifier ASE. Femtosecond preamplification and the use of a saturable absorber [4] allowed to have a μJ intensity level contrast improved femtosecond pulse to seed a CPA. Nonlinear optical filter have been investigated in optical fibers [5–7] with limited output power. Nonlinear contrast filters working in the mJ intensity level in a double CPA scheme have been proposed [8–10]. It is also possible to filter out the pedestal at the output of the laser system [11]. Most of these methods have not proved to be reliable enough for day-to-day use in a high-intensity femtosecond laser.

In this article, we report on the production of high-contrast femtosecond pulses reaching the requirement for Petawatt class laser with a measured contrast enhancement reducing the incoming pedestal to $10^{-10}$ time the main femtosecond pulse [12]. To achieve this, we are using a nonlinear solid-state cleaner for the contrast enhancement of femtosecond pulses. This filter is based on cross-polarized wave (XPW) generation in nonlinear crystals. It takes advantage of nonlinear properties associated with third-order tensor elements in $BaF_2$ cubic crystal to improve the temporal contrast of femtosecond laser pulses. Furthermore, the XPW filter does not affect the spectral shape or the phase of the cleaned pulse and it also acts as an efficient spatial filter.

We will first present theoretical consideration on contrast improvement and efficiency of the XPW contrast filter. Then experimental results about XPW filtering in the mJ energy range are displayed. Finally a way to make the experiment for XPW generation as efficient as the theoretical model predicts is presented.

## 2 Theory

XPW is a third-order nonlinear process; therefore XPW-generated signal presents a cubic dependence to the input intensity. In a first approximation, one can estimate the contrast improvement provided by the XPW setup by cubing the input contrast ($C_{\text{out}} \approx C_{\text{in}}^3$). Unfortunately we need to take into account the extinction ratio ($R$) of the polarizer–analyzer pair used in the setup. Leakage from the input pulse through the polarizer–analyzer pair ends up to be predominant compared to the XPW intrinsic contrast. Considering the leakage effect and making the hypothesis that the XPW generation process occurs in nondepleted regime, we obtain the output contrast $C_{\text{out}} \approx C_{\text{in}} \left[C_{\text{in}}^2 + R/\eta\right]$ with $\eta$ being the peak efficiency of the XPW generation [13]. Figure 1 is showing the dependence of the contrast improvement to the extinction ratio of the polarizer–analyzer pair. The 4,3db curve

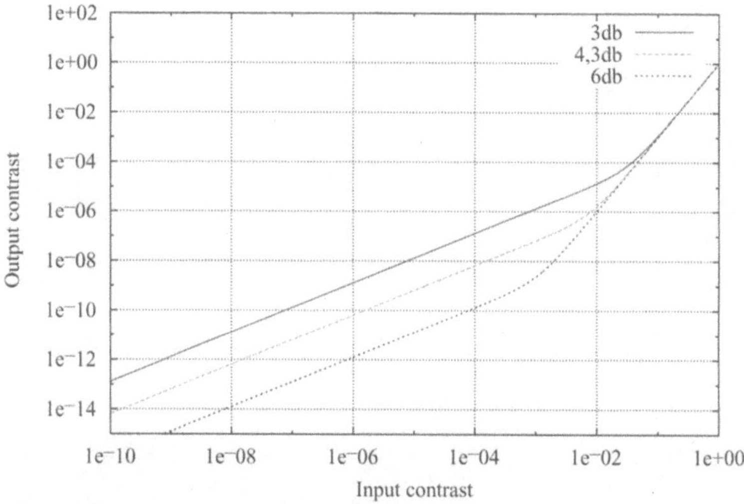


FIGURE 1. Calculation of contrast improvement made for three different extinction ratios of the polarizer–analyzer pair (i.e., 3db, 4,3db, and 6db).

corresponds to our experimental setup, and this curve is used for the calculation of the theoretical contrast for our laser plotted on Fig. 4. This extinction ratio can be expected from *out the shelf* polarizers. 6db curve is what you can expect from good polarizers. The 3db curve is there to show what to expect from setup with too many optical elements between the polarizers so that they are reducing the extinction ratio down to three orders of magnitude or less (like what can be expected from a multipass, nonlinear, elliptical polarization rotation setup).

Efficiency of the XPW is an important issue. We have already provided a full description of the nonlinear processes giving the XPW signal [14]. Figure 2 is displaying the XPW generation efficiency function of the input intensity calculated with those nonlinear processes. We took into account two types of pulse geometry as the XPW process is nonlinear with the intensity distribution of the pulse. If we consider a Gaussian beam in time and space, the maximum efficiency you get out of the XPW process is about 24%. If we considere a tophat beam in space (but Gaussian in time), the maximum efficiency as seen from Fig. 2 increases up to 38%.

## 3 XPW in the 1-mJ Range

XPW generation is a four-wave mixing process. A full description of this process was previously detailed for cubic and tetragonal crystals [12, 14–16] but it is important to point out that XPW generation is proportional to the product of $\chi^{(3)}_{xxxx}$ and the anisotropy ($\sigma = (\chi^{(3)}_{xxxx} - 2\chi^{(3)}_{xyyx} - \chi^{(3)}_{xxyy})/\chi^{(3)}_{xxxx}$ for cubic crystals). We

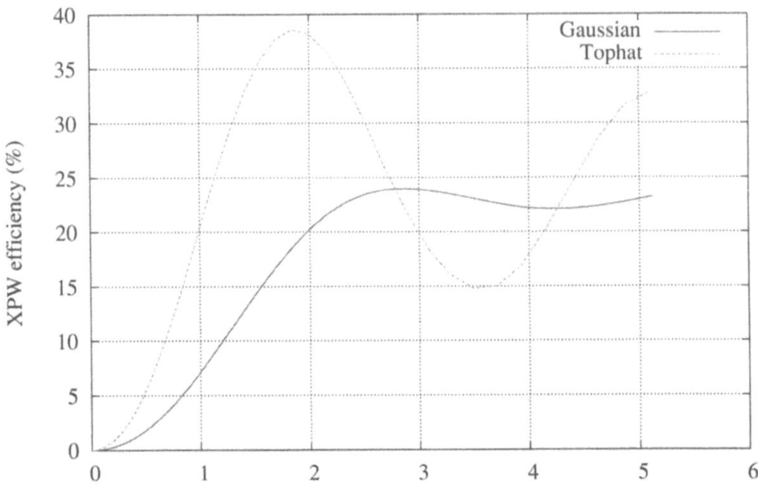


FIGURE 2. Theoretical efficiency of the XPW generation versus input intensity for a Gaussian and a tophat femtosecond pulse.

have used a $BaF_2$ crystal (m3m point group symmetry) as $\chi^{(3)}_{xxxx}$ value is moderate $(1, 59\ 10^{-22} m^2 V^{-2})$ but the anisotropy of $\chi^{(3)}$ is important $(\sigma = -1, 2)$ [15]. The XPW-generated wave has the same wavelength as the input pulse and a cubic dependence on the intensity facilitating an improvement in pulse temporal contrast.

As $BaF_2$ is isotropic with respect to linear optical properties, the process is characterized by perfect phase and group-velocity matching of two orthogonally polarized waves propagating along the z-axis. This property permits good efficiency of XPW generation and minimal pulse shape and spectral distortions.

Another advantage of $BaF_2$ is its transmission from the ultraviolet to the infrared. As its band gap energy is high (9.07 eV), multiphoton absorption is negligible. So the $\chi^{(3)}$ value should be almost constant in the visible and the near infrared, and XPW generation can be applied to various femtosecond laser wavelengths.

This promising system can also be adapted to other pulse energies by geometrical tuning, as the efficiency of the XPW generation is determined by the peak power intensity of the laser. We have demonstrated equal efficiency for input laser pulses from the microjoule to millijoule range.

Experiments were performed with 1 kHz repetition rate Ti:sapphire CPA laser. The laser system produces 42-fs pulses with a maximum energy of 2 mJ. The input pulse in the XPW setup is linearly polarized and focused by a 3-m focal length lens. A 2-mm thin $BaF_2$ crystal is placed after the focal point to optimize the conversion process by reaching the correct peak intensity level ($\sim 10^{12}\ W \cdot cm^{-2}$). The crystal is rotated at an optimized angle $\beta \sim 22.5°$ (the angle between the input polarization direction and its [100] axis) [15]. A second thin $BaF_2$ crystal can be inserted in the setup in a multipass configuration. Then an analyzer transmits the XPW-generated signal, suppressing the remaining from the input pulse with an extinction ratio of the polarizer–analyzer pair measured to be $5 \times 10^{-5}$.

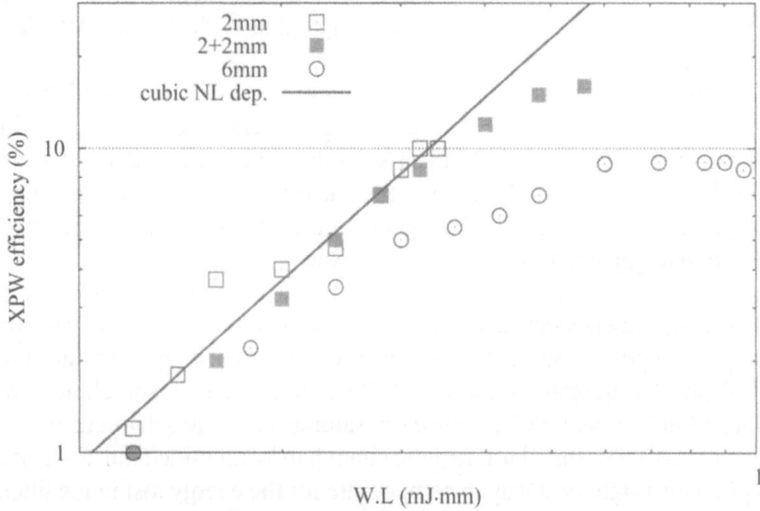


FIGURE 3. Measurement of the efficiency of the XPW generation versus (energy x crystal length) product. Measurements correspond to XPW with a thin 2-mm $BaF_2$ crystal, two imaged 2-mm crystal, and a thick 6-mm crystal.

As XPW is a nonlinear third order process, it is inherently not efficient. To overcome this problem, one has to work at the highest possible input intensity. We have been performing efficiency versus input intensity measurement up to the damage threshold of the crystal. Results are visible in Fig. 3 for a 2-mm thin $BaF_2$ crystal, for two 2-mm thin crystals, and for a thick 6-mm crystal. To efficiently compare the behavior of the various measured configurations, the efficiency is plotted relative to the product energy to crystal length.

The nonlinear response of the thin 2-mm crystal displayed in Fig. 3 corresponds to the expected quadratic dependence of the efficiency to the input intensity (due to the cubic dependence of a third-order process) up to the final measured point that is near damage threshold. The damage threshold is then the limiting effect for thin crystal XPW. This limit appears unfortunately before the theoretical efficiency limit of the XPW process calculated for Gaussian pulse displayed in Fig. 2.

When the input energy was 1.2 mJ, we succeeded in producing cleaned pulses with an energy of 120 μJ and low self-phase modulation. Thus the uncorrected energy transmission efficiency of the setup was 10%.

The 6-mm thick crystal follows the same nonlinear response. This nonlinear response saturates from efficiency as low as 5% giving a lower overall efficiency at high input intensity (Fig. 3). From our observations during experimentations with this thick crystal, it appears that the limiting factor is self-focusing along the propagation inside the crystal leading to early continuum generation and low damage threshold.

To overcome the early efficiency limitation due to the damage threshold in thin crystal and the saturation due to self-focusing of the pulse in thick crystal, and to reach the theoretical performance plotted on Fig. 2, we are using a two-thin-crystals scheme. This method proved to be efficient as we get the efficiency displayed in Fig. 3 directly comparable with the one for Gaussian pulse in Fig. 2.

There are various ways of using two or more thin crystals in the XPW setup. One is a double pass optical scheme allowing the beam spot on the crystal in the first pass to be imaged on the second pass. As the XPW process does not introduce group velocity dispersion, it preserves the phase matching between the input and the output pulse. The XPW pulses generated in each pass are then inherently phase matched and can interfere constructively.

Therefore, with limited intensity on each crystal, the double pass scheme allows for an overall efficiency that can reach the 24% limit plotted on Fig. 2. It overcomes the early damage threshold limit on each crystal that is then arising after the practical and theoretical saturation of the efficiency. It is somehow a way to simulate a thick crystal without the early saturation due to self-focusing.

The cleaned XPW signal is energetic enough to be amplified into a second CPA setup [8]. This might be a way to compensate for the energy lost in the filter, provided that the contrast is not excessively degraded during the second amplification.

We have estimated the contrast improvement by using a homemade high-dynamic range, third-order cross-correlator. We measured the XPW filtered and nonfiltered pulses with the same energy seeding the correlator ($\geq$100 μJ) to get comparable curves, as shown in Fig. 4. For this energy level, the correlator noise

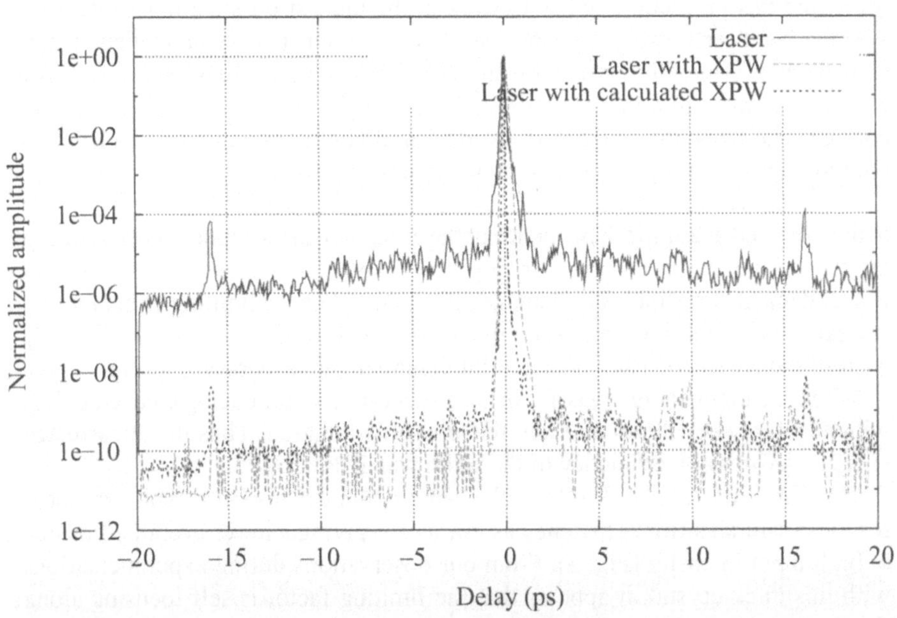


FIGURE 4. Third-order correlator traces of the attenuated down to 100 μJ laser pulse of the XPW-filtered laser pulse and the XPW theoretical contrast improvement applied to the input measurement.

is $10^{-11}$. The ASE intensity level of the input pulse is six orders of magnitude below the peak intensity of the main pulse.

As the XPW-generated pulse intensity profile exhibits a cubic dependence on the input profile, the pulse pedestal is drastically reduced. The $10^{-10}$ remaining ASE pedestal corresponds to a leakage of the incoming pulse pedestal through the analyzer. The calculated output contrast ratio by using the measured extinction (i.e., $5 \times 10^{-5}$) is $C_{\text{out}} \approx C_{\text{in}} R = 5 \times 10^{-11}$. It is very close to the measured one. For the conditions applicable for our experiment when $C_{\text{in}}^2 \ll R$, the output contrast ratio is proportionnal to the extinction ratio of the pair polarizer-analyser. The agreement between calculated and the experimental contrast ratio is demonstrated also in Fig. 4 where correlation function of the laser pulse, cleaned laser pulse, and theoretical predicted cleaned pulse are shown. We believe that XPW contrast can be brough down to $10^{-12}$ by using polarizers with a better extinction ratio (i.e., $10^{-6}$). Therefore, a Petawatt laser system based on the double CPA scheme will not suffer from the generated ASE in the second CPA.

Spectra measurements made for a XPW generation efficiency of 10% present no visible modulation defect, the spectral width is unchanged, and the spectrum exhibits a smooth Gaussian shape. As the beam is focused in air, at higher input-energy level self-phase modulation appears near the focal point and generates distortions of the filtered spectrum. For further developments, the use of a vacuum-packed setup will allow us to eliminate this effect.

Unlike for the nonlinear elliptical beam rotation technique [10], XPW generation does not require any specific input spectral phase. The output XPW pulse spectral phase presents no distortions or discontinuity. On the contrary, as XPW is filtering out post- and prepulses it is smoothing modulations on the input spectral phase. Spectral phase measurements are obtained with a homemade spectral phase interferometry for direct electric-field reconstruction (SPIDER).

An important characteristic for further amplification is the spatial quality of the cleaned beam, which is actually improved during the filtering [12]. Indeed, as the nonlinear effects occur in the far field, the highest spatial frequencies are removed. They are not intense enough to take part in the conversion process. So we can assert that our setup acts as an efficient temporal and spatial filter.

## 4 Conclusions

In conclusion, we demonstrate experimentally that multicrystal XPW scheme is a highly efficient setup for femtosecond pulse cleaning. XPW filter is achromatic, simple, and robust, as the nonlinear process occurs in a solid medium (i.e., cubic and $\chi^{(3)}$ anisotropic crystals, such as $BaF_2$). We have achieved an energy transmission of 24%, corresponding to the theoretical limit for Gaussian pulses. The temporal contrast has been measured to be improved by more than four orders of magnitude. Further improvement of the temporal contrast can be achieved by the selection of a polarizer–analyzer pair with a better extinction ratio as this extinction ratio is the main parameter for determining the overall contrast improvement of the XPW filter. Furthermore, the cleaned pulse does not exhibit any spectral phase distortions and the spatial profil is filtered out from high-frequency modulation from the input pulse. Those attractive performances allow us to believe that this technique will be useful in the design of future background-free Petawatt laser systems.

**Acknowledgments.** The authors thank the European Commission through the Suppression over High Dynamic Range of ASE at the Rising Edge of Ultraintense Femtosecond Pulses (SHARP), project, contract HPRI-CT-2001-50037.

N. Minkovski, S. Kourtev, and S. Saltiel acknowledge support from the European Commission LASERLAB (grant RII-CT-2003-506350) and from Bulgarian Science Fund (grant 1201).

## References

1. G. Chériaux, T. Planchon, F. Augé, G. Mourou, and J. P. Chambaret: Proceeding of Ultrafast Optics, **16**, Heidelberg: Springer-Verlag (2001).
2. G. N. Gibson, R. Klank, F. Gibson, and B. E. Bouma: Opt. Lett. **21**, 1055–1057 (1996).
3. A. Fernandez, T. Fuji, A. Poppe, A. Fürbach, F. Krausz, and A. Apolonski: Opt. Lett. **29**, 1366–1368 (2004).
4. J. Itatani, J. Faure, M. Nantel, G. Mourou, and S. Watanabe: Opt. Comm. **148**, 70–74 (1998).
5. D. Homoelle, A. L. Gaeta, V. Yanovsky, and G. Mourou: Opt. Lett. **27**, 1646–1648 (2002).
6. J. L. Tapié and G. Mourou: Shaping of clean. Opt. Lett. **17**, 136–138 (1992).
7. Y. Beaudoin, C. Y. Chien, J. S. Coe, J. L. Tapié, and G. Mourou: Opt. Lett. **17**, 865–867 (1992).
8. M. P. Kalashnikov, E. Risse, H. Schönnagel, A. Husakou, J. Herrmann, and W. Sandner: Opt. Express. **12**, 5088–5097 (2004).
9. A. Renault, F. Augé-Rochereau, T. Planchon, et al.: Opt. Comm., **248**, 535–541 (2005).
10. A. Jullien, F. Augé-Rochereau, G. Chériaux, et al.: Opt. Lett. **29**, 2184–2186 (2004).
11. G. Doumy, F. Quere, O. Gobert, et al.: Phys. Rev. E **69**, 026402 (2004).
12. A. Jullien, O. Albert, F. Burgy, et al.: Opt. Lett. **30**, 920–922 (2005).
13. A. Jullien, S. Kourtev, O. Albert, et al.: to be published.
14. A. Jullien, O. Albert, G. Chériaux et al.: J. Opt. Soc. Am. B **22**, 2635–2641 (2005).
15. N. Minkovski, G. I. Petrov, S. M. Saltiel, O. Albert, and J. Etchepare: J. Opt. Soc. Am. B **21**, 1659-1664 (2004).
16. N. Minkovski, S. M. Saltiel, G. I. Petrov, O. Albert, and J. Etchepare: Opt. Lett. **27**, 2025–2027 (2002).

# Part XVII

## OPCPA

# Infrared Optical Parametric Chirped Pulse Amplifier for High Harmonic Generation

T. Fuji,[1] N. Ishii,[1] Th. Metzger,[1] C. Y. Teisset,[1] L. Turi,[1] A. Baltuška,[1] N. Forget,[2] D. Kaplan,[2] A. Galvanauskas,[3] and F. Krausz[1,4]

[1] Max-Planck-Institut für Quantenoptik, Hans-Kopfermann-Strasse 1, D-85748, Garching, Germany
[2] Fastlite, Bat 403, Ecole Polytechnique, 91128 Palaiseau, France
[3] University of Michigan,College of Engineering, Department of Electrical Engineering and Computer Science, 1301 Beal Avenue Ann Arbor, 48109-2122, Michigan, U.S.A.
[4] Ludwig-Maximilians-Universität München, Am Coulombwall 1, D-85748, Garching, Germany
**tkf@mpq.mpg.de**

## 1 Introduction

Rapid advances in high-field physics achieved in recent years, most notably generation of isolated soft X-ray attosecond pulses, owe their success to the development of driver lasers with specific pulse properties. The latter include ultrahigh peak intensity, quasi-monocycle duration, and reliable control over the carrier-envelope phase (CEP) [1, 2]. Although the driver lasers currently employed in this research field operate nearly exclusively in the wavelength region of the Ti:sapphire gain (i.e. around 0.8 μm), a switching over to a longer, infrared (IR) wavelength would offer significant advantages. Because of the $\lambda^2$ scaling of the ponderomotive energy, the intensity of IR pulses needed to attain emission at a given X-ray photon energy could be substantially lowered in comparison with the 0.8-μm case [3–5]. This is expected to be extraordinarily helpful for up-scaling the X-ray frequency, decreasing the duration of X-ray attosecond pulses by at least a factor of $\lambda^{3/2}$, and suppressing undesired target preionization before the interaction with the strongest half-cycle of the laser pulse. From the standpoint of laser technology, the longer duration of the IR optical period reduces the number of cycles for a given pulse envelope and, therefore, relaxes the demand to the amplifier gain bandwidth, which in the case of 5-fs 0.8-μm pulses typically reaches the extreme >100 THz.

Several preliminary experiments in high-harmonic generation, using infrared (1~4 μm) ultrashort pulse driver laser were reported in the past [6,7]. However, due to a low pulse intensity, the achieved XUV photon energy was below ~75 eV.

Recently, development of optical parametric chirped pulse amplifier (OPCPA) [8,9] for generation of intense, phase-controlled, few-cycle pulses has progressed rapidly. However, the wavelength region of the amplifier has also been Ti:sapphire gain region [10–13]. It must be possible to apply the OPCPA concept to longer wavelength region. So far there have been a few reports concerning infrared OPCPA systems [9, 14]; however, these lasers do not satisfy required property for the application of high-field physics.

In this contribution, we describe an IR laser system based on OPCPA, which fulfills the requirements for high-field physics application in terms of spectral bandwidth, pulse duration, and CEP control. The laser system can be considered as a prototype of the next-generation attosecond pulse driver source.

## 2 Laser System

Figure 1 shows the schematic of the laser setup. The system is based on a two-stage OPCPA. The pump laser is based on a commercial 40-ps Nd:YLF amplifier laser system (HighQ Laser Production GmbH) producing 0.4 mJ at a 10-kHz repetition rate. The output of the laser is amplified in a homemade booster amplifier module based on a $\phi 4 \times 120$ mm Nd:YLF rod. After three passes through the module, the final output energy is ~2 mJ. The regenerative amplifier is electronically synchronized with a broadband Ti:sapphire seed oscillator. The spectrum generated from the Ti:sapphire oscillator is shown in Fig. 2a. We obtain the IR seed pulses by

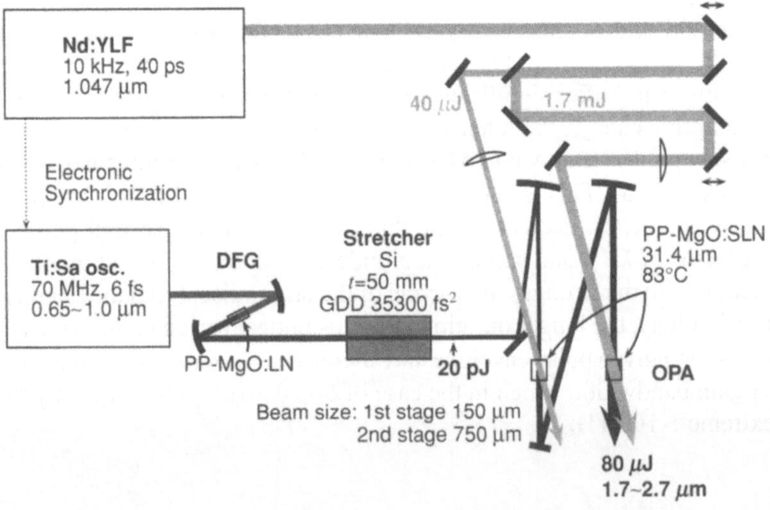


FIGURE 1. Schematic of the laser setup.

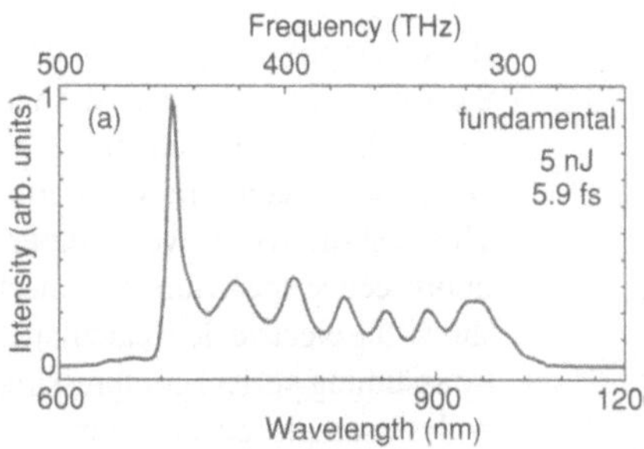


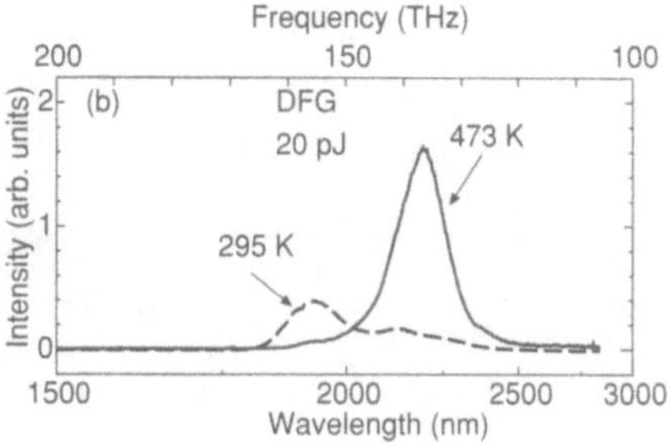


FIGURE 2. (a) The spectrum generated from the broadband all-chirped mirror Ti:sapphire laser. The energy is 5 nJ and the compressed pulse width becomes 5.9 fs. (b) The spectra of DFG of the Ti:sapphire laser output. The spectra are generated just by focusing the compressed laser pulse to 3-mm thick PP-MgO:LN with the QPM period of 13.9 μm. The dashed and solid curves show the spectra at room temperature and 200°C, respectively. The energy of the infrared pulse is 20 pJ at 200°C condition.

generating difference-frequency (DFG) radiation of the Ti:sapphire oscillator output in a 3-mm-long MgO doped periodically poled $LiNbO_3$ crystal (PP-MgO:LN, HC Photonics Ltd.) with a quasi-phase-matching period $\Lambda_{QPM} = 13.9$ μm. The crystal is heated to 200°C to suppress photorefraction. Figure 2b depicts the result of the frequency conversion of the 5-nJ 6-fs broadband pulses from a chirped-mirror Ti:sapphire oscillator at two different temperature of the crystal. Such DFG radiation has been shown to carry a constant CEP offset [15–17] and, therefore, provides a uniquely suitable source for CEP-controlled pulse amplification. Additionally, the DFG scheme here does not contain a delay line between high- and low-frequency components, otherwise the delay line should keep the stability within $\lambda/2$ to have stabilized CEP. Thus, we can expect better phase stability than the previous self-stablization schemes. The estimated energy of the infrared seed pulse is 20 pJ.

To achieve good temporal overlap between the picosecond pump and broadband seed pulses and thus ensure efficient pump energy conversion, the duration of the IR seed was stretched by an antireflection-coated silicon block. The total length of the Si bulk material was about 50 mm, which corresponds to a group delay dispersion of about 35300 $fs^2$ at 2.1 μm.

For our pump wavelength of 1.047 μm, the selected central wavelength of the parametric amplifier lies at the degeneracy point of the three-wave mixing, around which the broadest phase-matching bandwidth is expected. The measured spectrum of the super-fluorescence, which reveals the extent of the available gain bandwidth, is shown in Fig. 3 as a dash-dotted curve. The employed nonlinear crystals are a stoichiometrically grown PP-MgO:LN ($\Lambda_{QPM} = 31.4$ μm) heated to 83°C to minimize parasitic photorefractive effects. The lengths of the crystals are 3 mm for the first stage and 2 mm for the second stage. The pump beams were gently focused on the PP-MgO:LN with the $e^2$-level beam diameters of ∼150 μm for the first stage and ∼350 μm for the second stage. A small noncollinearity

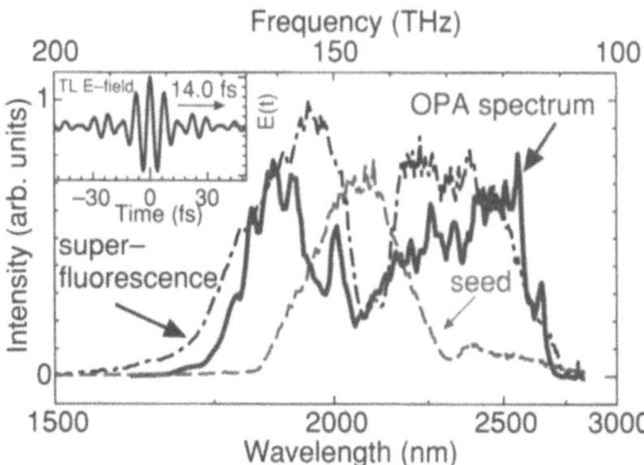


FIGURE 3. The solid curve shows amplified spectrum with the two-stage OPA. The dashed curve shows the seed pulse by DFG of Ti:sapphire oscillator. The dash-dotted curve is super-fluorescence spectrum. The inset shows the electric field calculated by assuming perfect compression with the amplified spectrum.

angle was introduced between the seed and the pump beams to enable spatial separation of the otherwise degenerated signal and idler waves and prevent instabilities caused by their mutual interference. At the first stage, the signal pulse is amplified up to 13.5 mW (1.35 μJ) with a 0.4-W (40-μJ) pump. At the second stage, the signal pulse is amplified up to 800 mW (80 μJ) with a 17-W (1.7-mJ) pump. The amplified spectrum is shown in Fig. 3 as solid line. The broadening of the spectrum compared to the seed spectrum is due to the saturated gain for the main part of the seed spectrum. A single-pass energy gain in excess of $\sim 10^6$ was obtained. However, this regime corresponded to a nonnegligible amount of background super-fluorescence that overwhelms the comparatively weak injected seed. Therefore, for further amplification, the signal has to be injected into subsequent parametric amplifier stages, which can be pumped correspondingly harder. As is evidenced by the gain profile (Fig. 3, dash-dotted curve), no significant gain narrowing is expected. Assuming ideal pulse compression, the measured amplified spectrum (Fig. 3, solid curve) corresponds to a 14.0-fs pulse, which carries the same number of cycles as a 5.1-fs pulse at 0.8 μm.

# 3 Compression Experiment

In order to check the possibility of compression, grating-based stretcher and compressor are implemented for the single-pass OPA. The generated spectrum is shown in Fig. 4a. The gratings are coated with aluminum and the grooves of the gratings is 150 l/mm. The throughput of the grating stretcher and the compressor is ~20%. The compressed pulse width measured with the autocorrelator based on two-photon-induced photocurrent [18] in a InGaAs detector. The measured interferometric autocorrelation trace is shown in Fig. 4b. The estimated pulse width is ~35 fs, which is ~5 cycles for 2.1 μm wavelength. By implementing adaptive optics, for example, a flexible mirror inside the grating-based stretcher, we expect to compress the pulse close to the transform-limited duration [19,20]. Additionally, the dispersion of the grating compressor is similar to bulk infrared

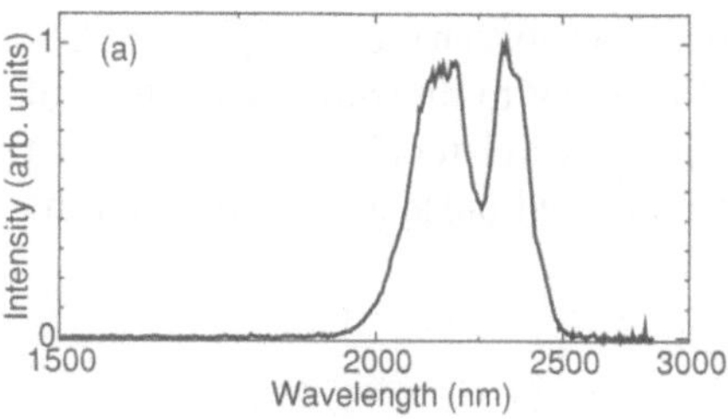


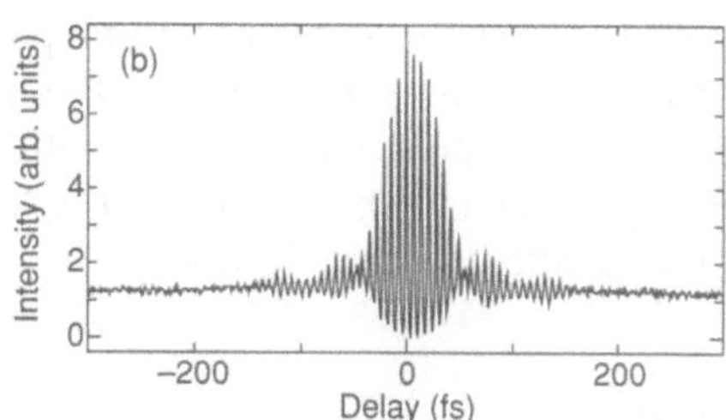


FIGURE 4. Compression experiment with grating-based stretcher and compressor. (a) The spectrum generated at the first stage of the OPA. (b) Measured interferometric autocorrelation trace with two-photon-induced photocurrent in a InGaAs detector.

fused silica, which has negative dispersion in that wavelength region. Replacing the grating compressor to an antireflection-coated infrared fused silica bulk must be helpful to increase the output energy dramatically.

For more precise compression, we use the programmable acousto-optic filter (DAZZLER, Fastlite) containing a 45-mm-long $TeO_2$ crystal as a stretcher and compressed by the antireflection coated 50-mm length Si block. The generated spectrum is shown in Fig. 5a. The measured interferometric autocorrelation trace is shown in Fig. 5b. When comparing to the calculated autocorrelation trace by assuming perfect compression, the pulse seems to be compressed down to theoretical limit, ~20 fs, or ~3 cycles. According to our knowledge, it is the first demonstration of few-cycle pulse generation in the several micron wavelength region.

With two-stage OPA, 0.4 mJ output has been achieved by 5 mJ pump energy; however, the contribution of parametric fluorescence background becomes significantly large (when the seed is blocked before the first stage, the power of the amplified spontaneous emission is only 10% less than that with injecting seed pulses) since the seed pulse energy is less than 1 pJ due to the low efficiency of broadband diffraction of gratings or DAZZLER (~10%). Nevertheless, more

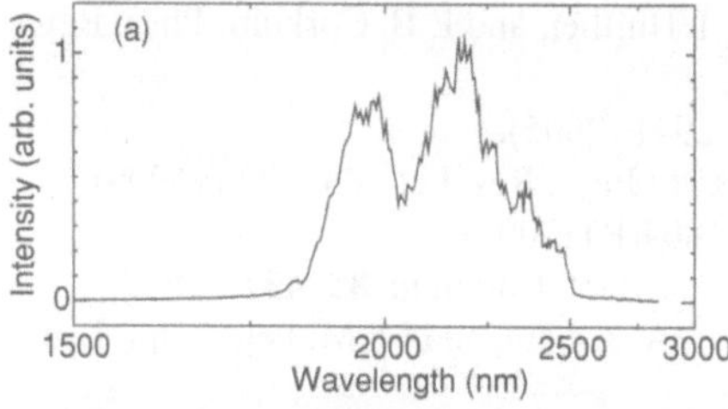


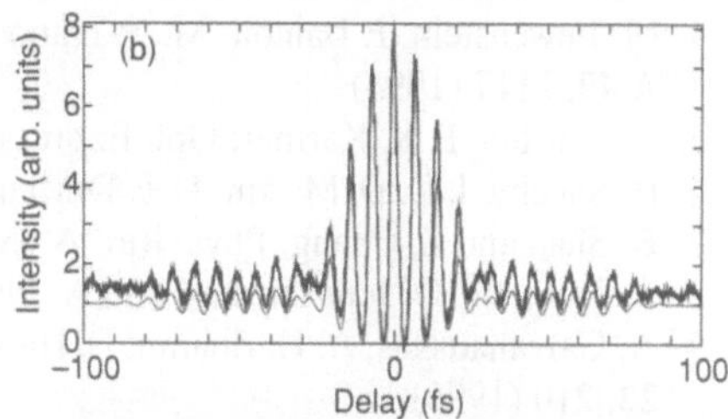


FIGURE 5. Compression experiment with DAZZLER stretcher and Si bulk compressor. (a) The spectrum generated at the first stage of the OPA. (b) The thick solid curve shows measured interferometric autocorrelation trace with two-photon-induced photocurrent in a InGaAs detector. The thin solid curve shows calculated autocorrelation trace assuming zero phase with the measured spectrum.

careful adjustment of temporal and spatial overlap between seed and pump pulses must still improve the situation. It is very important to measure the contrast of the amplified pulse quantitatively with high-order correlator [21]. We believe that it is possible to achieve >0.5-mJ, few-cycle infrared pulse generation with the scheme after the investigation.

## 4 Summary

In summary, we demonstrated a novel type of CEP-stable seed for a parametric amplifier that currently produces ~0.1-mJ 3-cycle pulses in the wavelength range of 1.5–3.0 μm. The laser system is directly scalable with the pump pulse energy. The reported scheme opens a promising route toward constructing a high-peak-power IR driver laser for high-field and attosecond applications. Additionally, the ultrashort infrared pulse could easily be converted to 5–10 μm wavelength, which is most interesting wavelength region for molecular vibrational spectroscopy, by difference frequency generation. We can expect few-cycle, several micro Joule mid-infrared pulses just by focusing the beam into a proper nonlinear crystal. It could be a quite unique light source for femtosecond molecular science.

**Acknowledgment.** We acknowledge HC Photonics for fruitful discussion concerning periodically poled crystals used in this research. This research was greatly motivated by a proposal of Prof. L. Dimauro from Ohio State University for improved attosecond high harmonic generation with an infrared driver.

## References

1. A. Baltuška, T. Udem, M. Uiberacker et al.: Nature **421**, 611 (2003).
2. A. Baltuška, M. Uiberacker, E. Goulielmakis et al.: IEEE J. Sel. Top. Quantum Electron. **9**, 972 (2003).
3. P. B. Corkum: Phys. Rev. Lett. **71**, 1994 (1993).
4. M. Lewenstein, P. Balcou, M. Y. Ivanov, A. L'Huillier, and P. B. Corkum: Phys. Rev. A **49**, 2117 (1994).
5. A. Gordon, F. X. Kärtner: Opt. Express **13**, 2941 (2005).
6. B. Sheehy, J. D. D. Martin, L. F. DiMauro et al.: Phys. Rev. Lett. **83**, 5270 (1999).
7. B. Shan and Z. Chang: Phys. Rev. A **65**, 011804(R) (2002).
8. A. Dubietis, G. Jonušauskas, and A. Piskarskas: Opt. Commun. **88**, 437 (1992).
9. A. Galvanauskas, A. Hariharan, D. Harter, M. A. Arbore, and M. M. Fejer: Opt. Lett. **23**, 210 (1998).
10. C. P. Hauri, P. Schlup, G. Arisholm, J. Biegert, and U. Keller: Opt. Lett. **29**, 1369 (2004).
11. R. T. Zinkstok, S. Witte, W. Hogervorst, and K. S. E. Eikema: Opt. Lett. **30**, 78 (2005).
12. N. Ishii, L. Turi, V. S. Yakovlev et al.: Opt. Lett. **30**, 567 (2005).
13. S. Witte, R. T. Zinkstok, W. Hogervorst, K. S. E. Eikema: Opt. Express **13**, 4903 (2005).

14. F. Rotermund, C. J. Yoon, V. Petrov et al.: Opt. Express **12**, 6421 (2004).
15. A. Baltuška, T. Fuji, and T. Kobayashi: Phys. Rev. Lett. **88**, 133901 (2002).
16. T. Fuji, A. Apolonski, and F. Krausz: Opt. Lett. **29**, 632 (2004).
17. C. Manzoni, G. Cerullo, and S. D. Silvestri: Opt. Lett. **29**, 2668 (2004).
18. J. K. Ranka, A. L. Gaeta, A. Baltuška, M. S. Pshenichnikov, and D. A. Wiersma: Opt. Lett. **22**, 1344 (1997).
19. G. Chériaux, O. Albert, V. Wänman, J. P. Chambaret, C. Félix, and G. Mourou: Opt. Lett. **26**, 169 (2001).
20. A. Baltuška, T. Fuji, and T. Kobayashi: Opt. Lett. **27**, 306 (2002).
21. F. Tavella, K. Schmid, N. Ishii, A. Marcinkevičius, L. Veisz, and F. Krausz: Appl. Phys. B **81**, 753 (2005).

# Double-passed, High-energy Quasiphase-matched Optical Parametric Chirped-pulse Amplifier

Igor Jovanovic, Nicolas Forget,* Curtis G. Brown, Christopher A. Ebbers, C. Le Blanc,* and C. P. J. Barty

Lawrence Livermore National Laboratory, Mail Code L-470, 7000 East Avenue, Livermore, California 94550, USA
* Laboratoire pour l'Utilisation des Lasers Intenses, École Polytechnique, Route de Saclay, 91128 Palaiseau cedex, France
jovanovic1@llnl.gov

## 1 Introduction

Quasiphase-matched (QPM) optical parametric chirped-pulse amplification (OPCPA) in periodically poled materials such as periodically poled $LiNbO_3$ (PPLN) and periodically poled $KTiOPO_4$ (PPKTP) has been shown to exhibit advantages over the OPCPA in bulk nonlinear crystals [1,2]. The use of the maximum material nonlinear coefficient results in ultrahigh gain with low pump peak power. Furthermore, the propagation of signal, pump, and idler beams along one of the crystal principal axes eliminates the birefringent walk-off, reduces angular sensitivity, and improves beam quality. Relatively high level of parasitic parametric fluorescence (PF) in QPM OPCPA represents an impediment for simple, single-stage, high-gain amplification of optical pulses from nanojoules to millijoules energies. PF in QPM is increased when compared to PF in critical phase matching in bulk crystals due to broader angular acceptance of the nonlinear conversion process. The PF reduces prepulse contrast and conversion efficiency by competition with the signal pulse for pump pulse energy. Previous experiments with QPM OPCPA have thus resulted in pulse energies limited to tens of microjoules. [3] Optical parametric amplification of a narrowband signal pulse in PPKTP utilizing two pump beams has been demonstrated at a millijoule level [4], but the conversion efficiency has been limited by low energy extraction of pump pulse in the first pass of amplification. Additionally, narrow spectral bandwidth was the result of operation far from signal-idler degeneracy.

Herein we present a novel double-pass, broad-bandwidth QPM OPCPA. Amplified signal energy of 1.2 mJ is produced in a single PPKTP crystal, utilizing a single 24-mJ pump pulse from a commercial pump laser. [5] To our knowledge, this is the highest energy demonstrated in QPM OPCPA. Double-passed QPM OPCPA exhibits high gain (>3 × 10$^6$), high prepulse contrast (>3 × 10$^7$), high energy stability (3% rms), and excellent beam quality. We additionally present a simple extension of QPM OPCPA to cascaded-optical parametric amplification (COPA) [6], resulting in, in principle, infinite prepulse contrast. This amplifier is highly suitable for a high-gain section of a high-energy OPCPA system, which has previously employed two or more crystals and higher pump energies, or a stand-alone, high-contrast preamplifier for a petawatt-class Nd:glass short-pulse laser. OPCPA front end systems for petawatt-class lasers are now used or planned at numerous laser facilities [7] and may benefit from utilization of this type of system.

## 2 Experimental Setup

Figure 1 depicts our experimental setup. The seed pulse centered at 1053 nm and with 6-nm FWHM bandwidth is generated by a mode-locked Yb:glass oscillator (High Q Laser IC-1053-200) operating at 77 MHz. The oscillator produces 200-fs pulses, which are subsequently stretched to 1.2-ns FWHM with conventional antiparallel grating pulse stretcher. The stretched pulse energy is 800 pJ. Pump pulses are generated by a commercial, injection-seeded Nd:YAG laser (Continuum Powerlite Plus), producing 532-nm pulses with 5.7-ns FWHM pulse duration in the pulse center with a spatiotemporal shape characteristic for an unstable resonator. The repetition rate of the pump laser is 10 Hz and its pulses are synchronized

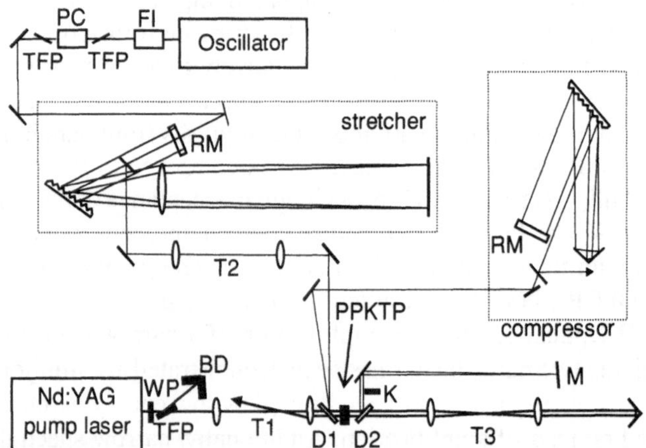


FIGURE 1. Schematic for millijoule OPCPA/COPA in PPKTP. FI, Faraday isolator; PC, Pockels cell; TFP, thin film polarizer; RM, roof mirror; T, telescope; WP, waveplate; D, dichroic beamsplitter; BD, beam dump; K, knife edge. Selection between OPCPA and COPA is accomplished by tilting the mirror M to direct the signal or idler pulse, respectively, into the second amplification pass.

with the oscillator pulses with a characteristic pulse-to-pulse jitter of 0.5 ns rms. Only a small amount (24 mJ) of the available pump energy from the pump laser is split from the pump laser and a circular aperture of ~7.5 mm diameter is inserted in the beam. The aperture is relay imaged using a 5:1 beam relay telescope onto a $1.5 \times 5 \times 7.5$ mm$^3$ PPKTP crystal (Raicol). The PPKTP crystal is periodically poled for a first-order, type-0 1053 nm + 1075 nm = 532 nm process and both its input and output surfaces are antireflection coated for 532 nm/1053 nm/1075 nm. A 3° wedge is used on the crystal back surface to prevent parasitic oscillation.

Seed and pump pulses are combined and injected into a PPKTP crystal with a small noncollinear angle of <0.5°. Seed pulses overfill the crystal aperture so that ~50% of the available seed energy is introduced into the crystal. After the first pass of amplification, the pump is separated from the signal and idler, using a dichroic beamsplitter and is fully relay imaged onto a roof mirror retro reflector. The roof mirror retro reflector displaces the pump beam laterally by 2 mm on the second (reverse) pass through the PPKTP crystal. Signal and idler pulses propagate freely without relay imaging to the second flat-mirror retro reflector. Idler pulses are separated from signal pulses using a knife edge (K).

Double-pass amplification scheme has a significant advantage over the single-pass amplification scheme due to the reduction of PF through propagation of the signal/idler pulse over extended distances, thus allowing angularly divergent PF to be separated from the main signal/idler pulse prior to second-pass amplification. This property allows high-energy, high-gain amplification with a fluorescence level orders-of-magnitude lower than previously. The removed fraction of PF can be increased by spatial filtering of the signal between two amplification passes.

# 3 Results

We have obtained 1.2 mJ of amplified signal energy by using 24 mJ of pump energy and a peak on-axis pump intensity of 170 MW/cm$^2$. Amplified spectrum exhibits spectral broadening to 12 nm FWHM (Fig. 2), which has been previously observed

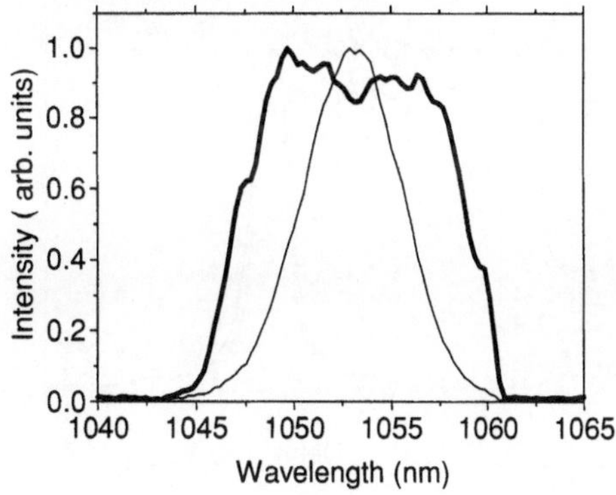


FIGURE 2. Oscillator (6 nm FWHM, thin line) and amplified signal spectra (12 nm FWHM, thick line). Spectral broadening is evident when OPCPA operates with high pump conversion efficiency.

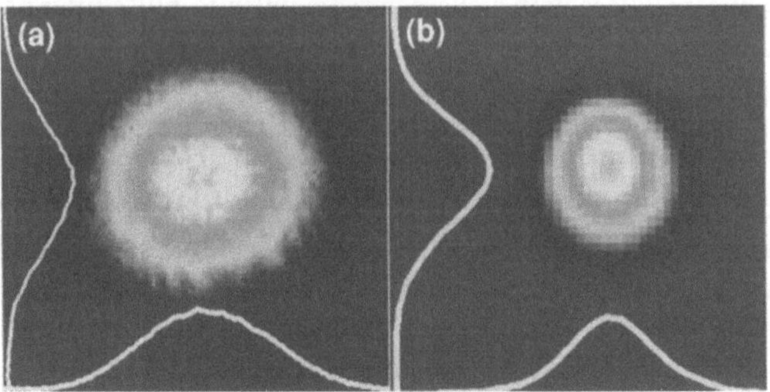


FIGURE 3. Amplified signal beam profiles: (a) free propagation for 1 m; (b) far field.

in saturated OPCPA. In Fig. 3 we show amplified signal-free propagation after 1 m and far field beam profiles. The amplified signal pulse has been recompressed and its intensity autocorrelation is shown in Fig. 4. The calculated transform-limited autocorrelation width is 280 fs FWHM, resulting in transform-limited pulse width of 147 fs FWHM. Nonidealities in the existing stretcher (spherical and chromatic aberrations) impose a ~200-fs aberration limit of the expansion-compression system. Pulse compression of QPM OPCPA-amplified pulse has been achieved at this aberration limit.

Prepulses in CPA systems can arise from preexisting oscillator pulse train or cavity- and etalon-produced pulses. We have measured the prepulse and fluorescence contrast of the amplified signal pulse after recompression and determined it to be $>3 \times 10^7$, limited by the experimental sensitivity of our measurement. The measured energy stability is 3% rms and is greatly enhanced when local pump depletion occurs.

When the idler pulse is removed in the signal/idler return line with the knife edge, the system operates in the normal optical parametric amplifier mode. When the signal pulse is removed instead of the idler pulse in the signal/idler return line, the system can operate in the cascaded-optical parametric amplifier (COPA) [6] mode. In this mode, the signal pulse is regenerated by difference frequency

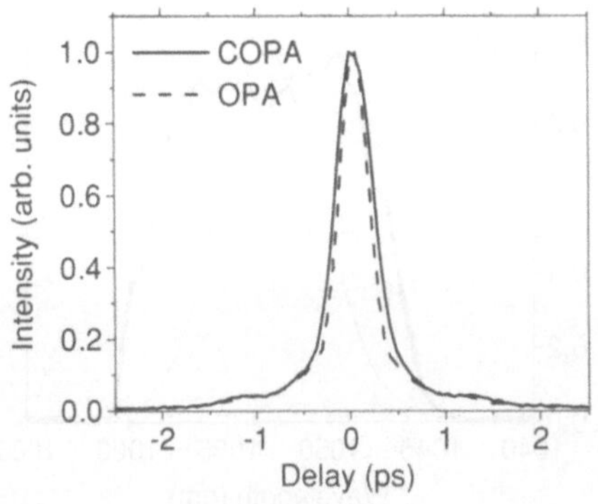


FIGURE 4. Experimental autocorrelation trace of the OPCPA recompressed pulse (dashed line, 350 fs FWHM) and COPA recompressed pulse (solid line, 422 fs FWHM).

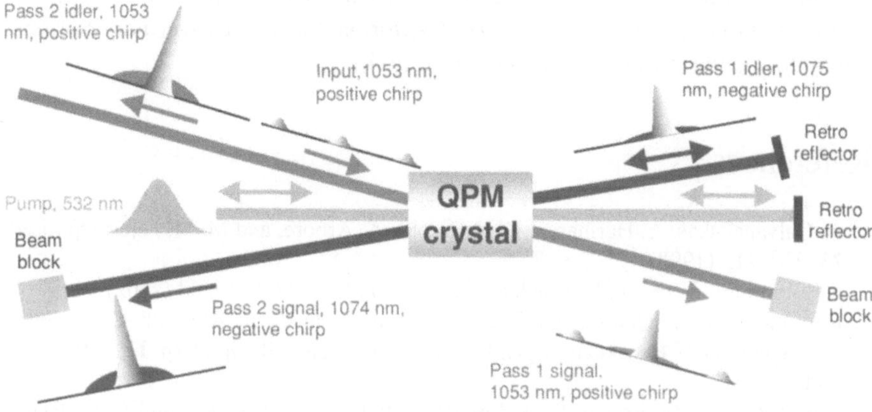


FIGURE 5. COPA in double-pass QPM OPCPA. In COPA, the unamplified oscillator pulse train is completely removed and the contrast is limited only by PF.

generation of the pump and idler in the second pass, resulting in extreme prepulse contrast enhancement (Fig. 5). Operation in the COPA mode takes advantage of two subsequent chirp reversals through double DFG in the nonlinear crystal and allows a complete removal of all other pulses from the mode-locked pulse train or optical background propagating on the signal axis outside of the temporal window defined by the pump pulse. COPA contrast is limited only by PF of the parametric process.

COPA-amplified pulse has been recompressed without any adjustments made to the pulse compressor used for OPCPA compression. The resulting autocorrelation trace is also shown in Fig. 4 and is compatible with the set of requirements for a front end of a petawatt-type Nd:glass laser system, which is typically limited to ~400-fs pulses.

# 4 Conclusion

In conclusion, we have demonstrated for the first time multipass QPM OPCPA and QPM COPA in PPKTP pumped by a commercial pump laser that produced record pulse energies of 1.2 mJ, spectrally broadened from 6 to 12 nm and subsequently compressed to ~200 fs (aberration limit of the system). Recompression to shorter pulse durations should be possible with higher precision expansion and compression components. Simple modification of the demonstrated scheme allows time multiplexing for higher energy extraction in situations when pump pulse is much longer than the signal pulse. We have not observed any crystal degradation in ~48 h of continuous operation at full energy, but additional work is required to identify the preferred nonlinear material with greatest resistance to photorefraction and other possible damage and/or degradation processes.

**Acknowledgments.** This work was performed under the auspices of the U.S. Department of Energy by the University of California, Lawrence Livermore National Laboratory under Contract No. W-7405-Eng-48.

## References

1. A. Galvanauskas, A. Hariharan, D. Harter, M. A. Arbore, and M. M. Fejer: Opt. Lett. **23**, 210–212 (1998).
2. F. Rotermund, V. Petrov, F. Noack et al.: Electron. Lett. **38**, 561–563 (2002).
3. I. Jovanovic, J. R. Schmidt, and C. A. Ebbers: Appl. Phys. Lett. **83**, 4125–4127 (2003).
4. A. Fragemann, V. Pasiskevicius, G. Karlsson, and F. Laurell: Opt. Exp. **11**, 1297–1302 (2003).
5. I. Jovanovic, C. G. Brown, C. A. Ebbers, C. P. J. Barty, N. Forget, and C. Le Blanc: Opt. Lett. **30**, 1036–1038 (2005).
6. B. Wattellier, I. Jovanovic, and C. P. J. Barty: Cascaded-optical parametric amplification for extreme contrast enhancement. In: *Conference on Lasers and Electro-Optics*, Technical Digest. Optical Society of America, Washington, DC (2002) p. 74.
7. OPCPA front end systems for petawatt-class lasers are used or under development at the Rutherford Appleton Laboratory, Osaka University, University of Rochester, École Polytechnique, Sandia National Laboratory, Lawrence Livermore National Laboratory, and other institutions.

# Control of Amplified Optical Parametric Fluorescence in Hybrid Chirped-pulse Amplification

Y. Hama,[1,2] K. Kondo,[1,2] H. Maeda,[1,2] A. Zoubir,[1,2] R. Kodama,[1,2] K. A. Tanaka,[1,2] and K. Mima[1,2]

[1] Graduate School of Engineering, Osaka University, 2-1 Yamadaoka, Suita, 565-0871, Japan
[2] Institute of Laser Engineering, Osaka University, 2-6 Yamadaoka, Suita, 565-0871, Japan

The amplified optical parametric fluorescence (AOPF) from optical parametric chirped pulse amplifier (OPCPA) was controlled by injecting a residual fundamental pulse of the seeded Q-switch Nd:YAG pumping laser as a quenching beam. The output pulse from the OPCPA was used as a seed pulse for Ti:sapphire chirped-pulse amplifier (CPA) system to generate up to 10 TW peak power with a high-contrast ratio. This intense laser pulse was focused on the supersonic Ar gas jet. High-energy electrons up to 30 MeV were observed.

## 1 Introduction

Recent technologies in ultrashort high peak power laser have opened various new fields. The focused intensity of these high peak power laser pulses can now reach to $10^{19}\ \mathrm{W \cdot cm^{-2}}$. Such lasers are capable of high energy charged particle generation, $\gamma$-ray generation, and fast ignition physics in ICF study, etc. The direct interaction between relativistic high field and over-dense materials is especially attractive. Here the development of high-contrast and high peak power laser system is essential for avoiding a generation of low-density preformed plasmas. To obtain a high-contrast pulse, the prepulse should be suppressed to low level. At this point of view, optical parametric chirped pulse amplification (OPCPA) is an attractive method.

In the conventional energy storage type laser amplifier, after amplification the main pulse is sliced out from amplified pulse train with Pockels cell and polarizer. Therefore the contrast ratio is determined by the performance of the pulse slicer. The typical contrast ratio is $10^6$ at most. In OPCPA, however, only the main pulse can be amplified by using short enough pump pulse. In this case, the amplification factor becomes the contrast ratio. Moreover, noncollinear OPCPA enables the wideband amplification over 100 nm in bandwidth [1,2]. Therefore, we can expect the development of a prepulse free ultrashort high peak power laser sytem.

## 2 Three-Stage OPCPA

The seed source was a self-mode-locked Ti:sapphire oscillator (Femtolasers Femtosource), which produces 10-fs pulses, with an energy of 6 nJ, a center wavelength of 770 nm, and a bandwidth of 77 nm. The output pulses from this oscillator propagated through a 4-pass Offner-type stretcher, which stretches the pulses to 500 ps. The bandwidth of the output pulse from this stretcher was limited to 55 nm (FWHM) by the allowance. These stretched pulses were delivered to the preamplifier stage composed of a three-stage noncollinear OPCPA amplifier and a one-stage double-pass Ti:sapphire amplifier. The schematic of this preamplifier is shown in Fig. 1. We used a commercial seeded Q-switched Nd:YAG laser (Spectra-Physics LAB-150) as a pump and a quench for this preamplifier. The pump pulse was 230 mJ/6 ns at 0.532 μm in wavelength at a 10-Hz repetition rate. In order to obtain a good spatial beam quality, the near-field pump beam profile was imaged on each β-barium borate (BBO) crystal. The BBO crystals, 15 mm in length and 5 mm × 5 mm in cross section, were cut at 26.5° for type I angular phase matching. The external noncollinear angle between the pump and the signal was carefully aligned to be 4.5° to obtain a wideband phase matching. The output energy after this three-stage OPA was 3 mJ.

However, if we block the seed pulse before this preamplifier, some AOPF is observed. In Fig. 2, the temporal profiles of amplified pulses with and without seed pulse are represented by a thin solid curve and a thick dotted curve, respectively. The pulse width of AOPF is 2.1 ns (FWHM). This profile represents the temporal

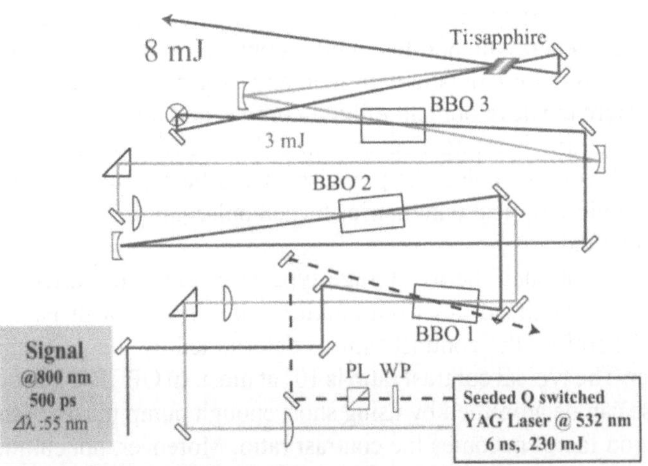


FIGURE 1. Schematic of the OPCPA/Ti:sapphire preamplifier in the hybrid system.

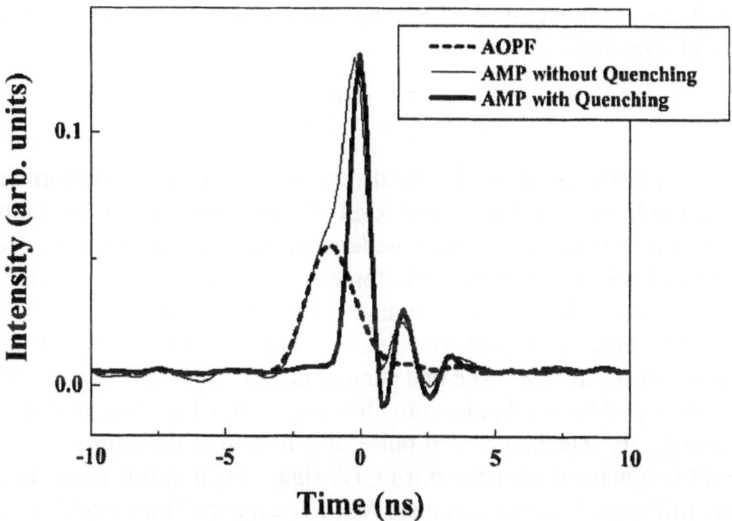


FIGURE 2. Temporal profiles of the amplified pulse at the OPCPA preamplifier stage with (thin solid line) and without (thick dotted line) a seed pulse. The thick solid line represents the temporal profile with quenching.

gain of OPA. The thin solid curve in Fig. 2 was obtained by injecting a seed pulse of 1.2 ns after the temporal peak of OPA. From this profile, we can clearly notice that there is some AOPF component preceding amplified main pulse (AMP) when the seed pulse is injected inadequately.

## 3 Control of AOPF

As is described in the first section, the prepulse arising from AOPF in OPA is a major problem for direct interaction experiments with a solid density material. In a study by Ross et al. [3], the level of AOPF in OPCPA is compared with that of amplified spontaneous emission (ASE) in CPA theoretically. The intensity of AOPF (IAOPF) can be estimated by the following equation:

$$I_{\mathrm{AOPF}} = \frac{h\nu G_0}{\sigma\tau}. \tag{1}$$

Here $h\nu$ is the photon energy, $G_0$ is the net small signal gain, $\sigma$ is the area of the diffraction limited spot, and $\tau$ is the coherence time of the wideband AOPF, which is given by

$$\tau = \frac{1}{\Delta\nu}, \tag{2}$$

where $\Delta\nu$ is the bandwidth of AOPF. Then,

$$\sigma = \frac{4(F\lambda)^2}{\pi}, \tag{3}$$

where $F$ is the $F$-number of the focusing optics and $\lambda$ is the center wavelength. Then Eq. (1) becomes

$$I_{\text{AOPF}} = \frac{\pi}{4F^2} \cdot \frac{h\nu \cdot \Delta\nu}{\lambda^2} \cdot G_0. \tag{4}$$

This intensity corresponds to the ideal case where OPCPA is performed with perfect spatial filtering and the initial level of OPF is assumed to be one photon ($h\nu$). From this estimation, it can be understood that there are two ways to keep the level of AOPF as low as possible for wideband OPA: (1) a decrease of $G_0$ while keeping enough output energy and (2) a tight spatial filtering. Generally, AOPF or ASE cannot be excluded easily since they have the same wavelength and the same timing as the AMP. The first attempt in reducing AOPF was to adjust the injection timing of the seed pulse into OPA amplifier, while maintaining enough output energy. By injecting a seed pulse of 1 ns before the gain peak of OPA, 3mJ could be obtained after the third OPA stage. Even in this case, the AOPF level was still large. In order to reduce AOPF to an even lower level, we actively reduced the gain of OPA by injecting a residual fundamental pulse of the seeded Q-switched YAG pumping laser as a quenching beam for OPF at the first stage OPA. The adequate noncollinear angle for 1.064 μm pulse was 0.2° greater than that for 0.8 μm in the crystal.

By analogy with injection locking of oscillators [4], the phase of the oscillating electric field in the crystal, polarized perpendicularly to that of the pump beam, is locked to that of the injected quenching beam, provided its intensity is higher than the intensity of AOPF. The quenching beam then experiences parametric amplification at 1.064 μm and begins to steal gain from the parametric amplification at 0.8 μm. According to Eq. (1), this decrease in $G_0$ also results in a decrease of the AOPF intensity at 0.8 μm. The quenching geometry is represented by dotted lines in Fig. 1. The power of the quenching beam was controlled by the combination of a polarizer and a half-wave plate, which are shown in Fig. 1. Figure 2 shows the temporal profile of AMP with quenching by a thick solid curve. The AOPF component shown in the temporal profile of the amplified seeded pulse when no quenching is applied completely vanishes in the presence of the quenching beam (see the linear scale plot of Fig. 2). Therefore, this method proves to be an efficient way of controlling AOPF in OPCPA systems.

## 4 10 TW OPCPA/Ti:sapphire Hybrid Laser System

As described in the previous section, in order to obtain enough energy with one pump source, a double-pass Ti:sapphire amplifier was inserted after the three-stage OPCPA to generate an output pulse of 8 mJ. This pulse is injected into a two-stage Ti:sapphire amplifier. The first stage is a triple-pass amplifier pumped with a total of 500 mJ from frequency-doubled Q-switched YAG laser. And the second stage, which is the final stage of the whole system, is also a triple-pass amplifier pumped with a total of 2.5 J. The output energy after the final amplifier

FIGURE 3. Output energy after the final Ti:sapphire amplifier in the hybrid system as a function of the quenching beam energy. Dots are with a seed, and triangles are without.

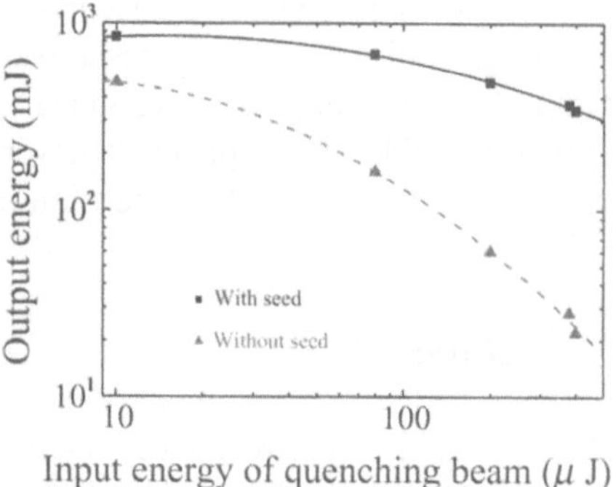


was over 1 J. We have measured the output energy while quenching the OPCPA. In Fig. 3, the output energy with and without a seed pulse was plotted as a function of the input energy of the quenching beam at the first OPA. By increasing the input energy of the quenching beam, the output energy decreases due to the decrease of the gain at the first OPA. However, the output energy with a seed decreases more slow than the output energy without a seed. The difference between the slopes of the curves in Fig. 3 is caused by the non-linear amplification in the final amplifier. At this stage, the amplification of AMP is in the saturation amplification region, although AOPF is in the small signal gain region with for a quenching power over 70 μJ. As a result, the contrast ratio between AMP and AOPF increases. The output pulse after the final amplifier is injected in the double-pass pulse compressor after expanding the effective beam size to 50 mm in diameter. The propagation efficiency from the final amplifier to the exit of the pulse compressor was measured to be 40% due to the small optics used in the expander and the diffraction efficiency of the gratings used in the pulse compressor. In Fig. 4, (a) the spectrum and (b) the auto-correlation trace of the final pulse are shown. The bandwidth is 33 nm FWHM without quenching. With quenching, although the bandwidth was almost the same, the peak was slightly shifted to the longer

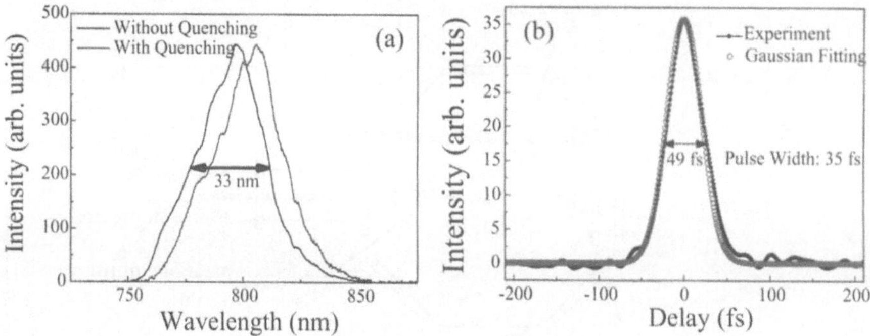


FIGURE 4. (a) Spectrum and (b) auto-correlation trace of the AMP after the pulse comoressor. The pulse width is estimated to be 35 fs FWHM.

wavelength side, which is considered to be caused by quenching with a longer wavelength pulse. From the measured auto-correlation trace in Fig. 4(b), the pulse width of the final output was estimated to be 35 fs FWHM, assuming a Gaussian temporal profile. The output peak power of this OPCPA/Ti:sapphire hybrid system was over 10 TW at a 10 Hz repetition rate.

## 5 Discussion

The contrast ratio between AMP itself and the amplified pulse trains before AMP was measured to be larger than $10^8$ with neutral density filters and a fast PIN photodiode detector. This high value can be easily understood from the principle of OPCPA. On the other hand, the contrast ratio between AMP and AOPF was lower, as shown in Fig. 3, and detectable AOPF was present after the three-stage OPA. However, AOPF cannot be compressed by the pulse compressor because of its incoherent phase. As a result, after the pulse compressor, the contrast ratio between the peak intensity of AMP and the instantaneous intensity of AOPF is estimated to increase by a factor equal to the compression ratio of $10^4$. If a seed is injected 1.2 ns after the gain peak of OPA, we obtained the temporal profile represented by the thin solid curve in Fig. 3. The apparent rising of AOPF can be observed, and is followed by the more rapid rising of AMP. In this case, the intensity of the AOPF is comparable to that of the AMP, and their contrast ratio is $10^4$ after the pulse compressor. However, by injecting a seed 1ns before the gain peak of the OPA, the instantaneous intensity level of AOPF was estimated to be lower than $10^{-2}$ of AMP before compression, with almost the same output energy. After compression, the contrast ratio was estimated to be $6 \times 10^6$ assuming a Gaussian temporal profile for AOPF. In this estimation, measured pulse widths were used. With quenching, the intensity ratio was improved 10 times, although the output energy decreased by half. From the same estimation, the contrast ratio between AMP and AOPF was increased to $5 \times 10^7$.

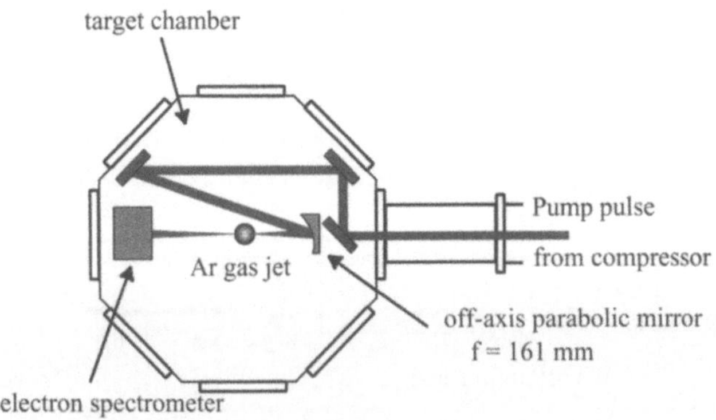


FIGURE 5. Experimental setup for enegetic electrons generation.

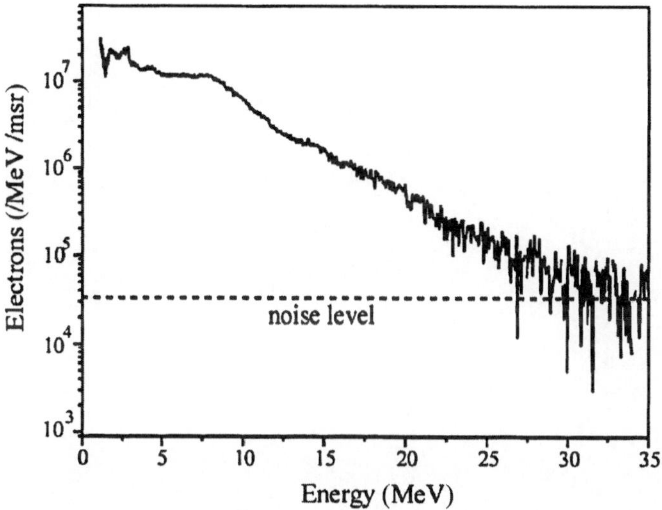


FIGURE 6. Electron energy spectrum for Ar gas jet.

## 6 Relativistic Electrons Generation

We have demonstrated a laser acceleration experiment with this laser system. Figure 5 shows experimental setup. The laser energy was 280 mJ. The propagation from the pulse compressor to the interaction chamber was in vacuum for avoiding a self-phase modulation in air. And this pulse was focused on Argon gas jet with an f/3.2 off axis parabolic mirror. The vacuum focal spot size was experimentally estimated to be 5 μm (FWHM). Then the intensity was evaluated to be higher than $2 \times 10^{19}$ W $\cdot$ cm$^{-2}$. The electrons were accelerated along the laser propagation axis. The energy spectrum was measured with the electron spectrometer (ESM). The magnetic field for dispersion was 0.5 T, which was produced by the permanent magnet. Each electron was bent to the corresponding direction. The electrons were detected by an imaging plate (Fuji BAS-SR2025)[5]. Figure 6 shows the energy spectrum which was obtained for the gas density of $4 \times 10^{18}$ cm$^{-3}$. We have observed energetic electrons up to 30 MeV.

## References


1. T. Kobayashi and A. Shirakawa: Appl. Phys. B 70, S239 (2000).
2. N. Ishii, L. Turi, V. S. Yakovlev, T. Fuji, F. Krausz, A. Baltuska, R. Butkus, G. Veitas, V. Smilgevicius, R. Danielins, and A. Piskarskas: Opt. Lett. 30, 567 (2005).
3. I. N. Ross, P. Matousek, M. Towrie, A. J. Langley, and J. L. Collier: Opt. Commun. 144, 125 (1997).
4. Siegman: Lasers, chapter 29.
5. K. A. Tanaka, T. Yabuuchi, T. Sato, R. Kodama, Y. Kitagawa, T. Takahashi, T. Ikeda, Y. Honda, and S. Okuda: Rev. Sci. Instrum. 76, 13507 (2005).

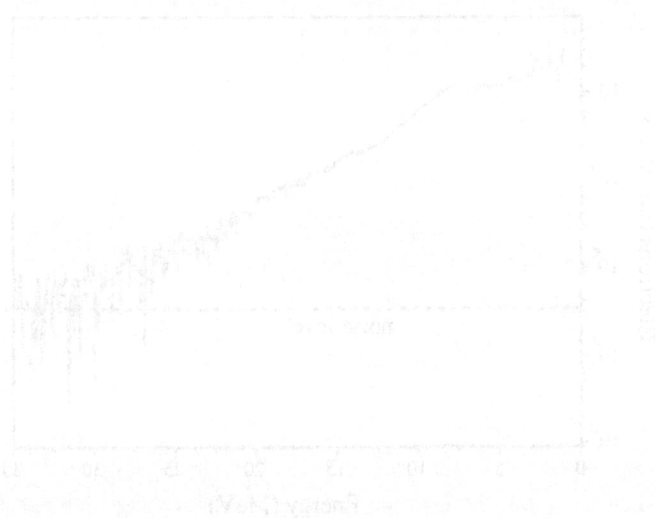




# Optical Synchronization for OPCPA Chains

C.Y. Teisset,[1] N. Ishii,[1] T. Fuji,[1] T. Metzger,[1] S. Köhler,[1]
R. Holzwarth,[1,2] A. Baltuška,[1] A.M. Zheltikov,[3] and F. Krausz[1,4]

[1] Max-Planck-Institut für Quantenoptik, Hans-Kopfermann-Straße 1, D-85748 Garching, Germany
catherine.teisset@mpq.mpg.de
[2] Menlo Systems GmbH, Am Klopferspitz 19, D-85152 Martinsried, Germany
[3] Physics Department, M.V. Lomonosov Moscow State University, Moscow 11992, Russia
[4] Ludwig-Maximilians-Universität München, Am Coulombwall 1, D-85748, Garching, Germany

**Summary.** We demonstrate two types of all-optical synchronization of two lasers based on dissimilar gain materials, that significantly simplify the scheme for optical parametric chirped-pulse amplification (OPCPA). Both methods lead to negligible timing jitter between the OPCPA seed and pump and eliminate one master oscillator and all synchronization electronics. In the first approach, a fraction of a broadband Ti:sapphire seed oscillator centered at 760 nm is frequency shifted in a photonic crystal fiber to enable synchronized seeding of a picosecond Nd:YAG pump laser. The seed radiation at 1064 nm is produced in the soliton regime which makes it inherently more intense and stable in comparison with other methods of frequency conversion. In the second approach, we further simplify the setup by employing direct optical seeding of a Nd:YLF amplifier from the NIR wing of our Ti:sapphire oscillator. Our work opens up the exciting possibility to use sub-picosecond pump pulse from highly efficient Yb-based amplifiers for jitterless parametric amplification of carrier-envelope phase stabilized pulses from Ti:sapphire oscillators.

**Key words:** optical parametric chirped pulse amplification, optical synchronization, photonic crystal fiber, ultrashort pulse

## 1 Introduction

Broadband, few-cycle, and frequently carrier-envelope phase (CEP-) controlled laser sources are urgently required in many applications of high-intensity and laser-field-sensitive physics [1]. Optical parametric chirped pulse amplification (OPCPA) [2, 3] offers a particularly promising route toward compact ultrashort ultrahigh-peak-power laser systems because of its very broad gain bandwidth, negligible thermal load on a nonlinear crystal, and an extremely high single-pass gain as compared to amplifiers based on laser gain media. High-intensity and high-energy sub-ps OPCPA pulses as well as sub-100-J few-cycle pulses have been demonstrated to date [4–7]. Recently, 10-fs TW-class multi-millijoule OPCPA systems were reported [8,9]. Unlike gain media with inversion storage, the parametric amplifier poses a severe challenge of ensuring strict synchronization between the pump and seed pulses. In this paper, we aim at finding a reliable and technologically attractive solution for passive synchronization between a picosecond laser amplifier and a femtosecond OPCPA seed oscillator that operates outside the gain region of the pump laser.

## 2 Known Pump-Seed Synchronization Technique

The problem of optical synchronization is easily solved in superfluorescence-seeded [10] and white-light-continuum-seeded parametric amplifiers [11–14], where the color-shifted seed pulse is derived directly from the pump pulse or from its (sub-) harmonics. In superfluorescence-seeded amplifiers, the seed originates from quantum noise. Consequently, the duration of the amplified pulse is near to the pump duration and the phase of the pulse may experience shot-to-shot variations. By contrast, amplifiers seeded with coherent white light are capable of reducing the duration of the amplified pulse down to several optical cycles, i.e., by orders of magnitude below the pump pulse duration [12, 13]. Nevertheless, these devices require short, typically sub-200-fs intense pulses for reliable white-light generation. Therefore, white-light-seeded few-cycle amplifiers are driven exclusively by chirped-pulse amplifiers (CPA) predominantly based on Ti:sapphire [11–14] and, more recently, on Yb-doped laser crystals [15]. Dependence on ultrashort pump lasers severely limits the practical power scalability of both superfluorescence- and white-light-seeded amplifiers. Superfluorescent devices are unsuitable for CEP-controlled amplification, whereas white-light-seeded amplifiers may exhibit CEP self-stabilization under special conditions [16, 17].

A straightforward solution to CEP-controlled parametric amplification, similar to the case of its laser-amplifier counterpart [18], is to use a prestabilized seed source [19, 20]. This idea was recently implemented by Hauri et al. [6], who demonstrated CEP stability preservation in an OPCPA seeded with a CEP-locked Ti:oscillator and pumped by the second harmonic of a Ti:sapphire amplifier that was also seeded by the same oscillator. While offering a very elegant solution to the problems of CEP control and optical synchronization, this system, nonetheless, has the power upscale constraints mentioned above. Scaling up the output of the CPA, Ti:sapphire pump laser becomes particularly unattractive because this laser in turn requires an upgradation of its own pump source—a green Q-switched ns laser.

In high-power OPCPAs, because of the heat disposal and pumping efficiency reasons, it is more economical to employ a stand-alone (multi-)kHz picosecond pump laser that is optimized for production of high-energy narrow-band pulses. This approach was used in Refs. [7–9], where a broadband Ti:sapphire oscillator, employed as OPCPA seed, was synchronized with the master oscillator of a picosecond pump laser based on Nd:YAG using a complex electronic servo loop [21]. State-of-the-art active synchronization loops [7], based on price GHz electronics, make it possible to keep the rms timing jitter between the seed and pump pulses on a sub-picosecond level. Whereas this locking precision satisfies well the demands of OPCPA pumped by Nd-based picosecond lasers (pulse duration range 20-100 ps), it is insufficient for use with subpicosecond Yb-based pump lasers.

As prospective OPCPA pump sources, Yb lasers [22] promise to be vastly superior to their Nd counterparts because of the shorter pulse duration, reduced parasitic heat, higher output power, and suitability for CPA operation. Using pump pulses in a 1-ps range, one can realize ultrabroadband saturated parametric gain in nonlinear crystals as thin as 1 mm [23] without running into problems with the B integral of the parametric amplifier. Shaped pulses with a bandwidth of several nanometers from Yb-based lasers are also advantageous in terms of maximizing OPCPA conversion efficiency, reducing superfluorescence background and avoiding gain narrowing. The desired near-rectangular pump pulse profile can be attained by temporal stretching and pulse wing shaping of the broadband pump pulse. The use of such chirped multi-ps pump pulses, however, will not lower the demand for femtosecond timing precision, because the latter is dictated by the steepness of the pump pulse edges. In this work, we set out to reduce the complexity of the laser and electronic setups in present-day OPCPAs based on a broadband seed oscillator and an external pump laser [7–9] and pave the way for the use of substantially shorter pump pulses in the near future. Below we survey different opportunities for passive optical seed-pump pulse synchronization in OPCPA and explain our chosen strategy.

## 3 Injection Seeding of the Pump Laser

We first consider two-color passively synchronized sources, the motivation behind this approach is the need to provide strong seed signals that are sufficiently above the amplified spontaneous emission (ASE) level in both the OPCPA and the injection-type pump laser. One possibility of realizing a two-color scheme is to employ self-stabilization of two mode-locked oscillators lasing at the seed and pump wavelengths, respectively, via a nonlinear optical (Kerr-lens) mechanism [24]. In terms of complexity, however, this method is not easier to implement than the active synchronization of two independent mode-locked oscillators. Another option to consider is a parametric or Raman type frequency converter that creates a narrowband spectral component at the wavelength of the picosecond amplifier. A suitably intense frequency-shifted pulse can be generated in a synchronously pumped optical parametric oscillator (OPO) driven by the broadband seed laser oscillator. Unfortunately, the OPO itself requires active cavity

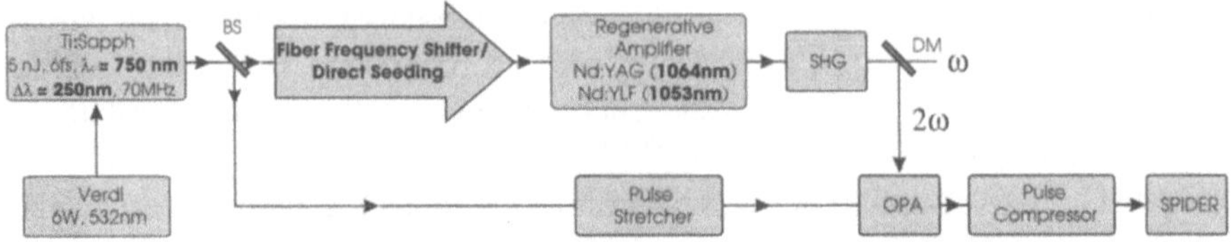


FIGURE 1. Injection seeding with and without PCF for all-optically synchronized OPCPA; BS, 50% beamsplitter; DM, dichroic mirror; SHG, second harmonic generation.

stabilization. In a cavity less alternative, a very weak frequency-shifted pulse was produced in a traveling wave superfluorescent parametric generator [25]. A separate two-stage cw-pumped parametric amplifier was used to boost this signal to the level suitable for seeding a laser amplifier.

In this work, we propose two straightforward solutions that combine the simplicity of passively synchronized OPCPA schemes using a single seed oscillator, with the flexibility and scalability of the actively synchronized two-source approach for optimizing independently the pump and seed lasers. Both solutions are illustrated in Fig. 1.

## 3.1 Frequency-Shifted Seeding

The first approach consists of a broadband Ti:Sapphire oscillator for seeding an OPCPA stage and of a frequency shifter for seeding a pulsed pump laser. In our setup, schematically presented in Fig. 1, the output of a 6-fs 5-nJ 70-MHz chirped-mirror Ti:Sapphire oscillator [26] is divided into two equal parts. One part is directed into an OPCPA stage and the other is frequency converted to obtain the seed for a picosecond Nd:YAG regenerative amplifier.

In the case of our unamplified oscillator pulses, parametric frequency shifters [25] become extremely inefficient because of the low pump-field intensity and group-velocity matching issues. Under these circumstances, soliton phenomena in optical fibers suggest an interesting alternative to standard strategies of frequency conversion. Optical solitons propagating in media with a nonlinear response experience reshaping and continuous frequency downshifting due to the Raman effect [27]. This phenomenon, called soliton self-frequency shift (SSFS) [28, 29], provides a convenient way of generating ultrashort pulses with a tunable carrier frequency. Photonic-crystal fibers (PCFs) [30] substantially enhance this nonlinear-optical process [31] due to a strong field confinement in a small-size fiber core and the possibility to tailor dispersion of guided modes by varying the fiber structure [32].

The key advantages of the SSFS-based strategy of frequency shifting originate from the intrinsic properties of Raman-shifted solitons [33]. In particular, the central frequency of the redshifted soliton can be tuned by varying the fiber length and the input pulse energy. Radiation energy carried by this signal is localized in the time domain within a short spike, dominating the temporal envelope of

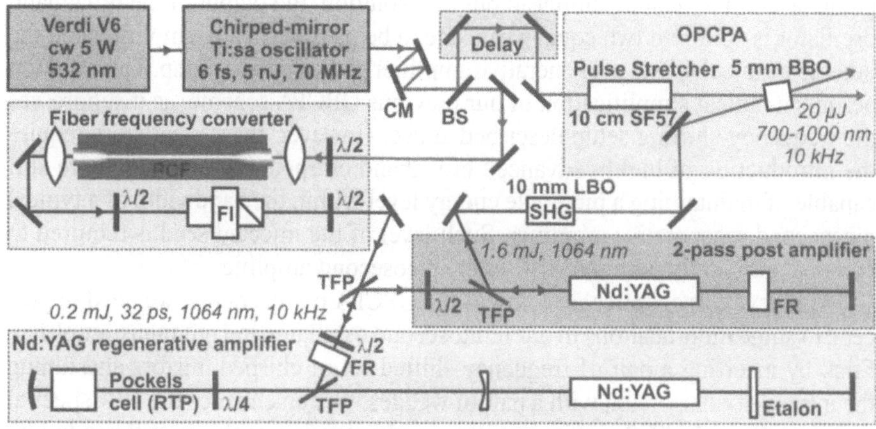


FIGURE 2. Layout of soliton-based synchronized OPCPA. PCF, photonic crystal fiber; TPF, thin-film polarizer; CM, chirped mirrors; BS, 50% beamsplitter; λ/2, half-wave plate; λ/4, quarter-wave plate; Nd:YAG, side-CW-pumped gain modules; FI, Faraday isolatorM; FR, Faraday rotator; DM, dichroic mirror; SHG, second harmonic generation.

radiation intensity in the fiber. Because of the anomalous group-velocity dispersion of the PCF, the red-shifted soliton becomes delayed and eventually isolated, both spectrally and temporally, with respect to the rest of the pump field. This isolation of the frequency-shifted soliton suppresses the interference between the solitonic part and the rest of the spectrum of radiation field. For a PCF with a characteristic length of tens of centimeters, the amplitude of the soliton spike in the temporal envelope of radiation intensity is typically an order of magnitude higher than the intensity of the remainder of the SPM-broadened pump field and Cherenkov-type emission, both of which are spread out in time. As a result, the frequency-shifted solitonic component is observed in experiments as the most stable part of the output spectrum [33], which is free of interference fringes, typical of the nonsolitonic part of the radiation field, including Cherenkov emission.

In the context of laser amplifier seeding, the combined spectrotemporal localization of the SSFS pulse makes it ideally suited for suppressing the ASE inside the laser cavity. In addition, the high amplitude stability of the SSFS pulse is very important for maintaining stability of the amplifier output at repetition rates that are well above the inverse population relaxation time of the laser gain medium. At such repetition rates, although the average output power can be saturated, the energy of each amplified pulse cannot be saturated individually. Therefore, stable and intense seed pulses are required.

## 3.2 Direct Seeding

Our second approach to all-optical synchronization relies on an octave-spanning Ti:Sapphire [34,35] oscillator, which is used for both seeding the OPCPA and its pump laser. In a manner similar to our first solution, the output of the broadband oscillator is split into two equal parts, one to be parametrically amplified and the other to be injected into a regenerative amplifier and serve as a pump. This solution permits a radical simplification of our previous OPCPA scheme, as it suppresses the frequency-shifting setup described above. However, these oscillators require the introduction of highly advanced broadband chirped mirrors, which are still capable of maintaining a picojoule energy level within the bandwidth of a typical picosecond regenerative amplifier. Such level of the injected seed is required to compete efficiently with the ASE in the picosecond amplifier.

To achieve direct optical seeding of the OPCPA pump source, we need to proceed to some modifications in our femtosecond oscillator and our pump amplifier. First, by inserting a pair of frequency-shifted set of chirped mirrors and tuning the intracavity dispersion with a pair of wedges, we can enhance the NIR spectral components of the Ti:Sapphire oscillator without reducing the overall bandwidth (Fig. 5a, solid curve) and the output power. Second, we build up a Nd:YLF regenerative amplifier, since it has a slightly blue-shifted gain material compared to that of Nd:YAG (Nd:YLF, $\lambda = 1053$ nm; Nd:YAG, $\lambda = 1064$ nm). Through these corrections in our OPCPA design, we can increase the spectral overlap between the pump and the seed lasers and provide the necessary picojoule seed to the regenerative amplifier.

## 4 Optically Synchronized OPCPA

We have investigated several commercial and research types of PCF that are capable of supporting the SSFS regime and selected the commercial fiber NL-PM-750 (Crystal Fibre A/S) for a clear solitonic feature in the NIR. The output spectrum of a 20-cm-long PCF injected with 2-nJ oscillator pulses is presented in Fig. 3.

The oscillator light was injected with $anf = 7.5$ mm aspheric lens. The central wavelength of the spectral soliton was fine-tuned by adjusting the waveplate in front of the PCF and optimizing the focusing. To improve mechanical stability of the PCF and avoid problems with optical damage, a monolithic fiber assembly was prepared by Menlo Systems GmbH. Larger-core-diameter end caps were spliced to the PCF and the end cap on the in-coupling side was wedge polished to avoid retroreflection into the Ti:Sapphire oscillator. The fiber assembly was fixed directly on the oscillator breadboard whereas the focusing and collimating lenses were mounded on xyz-translators. Mounting the fiber assembly in this way permits alignment-free operation for several weeks. A small adjustment of the in-coupling objective is only necessary upon the realignment of the oscillator.

The output of the PCF is directed into a homemade Nd:YAG regenerative amplifier (Fig. 1). To protect the fiber from a back reflection or leakage through the out-coupling polarizer of the regenerative amplifier, a Faraday isolator was introduced after the PCF. The estimated seed pulse energy contained within the

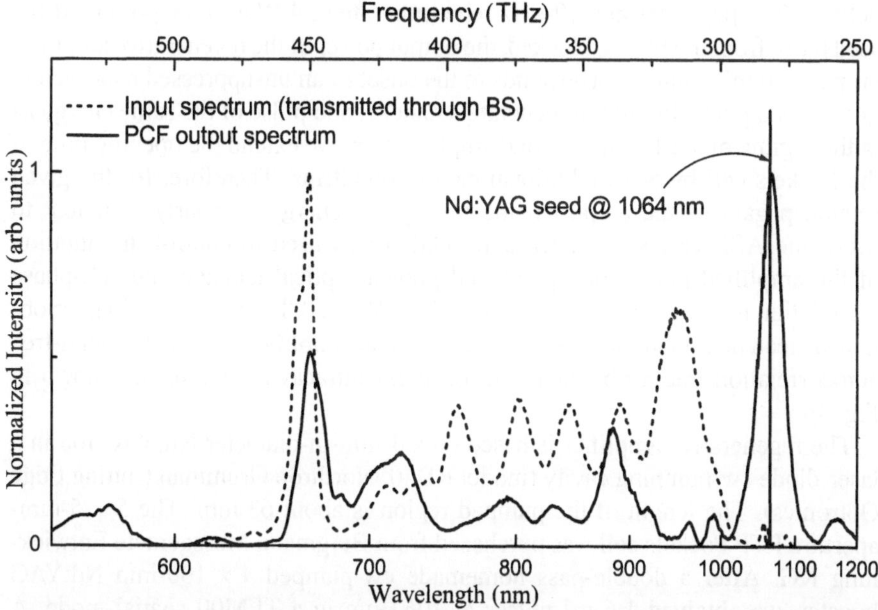


FIGURE 3. Frequency shift in the photonic crystal fiber. Solid curve shows an output spectrum optimized for injection seeding at 1064 nm. The input spectrum from the Ti:Sapphire oscillator is shown as dashed curved.

gain bandwidth of Nd:YAG is ~2 pJ, which proved to be sufficient to obtain clean pulse amplification and suppress the nanosecond Q-switched-pulse background.

This is evidenced by Fig. 4**a**, which shows the intracavity pulse train corresponding to the buildup of the amplified pulse in each cavity roundtrip. The regenerative amplifier operates in the regime of average power saturation and

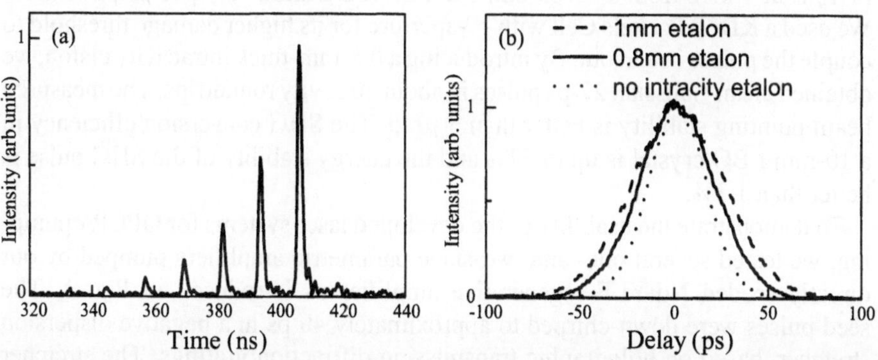


FIGURE 4. Performance of the regenerative amplifier seeded with an optical sotitonic pulse from a PCF: (**a**) intracavity pulse train of pulses amplified to the energy of 0.2 mJ; (**b**) background free autocorrelation with and without intracavity etalon for a fixed number of intracavity roundtrips.

delivers 240-μJ pulses at a 10-kHz repetition rate (2.4 W average power). If the seed pulse from the PCF is blocked, the output power of the regenerative amplifier drops to 50 mW, which corresponds to the onset of an unsuppressed nanosecond Q-switched pulse. In order to generate a nanosecond pulse of the same energy as in the regime of seeded picosecond amplification, we extend the opening time of the Pockels cell by 5–10 additional cavity roundtrips. Therefore, for the given roundtrip gain of the amplifier, the seed pulse energy is clearly sufficient to overcome ASE issues. An intracavity etalon was used to control the duration of the amplified picosecond pulse and prevent optical damage and self-phase modulation inside the regenerative amplifier. With a 0.8-mm-thick etalon, we obtained clean near-Gaussian 33-ps pulses from the amplifier. The background-free autocorrelation traces obtained with different intracavity etalons are shown in Fig. 4b.

The regenerative amplifier is based on a 4-mm-in-diameter Nd:YAG rod in a laser-diode cw-pumping cavity (model RD40, Northrop Grumman Cutting Edge Optronics). The length of the pumped region is about 63 mm. The 5 × 5-mm-aperture RTP Pockels cell was purchased from Bergmann Messgeraete Entwicklung KG. After a double-pass homemade cw-pumped 4 × 100-mm Nd:YAG booster, we obtained 1.6-mJ pulses at 1064 nm in a TEM00 spatial mode. A Faraday rotator was installed between the passes to compensate thermally induced birefringence of the YAG crystal and enable polarization out-coupling. 0.8-mJ pulses at 532 nm were obtained in a 10-mm-long critically phase-matched LBO crystal. The achieved 50% frequency doubling efficiency is an additional proof of the clean seeding and amplification process in our picosecond amplifier chain.

In the case of the directly seeded Nd:YLF amplifier, the energy of the oscillator pulses confined within the gain linewidth at 1053 nm was about 1 pJ. The seed energy was still sufficient to overcome the level of spontaneous emission in this lasing medium which has a lower gain cross section than Nd:YAG. The regenerative amplifier, based on the same amplifier head as that of Nd:YAG, delivers up to 4 mJ at 1 kHz from a 3×63 mm Nd:YLF rod without any postamplification. We used a KD*P Pockels Cell with 1”-aperture for its higher damage threshold to couple the pulses in and out. By introducing a 0.8-mm-thick intracavity etalon, we obtained clean Gaussian 27-ps pulses in about 40 cavity roundtrips. The measured beam pointing stability is better than 3 μrad. The SHG conversion efficiency in a 10-mm LBO crystal is up to 57% and the energy stability of the SHG pulse is better than 1.3%.

To demonstrate the usability of the developed laser systems for OPCPA pumping, we tested several one- and two-stage parametric amplifiers pumped by our directly seeded Nd:YLF regenerative amplifier as introduced in Fig. 1. The seed pulses were down-chirped to approximately 40 ps in a negative dispersion stretcher, based on holographic transmission diffraction gratings. The stretcher overchirped the seed pulses to compensate for the positive dispersion in the programmable acousto-optic filter (DAZZLER, Fastlite) and obtain an adequate temporal overlap between the pump (27 ps) and seed pulses (25 ps) at the OPA crystal. With the available SHG power of the Nd:YLF amplifier, we were able to generate

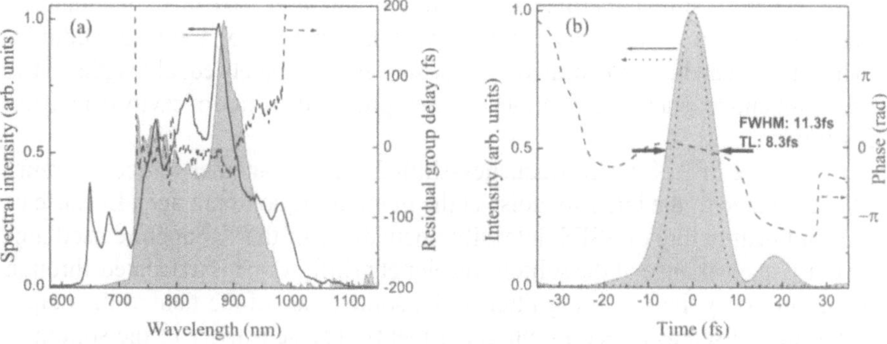


FIGURE 5. Results of chirped pulse amplification and recompression: (a) Ti:Sapphire oscillator spectrum (solid curve), spectrum of the amplified pulse (shaded contour), and its group delay measured with SPIDER (dashed curve); (b) recompressed pulse (shaded contour), phase (dashed curve), and the transformed-limited pulse (dotted curve).

broadband amplified signal pulses with energies up to 50 μJ for a single-stage OPA and 150 μJ in two stages in 4-5-mm-thick non-collinearly pumped BBO crystals. The two BBO crystals in the two-stage configuration were first pumped by 300 mW and then by 1.2 W to minimize the bandwidth narrowing through amplification. The spectrum and the group delay of a compressed 11.4-fs pulse, obtained in a saturated single-stage OPA, is presented in Fig. 5a and the corresponding pulse parameters are given in Fig. 5b. Remarkably, the performance of the broadband OPA shown in Fig. 5 is reliably reproduced each working day without any need to adjust the pump-seed delay, which has been previously impossible in the actively stabilized OPCPA setup [8].

## 5 Timing Jitter

The evaluated residual timing jitter between the seed and the pump pulses lies in the sub-30-fs domain and could not be measured by optical cross-correlation techniques available to us. The above mentioned results of parametric amplification, a type of cross-correlation measurement, reveal no traceable jitter. Below we discuss possible sources of residual timing jitter.

The most important contribution to the residual jitter in our system results from the cavity length drifts of the Ti:sapphire oscillator and of the regenerative amplifier. The short-term drift caused by mechanical instability can be disregarded, because the time span between the pulse injection into the regenerative amplifier and the Ti:Sapphire pulse injection into the OPCPA stage is merely 300–400 ns, well below the period of conceivable vibrations in a laboratory. The long-term drift is related to the thermal expansion of the cavities of the regenerative amplifier and the seed oscillator and depends on the ratio of the cavity lengths and the number of cavity roundtrips from the moment of seeding of the regenerative amplifier. With simple thermal stabilization of the laser base plate, it should be possible to control the cavity length to within a fraction of 1 μm. In addition, thermal drift can be compensated by choosing approximately equal lengths of the oscillator and regenerative amplifier cavities, provided the rate of cavity expansion is similar.

In the case of PCF-based frequency-shifted seeding, another source of timing jitter is related to the intensity noise at the input of the solitonic fiber. Because of the nonlinear nature of SSFS, intensity fluctuations of the Ti:Sapphire oscillator give rise to a variance of the soliton wavelength shift, which is translated, through the group delay, into a timing jitter of the soliton part of the field at the output of the fiber, used as a seed in our experiments. The sensitivity of the soliton to the variations of the input intensity was evaluated using a numerical model of SSFS in PCF [33] which uses the generalized nonlinear Schrödinger equation, including high-order dispersion effects and the Raman response of fused silica. According to these simulations, a 1% variation in the input pulse energy gives rise to additional spectral, $\delta\lambda$, and temporal, $\delta t$, shifts of the soliton estimated as $\delta t/\Delta t \simeq \delta\lambda/\Delta\lambda \simeq 0.7\%$, where $\Delta t$ and $\Delta\lambda$ are the time and wavelength Raman-induced shifts of the soliton in the absence of input energy variations. For the PCF length used in our experiments, $\Delta t$ is ~5 ps, which corresponds to $\delta t \simeq 30$ fs. Potentially, this timing jitter can be reduced through the optimization of the PCF dispersion profile. In addition, it must be pointed out that our jitter estimation is rather conservative because the pulse-energy noise of a Ti:Sapphire oscillator can be below 0.2% rms [36].

## 6 Conclusion

In conclusion, we have achieved a significant improvement in OPCPA synchronization as well as a substantial reduction of the complexity and cost of the entire system. Our frequency-shifting method opens the way to combine ubiquitous Nd-based picosecond amplifiers and emerging 1-ps-range Yb-based amplifiers with widely available several-nJ broadband Ti:sapphire seed oscillators, when no spectral overlap with the pump is achievable. Our simple all-optical seed-pump pulse synchronizations permit hassle-free parametric amplification of ultrashort light pulses and can be straightforwardly scaled according to the energy of the amplified pump pulse.

## References

1. T. Brabec and F. Krausz: Rev. Mod. Phys **72**, 545–591 (2000).
2. A. Dubietis, G. Jonušauskas, and A. Piskarskas: Opt. Commun. **88**, 437–440 (1992).
3. I. N. Ross, P. Matousek, M. Towrie, A. J. Langley, and J. L. Collier: Opt. Commun. **144**, 125–133 (1997).
4. X. Yang, Z. Xu, Y. Leng, et al.: Opt. Lett. **27**, 1135 (2002).

5. R. Butkus, R. Danielius, R. Dubietis, A. Piskarskas, and A. Stabinis: Appl. Phys. B **79**, 693–700 (2004).
6. C. P. Hauri, P. Schlup, G. Arisholm, J. Biegert, and U. Keller: Opt. Lett. **29**, 1369 (2004).
7. R. T. Zinkstok, S. Witte, W. Hogervorst, and K. S. E. Eikema: Opt. Lett. **30**, 78 (2004).
8. N. Ishii, L. Turi, V. S. Yakovlev, et al.: Opt. Lett. **30**, 567–569 (2005).
9. S. Witte, R. T. Zinkstok, W. Hogervorst, and K. S. E. Eikema: Opt. Exp. **13**, 4903 (2005).
10. G. Banfi, P. Di Trapani, R. Danielius, A. Piskarskas, R. Righini, and I. Santa: Opt. Lett. **18**, 1547–1579 (1993).
11. T. Sosnowski, P. B. Stephens, and T. B. Norris: Opt. Lett. **21**, 140–142 (1996).
12. G. Cerullo, M. Nisoli, S. Stagira, and S. De Silvestri: Opt. Lett. **23**, 1283 (1998).
13. A. Shirakawa, I. Sakane, M. Takasaka, and T. Kobayashi: Appl. Phys. Lett. **74**, 2268 (1999).
14. E. Riedle, M. Beutter, S. Lochbrunner, et al.: Appl. Phys. B. **71**, 457–465 (2000).
15. Fully integrated broadly tunable femtosecond Ytterbium system. Available at: www.lightcon.com.
16. A. Baltuǔska, T. Fuji, and T. Kobayashi: Phys. Rev. Lett. **88**, 133901 (2002).
17. C. Manzoni, G. Cerullo, and S. De Silvestri: Opt. Lett. **29**, 2668–2670 (2004).
18. A. Baltuška, T. Udem, M. Uiberacker, et al.: Nature **412**, 611–615 (2003).
19. D. J. Jones, S. A. Diddams, J. K. Ranka, et al.: Science **288**, 635–639 (2000).
20. A. Apolonski, A. Poppe, G. Tempea, et al.: Phys. Rev. Lett. **85**, 740–743 (2000).
21. M. J. W. Rodwell, D. M. Bloom, and K. J. Weingarten: IEEE J. Quantum Electron. **25**, 817 (1989).
22. W. F. Krupke: IEEE J. Sel. Topics Quantum Electron. **6**, 1287–1296 (2000).
23. A. Baltuška, T. Fuji, and T. Kobayashi, Opt. Lett. **27**, 306–308 (2002).
24. Z. Y. Wei, Y. Kobayashi, Z. G. Zhang, and K. Torizuka: Opt. Lett. **26**, 1806–1808 (2001).
25. H. Zheng, J. Wu, H. Xu, K. Wu, and E. Wu: Appl. Phys. B **79**, 837–839 (2004).
26. T. Fuji, A. Unterhuber, V. S. Yakovlev, et al.: Appl. Phys. B **77**, 125 (2003).
27. G. P. Agrawal: *Nonlinear Fiber Optics*. San Diego: Academic Press (2001).
28. F. M. Mitschke and L. F. Mollenauer: Opt. Lett. **11**, 659–661 (1986).
29. E. M. Dianov, A. Y. Karasik, P. V. Mamyshev, et al.: JETP Lett. **41**, 294–297 (1985).
30. Russell: Science **299**, 358–362 (2003).
31. X. Liu, C. Xu, W. H. Knox, et al.: Opt. Lett. **26**, 358–360 (2001).
32. W. H. Reeves, D. V. Skryabin, F. Biancalana, et al.: Nature **424**, 511–515 (2003).
33. E. E. Serebryannikov, A. M. Zheltikov, N. Ishii, et al.: Appl. Phys. B **81**, 579–584. (2005).
34. T. M. Fortier, D. J. Jones, and S. T. Cundiff: Opt. Lett. **28**, 2198–2200 (2003).
35. O. D. Mücke, R. Ell, A. Winter, et al.: Opt. Exp. **13**, 5163 (2005).
36. A. Poppe, L. Xu, F. Krausz, and C. Spielmann: IEEE J. Sel. Topics Quantum Electron. **4**, 179–184 (1998).

# OPCPA Systems for the Amplification of Ultrashort Pulses Up to Millijoules Level

A. Renault,[1,2] F. Augé-Rochereau,[1] R. Lopez-Martens,[1] and G. Chériaux[1]

[1] Laboratoire d'Optique Appliquée ENSTA, École Polytechnique, CNRS UMR 7639, 91761 Palaiseau cedex, France
[2] Amplitude Technologies, 2 rue du Bois Chaland, CE 2926, 91029 Evry cedex, France
**amandine.renault@ensta.fr**

**Abstract:** We present a quasicollinear near-degenerate optical parametric chirped pulse amplifier pumped at 1 kHz. Twice the initial spectral bandwidth and a single-pass gain of $2.6 \times 10^4$ were obtained after two stages of parametric amplification.

## 1 Introduction

The most recent developments in high field physics have demonstrated the necessity for ultrashort and ultraintense Ti:sapphire lasers to possess higher and higher specifications. In particular, to drive electronic processes in atoms and plasmas on the time scale of the optical cycle, researchers need to use terawatt-scale amplifiers providing 1 kHz sub-10-fs laser pulses, which are not available with conventional amplification techniques. Optical parametric chirped pulse amplification (OPCPA) is a good alternative to conventional Ti:sapphire amplifiers since it deals with extremely low prepulses, contrast preservation, and offers the possibility to amplify ultrashort pulses [1]. Indeed, thanks to a large gain bandwidth, and since there is only one passage through the nonlinear medium, there is no spectral gain narrowing. This is also a good way to reach the attosecond domain, thanks to the preservation of the carrier-envelop phase offset during the parametric process [2].

## 2 Ongoing Setup

We are currently investigating a degenerate quasi-collinear OPCPA configuration, for which the beam to amplify and the pump beam are produced by the same

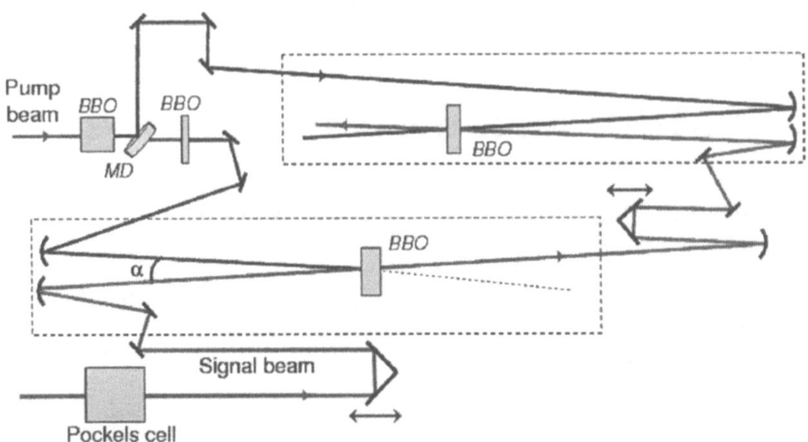


FIGURE 1. Design of the OPCPA setup with both a preamplifier (bottom) and a power amplifier (top).

source, at a kHz repetition rate. The 60-nm pulses generated by the Ti:sapphire oscillator at the central wavelength of 810 nm are split in two parts by a beam-splitter. Half of the beam is stretched to a few hundreds of femtoseconds and used as a seed for the OPCPA, while the half left seeds a CPA laser system and is then frequency doubled to pump the OPCPA. Inside a CPA system, the spectrum is cut to enhance the temporal duration of the pump, between 200 and 400 femtoseconds. The OPCPA consists of two BBO crystals, which are cut at 29.2° for type I phase matching (see Fig. 1). The first 3-mm long BBO crystal is used as a high-gain preamplifier, while the second (2-mm long) is used as a power-amplifier.

The pump of the power-amplifier is obtained by frequency doubling of the CPA output through a 5-mm long BBO crystal. The red part of the beam not converted in this crystal is frequency doubled in a 1-mm long BBO crystal so as to generate the blue pump for the preamplifier.

## 2.1 *Preamplifier Only*

We have first been working the preamplifier alone, below saturation and at gain saturation. The energy of the pump was 400 μJ at 400 nm. Measurement of the amplified signal spectrum shows that the spectral bandwidth is always preserved when we are working below saturation, and can even be broadened when we are working at saturation (see Fig. 2). The best results we obtained gave us a broadened spectrum of 120 nm, double the width of the spectrum produced by the oscillator.

The amplified signal has an energy of 2.5 μJ after this optical parametric preamplifer, which corresponds to a gain of $2.5 \times 10^3$.

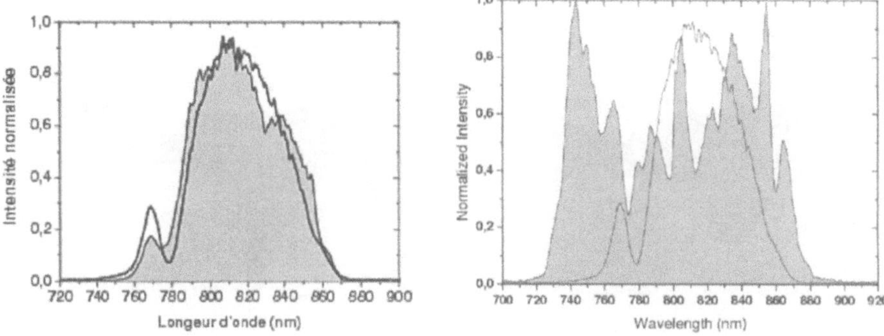


FIGURE 2. Spectrum of the amplified beam after the preamplifier, below saturation (left), and at saturation (right).

## *2.2 Both Preamplifier and Power Amplifier*

Next we have been working both preamplifier and power-amplifier below saturation (110 and 700 μJ of blue pump, respectively). The spectral bandwidth is lightly broadened during amplification, from 44 to 63 nm (see Fig. 3). The amplified signal energy was 1 μJ after the preamplifier and 26 μJ after the power-amplifier, which corresponds to a single-pass gain of $2.6 \times 10^4$. Spatial phase and spatial profile measurements with a Shack-Hartmann analyzer and a camera show that there is no significant spatial distortion during amplification (see Fig. 4).

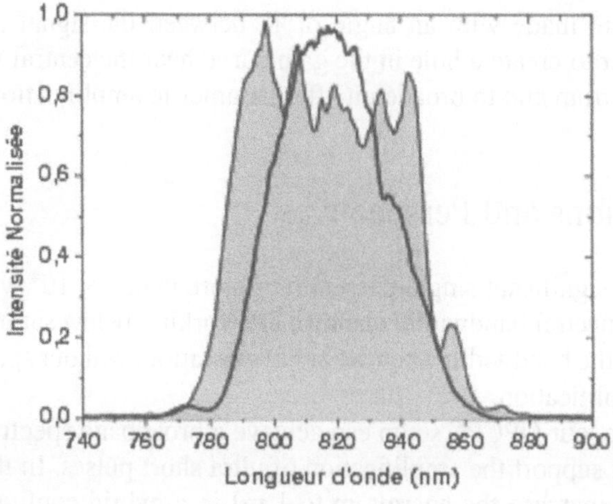


FIGURE 3. Spectrum of the amplified beam after both preamplifier and power amplifier, below saturation.

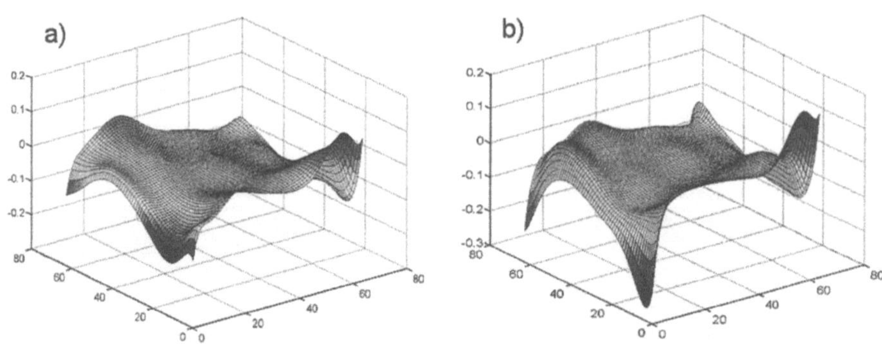


FIGURE 4. Shack–Hartmann measurement of the spatial phase after parametric amplification.

## 3 Theoretical Calculations of the Amplified Spectrum

We made calculations for the amplified spectrum when we are working on the OPA below saturation. Using the expression of the gain G (see Eq. (1)), and multiplying it with the experimental oscillator spectrum, we obtain the theoretical spectrum for the amplified beam. We observe a spectral broadening during the amplification process, and the experimental spectrum matches pretty well with the calculated spectrum (see Fig. 5).

$$G = 1 + \frac{{\gamma_o}^2}{\gamma^2} \sinh^2(\gamma z) \tag{1}$$

Here $\gamma$ is the gain coefficient and $z$ is the length of the nonlinear crystal. Calculations are made with an angle of 1° between the signal and the pump beams, in order to create a hole in the gain curve near the central wavelength of the signal spectrum and to broaden it after parametric amplification.

## 4 Conclusions and Perspectives

We obtained a significant single-pass gain of more than $2 \times 10^4$ with the preservation of the spectral bandwidth when we are working below saturation, and the broadening of the bandwidth when we are at saturation, without spatial distortion during the amplification.

Furthermore, our OPCPA setup can achieve a broadband spectral gain (>200 nm) and could support the amplification of ultra short pulses. In the next experiment we will increase the energy up to 1 mJ in a hybrid configuration with a single pass Ti:sapphire amplifier [3]. Furthermore, a stabilized phase oscillator will allow us to reach the few optical cycles regime, thanks to the preservation of the carrier-envelop phase offset in the parametric process.

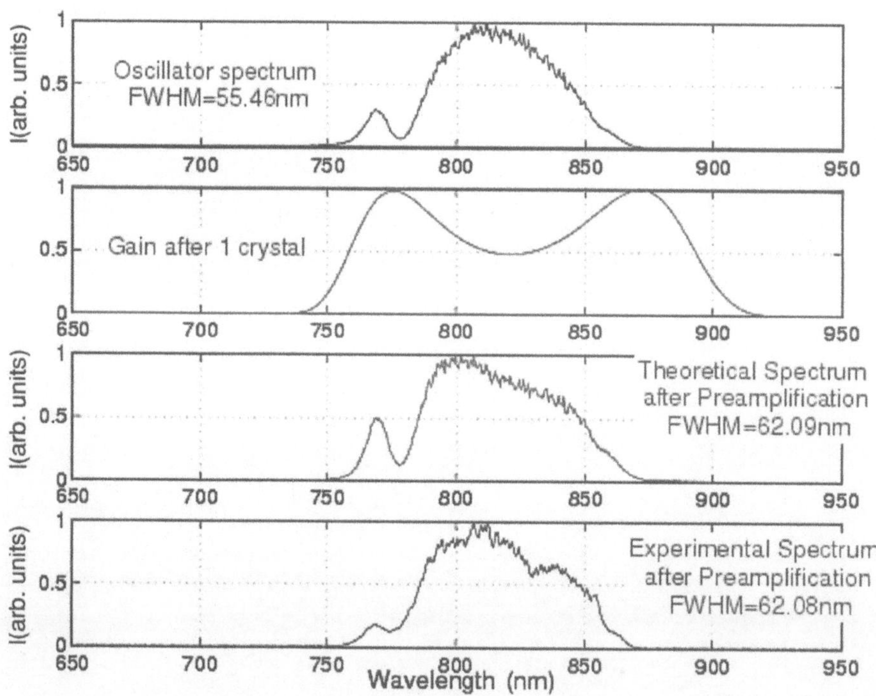


FIGURE 5. Theoretical spectrum after amplification in the preamplifier below saturation, compared to the experimental spectrum and the gain curve.

# References

1. I. N. Ross, P. Matousek, M. Towrie, A. J. Langley, and J. L. Collier: Opt.Comm. **144**, 125 (1997).
2. C. P. Hauri, P. Schlup, G. Arisholm, J. Biegert, and U. Keller: Opt. Lett. **29**, 1369 (2004).
3. I. Jovanovic, C. A. Ebbers, and C. P. Barty: Opt. Lett. **27**, 1622 (2002).

# A Novel Method of Ultrabroadband (white light) Femtosecond Optical Parametric Amplification

Chao-Kuei Lee,[1] Zing-Yung Zhang,[2] J. Y. Huang,[3] and Ci-Ling Pan[3]

[1] Institute of Electro-Optical Engineering, National Sun Yat-Sen University, Taiwan, ROC
chuckcklee@yahoo.com
[2] Department of Physics, Georgia Southern University, Statesboro, GA 30460, USA
[3] Department of Photonics and Institute of Optoelectronic Engineering, National Chiao Tung University, Taiwan, ROC

**Abstract:** In this paper, we have demonstrated, for the first time, a new and simple technique for generation ultrabroadband white light OPA. Over 300 nm bandwidth OPA output was generated.

## 1 Introduction

Ultrashort white light pulses have been employed to probe chemical intermediates in pump-probe spectroscopy [1]. Their application enables the detection of spectral changes in reactive systems over a large frequency range where multiple electronic states are involved. Moreover, simultaneous probing in a large spectral range avoids the problem of reconstructing the transient spectra from a set of subsequently performed experiments using transform-limited probe pulses, centered at a single-probe wavelength. However, the achievable pulse energies of single-filament white light sources at 1-kHz repetition rate are very low (typically <10 nJ) [2], which limits their applicability in ultrafast spectroscopy. The gain bandwidth of the OPA is predefined by the group-velocity mismatch (GVM) between the signal and idler pulses. The use of tilted-front pulses [3] and the noncollinear amplification geometry [4] offer new possibilities in extending the OPA gain bandwidth, and the combination of these methods in the meantime has became a widely recognized technique for the white light continuum amplification. Another important aspect of the OPA is the ability to be pumped by multiple beams. The multiple-beam pump was first introduced as a technique, which enabled to reduce the spatial bandwidth of the signal in the parametric generation process [5,6]. In the case of several noncollinear pump beams, the gain bandwidth of the OPA and the amplified spectral region are determined by the intersection geometry and the crystal orientation, i.e. phase-matching condition. However, complex alignment (NOPA, CPA, wavefront tilted, multistage) limits the performance and flexibility of ultrabroadband OPA. In this article, a new method to generate ultrabroad band OPA is demonstrated. Basically, it is a kind of modified multibeam pump OPA and single-shot configuration is implemented. Over 300 nm bandwidth output is performed.

## 2 Experimental Setup

Figure 1 shows the schematic of our ultrabroad band (white light) OPA-FROG. A regenerative amplified Ti:sapphire laser provides output energy of more than 1 mJ/pulse at 810 nm with pulse duration of 90 fs. The laser output is split into two parts: 90% of the energy is frequency-doubled to 405-nm. The SHG beam is then separated from the fundamental beam by dichroic mirrors and used to pump a type-I noncollinearly phase-matched BBO optical parametric amplifier (NOPA) with pump energy adjustable from 50–120 μJ. The remaining 10% of the Ti:sapphire laser beam is used to generate white light super continuum (WLS) with

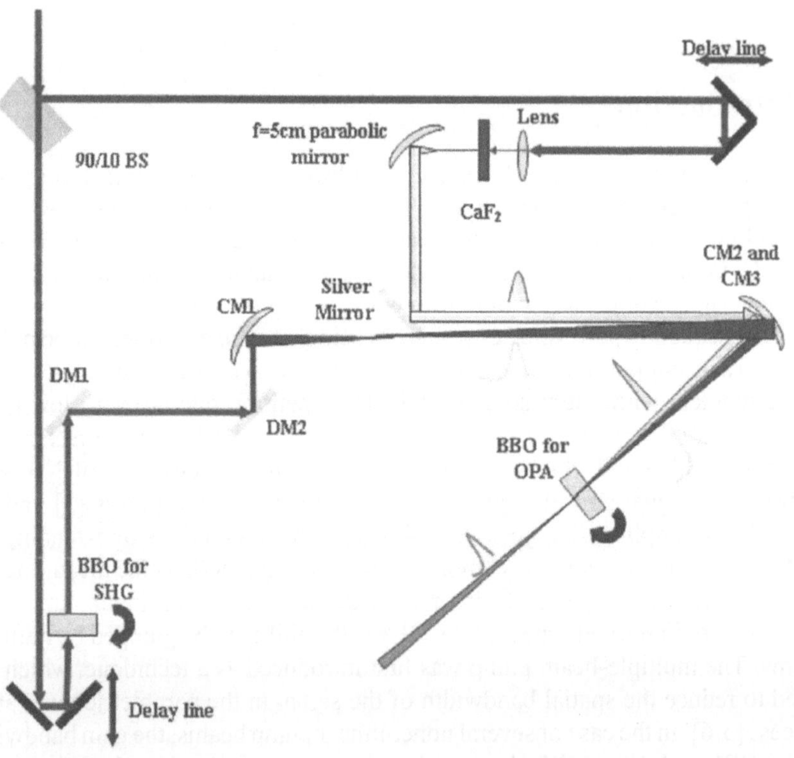


FIGURE 1. The schematic of our ultrabroadband (white light) OPA.

a 2-mm-thick CaF2 plate [2]. The WLS generated is collimated and then focused by a cylindrical mirror onto the 2-mm long BBO NOPA crystal cut at 29°. The SHG beam is then separated from the fundamental beam by two dichroic mirrors. The pump beam at 405-nm is horizontally focused with a cylindrical mirror to form a vertical line image. The WLS generated is collimated and then horizontally focused by a cylindrical mirror onto the BBO NOPA crystal as well. The pump and WLS are brought to the BBO crystal with a beam-crossing angle in the vertical direction. In a typical multibeam pump configuration [7], another pump beam with appropriately delay is necessary. In our design, a simple method was used to avoid the complex and additional optics. As shown in Fig. 1, a cylindrical mirror 1 was implemented after the dichroic mirror 2 to spread the beam around 2 cm in diameter at the cylindrical mirror 2 horizontally. The converging angle of the pump beam is around 16°. Beam diameter of pump beam before focusing is about 3 mm. Therefore, one can image several pump beams, for example 10 or more, was focused onto the BBO crystal. To solve the relative delay among pump beams and seeder from WLS resulting in that various pump beams will cause different relative optical delay, a single-shot scheme was introduced. Delay around, where $d$ is the beam diameter, and which denotes the intersecting angle between pump and seeder beam, is available. For example, 3 ps could be obtained for a beam with diameter of 5 mm and 5° intersecting angle. It should be enough for a WLS which was generated by focusing 50 fs pump.

## 3 Results and Discussion

Figure 2a shows the picture of our design. The WLS and 405 nm pump beam was focused horizontally onto the BBO crystal in the vertical direction. A clear OPA, see Fig. 2b was observed after adjusting the delay between WLS and 405 pump. The dispersive color bar like rainbow from near IR to green indicates that at least

(a)

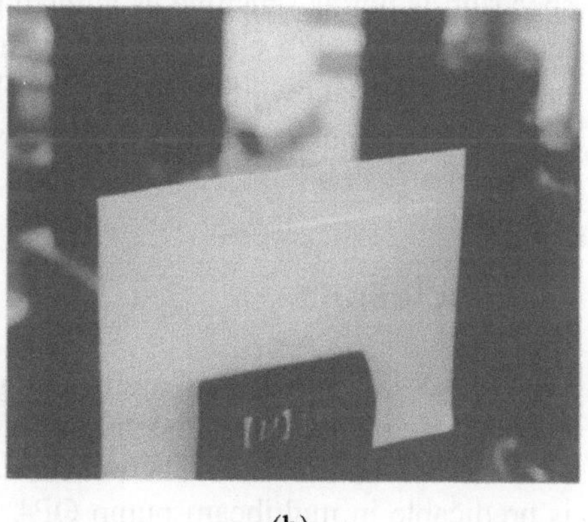

(b)

FIGURE 2. (a) Setup of our white light OPA and (b) OPA output.

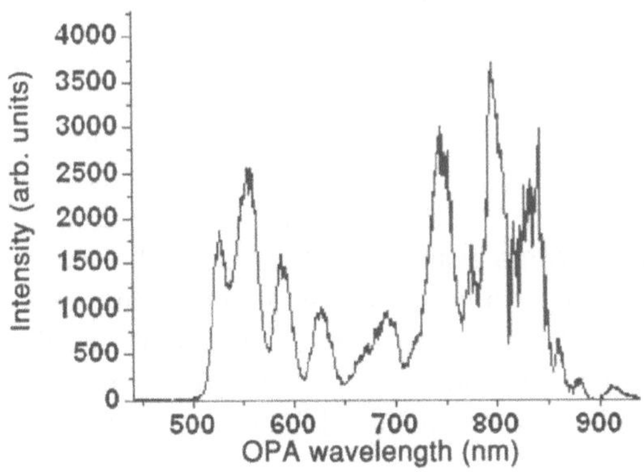


FIGURE 3. Over 300 nm bandwidth spectrum of OPA output at various BBOs angle (interacting angle = 6.5°).

100 nm bandwidth OPA was achieved. Tuning of the OPA was accomplished by varying the pump-seed delay, in order to temporal overlap different spectral regions of the white light, which are separated due to chirp, with the pump pulse. By changing the white light delay monotonically, the central frequency of the amplified pulses was tuned from the near IR to the blue. This result is in good agreement with Cerullo's work [8]. However, shorter wavelength output was achieved in our design. This is due to the longer delay between pump and seed that is available for single-shot configuration used in the OPA. Moreover, various intersecting angles were tested to obtained the most broaden bandwidth. Since GVM is a crucial factor influencing the values of gain and gain bandwidth [9], reduction of GVM can result in the longer interacting length, higher gain, and wider gain bandwidth. Figure 3 shows the spectrum of OPA whose intersecting angle is around 6.5°. Over 300 nm bandwidth pulses were generated and observed. We also measure the spectrum of OPA for various crystal angle and pump-seed delay. Similar bandwidth of pulse except the peak wavelength indicates that perfect phase matching was fulfilled at various wavelength for corresponding crystal angle. However, the spectrum is not as sensitive as adjusting the pump-seed delay. The peak wavelength does not shift clearly. The spectrum will first be cut in long wavelength part, the wider bandwidth of OPA was achieved thereafter, and finally the cut short wavelength output was generated when we tune pump-seed delay monotonically.

## 4 Conclusions

With noncollinear optical parametric generation and single-shot configuration between pump and seed beam, we demonstrated an ultrabroadband optical parametric amplification. Theoretical calculation reveals that broadening band is predicable in multibeam pump OPA. Over 300 nm bandwidth of output was achieved. Insensitive dependence of bandwidth on tuning the crystal angle and pump-seed delay was demonstrated. This is believed to be due to the noncollinear and single-shot configuration; and less than 3 fs could be expectable if appropriate pulse compression, such as group delay line, is implemented.

**Acknowledgment.** This work was supported by the Academic Excellence Program of the Ministry of Education under contract No. (91-E-FA08-1-4).

## References

1. R. L. Fork, C. V. Shank, C. Hirlimann, R. Yen, and W. J. Tomlinson: Opt. Lett. **8**, 1 (1983).
2. R. Huber, H. Satzger, W. Zinth, and J. Wachtveitl: Opt. Commun. **194**, 443 (2001).
3. P. Di Trapani, A. Andreoni, C. Solcia et al.: J. Opt. Soc. Am. B **12**, 2237 (1995).
4. P. Di Trapani, A. Andreoni, G. P. Ban. et al.: Phys. Rev. A **51**, 3164 (1995).
5. V. Smilgevicius and A. Stabinis: Opt. Commun. **106**, 69 (1994).
6. A. Baltuska, A. Berzanskis, R. Gadonas, A. Pugzlys, V. Smilgevicius, and A. Stabinis: Opt. Lett **20**, 2174 (1995).
7. A. Dubietis, R. Danielius, G. Tamosauskas, and A. Piskarskas: J. Opt. Soc. Am. B **15**, 1135 (1998).
8. G. Cerullo, M. Nisoli, and S. De Silvestri: Appl. Phys. Lett. **8**, 1 (1983).
9. G. Cerullo and S. De Silvestri: Rev. of Sci. Instruments **74**, 1 (2003).

# Study of Optical Parametric Chirped Pulse Amplification at 1064 and 780 nm

Xiaoyan Liang, Yuxin Leng, Ruxin Li, and Zhizhan Xu

State Key Laboratory of High Field Laser Physics, Shanghai Institute of Optics and Fine Mechanics, Chinese Academy of Sciences, PO Box 800-211, Shanghai 201800, China.
**liangxy@mail.siom.ac.cn**

We investigated high gain optical parametric chirped pulse amplifications at 1064 and 780 nm by using LBO crystals and periodically poled LiTaO3 (PPLT), pumped by synchronized frequency-doubled Nd:glass laser, Ti:sapphire laser, and Nd:YAG laser. Gain of $6 \times 10^{10}$ at 1064 nm with three-stage amplifiers, single pass gain of $3.7 \times 10^6$ at 780 nm using LBO, and single pass gain of $1.7 \times 10^6$ at 1064 nm in a PPLT have been demonstrated.

## 1 Introduction

Optical parametric chirped pulse amplification (OPCPA) has been regarded as a powerful scheme for obtaining high power laser pulse with extremely short duration, high contrast ratio, and good beam quality [1–4]. This scheme offers the possibility to implement ultrashort pulse and ultraintense laser with wide frequency tunable ability or at new lasing wavelength beyond the limitation imposed by the laser gain media. Moreover, a high repetition rate ultrashort pulse and ultraintense laser system can be realized due to the cancellation of the thermal effect in gain media, if a suitable pumping laser for OPCPA is ready. A number of research groups are pursuing the study of OPCPA both experimentally and theoretically [5]. We investigated and demonstrated high gain OPCPA at 1064 and 780 nm using LBO crystals and PPLT, pumped by precisely synchronized frequency-doubled Nd:glass laser, Ti:sapphire laser, and Nd:YAG laser.

## 2 OPCPA Laser at 1064 nm

We have developed a 1064 nm 16.7 TW/120 fs OPCPA laser system [6, 7], which was demonstrated with three stages of LBO nonlinear crystals as amplifiers pumped with frequency doubled Nd:glass laser at 532 nm in near-degenerative near-collinear phase matching geometry. One key feature of the laser system is that both the signal pulse and the pump pulse share the same Ti:sapphire oscillator operating at 1064 nm. This approach ensures the precise synchronization between the signal and the pump pulse of OPCPA. And the time jitter between signal and pump pulses was $<10$ ps. Since the optical parametric amplification occurred only in the temporally and spectrally overlapped range, the precise synchronization between the signal and the pump pulse is crucial for high efficient amplification. The narrow band nanosecond-duration pump laser pulse was obtained by temporally and spectrally shaping of the 100 fs pulse from the Ti:sapphire oscillator in a Nd:YAG regenerative amplifier before seeded to a cascade of Nd:YAG/Nd:glass amplifiers [8]. The 300 ps/18 nm chirped pulse at 1064 nm is amplified from 50 pJ to more than 3 J with the total energy gain of $6 \times 10^{10}$ in three stages of LBO OPCPA amplifiers, pumped with 12.4 J ns laser pulse at 532 nm in last amplifier. The energy conversion efficiency in final amplifier achieved 25.5%. After compressor, the measured pulse duration is 120 fs with energy of 2.01 J, which corresponds to a peak power of 16.7 TW.

Recently, we demonstrated a high-gain OPCPA preamplifier at 1064 nm using periodically poled LiTaO3 (PPLT) [9]. Our PPLT was poled by room temperature poling technique and the poling region was 15-mm long, 5-mm wide, and 0.5-mm thick. The poled period was $\Lambda = 7.81\ \mu\text{m}$, which satisfy type-0 QPM condition for 532 nm pumped degenerated optical parametric process at the temperature of 119.0°C, when photorefractive effect could be prevented effectively. We obtained single pass gain of $\sim 10^6$ and $\sim 10.3\%$ conversion efficiency, using low pump energy of 630μJ from a Nd:YAG laser synchronized with previous scheme. The maximum output signal energy is 35μJ, while the idler was $\sim$ 25μJ. There is no observed damage in crystal with pump intensity up to 800 $\text{MW/cm}^2$. The seed and amplified signal spectrum is shown in Fig. 1, with same spectral bandwidth

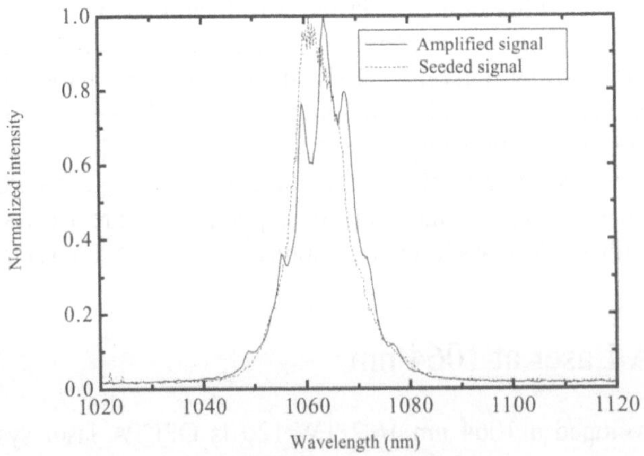


FIGURE 1. The signal spectrum of seed (dashed) and amplified (solid).

of 18 nm. And the amplified signal spectrum is almost same with the seed. The advantage of our system is that a high single pass gain was achieved with relatively low pump energies in microjoule range. Despite the disadvantage of low amplified energy in this experiment, such an amplifier can be used in low energy and high repetition rate OPCPA.

## 3 OPCPA Laser at 780 nm

OPCPA in nondegenerative noncollinear phase matching geometry was demonstrated and has led to high conversion efficiency at 800 nm, with precisely shaping the 532 nm pumped laser [10]. However, the gain bandwidth is very sensitive to the pump-signal angle [11]. And it results in the complex of optical system. A unique feature of the design is that the OPCPA preamplifier is pumped by a frequency-doubled spectra-shaped Ti:sapphire laser seeded by the same oscillator, and differs from other systems where the OPCPA is pumped by a separate laser system in nondegenerative noncollinear phase matching geometry. Compared with noncollinear geometry, simpler adjustment and better stability of the near-degenerative near-collinear geometry is more suitable for broadband OPCPA.

The near-collinear OPCPA preamplifier was tested experimentally and positive results were obtained [12]. High gain OPCPA at 780 nm in near-degenerative near-collinear phase matching geometry is demonstrated using LBO crystal for the first time. In our experiment, a ruled grating G (1700 l/mm) was used to narrow the spectrum of the pulse that came from the Ti:sapphire oscillator before it was injected into the Ti:sapphire regenerative amplifier. Moreover, a quartz etalon was inserted into the regenerative amplifier to stretch the pulse duration. By using these two pulse shaping techniques, the pulse duration and spectral bandwidth of the amplified pulse from the Ti:sapphire regenerative amplifier reached nanosecond level and ~0.38 nm (FWHM) at ~785 nm, respectively. The nonlinear crystal LBO was cut at $\theta = 90°$ and $\phi = 33.21°$, its size was $6 \times 6 \times 12$ mm$^3$. The stretched broadband chirped signal pulse near 780 nm is amplified up to be ~412 mJ with ~15 mJ/~270 ps pumping energy at ~4.5 GW/cm$^2$ pumping intensity, and the total gain is more than $3.7 \times 10^6$. As shown in Fig. 2a the bandwidth of the input chirped signal pulse was narrowed from the initial >40 to ~38 nm (FWHM) due to the grating stretcher. But the whole signal spectrum still covered from ~720 to ~820 nm. After OPA, the bandwidth of the amplified chirped signal pulse increased to ~71 nm, (see Fig. 2b). This OPCPA as the preamplifier will offer high gain without gain narrowing and promise high contrast ratio. It is very essential to replace the broadband regenerative amplification of chirped pulse amplification (CPA) by OPCPA.

## 4 Conclusions

In a conclusion, we investigated high gain optical parametric CPA at 1064 nm based on nonlinear optical crystals of LBO and PPLT, respectively. The pump

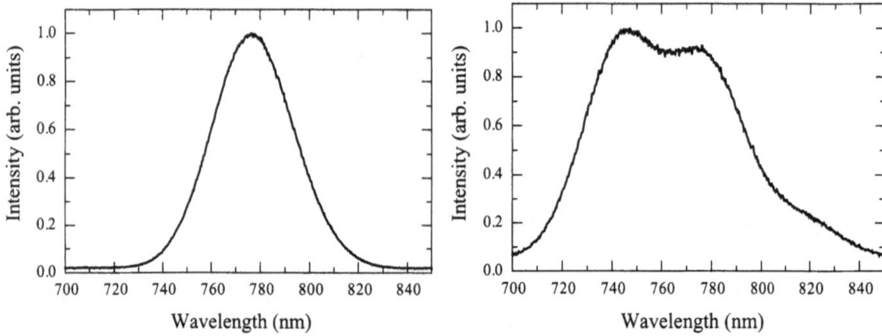


FIGURE 2. Spectrum of the chirped signal pulse. (a) The input signal pulse spectrum bandwidth is −38 nm (FWHM): (b) the amplified signal pulse spectrum in LBO-LOPCPA, bandwidth is −71 nm (FWHM).

source is precisely synchronized frequency-doubled Nd:YAG/Nd:glass laser, which provides the high gain and high conversion efficiency. From these results, we believe that PPLT is suitable for a high gain, lower pump, and high repetition rate OPCPA system, but LBO is more effective in high energy OPCPA laser. For OPCPA at 780 nm based on LBO crystal, the near-degenerative near-collinear phase matching geometry is also an effective method. The single pass gain of $3.7 \times 10^6$ was demonstrated. And this structure could be used as the replacement of the regenerative amplifier in a high intensity CPA laser system.

## References


1. A. Dubietis, G. Jonusauskas, and A. Piskarskas: Opt. Commun. **88**, 437 (1992).
2. I. N. Ross, P. Matousek, M. Towrie, A. J. Langley, and J. L. Collier: Opt. Commun. **144**, 125 (1997).
3. I. N. Ross, J. L. Collier, P. Matousek, et al.: Appl. Opt. **39**, 2422 (2000).
4. Y. Kitagawa, H. Fujita, R. Kodama, et al.: IEEE J. Quant. Electronics, **40**, 281 (2004).
5. R. Butkus, R. Danielius A. Dubietis, A. Piskarskas, and A. Stabinis: Appl. Phys. B**79**, 693 (2004).
6. Z. Xu, X. Yang, Y. Leng, et al.: Chin. Opt. Lett. 1, 24 (2003).
7. X. Yang, Z. Xu, Y. Leng, et al.: Opt. Lett. **27**, 1135 (2002).
8. Y. Leng, X. Yang, H. Lu, et al.: Opt. Eng. **43**, 2994 (2003).
9. B. Z. Zhao, X. Y. Liang, X. Leng, R. X. Li, Z. Z. Xu, X. P. Hu, and S. N. Zhu: IQEC/CLEO-PR, Tokyo, Japan (2005).
10. L. J. Waxer, V. Bagnoud, I. A. Begishev, M. J. Guardalben, J. Puth, and J. D. Zuegel: Opt. Lett. **28**, 1245 (2003).
11. X. Yang, Z. Xu, Z. Zhang, et al.: Appl. Phys. B **73**, 219 (2001).
12. Y. Leng, C. Wang, B. Z. Zhao, et al.: Opt. Eng. **44**, 1 (2005).


Printed in the United States of America

Springer